Mastering Quantum Mechanics

Zoltán Papp

Mastering Quantum Mechanics

Zoltán Papp
Department of Physics
California State University, Long Beach
Long Beach, CA, USA

ISBN 978-3-032-09010-2 ISBN 978-3-032-09011-9 (eBook)
https://doi.org/10.1007/978-3-032-09011-9

This Springer imprint is published by the registered company Springer Nature Switzerland AG
The registered company address is: Gewerbestrasse 11, 6330 Cham, Switzerland

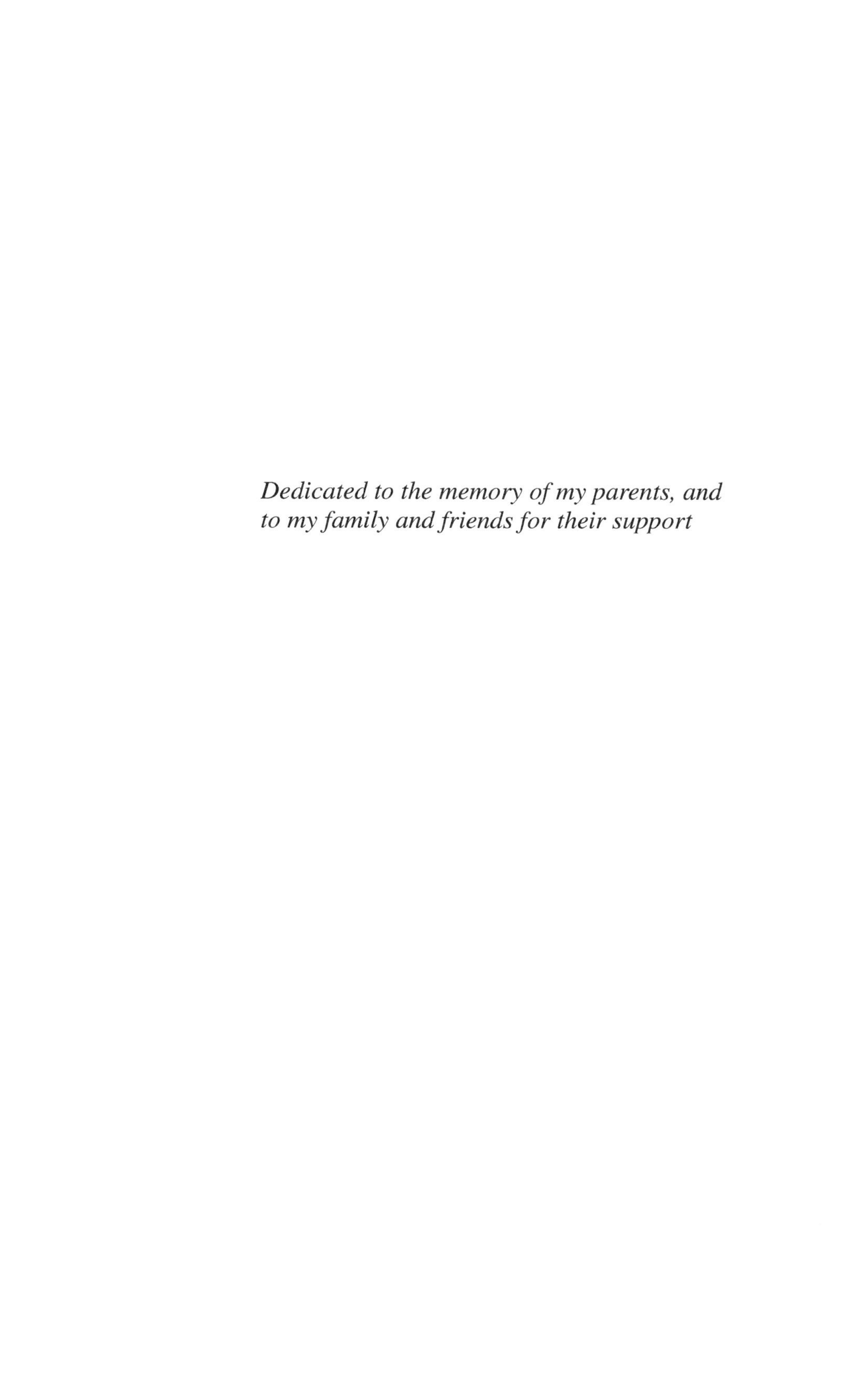

Dedicated to the memory of my parents, and to my family and friends for their support

Preface

This book is the result of the author's forty plus years of researching and teaching quantum mechanics. It is meant for students who have had some prior exposure to quantum mechanics and want to move to the next level. This book aims to lift those who are a little behind, and at the same time, give a new perspective for those who are more advanced.

The book covers material for a two-semester or a three-quarter course. The assumed prerequisites include some classical mechanics and electrodynamics. As some prior exposure to quantum mechanics is also assumed, experimental indications are not treated here, and the most basic topics are reviewed only briefly. On the other hand, important concepts that probably have been discussed before cannot be left out without risk of breaking the logic of this presentation. The book tries to cover the basic topics from different, possibly more advanced, perspectives.

The text starts with *Mathematical Foundations*, which provides a sound introduction to the required mathematics, like Hilbert spaces, Hermitian and unitary operators, invariant quantities, basis transformations, and functions of operators.

The next chapter is about the *Basics of Quantum Mechanics*. Here the postulates are presented, together with the various pictures, the representation of operators, uncertainty relations and some consequences of continuous and discrete symmetries.

A lot of physics can be explained by using the examples of *One-Dimensional Problems*. The general properties of the wave function are described in this chapter. The connection between bound-state and scattering quantities, together with time-dependent aspects, are established by using the technique of analytic continuation. Also, harmonic oscillators and periodic potential models are treated in detail.

The heavy industry of quantum mechanics starts with *Angular Momentum*. This chapter covers the quantization of angular momentum, orbital angular momentum, spin, rotation matrices, coupling schemes, rotation operators, spherical tensors and the Wigner-Eckart theorem.

The next chapter is about *Quantum Problems in Higher Dimensions*. The covered topics include the usual harmonic-oscillator and Coulomb problems, charged particles in a magnetic field, and the consequences of the gauge transformation of the electromagnetic potentials, including the Aharonov-Bohm effect with the quantization of magnetic flux.

In the chapter on *Identical Particles in Quantum Mechanics*, it is shown how the concept of indistinguishability of quantum particles leads to the symmetry of the wave function.

In *Approximation Methods for Time-Independent Problems*, a wide range of methods are presented, including the Hellmann-Feynman theorem, the Rayleigh-Schrodinger and Brillouin-Wigner perturbation methods, variational methods, the semiclassical approximation method, and related applications.

The *Time-Dependent Quantum Mechanics* starts with the time-dependent perturbation theory and applies it to constant and harmonic perturbations. Then sudden and adiabatic approximations with dynamical and geometric phases are presented, together with detailed analysis of two-state systems.

In the chapter about *Electromagnetic Radiation*, the electromagnetic field is quantized, which leads to the photon concept and the vacuum expectation of the field. Then atoms in a radiation field are considered, together with transition amplitudes for emission and absorption, the selection rules, and the natural broadening of spectral lines. Finally, the black-body radiation formulae are re-derived and the Casimir effect is discussed.

Scattering Theory is about exploring the Hamiltonian. Knowing a few bound states gives rather limited information about the physical system. To get the full picture, we need scattering information as well. The treatment of quantum scattering goes through asymptotic states and resolvent operators in order to derive the properties of the wave function from the asymptotic analysis of the integral equations. Then the phase shift, the Jost function, the time delay, and the Levinson theorem are introduced. The treatment of scattering theory cannot be complete without considering two additional topics: the Coulomb scattering and three-body problems using the Faddeev approach.

There are intimate relations between the two main pillars of modern physics, relativity and quantum mechanics. Their unification in *Relativistic Quantum Mechanics* presented us with the discovery of antiparticles, the electron spin, its coupling to the electromagnetic field, and so on.

In *Appendix*, some Fortran codes for solving simple bound and scattering state problems are presented.

This book should prepare students to take more advanced classes in the direction of many-body physics, quantum field theory, quantum optics, nuclear and particle physics, condensed-matter physics, and similar disciplines.

I am thankful to my friends and colleagues Benjamin P. Carter, Scott M. Keller, DerekWingard and Robert Woodhouse for critical reading of the manuscript and to Attila Csótó for his help in clarifying some items. While I was on sabbatical leave, colleagues Michael R. Peterson and Andreas Bill taught my class from this manuscript and provided feedback with pertinent suggestions. Special thanks is due to Sergey L. Yakovlev for his help and several constructive suggestions.

Long Beach, CA, USA
2025

Zoltán Papp

Competing Interests The author has no competing interests to declare that are relevant to the content of this manuscript.

Contents

Chapter 1
Mathematical Foundations

In this chapter we present the basic mathematical formalism needed for studying quantum mechanics. The formalism is rather abstract, but on the other hand it is very general. First, we introduce the Hilbert space and then linear operators together with their most basic properties.

1.1 Hilbert Space

1.1.1 Vector Spaces

The *field* of complex numbers, with addition and multiplication, is denoted by $\mathbb{C}$. A set V is called a *linear vector space* with respect to addition and multiplication by complex numbers if it satisfies, for any $\psi, \phi, \chi \in \mathbb{V}$ and scalars $a, b \in \mathbb{C}$, the following properties:

- addition: $\psi + \phi \in \mathbb{V}$,
- commutativity: $\psi + \phi = \phi + \psi$,
- associativity with respect to addition: $(\psi + \phi) + \chi = \psi + (\phi + \chi)$,
- zero vector: $0 \in \mathbb{V}$ such that $0 + \phi = \phi + 0 = \phi$,
- inverse vector: $-\phi \in V$ such that $(-\phi) + \phi = \phi + (-\phi) = 0$,
- linearity: $a\psi + b\phi \in \mathbb{V}$,
- distributivity: $a(\psi + \phi) = a\psi + a\phi$ and $(a + b)\psi = a\psi + b\psi$,
- associativity with respect to multiplication by scalars: $a(b\psi) = (ab)\psi$,
- unit and zero scalar: $1, 0 \in \mathbb{C}$ such that $1\psi = \psi$ and $0\psi = 0 \in \mathbb{V}$.

A set of n nonzero vectors $\phi_1, \phi_2, \ldots \phi_n \in \mathbb{V}$ are *linearly independent* if, for $a_i \in \mathbb{C}$, the equation

$$a_1\phi_1 + a_2\phi_2 + \cdots + a_n\phi_n = 0 \tag{1.1}$$

Z. Papp, *Mastering Quantum Mechanics*,
https://doi.org/10.1007/978-3-032-09011-9_1

is true only for $a_i = 0$ for all i. The *dimension of the vector space* is the maximal number N of independent vectors, which can be finite or infinite.

The set $\mathbb{V}_1 \subset \mathbb{V}$ is a vector subspace of $\mathbb{V}$ if it is closed with respect to vector operations, i.e. any linear combinations of vectors from $\mathbb{V}_1$ remain in $\mathbb{V}_1$. Vector spaces $\mathbb{V}_1$ and $\mathbb{V}_2$ are isomorphic if there is a one-to-one and onto mapping f from $\mathbb{V}_1$ to $\mathbb{V}_2$ which implies the property

$$f(a\phi_1 + b\phi_2) = af(\phi_1) + bf(\phi_2), \tag{1.2}$$

for all $\phi_1, \phi_2 \in \mathbb{V}_1$, $f(\phi_1), f(\phi_2) \in \mathbb{V}_2$ and for any $a, b \in \mathbb{C}$. Also, for every $\phi \in \mathbb{V}_1$ there is a $\psi \in \mathbb{V}_2$ such that $\psi = f(\phi)$.

1.1.2 Inner Product Spaces

If a vector space is augmented by an operation called the *inner product* or *scalar product*, we call them an *inner product space* or *scalar product space*. The inner product, denoted by the Dirac *bra-ket* symbols, is defined by the properties

- $\langle\psi|\phi\rangle \in \mathbb{C}$, such that $\langle\psi|\phi\rangle = \langle\phi|\psi\rangle^*$, i.e. the inner product of ψ and ϕ is equal to the complex conjugate of the inner product of ϕ and ψ.
- the inner product is linear with respect to the second argument

$$\langle\phi|a_1\psi_1 + a_2\psi_2\rangle = a_1\langle\phi|\psi_1\rangle + a_2\langle\phi|\psi_2\rangle, \tag{1.3}$$

  and anti-linear with respect to the first argument

$$\langle a_1\psi_1 + a_2\psi_2|\phi\rangle = a_1^*\langle\psi_1|\phi\rangle\rangle + a_2^*\langle\psi_2|\phi\rangle, \tag{1.4}$$

- the inner product is positive semi-definite

$$\langle\psi|\psi\rangle = \|\psi\|^2 \geq 0, \tag{1.5}$$

  where equality implies that $\psi \equiv 0$.

The symbol $\|\psi\| = \sqrt{\langle\psi|\psi\rangle}$ is called the *norm* of the vector. A vector space equipped with a norm is called a *normed vector space*. If $\langle\psi|\phi\rangle = 0$, we call the vectors ϕ and ψ *orthogonal*. If $\phi_1, \phi_2, \ldots \phi_N \in V$, where N is the dimension of the inner product space, are linearly independent, mutually orthogonal and $\|\phi_i\| = 1$ for each element, then we say that $\{\phi_i\}$ forms an *orthonormal basis*, or just *basis*. Any vector $|\psi\rangle \in V$ can be expressed in a unique way as a linear combination

$$|\psi\rangle = \sum_{i=1}^{N} a_i|\phi_i\rangle. \tag{1.6}$$

The inner product also satisfies the important inequalities:

$$|\langle\psi|\phi\rangle| \leq \|\psi\|\;\|\phi\| \qquad \text{(Schwarz or Cauchy-Bunyakovsky-Schwarz inequality)} \tag{1.7}$$

and

$$\|\psi+\phi\| \leq \|\psi\| + \|\phi\| \qquad \text{(triangular inequality)}\,. \tag{1.8}$$

These relations are the generalizations of the corresponding properties of Euclidean vectors $\boldsymbol{a}$, $\boldsymbol{b}$ and $\boldsymbol{c}$:

$$\boldsymbol{a}\cdot\boldsymbol{b} = a\,b\cos\theta \leq a\,b, \quad \text{and} \quad c \leq a+b \quad \text{if} \quad \boldsymbol{a}+\boldsymbol{b}+\boldsymbol{c}=0. \tag{1.9}$$

To prove the Schwarz inequality we define $\chi \in \mathbb{V}$ as

$$|\chi\rangle = |\phi\rangle - \alpha|\psi\rangle. \tag{1.10}$$

Then we find that

$$0 \leq \langle\chi|\chi\rangle = \langle(\phi-\alpha\psi)|(\phi-\alpha\psi)\rangle = \langle\phi|\phi\rangle - \alpha^*\langle\psi|\phi\rangle - \alpha\langle\phi|\psi\rangle + \alpha\alpha^*\langle\psi|\psi\rangle. \tag{1.11}$$

The right hand side has the minimum value with $\alpha = \langle\psi|\phi\rangle/\langle\psi|\psi\rangle$. This results in

$$0 \leq \langle\phi|\phi\rangle - \frac{\langle\psi|\phi\rangle^*}{\langle\psi|\psi\rangle}\langle\psi|\phi\rangle - \frac{\langle\psi|\phi\rangle}{\langle\psi|\psi\rangle}\langle\phi|\psi\rangle + \frac{\langle\psi|\phi\rangle}{\langle\psi|\psi\rangle}\frac{\langle\psi|\phi\rangle^*}{\langle\psi|\psi\rangle}\langle\psi|\psi\rangle. \tag{1.12}$$

Consequently

$$0 \leq \langle\phi|\phi\rangle - \frac{\langle\psi|\phi\rangle\langle\phi|\psi\rangle}{\langle\psi|\psi\rangle}, \tag{1.13}$$

and then

$$|\langle\psi|\phi\rangle| \leq \|\psi\|\;\|\phi\|. \tag{1.14}$$

To show the triangular inequality we calculate

$$\begin{aligned}\langle\phi+\psi|\phi+\psi\rangle = \|\phi\|^2 + \langle\phi|\psi\rangle + \langle\psi|\phi\rangle + \|\psi\|^2 &= \|\phi\|^2 + 2\Re(\langle\phi|\psi\rangle) + \|\psi\|^2 \\ &\leq \|\phi\|^2 + 2|\langle\phi|\psi\rangle| + \|\psi\|^2.\end{aligned} \tag{1.15}$$

Then, using the Schwarz inequality we obtain

$$\langle\phi+\psi|\phi+\psi\rangle \leq \|\phi\|^2 + 2\|\phi\|\;\|\psi\| + \|\psi\|^2 = (\|\phi\| + \|\psi\|)^2, \tag{1.16}$$

and thus

$$\|\psi+\phi\| \leq \|\psi\| + \|\phi\|. \tag{1.17}$$

1.1.3 Metric Spaces

A set $\mathbb{X}$ is a metric space if for every $\phi, \chi \in \mathbb{X}$, there is a function $d(\phi, \chi) : \mathbb{X} \times \mathbb{X} \to \mathbb{R}$, called the *metric*, or *distance*, with the properties:

- $d(\phi, \chi) > 0$ with any $\phi \neq \chi$,
- $d(\phi, \phi) = 0$,
- $d(\phi, \chi) = d(\chi, \phi)$,
- $d(\phi, \chi) \leq d(\phi, \psi) + d(\psi, \chi)$ (triangular inequality).

The infinite sequence $\phi_1, \phi_2, \ldots$ *converges* to ϕ if for any $\epsilon > 0$ there is an integer $N(\epsilon)$ such that for all $n > N(\epsilon)$ the distance $d(\phi, \phi_n) < \epsilon$. The series $\phi_1, \phi_2, \ldots$ is said to be a *Cauchy sequence* if for any $\epsilon > 0$ there exists an integer $N(\epsilon)$ such that for all $n, m > N(\epsilon)$, $d(\phi_n, \phi_m) < \epsilon$.

The metric space is *complete* if every Cauchy sequence converges to an element in $\mathbb{X}$. A normed vector space which is also complete is called a *Banach space*. A subset $\mathbb{S}$ of the metric space $\mathbb{X}$ is *dense* if for every element $\phi \in \mathbb{S}$ and for any $\epsilon > 0$, there is a $\chi \in \mathbb{S}$ such that $d(\chi, \phi) < \epsilon$.

1.1.4 Hilbert Space

It is easy to see that the norm is a metric

$$d(\phi, \chi) = \|\phi - \chi\|. \tag{1.18}$$

Therefore we can define convergence, completeness and denseness in terms of the norm. A complete inner product space with the norm as metric is called a *Hilbert space* $\mathcal{H}$. A Hilbert space admits a countable orthonormal basis, i.e. there exists a set of vectors $\{\phi_i\}$ that are linearly independent, mutually orthogonal $\langle\phi_i|\phi_j\rangle = \delta_{ij}$, their number equals the dimension of the Hilbert space, that can be finite or infinite, and any element of the Hilbert space can be expressed uniquely as a linear combination of the ϕ_i basis elements.

Examples

- The set of infinite dimensional vectors with finite norm

$$l_2 = \left\{\phi : \{\phi_k\}_{k\in N},\quad \|\phi\|^2 = \sum_k^{\infty} |\phi_k|^2 < \infty\right\}. \tag{1.19}$$

- The square-integrable functions in the configuration space $D \subseteq \mathbb{R}^d$

$$L_2(D) = \left\{ \psi : D \to \mathbb{C}, \ \ ||\psi||^2 = \int_D \mathrm{d}x \ \psi^*(x) w(x) \psi(x) < \infty \right\}, \tag{1.20}$$

where $w(x)$ is some non-negative weight function.

We can have a nice interpretation of the *bra-ket* formalism. The *ket* $|\psi\rangle$ denotes an element of the Hilbert space and the *bra* $\langle\phi|$ denotes a linear functional acting on a Hilbert space element. For example *bra* in L_2 means

$$\langle\phi| = \int_D \mathrm{d}x \ \phi^*(x) w(x) \cdot , \tag{1.21}$$

i.e. an integration with $\phi^*(x)$ and $w(x)$ is carried out with an arbitrary Hilbert space element. So the inner product is interpreted as

$$\langle\phi|\psi\rangle = \int_D \mathrm{d}x \ \phi^*(x) w(x) \cdot \psi(x), \tag{1.22}$$

i.e. the functional $\langle\phi|$ acts on the function $|\psi\rangle$ resulting in a complex number.

Similarly, in l_2 we have

$$\langle\phi| = \sum_k \phi_k^* \cdot \ , \tag{1.23}$$

and

$$\langle\phi|\psi\rangle = \sum_k \phi_k^* \cdot \psi_k . \tag{1.24}$$

The *ket* functions form the Hilbert space $\mathcal{H}$ and the *bra* functionals form the dual Hilbert space $\mathcal{H}^*$. These two Hilbert spaces are isomorphic, i.e. there is a one-to-one norm-preserving map between them (Riesz or Riesz-Fréchet representation theorem).

1.2 Linear Operators in Hilbert Space

An *operator* $\hat{A}$ that acts on the Hilbert space $\mathcal{H}$ transforms a *ket* $|\psi\rangle$ into another *ket* $|\psi'\rangle$

$$\hat{A}|\psi\rangle = |\psi'\rangle, \tag{1.25}$$

and similarly for a *bra*

$$\langle\psi|\hat{A} = \langle\psi'|. \tag{1.26}$$

The operators $\hat{A}$ and $\hat{B}$ are linear operators if for any complex number $c \in \mathbb{C}$ and vector $\psi \in \mathcal{H}$:

- $\hat{A}(c_1|\psi_1\rangle + c_2|\psi_2\rangle) = c_1\hat{A}|\psi_1\rangle + c_2\hat{A}|\psi_2\rangle$,
- $(\hat{A} + \hat{B})|\psi\rangle = \hat{A}|\psi\rangle + \hat{B}|\psi\rangle$,
- $(\hat{A}\hat{B})|\psi\rangle = \hat{A}(\hat{B}|\psi\rangle)$.

A linear operator $\hat{A}$ maps between normed spaces X and Y

$$A : \mathfrak{D}(\hat{A}) \subseteq X \to Y. \tag{1.27}$$

The linear subspace $\mathfrak{D}(\hat{A})$ on which $\hat{A}$ is defined is the *domain* of $\hat{A}$, and it is assumed to be dense. The *kernel*, or *null space*, of $\hat{A}$ is the subset which is mapped to zero

$$\mathrm{Ker}(\hat{A}) = \{\phi \in \mathfrak{D}(\hat{A}) \mid \hat{A}\phi = 0\} \subseteq X, \tag{1.28}$$

and the *range* of $\hat{A}$ is

$$\mathrm{Ran}(\hat{A}) = \{\hat{A}\phi \mid \phi \in \mathfrak{D}(\hat{A})\} = \hat{A}\mathfrak{D}(\hat{A}) \subseteq Y. \tag{1.29}$$

The operator $\hat{T}$ is *continuous* at ϕ_0 if for any $\epsilon > 0$ there is a $\delta(\epsilon) > 0$ such that $\|\hat{T}\phi - \hat{T}\phi_0\| < \epsilon$ if $\|\phi - \phi_0\| < \delta(\epsilon)$. A set $\mathbb{S}$ in a normed space is *bounded* if there is a constant C such that $\|\phi\| < C$ for every $\phi \in \mathbb{S}$. A transformation $\hat{T}$ is *bounded* if it maps each bounded set $\mathbb{S}$ into a bounded set. This means that $\|\hat{T}\phi\| < C\|\phi\|$ for every $\phi \in \mathbb{S}$.

1.2.1 Commutator Algebra

The symbol

$$[\hat{A}, \hat{B}] = \hat{A}\hat{B} - \hat{B}\hat{A} \tag{1.30}$$

is called the *commutator*, and the symbol

$$[\hat{A}, \hat{B}]_+ = \hat{A}\hat{B} + \hat{B}\hat{A} \tag{1.31}$$

is called the *anti-commutator*. Two operators *commute*, or are interchangeable, if their commutator is zero

$$[\hat{A}, \hat{B}] = \hat{A}\hat{B} - \hat{B}\hat{A} = 0, \quad \text{and then} \quad \hat{A}\hat{B} = \hat{B}\hat{A}. \tag{1.32}$$

The commutators have algebraic properties analogous to the Poisson-brackets:

- $[\hat{A}, \hat{B}] = -[\hat{B}, \hat{A}]$ (antisymmetry)
- $[\hat{A}, \hat{B} + \hat{C}] = [\hat{A}, \hat{B}] + [\hat{A}, \hat{C}]$ (linearity)

- $[\hat{A}, \hat{B}\hat{C}] = [\hat{A}, \hat{B}]\hat{C} + \hat{B}[\hat{A}, \hat{C}]$ (distributivity)
- $[\hat{A}, [\hat{B}, \hat{C}]] + [\hat{B}, [\hat{C}, \hat{A}]] + [\hat{C}, [\hat{A}, \hat{B}]] = 0$ (Jacobi identity).

1.2.2 The Identity Operator and Representations

The *identity operator*, denoted by I, maps each element of $\mathcal{H}$ into itself

$$I|\psi\rangle = |\psi\rangle. \tag{1.33}$$

An element $|\psi\rangle$ of the Hilbert space can be expressed in terms of orthonormal basis states $\{|\phi_i\rangle\}$

$$|\psi\rangle = \sum_i c_i|\phi_i\rangle. \tag{1.34}$$

Acting with $\langle\phi_j|$, we find

$$\langle\phi_j|\psi\rangle = \sum_i c_i\langle\phi_j|\phi_i\rangle = \sum_i c_i\delta_{ji} = c_j. \tag{1.35}$$

So,

$$|\psi\rangle = \sum_i \langle\phi_i|\psi\rangle|\phi_i\rangle = \sum_i |\phi_i\rangle\langle\phi_i|\psi\rangle, \tag{1.36}$$

which gives a representation of the identity operator in terms of the basis states

$$I = \sum_i |\phi_i\rangle\langle\phi_i|. \tag{1.37}$$

With identity operator expressed in terms of basis states we can get a basis *representation* of any element of the Hilbert space

$$|\psi\rangle = I|\psi\rangle = \sum_i |\phi_i\rangle\langle\phi_i|\cdot|\psi\rangle = \sum_i c_i|\phi_i\rangle, \tag{1.38}$$

where $c_i = \langle\phi_i|\psi\rangle$. There is a one-to-one and onto relation between $|\psi\rangle$ and the infinite dimensional vector $\{c_i\}$,

$$|\psi\rangle \mapsto \{c_i\}. \tag{1.39}$$

The vector $\{c_i\}$ is called the representation of the abstract Hilbert space vector $|\psi\rangle$ in the $\{\phi_i\}$ basis. Note that i can be a discrete or continuous index, or some combination of both.

We can also represent operators in a similar fashion:

$$\hat{A} = I\hat{A}I = \sum_{i,j} |\phi_i\rangle\langle\phi_i|\hat{A}|\phi_j\rangle\langle\phi_j| = \sum_{i,j} |\phi_i\rangle\underline{A}_{ij}\langle\phi_j|. \tag{1.40}$$

Here $\underline{A}_{ij}$ is a representation of the operator $\hat{A}$. If the basis index i is discrete, $\underline{A}_{ij}$ is a matrix, if continuous, it is a kernel of an integral operator, and the summation becomes an integration. This representation is also a one-to-one and onto relation:

$$\hat{A} \mapsto \underline{A}_{ij}. \tag{1.41}$$

The product of operators become

$$\hat{A}\hat{B} = I\hat{A}I\hat{B}I = \sum_{i,j,k} |\phi_i\rangle\langle\phi_i|\hat{A}|\phi_j\rangle\langle\phi_j|\hat{B}|\phi_k\rangle\langle\phi_k| = \sum_{i,j,k} |\phi_i\rangle\underline{A}_{ij}\underline{B}_{jk}\langle\phi_k|, \tag{1.42}$$

i.e. a product of matrices.

1.2.3 The Inverse Operator

The *inverse* operator is defined by

$$\hat{A}\hat{A}^{-1} = \hat{A}^{-1}\hat{A} = I. \tag{1.43}$$

If both $\hat{A}$ and $\hat{B}$ have inverses, then $(\hat{A}\hat{B})^{-1} = \hat{B}^{-1}\hat{A}^{-1}$.

For the inverse operator we find

$$\begin{aligned} I = I\hat{A}I\hat{A}^{-1}I &= \sum_{i,j,k} |\phi_i\rangle\langle\phi_i|\hat{A}|\phi_j\rangle\langle\phi_j|\hat{A}^{-1}|\phi_k\rangle\langle\phi_k| \\ &= \sum_{i,j,k} |\phi_i\rangle\underline{A}_{ij}\,\underline{A}^{-1}_{jk}\langle\phi_k| = \sum_{i} |\phi_i\rangle\langle\phi_i|, \end{aligned} \tag{1.44}$$

so it should be that $\sum_j \underline{A}_{ij}\,\underline{A}^{-1}_{jk} = \delta_{ik}$, which means that the matrix of the inverse operator $\hat{A}^{-1}$ is the inverse of the matrix $\underline{A}$.

1.2.4 Projection Operator

An operator is a *projection* if $\hat{P}^2 = \hat{P}$. We can easily see that a truncation of a unit operator is a projection operator

$$\hat{P} = \sum_{\{i\}} |\phi_i\rangle\langle\phi_i|, \tag{1.45}$$

where ϕ_i are the basis states and in the summation $\{i\}$ some basis states are left out. Indeed,

$$\hat{P}^2 = \left(\sum_{\{i\}} |\phi_i\rangle\langle\phi_i|\right)\left(\sum_{\{j\}} |\phi_j\rangle\langle\phi_j|\right) = \sum_{\{i,j\}} |\phi_i\rangle\langle\phi_i|\phi_j\rangle\langle\phi_j| = \sum_{\{i\}} |\phi_i\rangle\langle\phi_i| = \hat{P}. \tag{1.46}$$

This operator projects the elements of the Hilbert space onto the subspace spanned by $\{|\phi_i\rangle\}$, $i \in \{i\}$.

If two projections $\hat{P}_1$ and $\hat{P}_2$ commute, then $\hat{P}_1\hat{P}_2$ is also a projection

$$(\hat{P}_1\hat{P}_2)^2 = \hat{P}_1\hat{P}_2\hat{P}_1\hat{P}_2 = \hat{P}_1\hat{P}_1\hat{P}_2\hat{P}_2 = \hat{P}_1^2\hat{P}_2^2 = \hat{P}_1\hat{P}_2. \tag{1.47}$$

1.2.5 Adjoint Operators

A linear operator $\hat{A}^\dagger$ is *adjoint* or *Hermitian conjugate* to $\hat{A}$ if for any $\chi, \psi \in \mathcal{H}$

$$\langle\chi|\hat{A}\psi\rangle = \langle\hat{A}^\dagger\chi|\psi\rangle, \tag{1.48}$$

or in another form

$$\langle\chi|\hat{A}|\psi\rangle = \langle\psi|\hat{A}^\dagger|\chi\rangle^*. \tag{1.49}$$

The adjoint operation has the following properties:

- $(\alpha\hat{A})^\dagger = \alpha^*\hat{A}^\dagger$,
- $(\hat{A}^\dagger)^\dagger = \hat{A}$,
- $(\hat{A} + \hat{B})^\dagger = \hat{A}^\dagger + \hat{B}^\dagger$,
- $(\hat{A}\hat{B})^\dagger = \hat{B}^\dagger\hat{A}^\dagger$,
- $(\hat{A}|\psi\rangle)^\dagger = \langle\psi|\hat{A}^\dagger$,
- $(|\psi\rangle\langle\chi|)^\dagger = |\chi\rangle\langle\psi|$.

Considering Eq. (1.49) in a basis representation we find

$$\langle\chi|\hat{A}|\psi\rangle = \langle\chi|I\hat{A}I|\psi\rangle = \sum_{i,j}\langle\chi|\phi_i\rangle\langle\phi_i|\hat{A}|\phi_j\rangle\langle\phi_j|\psi\rangle \tag{1.50}$$

and

$$\begin{aligned}\langle\psi|\hat{A}^\dagger|\chi\rangle^* = \langle\psi|I\hat{A}^\dagger I|\chi\rangle^* &= \left[\sum_{i,j}\langle\psi|\phi_j\rangle\langle\phi_j|\hat{A}^\dagger|\phi_i\rangle\langle\phi_i|\chi\rangle\right]^* \\ &= \sum_{i,j}\langle\chi|\phi_i\rangle(\langle\phi_j|\hat{A}^\dagger|\phi_i\rangle)^*\langle\phi_j|\psi\rangle.\end{aligned} \tag{1.51}$$

These two expressions are, by definition, identical, therefore $\underline{A} = (\underline{A}^\dagger)^{T*}$ or $\underline{A}^\dagger = (\underline{A}^T)^*$. This means that the matrix of the adjoint operator $\hat{A}^\dagger$ is the transpose complex conjugate of the matrix of operator $\hat{A}$.

The linear operator $\hat{A}$ is *symmetric* if on its domain $\hat{A} = \hat{A}^\dagger$ and $\mathcal{D}(\hat{A}) \subset \mathcal{D}(\hat{A}^\dagger)$. If besides that $\mathcal{D}(\hat{A}) = \mathcal{D}(\hat{A}^\dagger)$ the operator is *self adjoint* or *Hermitian*. In a basis representation, the Hermitian operator becomes a Hermitian matrix

$$\underline{A} = (\underline{A}^T)^*. \tag{1.52}$$

The operator is *anti-Hermitian* or *skew-Hermitian* if

$$\hat{B} = -\hat{B}^\dagger. \tag{1.53}$$

An operator is *unitary* if it maps $\mathcal{H} \to \mathcal{H}$ one-to-one and onto such that

$$\hat{U}^\dagger = \hat{U}^{-1}, \quad \text{or} \quad \hat{U}\hat{U}^\dagger = \hat{U}^\dagger\hat{U} = I. \tag{1.54}$$

Unitary operators preserve the scalar product

$$\langle \hat{U}\phi | \hat{U}\psi \rangle = \langle \hat{U}^\dagger \hat{U}\phi | \psi \rangle = \langle \phi | \psi \rangle. \tag{1.55}$$

1.2.6 Function of Operators

A *function of an operator* can be defined by the corresponding power series

$$f(\hat{A}) = \sum_{n=0}^{\infty} a_n \hat{A}^n. \tag{1.56}$$

For example

$$\exp(\alpha\hat{A}) = \sum_{n=0}^{\infty} \frac{\alpha^n}{n!} \hat{A}^n. \tag{1.57}$$

Another example is the *resolvent* operator, which is an inverse operator

$$R(\lambda) = (\lambda - \hat{A})^{-1} = \frac{1}{\lambda}\left(1 + \frac{\hat{A}}{\lambda} + \left(\frac{\hat{A}}{\lambda}\right)^2 + \cdots\right), \tag{1.58}$$

where λ is a complex number.

If $[\hat{A}, \hat{B}] = 0$, then $[f(\hat{A}), \hat{B}] = 0$ and $[f(\hat{A}), g(\hat{B})] = 0$. However, if $[\hat{A}, \hat{B}] \neq 0$, $[f(\hat{A}), \hat{B}] \neq 0$ in general.

For example, $\exp(\hat{A})\hat{B}\exp(-\hat{A}) \neq \hat{B}$, but rather

$$\exp(\hat{A})\hat{B}\exp(-\hat{A}) = \hat{B} + [\hat{A}, \hat{B}] + \frac{1}{2!}[\hat{A}, [\hat{A}, \hat{B}]] + \frac{1}{3!}[\hat{A}, [\hat{A}, [\hat{A}, \hat{B}]]] + \cdots . \tag{1.59}$$

To prove this relation we may consider

$$f(\lambda) = \exp(\lambda\hat{A})\hat{B}\exp(-\lambda\hat{A}). \tag{1.60}$$

We can see that $f(0) = \hat{B}$. Then, taking the derivative, we obtain

$$\frac{\mathrm{d}f}{\mathrm{d}\lambda} = \hat{A}f(\lambda) - f(\lambda)\hat{A} = [\hat{A}, f(\lambda)], \tag{1.61}$$

and in a similar fashion

$$\frac{\mathrm{d}^2 f}{\mathrm{d}\lambda} = \left[\hat{A}, \frac{\mathrm{d}f}{\mathrm{d}\lambda}\right] = [\hat{A}, [\hat{A}, f(\lambda)]]. \tag{1.62}$$

This gives the formal power series

$$\begin{aligned} \exp(\lambda\hat{A})\hat{B}\exp(-\lambda\hat{A}) &= f(\lambda) = f(0) + \frac{f'(0)}{1!}\lambda + \frac{f''(0)}{2!}\lambda^2 + \frac{f'''(0)}{3!}\lambda^3 + \cdots \\ &= \hat{B} + [\hat{A}, \hat{B}]\lambda + \frac{1}{2!}[\hat{A}, [\hat{A}, \hat{B}]]\lambda^2 + \frac{1}{3!}[\hat{A}, [\hat{A}, [\hat{A}, \hat{B}]]]\lambda^3 + \cdots , \end{aligned} \tag{1.63}$$

and with $\lambda = 1$ we get the desired result of Eq. (1.59).

Furthermore, in general, $\exp(\hat{A})\exp(\hat{B}) \neq \exp(\hat{A} + \hat{B})$. What we can say, is that if

$$[\hat{A}, [\hat{A}, \hat{B}]] = [\hat{B}, [\hat{A}, \hat{B}]] = 0, \tag{1.64}$$

then

$$\exp(\hat{A})\exp(\hat{B}) = \exp(\hat{A} + \hat{B} + [\hat{A}, \hat{B}]/2) = \exp(\hat{A} + \hat{B})\exp([\hat{A}, \hat{B}]/2). \tag{1.65}$$

To show this, we define the function

$$f(\lambda) = \exp(\lambda\hat{A})\exp(\lambda\hat{B}). \tag{1.66}$$

Then taking the derivative and considering Eqs. (1.64) and (1.63), we have

$$\begin{aligned} \frac{\mathrm{d}f}{\mathrm{d}\lambda} &= \hat{A}\exp(\lambda\hat{A})\exp(\lambda\hat{B}) + \exp(\lambda\hat{A})\hat{B}\exp(\lambda\hat{B}) \\ &= \left[\hat{A} + \exp(\lambda\hat{A})\hat{B}\exp(-\lambda\hat{A})\right] f(\lambda) \\ &= \left(\hat{A} + \hat{B} + [\hat{A}, \hat{B}]\lambda\right) f(\lambda). \end{aligned} \tag{1.67}$$

By solving this differential equation for $f(\lambda)$ we find

$$f(\lambda) = \exp(\lambda\hat{A})\exp(\lambda\hat{B}) = \exp\left((\hat{A}+\hat{B})\lambda + [\hat{A},\hat{B}]\lambda^2/2\right), \tag{1.68}$$

and with $\lambda = 1$, we again find the desired result of Eq. (1.65). We can also easily verify

$$\exp(\hat{A}+\hat{B}) = \exp(\hat{A})\exp(\hat{B})\exp(-[\hat{A},\hat{B}]/2) = \exp(\hat{B})\exp(\hat{A})\exp([\hat{A},\hat{B}]/2). \tag{1.69}$$

These formulae are special cases of the Baker-Campbell-Hausdorff formula.

1.2.7 Eigenvalues and Eigenvectors

In general, the action of an operator on a state results in a different state

$$|\psi'\rangle = \hat{A}|\psi\rangle, \quad \text{where} \quad |\psi'\rangle \neq |\psi\rangle \quad \text{in general.} \tag{1.70}$$

It is a rather special case when the state vector remains invariant with respect to the action of $\hat{A}$, i.e. when $|\psi'\rangle \sim |\psi\rangle$, or

$$\hat{A}|\psi\rangle = \lambda|\psi\rangle. \tag{1.71}$$

These states are unique to operator $\hat{A}$. The state $|\psi\rangle$ is called the *eigenvector*, or *eigenstate*, and the complex number λ is the *eigenvalue*. The set of all eigenvalues $\{\lambda_i\}$ forms the *spectrum* of the operator. The eigenvalues are degenerate if $\lambda_i = \lambda_j$ for $i \neq j$.

An important property of *Hermitian operators* is that the eigenvalues are real and the eigenvectors are mutually orthogonal. Let us consider the eigenvalue equation

$$\hat{A}|\phi_n\rangle = \alpha_n|\phi_n\rangle. \tag{1.72}$$

Multiplying from the left by the *bra* eigenstate $\langle\phi_m|$ we obtain

$$\langle\phi_m|\hat{A}|\phi_n\rangle = \alpha_n\langle\phi_m|\phi_n\rangle. \tag{1.73}$$

Analogously, we can start from the adjoint relation

$$\langle\phi_m|\hat{A}^\dagger = \alpha_m^*\langle\phi_m|, \tag{1.74}$$

and applying to a vector on the right we get

$$\langle\phi_m|\hat{A}^\dagger|\phi_n\rangle = \alpha_n^*\langle\phi_m|\phi_n\rangle. \tag{1.75}$$

If the operator $\hat{A}$ is Hermitian, i.e. $\hat{A} = \hat{A}^\dagger$, the left sides of Eqs. (1.73) and (1.75) are the same, and subtracting them we find

$$0 = (\alpha_n - \alpha_m^*)\langle\phi_m|\phi_n\rangle. \tag{1.76}$$

If we take $m = n$, then $\langle\phi_n|\phi_n\rangle > 0$, consequently $\alpha_n = \alpha_n^*$, i.e. the eigenvalues are real. On the other hand, if $m \neq n$ and $\alpha_n \neq \alpha_m^*$, then $\langle\phi_m|\phi_n\rangle$ has to be zero, i.e. the eigenstates belonging to different eigenvalues are orthogonal.

In a similar fashion we can also show that the eigenvalues of an anti-Hermitian operator are either purely imaginary or zero.

The *eigenvalues* of a *unitary operator* are complex numbers with *moduli one* and if the eigenvalues are not degenerate, the eigenvectors are mutually *orthogonal*. Indeed, from

$$\hat{U}|\phi_n\rangle = u_n|\phi_n\rangle \quad \text{and} \quad \hat{U}|\phi_m\rangle = u_m|\phi_m\rangle, \tag{1.77}$$

we find

$$\langle\phi_m|\hat{U}^\dagger\hat{U}|\phi_n\rangle = u_m^* u_n\langle\phi_m|\phi_n\rangle = \langle\phi_m|\phi_n\rangle, \tag{1.78}$$

and consequently

$$(u_m^* u_n - 1)\langle\phi_m|\phi_n\rangle = 0. \tag{1.79}$$

So, if $m = n$, $\langle\phi_n|\phi_n\rangle > 0$, and thus

$$u_n^* u_n = |u_n|^2 = 1. \tag{1.80}$$

On the other hand, if $m \neq n$, $u_m \neq u_n$ and $(u_m^* u_n - 1) \neq 0$, then $\langle\phi_m|\phi_n\rangle = 0$, i.e. the eigenstates $|\phi_m\rangle$ and $|\phi_n\rangle$ are orthogonal.

The eigenvectors of a Hermitian operator form a basis in $\mathcal{H}$, i.e. the set of eigenvectors $\{\phi_i\}$, span the space of $\mathcal{H}$ such that

$$I = \sum_i |\phi_i\rangle\langle\phi_i|. \tag{1.81}$$

Using this representation we find that

$$\hat{A} = \hat{A}I = \sum_i \hat{A}|\phi_i\rangle\langle\phi_i| = \sum_i \alpha_i|\phi_i\rangle\langle\phi_i|. \tag{1.82}$$

This means that the operator is diagonal in its own eigenstate basis representation and the diagonal elements are the eigenvalues. Obviously

$$\hat{A}^2 = \sum_{i,j} \alpha_i|\phi_i\rangle\langle\phi_i| \cdot \alpha_j|\phi_j\rangle\langle\phi_j| = \sum_i \alpha_i^2|\phi_i\rangle\langle\phi_i|, \tag{1.83}$$

and similarly

$$\hat{A}^n = \sum_i \alpha_i^n |\phi_i\rangle\langle\phi_i|. \tag{1.84}$$

Consequently,

$$f(\hat{A}) = \sum_i f(\alpha_i)|\phi_i\rangle\langle\phi_i|. \tag{1.85}$$

If $\hat{A}$ and $\hat{B}$ are Hermitian operators and they commute, i.e. $[\hat{A}, \hat{B}] = \hat{A}\hat{B} - \hat{B}\hat{A} = 0$, then they possess *common eigenvectors*. If the null space of $\hat{A} - \lambda$ is one dimensional, each eigenvector of $\hat{A}$ is also an eigenvector of $\hat{B}$, and the other way around. So, we can construct a basis that diagonalizes both operators. Indeed, if a_n a nondegenerate eigenvalue,

$$\hat{A}|\phi_n\rangle = a_n|\phi_n\rangle, \tag{1.86}$$

then

$$\hat{B}\hat{A}|\phi_n\rangle = a_n(\hat{B}|\phi_n\rangle) = \hat{A}(\hat{B}|\phi_n\rangle), \tag{1.87}$$

showing that $\hat{B}|\phi_n\rangle$ is an eigenvector of $\hat{A}$ belonging to the eigenvalue a_n. But since eigenvectors are unique, except for a normalization factor,

$$\hat{B}|\phi_n\rangle \sim |\phi_n\rangle, \tag{1.88}$$

or

$$\hat{B}|\phi_n\rangle = b_n|\phi_n\rangle \tag{1.89}$$

for some b_n. This means that $|\phi_n\rangle$ is an eigenvector of $\hat{B}$ with eigenvalue b_n.

1.2.8 Cauchy Integral Representation of Functions of Operators

We may recall the Cauchy integral representation of complex functions

$$f(z) = \frac{1}{2\pi i}\oint_\gamma \frac{f(\lambda)}{\lambda - z}\mathrm{d}\lambda = \frac{1}{2\pi i}\oint_\gamma f(\lambda)(\lambda - z)^{-1}\mathrm{d}\lambda, \tag{1.90}$$

where γ is a contour in counterclockwise direction around z, and f is analytic on the area encircled by γ. There is an analogous representation for functions of Hermitian operators

$$f(\hat{A}) = \frac{1}{2\pi i}\oint_C \mathrm{d}\lambda f(\lambda)(\lambda - \hat{A})^{-1}, \tag{1.91}$$

where the integration goes in a counterclockwise direction around the perimeter of C that covers the spectrum of $\hat{A}$, and $f(\lambda)$ is analytic on C, including the perimeter.

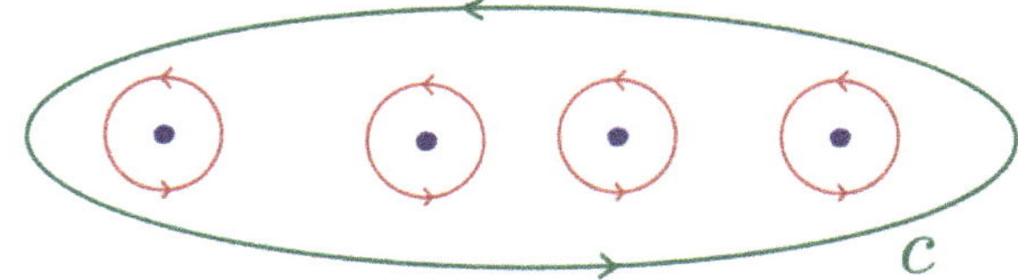

Fig. 1.1 Contour C encircles the spectrum of $\hat{A}$ and by analytic deformation it shrinks to the individual poles

Let $\{a_j\}$ and $\{\phi_j\}$ be the eigenvalues and the corresponding normalized eigenvectors of $\hat{A}$, respectively. The resolvent operator is a function of $\hat{A}$, therefore its action on the eigenstate gives

$$(\lambda - \hat{A})^{-1}|\phi_j\rangle = \frac{1}{\lambda - a_j}|\phi_j\rangle. \tag{1.92}$$

If C'_{a_j} is a small circle around the pole with path $\lambda = a_j + r_0 \exp(i\varphi)$, then $\mathrm{d}\lambda = ir_0 \exp(i\varphi)\mathrm{d}\varphi$, and we find

$$\begin{aligned}\frac{1}{2\pi i}\oint_{C'_{a_j}} \mathrm{d}\lambda(\lambda - \hat{A})^{-1}|\phi_j\rangle\langle\phi_j| &= \frac{1}{2\pi i}\oint_{C'_{a_j}} \mathrm{d}\lambda \frac{1}{\lambda - a_j}|\phi_j\rangle\langle\phi_j| \\ &= \frac{1}{2\pi i}\int_0^{2\pi} \mathrm{d}\varphi \frac{ir_0 \exp(i\varphi)}{a_j + r_0 \exp(i\varphi) - a_j}|\phi_j\rangle\langle\phi_j| = |\phi_j\rangle\langle\phi_j|.\end{aligned} \tag{1.93}$$

Recalling that for a projection operator $\hat{P}^2 = \hat{P}$, we can see that a small circle contour integral of the resolvent operator around an eigenvalue singularity is a projection onto the corresponding eigenstate

$$\frac{1}{2\pi i}\oint_{C'_{a_j}} \mathrm{d}\lambda(\lambda - \hat{A})^{-1} = |\phi_j\rangle\langle\phi_j| = \hat{P}_{a_j}. \tag{1.94}$$

In the representation of functions of operators in Eq. (1.91), $f(\lambda)$ is analytic on the region enclosed by C, which contains all the eigenvalues of $\hat{A}$. According to Cauchy's integral theorem of complex analysis, we can deform the contour C analytically such that the integral is equal to the sum of contour integrals over the small circles enclosing the eigenvalues of $\hat{A}$ (see Fig. 1.1). Consequently,

$$f(\hat{A}) = \frac{1}{2\pi i}\oint_C \mathrm{d}\lambda f(\lambda)(\lambda I - \hat{A})^{-1} = \sum_j \frac{1}{2\pi i}\oint_{C_{a_j}} \mathrm{d}\lambda f(\lambda)(\lambda I - \hat{A})^{-1} = \sum_j f(a_j)|\phi_j\rangle\langle\phi_j|. \tag{1.95}$$

We have found that Eq. (1.91) is indeed another expression for functions of operators. Taking $f \equiv 1$ we obtain a completeness relation for the eigenstates of Hermitian operators

$$I = \sum_j |\phi_j\rangle\langle\phi_j|. \tag{1.96}$$

1.2.9 Unitary Transformations

Let us consider two sets of basis states, $\{|\phi_n\rangle\}$ and $\{|\chi_n\rangle\}$, and define a one-to-one and onto mapping between them by the operator $\hat{U}$

$$\hat{U}|\phi_n\rangle = |\chi_n\rangle \quad \text{and} \quad \langle\phi_n|\hat{U}^\dagger = \langle\chi_n|. \tag{1.97}$$

Now,

$$\hat{U} = \hat{U}I = \hat{U}\sum_n |\phi_n\rangle\langle\phi_n| = \sum_n |\chi_n\rangle\langle\phi_n| \tag{1.98}$$

and

$$\hat{U}^\dagger = I\hat{U}^\dagger = \sum_n |\phi_n\rangle\langle\phi_n|\hat{U}^\dagger = \sum_n |\phi_n\rangle\langle\chi_n|. \tag{1.99}$$

Then,

$$\hat{U}\hat{U}^\dagger = \sum_{n,m} |\chi_n\rangle\langle\phi_n|\phi_m\rangle\langle\chi_m| = \sum_n |\chi_n\rangle\langle\chi_n| = I, \tag{1.100}$$

and

$$\hat{U}^\dagger\hat{U} = \sum_{n,m} |\phi_n\rangle\langle\chi_n|\chi_m\rangle\langle\phi_m| = \sum_n |\phi_n\rangle\langle\phi_n| = I, \tag{1.101}$$

i.e. $\hat{U} = \sum_n |\phi_n\rangle\langle\chi_n|$ is a unitary operator, and the mapping between basis states is a *unitary transformation*.

Unitary transformations, in general, are performed by unitary operators

$$|\psi'\rangle = \hat{U}|\psi\rangle \quad \text{and} \quad \langle\phi'| = \langle\phi|\hat{U}^\dagger. \tag{1.102}$$

Unitary transformations also transform operators. If

$$|\phi\rangle = \hat{A}|\psi\rangle \tag{1.103}$$

then

$$|\phi'\rangle = \hat{U}|\phi\rangle = \hat{U}\hat{A}\hat{U}^\dagger\hat{U}|\psi\rangle = \hat{U}\hat{A}\hat{U}^\dagger|\psi'\rangle = \hat{A}'|\psi'\rangle, \tag{1.104}$$

i.e. operators should transform according to

$$\hat{A}' = \hat{U}\hat{A}\hat{U}^\dagger. \tag{1.105}$$

We can notice that

$$\hat{A}'^2 = \hat{U}\hat{A}\hat{U}^\dagger\hat{U}\hat{A}\hat{U}^\dagger = \hat{U}\hat{A}^2\hat{U}^\dagger, \tag{1.106}$$

and therefore

$$f(\hat{A}') = \hat{U}f(\hat{A})\hat{U}^\dagger. \tag{1.107}$$

We can easily see that if $\hat{A}$ is Hermitian, $\hat{A}'$ is also Hermitian

$$(\hat{A}')^\dagger = (\hat{U}\hat{A}\hat{U}^\dagger)^\dagger = \hat{U}\hat{A}^\dagger\hat{U}^\dagger = \hat{U}\hat{A}\hat{U}^\dagger = \hat{A}', \tag{1.108}$$

i.e. unitary transformations preserve Hermiticity.

Unitary transformations leave the eigenvalues unchanged. Indeed, if $\hat{A}|\psi_n\rangle = a_n|\psi_n\rangle$, then

$$\hat{A}'|\psi_n'\rangle = (\hat{U}\hat{A}\hat{U}^\dagger)(\hat{U}|\psi_n\rangle) = \hat{U}\hat{A}|\psi_n\rangle = \hat{U}a_n|\psi_n\rangle = a_n\hat{U}|\psi_n\rangle = a_n|\psi_n'\rangle. \tag{1.109}$$

Also, if $[\hat{A}, \hat{B}] = cI$ with $c \in \mathbb{C}$, i.e. the commutator is a complex number, then it is also preserved by unitary transformations

$$\begin{aligned}[\hat{A}', \hat{B}'] =&[\hat{U}\hat{A}\hat{U}^\dagger, \hat{U}\hat{B}\hat{U}^\dagger] = \hat{U}\hat{A}\hat{U}^\dagger\hat{U}\hat{B}\hat{U}^\dagger - \hat{U}\hat{B}\hat{U}^\dagger\hat{U}\hat{A}\hat{U}^\dagger \\ =&\hat{U}[\hat{A}, \hat{B}]\hat{U}^\dagger = \hat{U}cI\hat{U}^\dagger = cI = [\hat{A}, \hat{B}].\end{aligned} \tag{1.110}$$

Furthermore, the matrix elements and the scalar products are also preserved

$$\langle\phi'|\hat{A}'|\psi'\rangle = \langle\hat{U}\phi|\hat{U}\hat{A}\hat{U}^\dagger|\hat{U}\psi\rangle = \langle\phi|\hat{U}^\dagger\hat{U}\hat{A}\hat{U}^\dagger\hat{U}|\psi\rangle = \langle\phi|\hat{A}|\psi\rangle. \tag{1.111}$$

Taking $\hat{A} = I$ we obtain the preservation of the scalar product and the norm.

1.2.10 *Infinitesimal Unitary Transformations*

Infinitesimal operators are just infinitesimally different from the unit operator. The operator

$$\hat{U}_\varepsilon = I + i\varepsilon\hat{G} \tag{1.112}$$

is an infinitesimal unitary operator if ε is real and infinitesimal, while the generator $\hat{G}$ is Hermitian. Indeed,

$$\hat{U}_\varepsilon\hat{U}_\varepsilon^\dagger = (I + i\varepsilon\hat{G})(I + i\varepsilon\hat{G})^\dagger = (I + i\varepsilon\hat{G})(I - i\varepsilon\hat{G}^\dagger) \approx I + i\varepsilon(\hat{G} - \hat{G}^\dagger) \approx I, \tag{1.113}$$

since $\hat{G} = \hat{G}^\dagger$ and we can neglect the ε^2 term for infinitesimal ε.

The transformation of the states generated by $\hat{U}_\varepsilon$ is infinitesimal as well,

$$|\psi'\rangle = (I + i\varepsilon\hat{G})|\psi\rangle = |\psi\rangle + i\varepsilon\hat{G}|\psi\rangle = |\psi\rangle + |\delta\psi\rangle, \tag{1.114}$$

and the transformation of operators is also infinitesimal,

$$\begin{aligned}\hat{A}' = (I + i\varepsilon\hat{G})\hat{A}(I + i\varepsilon\hat{G})^\dagger &= (I + i\varepsilon\hat{G})\hat{A}(I - i\varepsilon\hat{G}) \approx \hat{A} + i\varepsilon(\hat{G}\hat{A} - \hat{A}\hat{G}) \\ &\approx \hat{A} + i\varepsilon[\hat{G}, \hat{A}].\end{aligned} \tag{1.115}$$

We can see that if the generator $\hat{G}$ commutes with $\hat{A}$, $[\hat{G}, \hat{A}] = 0$, then $\hat{A}' = \hat{A}$, i.e. $\hat{A}$ is preserved by the infinitesimal unitary transformation.

Infinitely many consecutive infinitesimal unitary transformations results in a finite transformation

$$\hat{U}_\alpha(\hat{G}) = \lim_{N\to\infty} \prod_{k=1}^{N} \left(I + i\frac{\alpha}{N}\hat{G}\right) = \lim_{N\to\infty} \left(I + i\frac{\alpha}{N}\hat{G}\right)^N = \exp(i\alpha\hat{G}), \tag{1.116}$$

which is again unitary, provided α is real and $\hat{G}$ is Hermitian, since

$$\left[\exp(i\alpha\hat{G})\right]^\dagger = \exp(-i\alpha\hat{G}) = \left[\exp(i\alpha\hat{G})\right]^{-1}. \tag{1.117}$$

The operator $\hat{A}$ is transformed with respect to $\hat{U}_\alpha(\hat{G})$ as

$$\begin{aligned} \hat{A}' &= \hat{U}_\alpha(\hat{G})\hat{A}\hat{U}_\alpha^\dagger(\hat{G}) = \exp(i\alpha\hat{G})\hat{A}\exp(-i\alpha\hat{G}) \\ &= \hat{A} + i\alpha[\hat{G}, \hat{A}] + \frac{(i\alpha)^2}{2!}[\hat{G}, [\hat{G}, \hat{A}]] + \cdots . \end{aligned} \tag{1.118}$$

Here again, we find that the operator is invariant with respect to the unitary transformation $\hat{U}_\alpha(\hat{G})$, and $\hat{A}' = \hat{A}$, if $[\hat{G}, \hat{A}] = 0$.

1.2.11 Invariant Quantities

We have found before that the eigenvalues are invariant with respect to unitary transformations. The *trace* is also an invariant. The trace of an operator is defined with the help of a basis $\{\phi_i\}$

$$\mathrm{Tr}(\hat{A}) = \sum_n \langle\phi_n|\hat{A}|\phi_n\rangle < \infty. \tag{1.119}$$

Note that the trace is defined only if the sum is finite. We can see that

$$\begin{aligned} \mathrm{Tr}(\hat{A}\hat{B}) &= \sum_n \langle\phi_n|\hat{A}\hat{B}|\phi_n\rangle = \sum_{n,m} \langle\phi_n|\hat{A}|\phi_m\rangle\langle\phi_m|\hat{B}|\phi_n\rangle = \sum_{n,m} \langle\phi_m|\hat{B}|\phi_n\rangle\langle\phi_n|\hat{A}|\phi_m\rangle \\ &= \sum_m \langle\phi_m|\hat{B}\hat{A}|\phi_m\rangle = \mathrm{Tr}(\hat{B}\hat{A}). \end{aligned} \tag{1.120}$$

Therefore,

$$\mathrm{Tr}(\hat{A}') = \mathrm{Tr}(\hat{U}\hat{A}\hat{U}^\dagger) = \mathrm{Tr}(\hat{U}^\dagger\hat{U}\hat{A}) = \mathrm{Tr}(\hat{A}), \tag{1.121}$$

i.e. the trace is invariant with respect to unitary transformations, and consequently it does not depend on the particular basis representation.

Similarly, the *determinant* of an operator is defined by the determinant of its matrix representation

$$\det(\hat{A}) = \det(\underline{A}) < \infty. \tag{1.122}$$

The determinant is also invariant with respect to unitary transformations

$$\begin{aligned}\det(\hat{A}') &= \det(\hat{U}\hat{A}\hat{U}^{\dagger}) = \det(\hat{U})\det(\hat{A})\det(\hat{U}^{\dagger}) = \det(\hat{U})\det(\hat{A})\det(\hat{U}^{-1})\\ &= \det(\hat{A}),\end{aligned} \tag{1.123}$$

and consequently, it is also independent of the particular basis representation.

1.3 Continuous Basis

So far we have considered bases with discrete index n. If the basis has continuously infinite dimension, the discrete n becomes a continuous variable λ and the summation becomes an integration

$$I = \int d\lambda \; |\lambda\rangle\langle\lambda|. \tag{1.124}$$

The representation of an arbitrary vector is given by

$$|\psi\rangle = I|\psi\rangle = \int d\lambda \; |\lambda\rangle\langle\lambda||\psi\rangle, \tag{1.125}$$

and consequently, we have

$$\langle\lambda'|\psi\rangle = \int d\lambda \; \langle\lambda'|\lambda\rangle\langle\lambda||\psi\rangle. \tag{1.126}$$

We can see that $\langle\lambda'|\lambda\rangle$ must be a δ function

$$\langle\lambda'|\lambda\rangle = \delta(\lambda' - \lambda), \tag{1.127}$$

because only the δ function can pick one value of a function through integrating over the whole interval. Then, Eq. (1.126) becomes

$$\psi(\lambda') = \int d\lambda \; \delta(\lambda' - \lambda)\psi(\lambda). \tag{1.128}$$

The inner product of states in a continuous basis representation is given by

$$\langle\phi|\psi\rangle = \langle\phi|I|\psi\rangle = \int d\lambda \; \langle\phi|\lambda\rangle\langle\lambda|\psi\rangle = \int d\lambda \; \phi^*(\lambda)\,\psi(\lambda), \tag{1.129}$$

while for the matrix elements of an operator, we have

$$\langle\phi|\hat{A}|\psi\rangle = \langle\phi|I\hat{A}I|\psi\rangle = \int d\lambda\, d\lambda'\, \langle\phi|\lambda\rangle\langle\lambda|\hat{A}|\lambda'\rangle\langle\lambda'|\psi\rangle = \int d\lambda\, d\lambda'\, \phi^*(\lambda)\, A(\lambda, \lambda')\, \psi(\lambda'). \tag{1.130}$$

We can see that the abstract *ket* $|\psi\rangle$ in a discrete basis representation becomes a finite or an infinite dimensional vector $(\psi_1, \psi_2, \ldots)$, while in a continuous basis representation it becomes a wave function $\psi(\lambda)$. The operator $\hat{A}$ in a discrete basis representation is a finite or an infinite dimensional matrix $\underline{A}_{nn'}$, while in a continuous basis representation it is the kernel of an integral operator $A(\lambda, \lambda')$.

1.3.1 Configuration Space Representation

The position operator $\hat{x}$ and its eigenvector $|x\rangle$ are defined by

$$\hat{x}|x\rangle = x|x\rangle, \tag{1.131}$$

where the eigenvalue x is a number marking the position. This relation provides us with the resolution of the identity

$$I = \int dx\, |x\rangle\langle x|, \tag{1.132}$$

which leads to the position representation of the state

$$|\psi\rangle = I|\psi\rangle = \int dx\, |x\rangle\langle x|\psi\rangle = \int dx\, \psi(x)|x\rangle, \tag{1.133}$$

where $\langle x|\psi\rangle = \psi(x)$ is the wave function in configuration space, or coordinate, representation. For a matrix element of an operator we obtain

$$\langle\phi|\hat{A}|\psi\rangle = \int dx\, dx'\, \langle\phi|x\rangle\langle x|\hat{A}|x'\rangle\langle x'|\psi\rangle = \int dx\, dx'\, \phi^*(x)\, A(x, x')\, \psi(x'), \tag{1.134}$$

and the norm comes as a special case

$$\langle\psi|\psi\rangle = \int dx\, \langle\psi|x\rangle\langle x|\psi\rangle = \int dx\, \psi^*(x)\, \psi(x) = \int dx\, |\psi(x)|^2. \tag{1.135}$$

1.3.2 Momentum Space Representation

The momentum space representation is completely analogous. Here the basis states are the eigenstates of the momentum operator

$$\hat{p}_x|p_x\rangle = p_x|p_x\rangle, \tag{1.136}$$

where p_x is the numerical value of the x component of the momentum. The corresponding momentum space formulas are as follows:

$$I = \int \mathrm{d}p_x \, |p_x\rangle\langle p_x|, \tag{1.137}$$

$$|\psi\rangle = I|\psi\rangle = \int \mathrm{d}p_x \, |p_x\rangle\langle p_x|\psi\rangle, \tag{1.138}$$

$$\langle\phi|\hat{A}|\psi\rangle = \int \mathrm{d}p_x \, \mathrm{d}p'_x \, \langle\phi|p_x\rangle\langle p_x|\hat{A}|p'_x\rangle\langle p'_x|\psi\rangle = \int \mathrm{d}p_x \, \mathrm{d}p'_x \, \phi^*(p_x) \, A(p_x, p'_x) \, \psi(p'_x), \tag{1.139}$$

$$\langle\psi|\psi\rangle = \int \mathrm{d}p_x \, \langle\psi|p_x\rangle\langle p_x|\psi\rangle = \int \mathrm{d}p_x \, \psi^*(p_x) \, \psi(p_x) = \int \mathrm{d}p_x \, |\psi(p_x)|^2, \tag{1.140}$$

where $\langle p_x|\psi\rangle = \psi(p_x)$ is the momentum space wave function.

All representations are equivalent. Use of a specific representation depends on the nature of the problem and personal preferences. Since the configuration space representation is closer to our physical intuition, this is used most of the time, but momentum space representation also has its definite role in specific problems.

Chapter 2
Basics of Quantum Mechanics

2.1 Postulates of Quantum Mechanics

Quantum mechanics is defined by a set of assumptions, axioms, or postulates. These assumptions are not questioned; they are validated by the results we obtain by using them. The postulates presented here are referred to as the *Copenhagen Interpretation of Quantum Mechanics*, worked out mostly by Niels Bohr, Werner Heisenberg and Max Born.

Non-relativistic quantum mechanics is based on the following postulates:

1. *Postulate on the state*: The quantum state of a physical system at any time t is completely described by a normalized Hilbert space vector $|\psi(t)\rangle \in \mathcal{H}$.
2. *Observables*: Physically measurable quantities, called observables, are represented by Hermitian operators. We assume that the eigenstates of each observable form a complete set in $\mathcal{H}$.
3. *Measurement*: The outcome of a measurement of an observable $\hat{A}$ is one of the a_n eigenvalues of the corresponding Hermitian operator

$$\hat{A}|\phi_n\rangle = a_n|\phi_n\rangle, \tag{2.1}$$

where $|\phi_n\rangle$ is the normalized eigenstate. If the result of a measurement is a_n, the system, immediately after measurement, is given by

$$|\psi\rangle_{after} = |\phi_n\rangle\langle\phi_n \mid \psi\rangle_{before}. \tag{2.2}$$

4. *Probabilistic outcome*: The probability of obtaining an eigenvalue a_n is given by

$$P(a_n) = |\langle\phi_n|\psi(t)\rangle|^2, \tag{2.3}$$

if a_n is a non-degenerate eigenvalue. If a_n is f-fold degenerate, then

Z. Papp, *Mastering Quantum Mechanics*,
https://doi.org/10.1007/978-3-032-09011-9_2

$$P(a_n) = \sum_{j=1}^{f} |\langle \phi_n^{(j)} | \psi(t) \rangle|^2, \tag{2.4}$$

and

$$\frac{\mathrm{d}P(a)}{\mathrm{d}a} = |\psi(a)|^2, \tag{2.5}$$

if a is a continuous eigenvalue. The expectation value of a measurement is the average of all possible outcomes, weighted by their respective likelihood. If the system is in the quantum state $|\psi\rangle$, the expectation value of the measurement of a physical quantity described by the operator $\hat{A}$ is given by

$$\langle A \rangle = \langle \psi | \hat{A} | \psi \rangle = \sum_j a_j |\langle \psi | \phi_j \rangle|^2, \tag{2.6}$$

where a_j and $|\phi_j\rangle$ are the eigenvalues and corresponding eigenvectors of $\hat{A}$, respectively.

5. *Time evolution*: The time evolution of the quantum system is governed by the time-dependent Schrödinger equation

$$i\hbar \frac{\mathrm{d}}{\mathrm{d}t} |\psi(t)\rangle = \hat{H} |\psi(t)\rangle, \tag{2.7}$$

where $\hat{H}$ is the Hamiltonian.

2.2 Properties of Quantum States

2.2.1 Superposition Principle

The *superposition principle* asserts that if $|\psi_1\rangle$ and $|\psi_2\rangle$ are two possible states, then any linear combination of them

$$|\psi\rangle = c_1 |\psi_1\rangle + c_2 |\psi_2\rangle, \tag{2.8}$$

where c_1 and c_2 are complex numbers with $|c_1|^2 + |c_2|^2 \neq 0$, would also be a possible quantum state. The converse is also true, any quantum state can be represented as a linear combination of two or more other distinct states.

The superposition principle is realized by the fact that we model quantum state vectors in Hilbert space, which is a linear vector space. This is also in agreement with the linearity of the Schrödinger equation. Any linear combination of the solutions is also a solution.

2.2.2 Time Evolution in Quantum Mechanics

In quantum mechanics the physical quantities are associated with operators that act on the elements of the Hilbert space. The measured values are the matrix elements of the operator and both the operator and the states are generally time-dependent

$$\langle\psi(t)|\hat{A}(t)|\psi(t)\rangle. \tag{2.9}$$

In dealing with time dependencies, there are three major *pictures* of quantum mechanics.

Schrödinger Picture

The most frequently used picture is the *Schrödinger picture*. Here the time-dependence of the wave function $|\psi(t)\rangle$ is governed by the Schrödinger equation

$$i\hbar\frac{\mathrm{d}}{\mathrm{d}t}|\psi(t)\rangle = \hat{H}(t)|\psi(t)\rangle, \tag{2.10}$$

and the solution can be written in the form

$$|\psi(t)\rangle = \hat{U}(t)|\psi(0)\rangle. \tag{2.11}$$

The time evolution operator $\hat{U}(t)$ satisfies the equation

$$i\hbar\frac{\mathrm{d}}{\mathrm{d}t}\hat{U}(t) = \hat{H}(t)\hat{U}(t), \tag{2.12}$$

and the solution, for the special case of a time independent Hamiltonian, is given by

$$\hat{U}(t) = \exp\Big(-i\hat{H}t/\hbar\Big). \tag{2.13}$$

The matrix element of an operator between states is given by $\langle\psi(t)|\hat{A}(t)|\phi(t)\rangle$. If the operator $\hat{A}(t)$ has explicit time dependence, then

$$\begin{aligned}\frac{\mathrm{d}}{\mathrm{d}t}\langle\psi(t)|\hat{A}(t)|\phi(t)\rangle &= \langle\frac{\mathrm{d}}{\mathrm{d}t}\psi(t)|\hat{A}(t)|\phi(t)\rangle + \langle\psi(t)|\hat{A}(t)|\frac{\mathrm{d}}{\mathrm{d}t}\phi(t)\rangle + \langle\psi(t)|\frac{\partial}{\partial t}\hat{A}(t)|\phi(t)\rangle\\ &= -\frac{1}{i\hbar}\langle\psi(t)|\hat{H}(t)\hat{A}(t)|\phi(t)\rangle + \frac{1}{i\hbar}\langle\psi(t)|\hat{A}(t)\hat{H}(t)|\phi(t)\rangle\\ &+ \langle\psi(t)|\frac{\partial}{\partial t}\hat{A}(t)|\phi(t)\rangle\\ &= \frac{1}{i\hbar}\langle\psi(t)|\Big[\hat{A}(t),\hat{H}(t)\Big]|\phi(t)\rangle + \langle\psi(t)|\frac{\partial}{\partial t}\hat{A}(t)|\phi(t)\rangle.\end{aligned} \tag{2.14}$$

If we take $\hat{A}(t) \equiv I$, we find that

$$\frac{\mathrm{d}}{\mathrm{d}t}\langle\psi(t)|\phi(t)\rangle = 0, \tag{2.15}$$

i.e. the overlap $\langle\psi(t)|\phi(t)\rangle$ and the norm $\|\psi\| = \sqrt{\langle\psi(t)|\psi(t)\rangle}$ are conserved. We can see that if the operator $\hat{A}$ does not depend explicitly on time, i.e. $\partial\hat{A}/\partial t = 0$, and it commutes with the Hamiltonian, $\left[\hat{A}, \hat{H}\right] = 0$, then the expectation value $\langle\hat{A}\rangle = \langle\psi(t)|\hat{A}|\psi(t)\rangle$ is a constant of motion.

We have found that the overlap and the norm is conserved, therefore

$$\begin{aligned}\langle\psi(t)|\phi(t)\rangle =&\langle\hat{U}(t,t_0)\psi(t_0)|\hat{U}(t,t_0)\phi(t_0)\rangle = \langle\psi(t_0)|\hat{U}^\dagger(t,t_0)\hat{U}(t,t_0)|\phi(t_0)\rangle\\ =&\langle\psi(t_0)|\phi(t_0)\rangle,\end{aligned} \tag{2.16}$$

i.e. $\hat{U}^\dagger(t,t_0)\hat{U}(t,t_0) = I$ and $\hat{U}$ is unitary. The time evolution operators form a group. The group contains a unit element and an inverse element

$$\hat{U}(t,t) = 1, \quad \hat{U}^\dagger(t,t_0) = \hat{U}^{-1}(t,t_0) = \hat{U}(t_0,t), \tag{2.17}$$

and the product is also an element of the group

$$\hat{U}(t_1,t_2)\hat{U}(t_2,t_3) = \hat{U}(t_1,t_3) \quad \text{for} \quad t_3 < t_2 < t_1. \tag{2.18}$$

Heisenberg Picture

In the *Heisenberg picture*, the time-dependence of the wave function is transformed out

$$|\psi(t)\rangle_H = \hat{U}^\dagger(t)|\psi(t)\rangle = |\psi(0)\rangle. \tag{2.19}$$

Since the physics does not depend on how we picture quantum mechanics, we should have

$$\langle\psi(t)|\hat{A}(t)|\psi(t)\rangle = \langle\psi(t)|\hat{U}\hat{U}^\dagger\hat{A}(t)\hat{U}\hat{U}^\dagger|\psi(t)\rangle = {}_H\langle\psi(t)|\hat{A}_H(t)|\psi(t)\rangle_H, \tag{2.20}$$

where

$$\hat{A}_H(t) = \hat{U}^\dagger(t)\hat{A}(t)\hat{U}(t). \tag{2.21}$$

Since $\hat{U}(t)$ is a function of $\hat{H}$, they commute, and therefore

$$\hat{H}_H = \hat{U}^\dagger(t)\hat{H}\hat{U}(t) = \hat{H}. \tag{2.22}$$

So, at $t = 0$ we have

$$|\psi(0)\rangle_H = |\psi(0)\rangle \qquad \text{and} \qquad \hat{A}_H(0) = \hat{A}. \tag{2.23}$$

If we assume that the Schrödinger picture operator $\hat{A}$ is time independent, i.e. $\hat{A}(\not{t})$, we find for the time evolution in the Heisenberg picture

$$\begin{aligned}\frac{\mathrm{d}}{\mathrm{d}t}\hat{A}_H =&\frac{\mathrm{d}\hat{U}^\dagger}{\mathrm{d}t}\hat{A}\hat{U} + \hat{U}^\dagger\hat{A}\frac{\mathrm{d}\hat{U}}{\mathrm{d}t} = -\frac{1}{i\hbar}\hat{U}^\dagger\hat{H}\hat{U}\hat{U}^\dagger\hat{A}\hat{U} + \frac{1}{i\hbar}\hat{U}^\dagger\hat{A}\hat{U}\hat{U}^\dagger\hat{H}\hat{U} \\ =&\frac{1}{i\hbar}[\hat{A}_H, \hat{H}_H] = \frac{1}{i\hbar}[\hat{A}_H, \hat{H}].\end{aligned} \tag{2.24}$$

We again find that when $\hat{A}$ does not have explicit time dependence and commutes with the Hamiltonian, then $\hat{A}_H$ also commutes with $\hat{H}$, and consequently $\hat{A}$ is a constant of motion.

Interaction or Dirac Picture

In this approach we assume that the Hamiltonian is split into time independent $\hat{H}_0$ and time dependent $\hat{V}(t)$

$$\hat{H} = \hat{H}_0 + \hat{V}(t). \tag{2.25}$$

We define the *interaction picture* state by

$$|\psi(t)\rangle_I = \exp\Big(i\hat{H}_0 t/\hbar\Big)|\psi(t)\rangle. \tag{2.26}$$

So, just like in the Heisenberg picture, we transform away the time dependence of the state, but here only a part of it, which is due to $\hat{H}_0$. At $t = 0$, we have $|\psi(0)\rangle_I = |\psi(0)\rangle$.

The time dependence of the state is described by the interaction picture Schrödinger equation

$$\begin{aligned}i\hbar\frac{\mathrm{d}}{\mathrm{d}t}|\psi(t)\rangle_I &= -\hat{H}_0 \exp\Big(i\hat{H}_0 t/\hbar\Big)|\psi(t)\rangle + \exp\Big(i\hat{H}_0 t/\hbar\Big)\Big(i\hbar\frac{\partial}{\partial t}|\psi(t)\rangle\Big) \\ &= -\hat{H}_0 \exp\Big(i\hat{H}_0 t/\hbar\Big)|\psi(t)\rangle + \exp\Big(i\hat{H}_0 t/\hbar\Big)(\hat{H}_0 + \hat{V}(t))|\psi(t)\rangle \\ &= \exp\Big(i\hat{H}_0 t/\hbar\Big)\hat{V}(t)\exp\Big(-i\hat{H}_0 t/\hbar\Big)\exp\Big(i\hat{H}_0 t/\hbar\Big)|\psi(t)\rangle \\ &= \hat{V}_I(t)|\psi(t)\rangle_I,\end{aligned} \tag{2.27}$$

where

$$\hat{V}_I(t) = \exp\Big(i\hat{H}_0 t/\hbar\Big)\hat{V}(t)\exp\Big(-i\hat{H}_0 t/\hbar\Big). \tag{2.28}$$

In general, the interaction picture operator is defined by

$$\hat{A}_I(t) = \exp\Big(i\hat{H}_0 t/\hbar\Big)\hat{A}(t)\exp\Big(-i\hat{H}_0 t/\hbar\Big). \tag{2.29}$$

If $\hat{A}$ does not have explicit time dependence, i.e. $\hat{A}(\not{t})$, we find

$$\begin{aligned}\frac{\mathrm{d}}{\mathrm{d}t}\hat{A}_I &= \frac{i}{\hbar}\hat{H}_0 \exp\left(i\hat{H}_0 t/\hbar\right)\hat{A}\exp\left(-i\hat{H}_0 t/\hbar\right) \\ &\quad - \frac{i}{\hbar}\exp\left(i\hat{H}_0 t/\hbar\right)\hat{A}\hat{H}_0 \exp\left(-i\hat{H}_0 t/\hbar\right) \\ &= \frac{1}{i\hbar}(\hat{A}_I\hat{H}_0 - \hat{H}_0\hat{A}_I) = \frac{1}{i\hbar}[\hat{A}_I, \hat{H}_0] = \frac{1}{i\hbar}[\hat{A}_I, \hat{H}_{0I}],\end{aligned} \tag{2.30}$$

and if $\hat{A}_I$ commutes with $\hat{H}_0$, $\hat{A}_I$ is a constant of motion.

2.2.3 Stationary Time Evolution

If the Hamiltonian does not depend on time, we can assume that the wave function is a product of a time-dependent and a time-independent parts

$$|\psi(t)\rangle = f(t)|\psi\rangle. \tag{2.31}$$

Substituting into the time dependent Schrödinger equation we find

$$i\hbar\left(\frac{\mathrm{d}f(t)}{\mathrm{d}t}\right)|\psi\rangle = f(t)(\hat{H}|\psi\rangle), \tag{2.32}$$

or

$$\frac{i\hbar}{f(t)}\left(\frac{\mathrm{d}f(t)}{\mathrm{d}t}\right)|\psi\rangle = \hat{H}|\psi\rangle. \tag{2.33}$$

We can see that the time-dependent left side is equal to the time-independent right side for all values of time t. This is possible only if the time dependent part is a constant. Therefore,

$$\frac{i\hbar}{f(t)}\left(\frac{\mathrm{d}f(t)}{\mathrm{d}t}\right) = E \quad \text{and} \quad \hat{H}|\psi\rangle = E|\psi\rangle, \tag{2.34}$$

with E constant. Consequently,

$$f(t) = \exp(-iEt/\hbar) \tag{2.35}$$

and

$$|\psi(t)\rangle = \exp(-iEt/\hbar)|\psi\rangle. \tag{2.36}$$

The Eq. (2.34) is the time-independent Schrödinger equation with E as the energy eigenvalue. The time dependence of the stationary state is governed by E. If the state is a superposition of states with different energies, the time evolution is a linear combination of stationary time evolutions

$$|\psi(t)\rangle = \sum_n c_n \exp(-iE_n t/\hbar)|\psi_n\rangle, \tag{2.37}$$

with coefficients c_n, which are determined by the initial conditions.

2.3 Representation of Observables by Operators

2.3.1 Correspondence Principle

The guiding principle of representing the quantum mechanical observables by operators is the *correspondence principle*. Physical quantities in quantum mechanics should converge to their classical mechanical counterparts in the macroscopic limit. The time evolution of classical mechanical quantities is given by the Poisson bracket with the Hamilton function

$$\frac{\mathrm{d}A}{\mathrm{d}t} = \{A, H\} + \frac{\partial A}{\partial t}. \tag{2.38}$$

This relation is analogous to the time dependence of the quantum mechanical expectation value in Eq. (2.14)

$$\frac{\mathrm{d}}{\mathrm{d}t}\left\langle \hat{A} \right\rangle = \frac{1}{i\hbar}\left\langle [\hat{A}, \hat{H}] \right\rangle + \left\langle \frac{\partial \hat{A}}{\partial t} \right\rangle. \tag{2.39}$$

We should also remember that the commutator algebra is analogous to the algebra of Poisson brackets. These suggest that operators in quantum mechanics should follow commutation rules analogous to the classical mechanical Poisson bracket rules,

$$\{A, B\} \longrightarrow \frac{1}{i\hbar}\left[\hat{A}, \hat{B}\right]. \tag{2.40}$$

In classical mechanics, the cartesian position and momentum vector coordinates are denoted by $\boldsymbol{r}(x, y, z)$ and $\boldsymbol{p}(p_x, p_y, p_z)$, respectively, and their components follow the Poisson bracket relations

$$\{r_i, r_j\} = 0, \quad \{p_i, p_j\} = 0, \quad \{r_i, p_j\} = \delta_{ij}. \tag{2.41}$$

In quantum mechanics, we postulate analogous commutation relations

$$[\hat{r}_i, \hat{r}_j] = 0, \quad [\hat{p}_i, \hat{p}_j] = 0, \quad [\hat{r}_i, \hat{p}_j] = i\hbar\delta_{ij}. \tag{2.42}$$

If a physical variable in classical mechanics is a function of $\boldsymbol{r}$ and $\boldsymbol{p}$, $f(\boldsymbol{r}, \boldsymbol{p})$, the corresponding quantum analog is the Hermitian operator $\hat{f}(\hat{\boldsymbol{r}}, \hat{\boldsymbol{p}})$. Consequently, the

Hamilton operator of a particle with mass m moving in a potential is given by

$$\hat{H} = \frac{\hat{\boldsymbol{p}}^2}{2m} + \hat{V}(\hat{\boldsymbol{r}}). \tag{2.43}$$

The angular momentum operator of a single particle is given by

$$\hat{\boldsymbol{L}} = \hat{\boldsymbol{r}} \times \hat{\boldsymbol{p}}, \tag{2.44}$$

which, in cartesian coordinates, amounts to

$$\hat{L}_x = \hat{y}\hat{p}_z - \hat{z}\hat{p}_y, \quad \hat{L}_y = \hat{z}\hat{p}_x - \hat{x}\hat{p}_z, \quad \hat{L}_z = \hat{x}\hat{p}_y - \hat{y}\hat{p}_x. \tag{2.45}$$

From the commutation relations of coordinates and momenta, we can easily derive the commutation relation between the components of the angular momentum operator

$$[\hat{L}_x, \hat{L}_y] = i\hbar\hat{L}_z, \quad [\hat{L}_y, \hat{L}_z] = i\hbar\hat{L}_x, \quad [\hat{L}_z, \hat{L}_x] = i\hbar\hat{L}_y. \tag{2.46}$$

These results are in complete agreement with the corresponding classical mechanical Poisson bracket relations

$$\{L_i, L_j\} = \epsilon_{ijk}L_k \quad \text{and} \quad \{\boldsymbol{L}^2, L_i\} = 0. \tag{2.47}$$

2.3.2 *Momentum Operator in Configuration Space Representation*

According to the correspondence principle, the commutation relation between position and momentum is given by

$$\left[\hat{x}, \hat{p}_x\right] = i\hbar. \tag{2.48}$$

Applying this to an arbitrary state in configuration space, we have

$$i\hbar\phi(x) = \left[\hat{x}, \hat{p}_x\right]\phi(x) = \hat{x}\hat{p}_x\phi(x) - \hat{p}_x\hat{x}\phi(x) = \hat{x}\hat{p}_x\phi(x) - \hat{p}_x x\phi(x). \tag{2.49}$$

We know from classical mechanics, that the momentum operator is the infinitesimal generator of the spatial translation. Ergo, it should be proportional to $\partial/\partial x$, i.e. $\hat{p}_x = \alpha\partial/\partial x$. Putting this into the previous relation we find

$$i\hbar\phi = x\alpha\partial\phi/\partial x - \alpha\partial x\phi/\partial x = -\alpha\phi, \tag{2.50}$$

then $\alpha = -i\hbar$ and

$$\hat{p}_x = -i\hbar\frac{\partial}{\partial x}. \tag{2.51}$$

The kinetic energy operator is given by

$$\hat{K} = \frac{\hat{p}_x^2}{2m} = -\frac{\hbar^2}{2m}\frac{\partial^2}{\partial x^2}, \tag{2.52}$$

or, in three spatial dimensions we have

$$\hat{K} = \frac{\hat{\boldsymbol{p}}^2}{2m} = -\frac{\hbar^2}{2m}\left(\frac{\partial^2}{\partial x^2} + \frac{\partial^2}{\partial y^2} + \frac{\partial^2}{\partial z^2}\right) = -\frac{\hbar^2}{2m}\Delta\,. \tag{2.53}$$

We can also determine the configuration space eigenstates $\phi_p(x) = \langle x|p_x\rangle$ of the momentum operator

$$\hat{p}_x\phi_p(x) = -i\hbar\frac{\partial}{\partial x}\phi_p(x) = p_x\phi_p(x). \tag{2.54}$$

We see immediately that

$$\langle x|p_x\rangle = \phi_p(x) = C\exp(ip_x x/\hbar), \tag{2.55}$$

or, if we introduce the wave number $k_x = p_x/\hbar$, we have

$$\langle x|k_x\rangle = \phi_k(x) = C'\exp(ik_x x). \tag{2.56}$$

The eigenstates of the momentum operator are normalized to the δ function

$$\langle p_x'|p_x\rangle = \delta(p_x' - p_x) = \int_{-\infty}^{\infty} \mathrm{d}x\,\langle p_x'|x\rangle\,\langle x|p_x\rangle = |C|^2\int_{-\infty}^{\infty} \mathrm{d}x\;\exp(-ip_x'x/\hbar)\,\exp(ip_x x/\hbar), \tag{2.57}$$

and from the representation

$$\delta(x-a) = \frac{1}{2\pi}\int_{-\infty}^{\infty} \mathrm{d}y\,\exp[\pm i(x-a)y] \tag{2.58}$$

we infer that

$$\langle x|p_x\rangle = \phi_p(x) = \frac{1}{\sqrt{2\pi\hbar}}\exp(ip_x x/\hbar). \tag{2.59}$$

A similar normalization of $\phi_k(x)$ to $\delta(k_x' - k_x)$ results in

$$\langle x|k_x\rangle = \phi_k(x) = \frac{1}{\sqrt{2\pi}}\exp(ik_x x). \tag{2.60}$$

The corresponding three-dimensional plane waves are given by

$$\langle \boldsymbol{r}|\boldsymbol{p}\rangle = \phi_{\boldsymbol{p}}(\boldsymbol{r}) = \frac{1}{(2\pi\hbar)^{3/2}}\exp(i\boldsymbol{p}\boldsymbol{r}/\hbar) \tag{2.61}$$

and

$$\langle \boldsymbol{r}|\boldsymbol{k}\rangle = \phi_{\boldsymbol{k}}(\boldsymbol{r}) = \frac{1}{(2\pi)^{3/2}}\exp(i\boldsymbol{k}\boldsymbol{r}), \tag{2.62}$$

respectively.

We can easily prove the following commutation relations

$$\begin{aligned} &\left[\hat{x}^n, \hat{p}_x\right] = i\hbar n\hat{x}^{n-1}, \quad \left[\hat{x}, \hat{p}_x^n\right] = i\hbar n\hat{p}_x^{n-1}, \\ &\left[f(\hat{x}), \hat{p}_x\right] = i\hbar\frac{\partial}{\partial\hat{x}}f(\hat{x}), \quad \left[\hat{x}, f(\hat{p}_x)\right] = i\hbar\frac{\partial}{\partial\hat{p}_x}f(\hat{p}_x), \end{aligned} \tag{2.63}$$

and we can infer the position operator in momentum representation

$$\hat{x} = i\hbar\frac{\partial}{\partial p_x}. \tag{2.64}$$

2.3.3 Probability Current

The time dependent Schrödinger equation for a particle moving in a real potential $V(x)$ is given by

$$i\hbar\frac{\mathrm{d}\psi}{\mathrm{d}t} = -\frac{\hbar^2}{2m}\frac{\partial^2\psi}{\partial x^2} + V(x)\psi, \tag{2.65}$$

and its complex conjugate reads

$$-i\hbar\frac{\mathrm{d}\psi^*}{\mathrm{d}t} = -\frac{\hbar^2}{2m}\frac{\partial^2\psi^*}{\partial x^2} + V(x)\psi^*. \tag{2.66}$$

If we multiply the first equation by ψ^*, the second one by ψ, and subtract them, we obtain

$$i\hbar\frac{\mathrm{d}}{\mathrm{d}t}(\psi^*\psi) = -\frac{\hbar^2}{2m}\left(\psi^*\frac{\partial^2\psi}{\partial x^2} - \psi\frac{\partial^2\psi^*}{\partial x^2}\right). \tag{2.67}$$

By introducing the probability density

$$\rho(x,t) = \psi^*(x,t)\psi(x,t) \tag{2.68}$$

and probability current

$$j(x,t) = \frac{i\hbar}{2m}\left(\psi\frac{\partial\psi^*}{\partial x} - \psi^*\frac{\partial\psi}{\partial x}\right) \tag{2.69}$$

we can cast Eq. (2.67) into the form

$$\frac{\mathrm{d}\rho(x,t)}{\mathrm{d}t} + \frac{\partial j(x,t)}{\partial x} = 0. \tag{2.70}$$

This equation expresses the conservation of probability. The corresponding three dimensional formulae are given by

$$\rho(\boldsymbol{r},t) = \psi^*(\boldsymbol{r},t)\psi(\boldsymbol{r},t), \quad \boldsymbol{j}(\boldsymbol{r},t) = \frac{i\hbar}{2m}\left(\psi\nabla\psi^* - \psi^*\nabla\psi\right), \tag{2.71}$$

and

$$\frac{\mathrm{d}\rho(\boldsymbol{r},t)}{\mathrm{d}t} + \nabla\cdot\boldsymbol{j}(\boldsymbol{r},t) = 0. \tag{2.72}$$

2.3.4 Consecutive Measurement of Observables

Two operators $\hat{A}$ and $\hat{B}$ are said to be compatible if they commute, i.e. $[\hat{A}, \hat{B}] = 0$. We have seen before that in this case they share common eigenvectors

$$\hat{A}|\psi_n\rangle = a_n|\psi_n\rangle \quad \text{and} \quad \hat{B}|\psi_n\rangle = b_n|\psi_n\rangle. \tag{2.73}$$

Let us examine the outcome of two consecutive measurements on state $|\psi\rangle$.

- *$\hat{A}$ and $\hat{B}$ are not compatible*. By applying $\hat{A}$ first and then $\hat{B}$, we find

$$|\psi\rangle \xrightarrow{\hat{A}} |\psi_{n_a}^{(a)}\rangle\langle\psi_{n_a}^{(a)}|\psi\rangle \xrightarrow{\hat{B}} |\psi_{n_b}^{(b)}\rangle\langle\psi_{n_b}^{(b)}|\psi_{n_a}^{(a)}\rangle\langle\psi_{n_a}^{(a)}|\psi\rangle. \tag{2.74}$$

So, the final state is an eigenstate of $\hat{B}$.
If we apply $\hat{B}$ first, and then $\hat{A}$, we obtain

$$|\psi\rangle \xrightarrow{\hat{B}} |\psi_{n_b}^{(b)}\rangle\langle\psi_{n_b}^{(b)}|\psi\rangle \xrightarrow{\hat{A}} |\psi_{n_a}^{(a)}\rangle\langle\psi_{n_a}^{(a)}|\psi_{n_b}^{(b)}\rangle\langle\psi_{n_b}^{(b)}|\psi\rangle, \tag{2.75}$$

and the final state is an eigenstate of $\hat{A}$.
Operators $\hat{A}$ and $\hat{B}$, in general, do not share eigenstates. Therefore, the outcome of two consecutive measurements depends on the sequence in which the operators have been applied.
- *$\hat{A}$ and $\hat{B}$ are compatible*. Here, by applying $\hat{A}$, and then $\hat{B}$, gives

$$|\psi\rangle \xrightarrow{\hat{A}} |\psi_{n_a}^{(a)}\rangle\langle\psi_{n_a}^{(a)}|\psi\rangle \xrightarrow{\hat{B}} |\psi_{n_a}^{(a)}\rangle\langle\psi_{n_a}^{(a)}|\psi\rangle. \tag{2.76}$$

Now, since $|\psi_{n_a}^{(a)}\rangle$ is also an eigenstate of $\hat{B}$, applying $\hat{B}$ does not change the state. Therefore, either sequence of applications leads to one of their common eigenstates.

So, if $\hat{A}$ and $\hat{B}$ do not commute, they cannot be measured simultaneously and the sequence matters. On the other hand, compatible observables can be measured simultaneously with arbitrary accuracy.

What if operator $\hat{A}$ has degenerate eigenvalues? In this case, measuring an eigenvalue does not specify the eigenstate. Then we should measure with $\hat{B}$, which should be compatible with $\hat{A}$. This measurement may resolve this degeneracy since only a subset of the degenerate eigenstates of $\hat{A}$ are going to be eigenstates of $\hat{B}$. If this eigenvalue is still a degenerate eigenvalue of $\hat{B}$, to resolve the degeneracy, we need to apply another operator $\hat{C}$, which should commute with both $\hat{A}$ and $\hat{B}$. We should keep doing this until the degeneracy is removed for all eigenstates. The corresponding set of operators form a *complete set of commuting observables*. Their eigenvalues uniquely determine the system. For example, in the problem of a particle moving in a spherical potential, $\hat{H}$, $\hat{L}^2$ and $\hat{L}_z$ form a complete set of commuting observables.

We have found that if two or more operators mutually commute, we can find for them a complete set of simultaneous eigenvectors. The converse is also true: if two operators $\hat{A}$ and $\hat{B}$ have a complete set of simultaneous eigenvectors, then these operators commute. Assume, we have a complete set of eigenvectors $|n, \nu\rangle$ which are eigenvectors of $\hat{A}$ and $\hat{B}$ with eigenvalues a_n and b_ν, respectively, such that

$$\hat{A}|n, \nu\rangle = a_n|n, \nu\rangle \quad \text{and} \quad \hat{B}|n, \nu\rangle = b_\nu|n, \nu\rangle. \tag{2.77}$$

Operating with $\hat{B}$ on the first equation and $\hat{A}$ on the second one gives

$$\hat{B}\hat{A}|n, \nu\rangle = a_n b_\nu|n, \nu\rangle \quad \text{and} \quad \hat{A}\hat{B}|n, \nu\rangle = b_\nu a_n|n, \nu\rangle, \tag{2.78}$$

and subtracting, we find

$$(\hat{A}\hat{B} - \hat{B}\hat{A})|n, \nu\rangle = 0. \tag{2.79}$$

Since this result is for a complete set of vectors $|n, \nu\rangle$, this must be true for any linear combinations, and therefore for any vector in that Hilbert space. Hence

$$\hat{A}\hat{B} - \hat{B}\hat{A} = [\hat{A}, \hat{B}] = 0, \tag{2.80}$$

i.e. $\hat{A}$ and $\hat{B}$ commute.

2.3.5 Uncertainty of Measurements

If we measure observables $\hat{A}$ and $\hat{B}$ repeatedly on a state $|\psi\rangle$, prepared in the same way, we find the expectation values

$$\langle A\rangle = \langle\psi|\hat{A}|\psi\rangle \quad \text{and} \quad \langle B\rangle = \langle\psi|\hat{B}|\psi\rangle. \tag{2.81}$$

It is interesting to see what their uncertainties are, i.e. what are the average deviations from the expectation values

$$\Delta A = \sqrt{\langle(\hat{A} - \langle A\rangle)^2\rangle} = \sqrt{\langle\hat{A}^2 - 2\hat{A}\langle A\rangle + \langle A\rangle^2\rangle} = \sqrt{\langle\hat{A}^2\rangle - \langle\hat{A}\rangle^2}. \tag{2.82}$$

The *Heisenberg uncertainty relation* sets up a limit between the uncertainties of simultaneous measurement of physical quantities.

Let us define the operator

$$\hat{C} = (\hat{A} - \langle\hat{A}\rangle) + i\alpha(\hat{B} - \langle\hat{B}\rangle), \tag{2.83}$$

where α is a real number. Since $\hat{C}|\psi\rangle \in \mathcal{H}$, we find

$$0 \le \langle\psi|\hat{C}^\dagger\hat{C}|\psi\rangle = \langle\psi|[(\hat{A} - \langle\hat{A}\rangle) - i\alpha(\hat{B} - \langle\hat{B}\rangle)][(\hat{A} - \langle\hat{A}\rangle) + i\alpha(\hat{B} - \langle\hat{B}\rangle)]|\psi\rangle. \tag{2.84}$$

Expanding out, we obtain

$$\langle(\hat{A} - \langle A\rangle)^2\rangle + i\alpha\langle(\hat{A}\hat{B} - \hat{B}\hat{A})\rangle + \alpha^2\langle(\hat{B} - \langle B\rangle)^2\rangle \ge 0, \tag{2.85}$$

or

$$(\Delta A)^2 + i\alpha\langle[\hat{A}, \hat{B}]\rangle + \alpha^2(\Delta B)^2 \ge 0. \tag{2.86}$$

This is valid for any value of α. However, the left hand side has minimum at

$$\alpha = -i\frac{\langle[\hat{A}, \hat{B}]\rangle}{2(\Delta B)^2}. \tag{2.87}$$

The minimum value is

$$(\Delta A)^2 + \frac{\langle[\hat{A}, \hat{B}]\rangle^2}{2(\Delta B)^2} - \frac{\langle[\hat{A}, \hat{B}]\rangle^2}{4(\Delta B)^4}(\Delta B)^2 = (\Delta A)^2 + \frac{\langle[\hat{A}, \hat{B}]\rangle^2}{4(\Delta B)^2} \ge 0, \tag{2.88}$$

which gives

$$\Delta A^2 \cdot \Delta B^2 \ge \frac{1}{4}\langle i[\hat{A}, \hat{B}]\rangle^2, \tag{2.89}$$

or

$$\Delta A \cdot \Delta B \ge \frac{1}{2}\langle \pm i[\hat{A}, \hat{B}]\rangle, \tag{2.90}$$

or rather

$$\Delta A \cdot \Delta B \ge \frac{1}{2}|\langle[\hat{A}, \hat{B}]\rangle|. \tag{2.91}$$

This is the famous *Heisenberg uncertainty relation*. It has been derived originally for position and momentum operators with the help of Fourier transforms. The present, more general derivation relates the uncertainty of non-commuting operators, and it has been derived by Robertson and Schrödinger. If we take $\hat{A} = \hat{x}_i$ and $\hat{B} = \hat{p}_j$, then $[\hat{x}_i, \hat{p}_j] = i\hbar\delta_{ij}$, we get the uncertainty relation of measuring the position and momentum simultaneously

$$\Delta x_i \cdot \Delta p_j \geq \frac{\hbar}{2}\delta_{ij}. \tag{2.92}$$

Minimal Uncertainty State

We can see from Eq. (2.84) that with a proper choice of $|\psi\rangle$, we can achieve equality if

$$[(\hat{A} - \langle\hat{A}\rangle) + i\alpha(\hat{B} - \langle\hat{B}\rangle)]|\psi\rangle = 0. \tag{2.93}$$

This relation defines the minimal uncertainty state. From Eqs. (2.87) and (2.90) we find that

$$i\alpha = \frac{\Delta A}{\Delta B}. \tag{2.94}$$

For $\hat{x}$ and $\hat{p}_x$ this state is defined by the differential equation

$$\left(-i\hbar\frac{\partial}{\partial x} - p_0\right)\psi = -\frac{\Delta p}{\Delta x}(x - x_0)\psi, \tag{2.95}$$

where $p_0 = \langle p\rangle$ and $x_0 = \langle x\rangle$. The solution is the normalized Gaussian wave packet

$$\psi = \frac{1}{(2\pi\sigma^2)^{1/4}}\exp\left(-\frac{(x - x_0)^2}{4\sigma^2} + i\frac{p_0}{\hbar}x\right), \tag{2.96}$$

where $\sigma^2 = \hbar\Delta x/(2\Delta p)$.

Angular Momentum Uncertainty Relations

The commutation relations between the components of the angular momentum operator lead to the corresponding uncertainty relations. If we take $\hat{A} = \hat{J}_x$ and $\hat{B} = \hat{J}_y$, and considering that $\left[\hat{J}_x, \hat{J}_y\right] = i\hbar\hat{J}_z$, we get

$$\Delta J_x \cdot \Delta J_y \geq \frac{\hbar}{2}|\langle\hat{J}_z\rangle|. \tag{2.97}$$

On the other hand, there exists no uncertainty relation between angle ϕ and $\hat{L}_z$, although they are canonically conjugate variables. One would define their action on an arbitrary function $f(\phi)$ as

$$\hat{L}_z f(\phi) = -i\hbar \frac{\partial}{\partial \phi} f(\phi) \quad \text{and} \quad \hat{\phi} f(\phi) = \phi f(\phi), \tag{2.98}$$

with ϕ being an angle and $f(\phi + 2\pi) = f(\phi)$. However, the function $\phi f(\phi)$ does not satisfy this periodicity requirement. The action of $\hat{\phi}$ on a legitimate function $f(\phi)$ leads to an illegitimate function $\phi f(\phi)$, and therefore $\hat{\phi}$ is not self-adjoint. What is possible in this regard is the use of the commutation relations

$$\left[\sin\phi, \hat{L}_z\right] = i\hbar\cos\phi \quad \text{and} \quad \left[\cos\phi, \hat{L}_z\right] = -i\hbar\sin\phi, \tag{2.99}$$

which results in

$$\Delta L_z \cdot \Delta \sin\phi \geq \frac{\hbar}{2} |\langle \cos\phi \rangle| \quad \text{and} \quad \Delta L_z \cdot \Delta \cos\phi \geq \frac{\hbar}{2} |\langle \sin\phi \rangle|. \tag{2.100}$$

Time-Energy Uncertainty Relation

While the time-energy uncertainty relation is referred to quite often in physical arguments, this relation cannot be derived directly from the above considerations. In quantum mechanics, time is a parameter and it cannot be represented by a Hermitian operator. Time evolution is generated by the Hamiltonian, or the energy operator, in the same manner as the spatial displacement is generated by the momentum operator. The latter leads to the commutation relation between position and momentum. It would be tempting to conjecture an analogous commutation relation between the energy and time

$$[t, E] = -i\hbar, \tag{2.101}$$

and identify the energy operator as

$$E = H = i\hbar \frac{\partial}{\partial t}. \tag{2.102}$$

Indeed, with this correspondence, Eq. (2.101) reads

$$[t, E]\,|\phi\rangle = t\, i\hbar \frac{\partial}{\partial t} |\phi\rangle - i\hbar \frac{\partial}{\partial t} (t|\phi\rangle) = -i\hbar |\phi\rangle, \tag{2.103}$$

where $|\phi\rangle \in \mathcal{H}$ is arbitrary.

Let us assume that $|\psi_0\rangle$ is the ground state of a physical system described by the time-independent Hamiltonian

$$\hat{H}|\psi_0\rangle = E_0|\psi_0\rangle. \tag{2.104}$$

Let us take the operator $\hat{B} = \exp(i\omega t)$, with real ω, which just changes the phase of a state. From Eqs. (2.101) and (2.102), it follows that

$$\left[\hat{H}, \hat{B}\right] = \left[i\hbar\frac{\partial}{\partial t}, \exp(i\omega t)\right] = -\hbar\omega\hat{B}. \tag{2.105}$$

Consequently,

$$\hat{H}\hat{B}|\psi_0\rangle = \hat{B}\hat{H}|\psi_0\rangle - \hbar\omega\hat{B}|\psi_0\rangle, \quad \text{or} \quad \hat{H}(\hat{B}|\psi_0\rangle) = (E_0 - \hbar\omega)(\hat{B}|\psi_0\rangle). \tag{2.106}$$

So, $\hat{B}|\psi_0\rangle$ would be an eigenstate of $\hat{H}$ with eigenvalue $E_0 - \hbar\omega$, which is lower then E_0. Then E_0 would not be the ground state of the system for any $t \neq 0$. All these controversies came from the commutation relation Eq. (2.101) when we assumed that the time is represented by a Hermitian operator. So, time in quantum mechanics can only be a parameter.

What we can do, however, is to take $\hat{B} = \hat{H}$ in Eq. (2.81) and obtain

$$\Delta A \cdot \Delta H \geq \frac{1}{2}|\langle[\hat{A}, \hat{H}]\rangle| = \frac{1}{2}|i\hbar\frac{\mathrm{d}}{\mathrm{d}t}\langle\hat{A}\rangle|. \tag{2.107}$$

By defining

$$\Delta\tau = \frac{\Delta A}{\left|\frac{\mathrm{d}}{\mathrm{d}t}\left\langle\hat{A}\right\rangle\right|}, \tag{2.108}$$

we obtain with $\Delta H = \Delta E$ that

$$\Delta\tau \cdot \Delta E \geq \frac{\hbar}{2}. \tag{2.109}$$

Here $\Delta\tau$ is not the uncertainty of the time parameter, but rather it represents some characteristic time interval of the system. It measures the time interval needed for the change of the physical variable A to become significant with respect to its uncertainty. Thus, $\Delta\tau$ is roughly the time needed to observe an appreciable change in the value of the observable $\hat{A}$. If the system is in a stationary eigenstate of the Hamiltonian, the energy is well defined and $\Delta E = 0$. Consequently $\Delta\tau \to \infty$, i.e. no change happens even after an infinitely long time. If the system is a linear combination of various energy eigenstates $\Delta E \neq 0$. Then, during a finite interval $\Delta\tau$, the different energy eigenstates evolve differently in time, destroying the original configuration.

As an example, we consider a wave packet of width Δx and group velocity v_g. The time interval measured as it crosses a certain point is

$$\Delta\tau \simeq \frac{\Delta x}{v_g}. \tag{2.110}$$

On the other hand, since the momentum has Δp_x uncertainty, its energy also has some uncertainty

$$\Delta E \simeq \frac{\partial E}{\partial p_x}\Delta p_x = v_g\Delta p_x. \tag{2.111}$$

So,

$$\Delta\tau \cdot \Delta E \simeq \Delta x \cdot \Delta p_x \geq \frac{\hbar}{2}, \tag{2.112}$$

i.e. the time-energy uncertainty can be interpreted as the uncertainty of the passing time and the energy spread.

Another application of the time-energy uncertainty relation is the radioactive decay. The observed energy spread Γ of the decayed particles and the mean lifetime τ follow the rule

$$\tau \cdot \Gamma \approx \hbar. \tag{2.113}$$

Similarly, in case of transition between quantum levels due to persistent external perturbations we have

$$T \cdot \Delta E \approx \hbar, \tag{2.114}$$

where T is the duration of the perturbation and ΔE is the separation between the levels.

2.3.6 *Ehrenfest Theorems*

Let us calculate the time evaluation of the expectation values of the position and momentum operators. From Eq. (2.14) and with the help of the commutation relation $[\hat{\boldsymbol{r}}, \hat{\boldsymbol{p}}^2] = 2i\hbar\hat{\boldsymbol{p}}$, we obtain

$$\frac{d}{dt}\langle\hat{\boldsymbol{r}}\rangle = \frac{1}{i\hbar}\langle[\hat{\boldsymbol{r}}, \hat{H}]\rangle = \frac{1}{i\hbar}\langle[\hat{\boldsymbol{r}}, \hat{\boldsymbol{p}}^2/(2m) + \hat{V}(\hat{\boldsymbol{r}})]\rangle = \frac{1}{i\hbar}\langle[\hat{\boldsymbol{r}}, \hat{\boldsymbol{p}}^2/2m]\rangle = \frac{\langle\hat{\boldsymbol{p}}\rangle}{m}. \tag{2.115}$$

Then, by using the relation $[\hat{\boldsymbol{p}}, \hat{V}(\hat{\boldsymbol{r}})] = -i\hbar\nabla\hat{V}(\hat{\boldsymbol{r}})$, we find

$$\frac{d}{dt}\langle\hat{\boldsymbol{p}}\rangle = \frac{1}{i\hbar}\langle[\hat{\boldsymbol{p}}, \hat{H}]\rangle = \frac{1}{i\hbar}\langle[\hat{\boldsymbol{p}}, \hat{\boldsymbol{p}}^2/(2m) + \hat{V}(\hat{\boldsymbol{r}})]\rangle = \frac{1}{i\hbar}\langle[\hat{\boldsymbol{p}}, \hat{V}(\hat{\boldsymbol{r}})]\rangle = -\langle\nabla\hat{V}(\hat{\boldsymbol{r}})\rangle. \tag{2.116}$$

So, we have recovered the Newtonian equations of classical mechanics as expectation values of the corresponding quantum mechanical operators.

2.4 Density Operator

So far we have considered states which are defined by a complete set of commuting observables. These states are the *pure states*. They provide the maximal possible knowledge we can acquire about the system. A measurement gives a definite value of all the physical properties that are in the complete set of commuting observables. For example, a spin-1/2 particle can be in a pure spin-up $|+\rangle$ state. However, we may have a physical situation where the spin orientation is not set and the system is in the state that is a linear combination of spin-up $|+\rangle$ and spin-down $|-\rangle$ states. This is a *mixed state*. Here we are experiencing the lack of knowledge of the physical system. We may have some knowledge, however, on the probability about how the mixed state is combined by other states.

We assume that the physical system is a combination of states $\{|\alpha\rangle\}$, with p_α probability finding the system in state $|\alpha\rangle$,

$$0 \leq p_\alpha \leq 1, \quad \sum_\alpha p_\alpha = 1. \tag{2.117}$$

The expectation value of an observable $\hat{O}$ can be defined as

$$\langle \hat{O} \rangle = \sum_\alpha p_\alpha \langle \alpha | \hat{O} | \alpha \rangle. \tag{2.118}$$

With the projector $\hat{P}_\alpha = |\alpha\rangle\langle\alpha|$ we have

$$\mathrm{Tr}(\hat{P}_\alpha \hat{O}) = \sum_n \langle \phi_n | \hat{P}_\alpha \hat{O} | \phi_n \rangle = \sum_n \langle \phi_n | \alpha \rangle \langle \alpha | \hat{O} | \phi_n \rangle = \sum_n \langle \alpha | \hat{O} | \phi_n \rangle \langle \phi_n | \alpha \rangle = \langle \alpha | \hat{O} | \alpha \rangle. \tag{2.119}$$

where the states $\{|\phi_n\rangle\}$ form a basis. With that, we can write the expectation value

$$\langle \hat{O} \rangle = \sum_\alpha p_\alpha \mathrm{Tr}(\hat{P}_\alpha \hat{O}) = \mathrm{Tr}\left(\sum_\alpha p_\alpha \hat{P}_\alpha \hat{O} \right), \tag{2.120}$$

and, with the help of *density operator*

$$\hat{\rho} = \sum_\alpha p_\alpha \hat{P}_\alpha = \sum_\alpha p_\alpha |\alpha\rangle\langle\alpha|, \tag{2.121}$$

we have

$$\langle \hat{O} \rangle = \mathrm{Tr}(\hat{\rho}\hat{O}). \tag{2.122}$$

For a pure state $|\beta\rangle$, the formalism falls back to the known results

$$p_\beta = 1, \quad \hat{\rho} = \hat{P}_\beta = |\beta\rangle\langle\beta|, \quad \langle \hat{O} \rangle = \langle \beta | \hat{O} | \beta \rangle. \tag{2.123}$$

We can easily establish the following relations:

- $\hat{\rho}^\dagger = \hat{\rho}$,
- $\mathrm{Tr}(\hat{\rho}) = \langle I \rangle = 1$,
- $\langle \psi | \hat{\rho} | \psi \rangle = \sum_\alpha p_\alpha |\langle \psi | \alpha \rangle|^2 \geq 0$,
- $\mathrm{Tr}(\hat{\rho}^2) = \langle \hat{\rho} \rangle \leq \sum_\alpha p_\alpha \leq 1$, and $\mathrm{Tr}(\hat{\rho}^2) = 1$ if and only if $|\alpha\rangle$ is a pure state.

The time evolution of state $|\alpha\rangle$ is given by the Schrödinger equation

$$i\hbar\frac{\mathrm{d}}{\mathrm{d}t}|\alpha\rangle = \hat{H}|\alpha\rangle, \tag{2.124}$$

where $\hat{H}$ is the Hamiltonian of the system. Similarly,

$$-i\hbar\frac{\mathrm{d}}{\mathrm{d}t}\langle\alpha| = \langle\alpha|\hat{H}. \tag{2.125}$$

Taking the time derivative of Eq. (2.121) and assuming that the probabilities p_α are independent of time, we find

$$i\hbar\frac{\mathrm{d}}{\mathrm{d}t}\hat{\rho} = i\hbar\sum_\alpha p_\alpha\left(\frac{\mathrm{d}|\alpha\rangle}{\mathrm{d}t}\langle\alpha| + |\alpha\rangle\frac{\mathrm{d}\langle\alpha|}{\mathrm{d}t}\right) = \hat{H}\hat{\rho} - \hat{\rho}\hat{H} = \left[\hat{H},\hat{\rho}\right]. \tag{2.126}$$

As an example, we consider spin-1/2 states $|+\rangle$ and $|-\rangle$ with probabilities p_+ and p_-, respectively, such that $p_+ + p_- = 1$. In the basis $\{|+\rangle, |-\rangle\}$ we find

$$\hat{P}_+ = \begin{pmatrix} 1 & 0 \\ 0 & 0 \end{pmatrix}, \quad \hat{P}_- = \begin{pmatrix} 0 & 0 \\ 0 & 1 \end{pmatrix}, \quad \text{and} \quad \hat{\rho} = \begin{pmatrix} p_+ & 0 \\ 0 & p_- \end{pmatrix}, \quad \text{with} \quad \mathrm{Tr}(\hat{\rho}) = 1. \tag{2.127}$$

As a consequence,

$$\langle\hat{S}_z\rangle = \frac{\hbar}{2}\mathrm{Tr}(\hat{\rho}\sigma_z) = \frac{\hbar}{2}\mathrm{Tr}\begin{pmatrix} p_+ & 0 \\ 0 & -p_- \end{pmatrix} = \frac{\hbar}{2}(p_+ - p_-). \tag{2.128}$$

If the state is prepared such that $p_+ = p_- = 1/2$ and $\hat{\rho} = I/2$, we find for the spin orientation in the direction $\boldsymbol{e}_n$

$$\langle\hat{\boldsymbol{S}}\cdot\boldsymbol{e}_n\rangle = \frac{\hbar}{2}\frac{1}{2}\mathrm{Tr}(\boldsymbol{\sigma}\cdot\boldsymbol{e}_n) = 0. \tag{2.129}$$

Specifically,

$$\langle\hat{S}_x\rangle = \mathrm{Tr}\left[\frac{1}{2}\begin{pmatrix} 1 & 0 \\ 0 & 1 \end{pmatrix}\frac{\hbar}{2}\begin{pmatrix} 0 & 1 \\ 1 & 0 \end{pmatrix}\right] = \frac{\hbar}{4}\mathrm{Tr}\begin{pmatrix} 0 & 1 \\ 1 & 0 \end{pmatrix} = 0. \tag{2.130}$$

In another example the spin state is along the direction $\boldsymbol{e}_\varphi = (\cos\varphi, \sin\varphi, 0)$,

$$|\varphi\rangle = \frac{1}{\sqrt{2}}\left(|+\rangle + e^{i\varphi}|-\rangle\right). \tag{2.131}$$

Then,

$$\hat{\rho} = P_\varphi = |\varphi\rangle\langle\varphi| = \frac{1}{2}\begin{pmatrix} 1 & e^{-i\varphi} \\ e^{i\varphi} & 1 \end{pmatrix}. \tag{2.132}$$

We can see that this state is a pure state, since

$$\hat{\rho}^2 = \hat{\rho} \quad \text{and} \quad \mathrm{Tr}\big(\hat{\rho}^2\big) = 1. \tag{2.133}$$

In this case, the measurement for the x-component of the spin yields

$$\langle \hat{S}_x \rangle = \mathrm{Tr}\left[\frac{1}{2}\begin{pmatrix} 1 & e^{-i\varphi} \\ e^{i\varphi} & 1 \end{pmatrix}\frac{\hbar}{2}\begin{pmatrix} 0 & 1 \\ 1 & 0 \end{pmatrix}\right] = \frac{\hbar}{4}\mathrm{Tr}\begin{pmatrix} e^{-i\varphi} & 1 \\ 1 & e^{i\varphi} \end{pmatrix} = \frac{\hbar}{2}\cos\varphi. \tag{2.134}$$

2.5 Symmetries and Conservation Laws

2.5.1 Continuous Symmetries

We have seen before that

$$\hat{U}_\epsilon(\hat{G}) = I + i\epsilon\hat{G} \tag{2.135}$$

is an infinitesimal unitary transformation if ϵ is real and infinitesimal, and the generator $\hat{G}$ is Hermitian. It transforms the state as

$$|\psi'\rangle = (I + i\epsilon\hat{G})|\psi\rangle = |\psi\rangle + |\delta\psi\rangle, \tag{2.136}$$

and the operator as

$$\hat{A}' = (I + i\epsilon\hat{G})\hat{A}(I - i\epsilon\hat{G}) \approx \hat{A} + i\epsilon\left[\hat{G}, \hat{A}\right]. \tag{2.137}$$

We can see that the operator is unchanged if it commutes with the generator $\hat{G}$.

Now we consider the symmetry with respect to time and spatial translations. The consequences of space rotation will be discussed in a separate chapter.

Time Evolution

Let us take

$$\hat{G} = \hat{H}/\hbar, \quad \epsilon = -\delta t, \quad \text{and then} \quad \hat{U}_{\delta t}(\hat{H}) = I - i\delta t\, \hat{H}/\hbar. \tag{2.138}$$

Acting on $|\psi\rangle$ we find

$$\begin{aligned} |\psi'\rangle &= \left(I - i\delta t\, \hat{H}/\hbar\right)|\psi(t)\rangle = |\psi(t)\rangle - i\delta t\, \hat{H}/\hbar|\psi(t)\rangle \\ &= |\psi(t)\rangle + \delta t\frac{\mathrm{d}}{\mathrm{d}t}|\psi\rangle = |\psi(t + \delta t)\rangle. \end{aligned} \tag{2.139}$$

This means that just like in classical mechanics, the Hamiltonian is the generator for the time evolution of the system.

If $\hat{H}$ does not depend explicitly on time, the finite time evolution operator is given by

$$\hat{U}(\tau) = \exp\Big(- i\hat{H}\tau/\hbar\Big), \tag{2.140}$$

and its action on the state gives

$$\hat{U}(\tau)|\psi(t)\rangle = \exp\Big(- i\hat{H}\tau/\hbar\Big)|\psi(t)\rangle = |\psi(t+\tau)\rangle. \tag{2.141}$$

We can see that the Hamiltonian is invariant with respect to infinitesimal time evolution transformation

$$\hat{H}' = \hat{U}_{\delta t}\hat{H}\hat{U}^{\dagger}_{\delta t} = \hat{H} - i\delta t\Big[\hat{H}, \hat{H}\Big] = \hat{H}. \tag{2.142}$$

If the $\hat{H}$ does not have explicit time dependence the energy is conserved

$$\frac{\mathrm{d}}{\mathrm{d}t}\langle\hat{H}\rangle = \frac{i}{\hbar}\langle[\hat{H}, \hat{H}]\rangle + \langle\frac{\partial H}{\partial t}\rangle = 0. \tag{2.143}$$

Spatial Translation

Let us take now

$$\hat{G} = \hat{p}_x/\hbar, \quad \epsilon = \delta x, \quad \text{and then} \quad \hat{U}_{\delta x}(\hat{p}_x) = I + i\,\delta x\,\hat{p}_x/\hbar. \tag{2.144}$$

Acting on $|\psi\rangle$, we find

$$\begin{aligned} |\psi'\rangle &= \big(I + i\,\delta x\,\hat{p}_x/\hbar\big)\,|\psi\rangle = |\psi(x)\rangle + i\,\delta x\,\hat{p}_x/\hbar|\psi(x)\rangle \\ &= |\psi(x)\rangle + \delta x\frac{\partial}{\partial x}|\psi(x)\rangle = |\psi(x+\delta x)\rangle. \end{aligned} \tag{2.145}$$

We have found again, that just like in classical mechanics, that $\hat{p}$ is the generator of the spatial translation. For the operator $\hat{x}$ we find

$$\hat{x}' = (I + i\,\delta x\,\hat{p}_x/\hbar)\hat{x}(I - i\,\delta x\,\hat{p}_x/\hbar) \approx \hat{x} + i\delta t[\hat{H}, \hat{H}] = \hat{H}. \tag{2.146}$$

The finite spatial translation operator is given by

$$\hat{U}(a) = \exp(i\,a\,\hat{p}_x/\hbar), \tag{2.147}$$

and its action gives

$$\hat{x}' = \hat{U}(a)\hat{x}\hat{U}^{\dagger}(a) = \hat{x} + i\,a\left[\hat{p}_x, \hat{x}\right]/\hbar = \hat{x} + a, \tag{2.148}$$

and

$$\hat{U}(a)|\psi(x)\rangle = \exp(i\,a\,\hat{p}_x/\hbar)|\psi(x)\rangle = |\psi(x+a)\rangle. \tag{2.149}$$

If the Hamiltonian does not depend on the spatial coordinates, it commutes with the linear momentum operator, $[\hat{H}, \hat{\boldsymbol{p}}] = 0$. Then, the Hamiltonian is invariant with respect to spatial translation

$$\hat{H}' = \exp(i\,\boldsymbol{a}\hat{\boldsymbol{p}}/\hbar)\hat{H}\exp(-i\,\boldsymbol{a}\hat{\boldsymbol{p}}/\hbar) = \exp(i\,\boldsymbol{a}\hat{\boldsymbol{p}}/\hbar)\exp(i\,\boldsymbol{a}\hat{\boldsymbol{p}}/\hbar)\hat{H} = \hat{H}. \tag{2.150}$$

Since the momentum operator $\hat{\boldsymbol{p}}$ commutes with the Hamiltonian, $\hat{\boldsymbol{p}}$ is a constant of motion

$$\frac{\mathrm{d}}{\mathrm{d}t}\langle\hat{\boldsymbol{p}}\rangle = \frac{i}{\hbar}\langle[\hat{\boldsymbol{p}}, \hat{H}]\rangle + \langle\frac{\partial\hat{\boldsymbol{p}}}{\partial t}\rangle = 0, \tag{2.151}$$

i.e. the momentum is conserved.

2.5.2 Discrete Symmetries

Parity

Parity, or spatial reflection, is defined by

$$\hat{\mathcal{P}}|x\rangle = |-x\rangle, \quad \text{or} \quad \hat{\mathcal{P}}|\boldsymbol{r}\rangle = |-\boldsymbol{r}\rangle. \tag{2.152}$$

The operator $\hat{\mathcal{P}}$ is Hermitian since

$$\int \mathrm{d}x\phi^*(x)\hat{\mathcal{P}}\psi(x) = \int \mathrm{d}x\phi^*(x)\psi(-x) = \int \mathrm{d}x\phi^*(-x)\psi(x) = \int \mathrm{d}x[\hat{\mathcal{P}}\phi(x)]^*\psi(x). \tag{2.153}$$

Multiplying Eq. (2.152) from the left by $\hat{\mathcal{P}}$, we find

$$\hat{\mathcal{P}}^2 = I. \tag{2.154}$$

This means that $\hat{\mathcal{P}} = \hat{\mathcal{P}}^{\dagger} = \hat{\mathcal{P}}^{-1}$, i.e. $\hat{\mathcal{P}}$ is unitary with eigenvalues ± 1. The eigenvalues are discrete, therefore the parity transformation is a discrete transformation. If the state $|\psi\rangle$ has definite parity, we can write

$$\hat{\mathcal{P}}|\psi_{\pm}\rangle = \pi_{\psi}|\psi_{\pm}\rangle, \tag{2.155}$$

with $\pi_{\psi} = \pm 1$. The states with $\pi_{\psi} = +1$ are called even parity, or symmetric, while with $\pi_{\psi} = -1$ are called odd parity, or antisymmetric, states. The states $|\psi_{+}\rangle$ and $|\psi_{-}\rangle$ are orthogonal, since they belong to different eigenstates of a Hermitian operator

$$\langle\psi_+|\psi_-\rangle = \langle\psi_+|\hat{\mathcal{P}}^2|\psi_-\rangle = \langle\hat{\mathcal{P}}\psi_+|\hat{\mathcal{P}}\psi_-\rangle = -\langle\psi_+|\psi_-\rangle = 0. \tag{2.156}$$

Since $\hat{\mathcal{P}}$ is unitary, the operator $\hat{x}$ transforms as $x' = \hat{\mathcal{P}}x\hat{\mathcal{P}}$. From $\langle x'|\hat{x}|x\rangle = x\langle x'|x\rangle = x\delta(x'-x)$, and considering that

$$\langle x'|\hat{x}'|x\rangle = \langle x'|\hat{\mathcal{P}}\hat{x}\hat{\mathcal{P}}|x\rangle = \langle -x'|\hat{x}|-x\rangle = -x\delta(x'-x), \tag{2.157}$$

we can conclude that

$$\hat{\mathcal{P}}\hat{x}\hat{\mathcal{P}} = -\hat{x}, \quad \text{or} \quad \hat{\mathcal{P}}\hat{\boldsymbol{r}}\hat{\mathcal{P}} = -\hat{\boldsymbol{r}}. \tag{2.158}$$

Since $\mathcal{P}$ is a unitary transformation, the coordinate-momentum commutation relation should be invariant

$$i\hbar = \hat{\mathcal{P}}i\hbar\hat{\mathcal{P}} = \hat{\mathcal{P}}\left[\hat{x}, \hat{p}_x\right]\hat{\mathcal{P}} = \hat{\mathcal{P}}\hat{x}\hat{\mathcal{P}}\hat{\mathcal{P}}\hat{p}_x\hat{\mathcal{P}} - \hat{\mathcal{P}}\hat{p}_x\hat{\mathcal{P}}\hat{\mathcal{P}}\hat{x}\hat{\mathcal{P}} = (-\hat{x})\hat{\mathcal{P}}\hat{p}_x\hat{\mathcal{P}} - \hat{\mathcal{P}}\hat{p}_x\hat{\mathcal{P}}(-\hat{x}), \tag{2.159}$$

therefore

$$\hat{\mathcal{P}}\hat{p}_x\hat{\mathcal{P}} = -\hat{p}_x, \quad \text{or} \quad \hat{\mathcal{P}}\hat{\boldsymbol{p}}\hat{\mathcal{P}} = -\hat{\boldsymbol{p}}. \tag{2.160}$$

These relations can also serve as a representation independent definition of the parity operator.

The orbital angular momentum is defined by $\hat{\boldsymbol{L}} = \hat{\boldsymbol{r}} \times \hat{\boldsymbol{p}}$. Consequently,

$$\hat{\mathcal{P}}\hat{\boldsymbol{L}}\hat{\mathcal{P}} = \hat{\mathcal{P}}\hat{\boldsymbol{r}}\hat{\mathcal{P}} \times \hat{\mathcal{P}}\hat{\boldsymbol{p}}\hat{\mathcal{P}} = (-\hat{\boldsymbol{r}}) \times (-\hat{\boldsymbol{p}}) = \hat{\boldsymbol{L}}. \tag{2.161}$$

The spin is another form of angular momentum, therefore we should take

$$\hat{\mathcal{P}}\hat{\boldsymbol{S}}\hat{\mathcal{P}} = \hat{\boldsymbol{S}}, \tag{2.162}$$

and then, for a general angular momentum, we should have

$$\hat{\mathcal{P}}\hat{\boldsymbol{J}}\hat{\mathcal{P}} = \hat{\boldsymbol{J}}. \tag{2.163}$$

These type of operators behave like vectors with respect to rotation but do not change sign with respect to parity transformations. They are called *pseudo-vector*, or *axial vector*, operators.

The scalar product $\hat{\boldsymbol{p}}\hat{\boldsymbol{r}}$ is scalar, which is invariant with respect to parity transformation

$$\hat{\mathcal{P}}(\hat{\boldsymbol{p}}\hat{\boldsymbol{r}})\hat{\mathcal{P}} = \hat{\boldsymbol{p}}\hat{\boldsymbol{r}}. \tag{2.164}$$

On the other hand, other scalar products with spin changes sign

$$\hat{\mathcal{P}}(\hat{\boldsymbol{p}}\hat{\boldsymbol{S}})\hat{\mathcal{P}} = -\hat{\boldsymbol{p}}\hat{\boldsymbol{S}} \quad \text{and} \quad \hat{\mathcal{P}}(\hat{\boldsymbol{r}}\hat{\boldsymbol{S}})\hat{\mathcal{P}} = -\hat{\boldsymbol{r}}\hat{\boldsymbol{S}}. \tag{2.165}$$

These type of operators are called *pseudo-scalar* operators.

In the Hamiltonians

$$\hat{H} = \frac{\hat{p}_x^2}{2m} + \hat{V}(x), \quad \text{or} \quad \hat{H} = \frac{\hat{\boldsymbol{p}}^2}{2m} + \hat{V}(\boldsymbol{r}), \tag{2.166}$$

the kinetic energy is scalar, $[\hat{\boldsymbol{p}}^2, \hat{\mathcal{P}}] = 0$. Then, if $\hat{V}(x) = \hat{V}(-x)$, or $\hat{V}(\boldsymbol{r}) = \hat{V}(-\boldsymbol{r})$, respectively, the Hamiltonians are parity invariant

$$[\hat{\mathcal{P}}, \hat{H}] = 0. \tag{2.167}$$

If the parity operator commutes with the Hamiltonian, it is a constant of motion and the parity is conserved. Then, the eigenstates of the Hamiltonians have definite parity, even or odd, and this property of the states are conserved.

Time Reversal

In classical mechanics, the time reversal transforms $(\boldsymbol{r}, \boldsymbol{p})$ to $(\boldsymbol{r}, -\boldsymbol{p})$. It leaves the position unchanged but reverses the momentum. The corresponding quantum mechanical operator should transform $(|\boldsymbol{r}\rangle, |\boldsymbol{p}\rangle)$ to $(|\boldsymbol{r}\rangle, |-\boldsymbol{p}\rangle)$. This mapping cannot be done with a linear operator. The only linear operator which maps $|\boldsymbol{r}\rangle \mapsto |\boldsymbol{r}\rangle$ is the unit operator. The unit operator, however, does not map $|\boldsymbol{p}\rangle \mapsto |-\boldsymbol{p}\rangle$.

To determine the quantum mechanical operator for the time reversal transformation, we can find some guidance by considering the time-dependent Schrödinger equation with a real-valued time-independent potential

$$i\hbar\frac{\mathrm{d}}{\mathrm{d}t}\psi(\boldsymbol{r}, t) = \left(-\frac{\hbar^2}{2m}\Delta + V\right)\psi(\boldsymbol{r}, t). \tag{2.168}$$

Under the time reversal transformation,

$$t \to -t, \tag{2.169}$$

the Schrödinger equation becomes

$$-i\hbar\frac{\mathrm{d}}{\mathrm{d}t}\psi(\boldsymbol{r}, -t) = \left(-\frac{\hbar^2}{2m}\Delta + V\right)\psi(\boldsymbol{r}, -t). \tag{2.170}$$

We can see that the transformed wave function $\psi(\boldsymbol{r}, -t)$ does not satisfy the original Schrödinger equation, but rather the complex conjugated one

$$-i\hbar\frac{\mathrm{d}}{\mathrm{d}t}\psi^*(\boldsymbol{r}, t) = \left(-\frac{\hbar^2}{2m}\Delta + V\right)\psi^*(\boldsymbol{r}, t). \tag{2.171}$$

So, the wave function under the time reversal operation transforms as

$$\psi(t) \rightarrow \psi^*(-t). \tag{2.172}$$

Let us denote the time reversal operator by $\hat{\mathcal{T}}$. Since $\boldsymbol{r}$ is not changed by $\hat{\mathcal{T}}$, we must have

$$\hat{\mathcal{T}}\hat{\boldsymbol{r}}|\psi\rangle = \hat{\mathcal{T}}\hat{\boldsymbol{r}}\hat{\mathcal{T}}^{-1}\hat{\mathcal{T}}|\psi\rangle = \hat{\boldsymbol{r}}\hat{\mathcal{T}}|\psi\rangle. \tag{2.173}$$

Consequently,

$$\hat{\mathcal{T}}\hat{\boldsymbol{r}}\hat{\mathcal{T}}^{-1} = \hat{\boldsymbol{r}}. \tag{2.174}$$

The same transformation reverses the direction of the momentum operator

$$\hat{\mathcal{T}}\hat{\boldsymbol{p}}|\psi\rangle = \hat{\mathcal{T}}\hat{\boldsymbol{p}}\hat{\mathcal{T}}^{-1}\hat{\mathcal{T}}|\psi\rangle = -\hat{\boldsymbol{p}}\hat{\mathcal{T}}|\psi\rangle, \tag{2.175}$$

i.e.

$$\hat{\mathcal{T}}\hat{\boldsymbol{p}}\hat{\mathcal{T}}^{-1} = -\hat{\boldsymbol{p}}. \tag{2.176}$$

Then, the orbital angular momentum should transform as

$$\hat{\mathcal{T}}\hat{\boldsymbol{L}}\hat{\mathcal{T}}^{-1} = -\hat{\boldsymbol{L}}, \tag{2.177}$$

and so should the spin and the angular momentum in general

$$\hat{\mathcal{T}}\hat{\boldsymbol{S}}\hat{\mathcal{T}}^{-1} = -\hat{\boldsymbol{S}} \quad \text{and} \quad \hat{\mathcal{T}}\hat{\boldsymbol{J}}\hat{\mathcal{T}}^{-1} = -\hat{\boldsymbol{J}}. \tag{2.178}$$

The above transformation of the spin is not obvious, because the spin is an internal property of particles. However, if we associate the spin with loop currents, the time reversal changes the direction of the current, and thus the direction of the spin.

If we apply $\hat{\mathcal{T}}$ to the commutation relation

$$\left[\hat{x}, \hat{p}_x\right] = i\hbar, \tag{2.179}$$

we get

$$\hat{\mathcal{T}}\hat{x}\hat{\mathcal{T}}^{-1}\hat{\mathcal{T}}\hat{p}_x\hat{\mathcal{T}}^{-1} - \hat{\mathcal{T}}\hat{p}_x\hat{\mathcal{T}}^{-1}\hat{\mathcal{T}}\hat{x}\hat{\mathcal{T}}^{-1} = \hat{\mathcal{T}}i\hbar\hat{\mathcal{T}}^{-1}. \tag{2.180}$$

Then, utilizing Eqs. (2.174) and (2.176), we have

$$\hat{x}(-\hat{p}_x) - (-\hat{p}_x)\hat{x} = -\left[\hat{x}, \hat{p}_x\right] = \hat{\mathcal{T}}i\hbar\hat{\mathcal{T}}^{-1}, \tag{2.181}$$

which is compatible with the original commutation relation if

$$\hat{\mathcal{T}}i\hat{\mathcal{T}}^{-1} = -i = i^*. \tag{2.182}$$

Clearly this operator cannot be linear, it is an anti-linear operator. Anti-linear operators are similar to the linear operators, but act differently on a linear combination of states

$$\hat{A}(c_1|\psi_1\rangle + c_2|\psi_2\rangle) = c_1^*\hat{A}|\psi_1\rangle + c_2^*\hat{A}|\psi_2\rangle. \tag{2.183}$$

The product of two anti-linear operators is linear. An operator $\hat{A}$ is anti-unitary if it is anti-linear, its inverse $\hat{A}^{-1}$ exists, and has the property $\langle\hat{A}\phi|\hat{A}\psi\rangle = \langle\phi|\psi\rangle^* = \langle\psi|\phi\rangle$. As a consequence, anti-unitary operators preserve the norm $|\langle\hat{A}\phi|\hat{A}\psi\rangle| = |\langle\phi|\psi\rangle^*| = |\langle\phi|\psi\rangle|$, just like the unitary operators. So, any mapping of the Hilbert space onto itself that preserves the norm can be performed by either a unitary or an anti-unitary operator. This is *Wigner's theorem.*

Clearly, the time reversal operator $\hat{\mathcal{T}}$ is anti-unitary. The above commutation relations are invariant with respect to any linear similarity transformation, therefore $\hat{\mathcal{T}}$ can be written as

$$\hat{\mathcal{T}} = \hat{U}\hat{K}, \tag{2.184}$$

where $\hat{U}$ is unitary and $\hat{K}$ just conjugates complex numbers. Then $\hat{K}^2 = I$ and $\hat{K}^{-1} = \hat{K}$, and consequently, $\hat{\mathcal{T}}^{-1} = \hat{K}\hat{U}^\dagger$.

The Hamiltonian is invariant with respect to the time reversal if

$$\hat{\mathcal{T}}\hat{H}\hat{\mathcal{T}}^{-1} = \hat{H}, \quad \text{or} \quad [\hat{\mathcal{T}}, \hat{H}] = 0. \tag{2.185}$$

This invariance, however, does not involve any conserved quantity. The conservation property is valid only for unitary operators. However, if $\hat{H}$ is invariant with respect to time reversal and the eigenstate $|\psi\rangle$ is non-degenerate, then we find

$$\hat{H}\hat{\mathcal{T}}|\psi\rangle = \hat{\mathcal{T}}\hat{H}|\psi\rangle = \hat{\mathcal{T}}E|\psi\rangle = E(\hat{\mathcal{T}}|\psi\rangle), \tag{2.186}$$

i.e. $\hat{\mathcal{T}}|\psi\rangle$ is also an eigenstate of $\hat{H}$ with eigenvalue E. Since the state is not degenerate, we must have

$$\hat{\mathcal{T}}|\psi\rangle = \alpha|\psi\rangle. \tag{2.187}$$

This, in coordinate representation, reads

$$\langle x|\hat{\mathcal{T}}\psi\rangle = \langle x|\psi\rangle^* = \alpha\langle x|\psi\rangle, \tag{2.188}$$

since $\hat{\mathcal{T}}$ has a complex conjugation. We have found that in coordinate representation the wave function is proportional to its complex conjugate, so $\langle x|\psi\rangle$ can be taken as real.

Two successive applications of time reversal leaves the physical situation unchanged. Therefore $\hat{\mathcal{T}}^2|\psi\rangle$ and $|\psi\rangle$ should describe the same quantum state, i.e.

$$\hat{\mathcal{T}}^2|\psi\rangle = \alpha|\psi\rangle, \tag{2.189}$$

with $|\alpha| = 1$, since the state should be normalized. Applying this relation to a time reversed state, we find

$$\hat{\mathcal{T}}^2(\hat{\mathcal{T}}|\psi\rangle) = \hat{\mathcal{T}}(\hat{\mathcal{T}}^2|\psi\rangle) = \hat{\mathcal{T}}(\alpha|\psi\rangle) = \alpha^*(\hat{\mathcal{T}}|\psi\rangle), \tag{2.190}$$

so $\alpha = \alpha^*$. Consequently $\alpha = \pm 1$ and

$$\hat{\mathcal{T}}^2|\psi\rangle = \pm|\psi\rangle. \tag{2.191}$$

As the physical situation is not changed by two consecutive time reversals, the matrix elements should not change either

$$\langle\hat{\mathcal{T}}\psi|\psi\rangle = \langle\hat{\mathcal{T}}\psi|\hat{\mathcal{T}}^2\psi\rangle = \alpha\langle\hat{\mathcal{T}}\psi|\psi\rangle. \tag{2.192}$$

This result means that if the Hamiltonian is invariant with respect to time reversal and $\alpha = -1$, the states $|\psi\rangle$ and $\hat{\mathcal{T}}|\psi\rangle$ have to be orthogonal. They are linearly independent, yet they have the same energy. So, these states are degenerate. This is *Kramer's theorem.*

Charge Conjugation

Charge conjugation converts a particle state $|\psi\rangle$ to its counterpart with opposite charge $|\bar{\psi}\rangle$

$$\hat{C}|\psi\rangle = |\bar{\psi}\rangle, \tag{2.193}$$

and it reverses all the internal quantum numbers. This transformation maps normalized Hilbert space vectors into normalized Hilbert space vectors, so it has to be unitary, i.e. $\hat{C}\hat{C}^\dagger = 1$. Furthermore, repeated applications give

$$\hat{C}^2|\psi\rangle = \hat{C}|\bar{\psi}\rangle = |\psi\rangle. \tag{2.194}$$

Thus $\hat{C}^2 = 1$ and $\hat{C} = \hat{C}^{-1}$; therefore, $\hat{C} = \hat{C}^\dagger$, i.e. $\hat{C}$ is Hermitian and consequently an observable.

The eigenvalue problem for charge conjugation reads

$$\hat{C}|\psi\rangle = \eta_C|\psi\rangle. \tag{2.195}$$

Since

$$\hat{C}^2|\psi\rangle = \eta_C\hat{C}|\psi\rangle = \eta_C^2|\psi\rangle = |\psi\rangle, \tag{2.196}$$

we find that $\eta_C = \pm 1$, which is called *C-parity* or *charge parity*. For a particle to be in a C-parity eigenstate, it is required that $|\psi\rangle$ and $\hat{C}|\psi\rangle$ posses the same set of quantum numbers. They should be truly neutral particles, with all quantum charges and magnetic moments equal to zero. Only the photon γ and particle-antiparticle bound states, like the neutral π^0 and η particles, which are quark-antiquark states, are

C-parity eigenstates. The electromagnetic field changes sign under the $\hat{C}$ transformation, therefore its quantum number should have negative C-parity, i.e. $\eta_C(\gamma) = -1$ for the photon.

Chapter 3
One-Dimensional Problems

In this chapter we study quantum mechanical problems in one spatial dimension. We show how quantum mechanics works for relatively simple systems and we introduce methods and concepts that can be applied for more complicated cases. The Hamiltonian in coordinate representation with $\hat{p}_x = -i\hbar \mathrm{d}/\mathrm{d}x$ is given by

$$\hat{H} = \frac{\hat{p}_x^2}{2m} + \hat{V}(x) = -\frac{\hbar^2}{2m}\frac{\mathrm{d}^2}{\mathrm{d}x^2} + V(x). \tag{3.1}$$

The coordinate x can range from $-\infty$ to ∞, but in some cases, it can also be restricted to finite or semi-infinite intervals.

3.1 Properties of the Wave Function

3.1.1 Boundary Condition

In one dimension, the inner product is given by

$$\langle \phi | \psi \rangle = \int_a^b \mathrm{d}x \, \phi^*(x)\psi(x), \tag{3.2}$$

where $a < b$, and they can be finite or infinite. We assume that the wave function ψ satisfies the stationary Schrödinger equation

$$\hat{H}\psi = -\frac{\hbar^2}{2m}\psi'' + V\psi = E_\psi \psi. \tag{3.3}$$

Since $\hat{H}$ is Hermitian, with another solution ϕ, we immediately have

Z. Papp, *Mastering Quantum Mechanics*,
https://doi.org/10.1007/978-3-032-09011-9_3

$$\langle\phi|\hat{H}\psi\rangle - \langle\hat{H}\phi|\psi\rangle = -\frac{\hbar^2}{2m}\int_a^b \mathrm{d}x(\phi^*\psi'' - \phi^{*\prime\prime}\psi) = -\frac{\hbar^2}{2m}(\phi^*\psi' - \phi^{*\prime}\psi)|_a^b = 0. \quad (3.4)$$

This condition is satisfied if one of the following boundary conditions is satisfied by the solutions.

1. *Dirichlet boundary condition*:

$$\psi(a) = \psi(b) = 0. \quad (3.5)$$

2. *Neumann boundary condition*:

$$\psi'(a) = \psi'(b) = 0. \quad (3.6)$$

3. *Robin boundary condition*:

$$\psi'(a) = \alpha_a\psi(a), \quad \psi'(b) = \alpha_b\psi(b). \quad (3.7)$$

   This is a generalization of the previous two boundary conditions. If $\alpha_{a,b} \to 0$, we recover the Neumann condition, while if $\alpha_{a,b} \to \infty$, we obtain the Dirichlet condition.
4. *Periodic boundary condition*:

$$\psi(b) = \psi(a), \quad \psi'(b) = \psi'(a). \quad (3.8)$$

3.1.2 Continuity of the Wave Function

The Schrödinger equation is given by

$$-\psi'' - (\epsilon - v)\psi = 0, \quad (3.9)$$

where $\epsilon = 2mE/\hbar^2$ and $v = 2mV/\hbar^2$ are real. So, if ψ is complex, then both the real and the imaginary parts are solutions. Consequently we can choose, if convenient, ψ to be real.

If, in Eq. (3.9) the potential $V(x)$ is continuous, then ψ'' is continuous, and so is ψ' and ψ. However, if at $c \in [a, b]$, the potential is not continuous, ψ'' is not continuous either, and there is no classical solution. But as the Hamiltonian is Hermitian, the solution can be defined in terms of distribution

$$\langle -\phi'' - (\epsilon - v)\phi|\psi\rangle = 0, \quad (3.10)$$

where $\phi \in C_0^\infty(a, b)$, i.e. infinitely differentiable function vanishing at the boundaries on the interval $[a, b]$. The left hand side of Eq. (3.10) reads

$$\int_a^b \mathrm{d}x\,[-\phi^{*\prime\prime}(x) + v(x)\phi^*(x) - \epsilon\phi^*(x)]\psi(x) = 0. \tag{3.11}$$

If we split the integral and integrate by parts in terms of the second derivative, we obtain

$$\begin{aligned} 0 = &\int_a^c \mathrm{d}x\,[-\phi^{*\prime\prime}(x) + v(x)\phi^*(x) - \epsilon\phi^*(x)]\psi(x) \\ &+ \int_c^b \mathrm{d}x\,[-\phi^{*\prime\prime}(x) + v(x)\phi^*(x) - \epsilon\phi^*(x)]\psi(x) \\ = &\int_a^c \mathrm{d}x\,\phi^*(x)[-\psi''(x) + v(x)\psi(x) - \epsilon\psi(x)] \\ &+ \int_c^b \mathrm{d}x\,\phi^*(x)[-\psi''(x) + v(x)\psi(x) - \epsilon\psi(x)] \\ &+ \phi^{*\prime}(c)[\psi(c-0) - \psi(c+0)] - \phi^*(c)[\psi'(c-0) - \psi'(c+0)]. \end{aligned} \tag{3.12}$$

Since ϕ is an infinitely differentiable arbitrary function, all terms should separately vanish, resulting in the boundary conditions for the wave function

$$\begin{aligned} -\psi''(x) + v(x)\psi(x) - \epsilon\psi(x) = 0, &\quad x \in [a, c), \\ -\psi''(x) + v(x)\psi(x) - \epsilon\psi(x) = 0, &\quad x \in (c, b], \\ \psi(c-0) = \psi(c+0), & \\ \psi'(c-0) = \psi'(c+0), & \end{aligned} \tag{3.13}$$

i.e. ψ should satisfy the Schrödinger equation in each of the subintervals with the boundary condition that ψ and ψ' should be continuous at the discontinuity of the potential.

3.1.3 Parity-Symmetric Potential

The parity transformation amounts to mirroring of the coordinates, $\hat{\mathcal{P}}x = -x$ and $\hat{\mathcal{P}}\hat{p}_x = -\hat{p}_x$. Therefore $\hat{\mathcal{P}}\hat{p}_x^2 = (-\hat{p}_x)^2 = \hat{p}_x^2$. If the potential is such that $\hat{\mathcal{P}}V(x) = V(-x) = V(x)$, the Hamiltonian is invariant with respect to the parity transformation, $[\hat{H}, \hat{\mathcal{P}}] = 0$. Then $\hat{H}$ and $\hat{\mathcal{P}}$ must have common eigenstates. The solution must have a definite parity, which should be even or odd, symmetric or antisymmetric. Figure 3.1 shows typical even and odd parity wave functions.

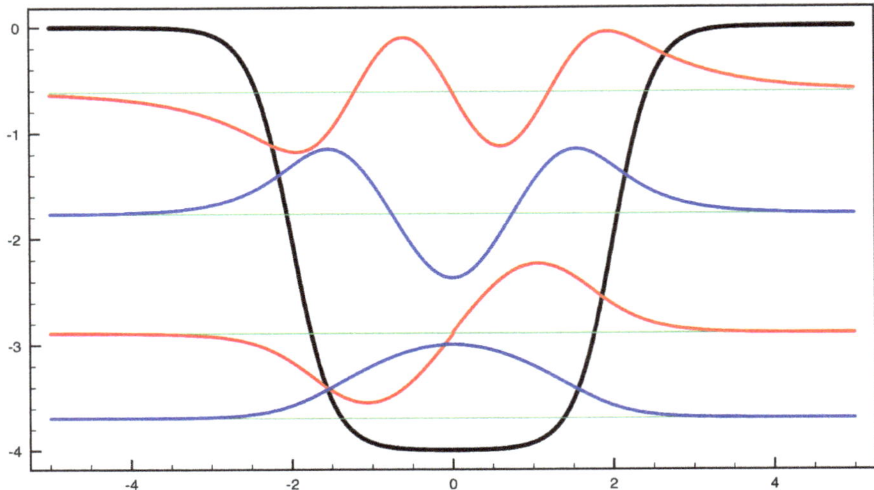

Fig. 3.1 Schematic picture of even and odd parity wave functions in a potential with the corresponding energy levels. The ground state wave function has no node, the states with higher energy has more nodes and stretches further out

3.1.4 Curvature of the Wave Function

The $E = V(x)$ point is the classical "turning point". If $E > V(x)$, the kinetic energy is positive and the region is a classically allowed region. Then

$$\frac{\psi''(x)}{\psi(x)} = -\frac{2m}{\hbar^2}(E - V(x)) < 0, \tag{3.14}$$

and $\psi(x)$ is concave, bending toward the x axis. The wave function oscillates with local wave number $k(x) = \sqrt{2m(E - V(x))/\hbar^2}$.

If $E < V(x)$, the kinetic energy is negative, corresponding to a classically forbidden region, and we have

$$\frac{\psi''(x)}{\psi(x)} = -\frac{2m}{\hbar^2}(E - V(x)) > 0. \tag{3.15}$$

This means that $\psi(x)$ is convex, bending away form the x axis, and the wave function is exponentially decaying with the rate $\kappa(x) = \sqrt{2m|E - V(x)|/\hbar^2}$. These behaviors are nicely visible in Fig. 3.1. As the energy line crosses the potential curve and the region becomes classically forbidden, the curvature of the wave function changes from concave to convex.

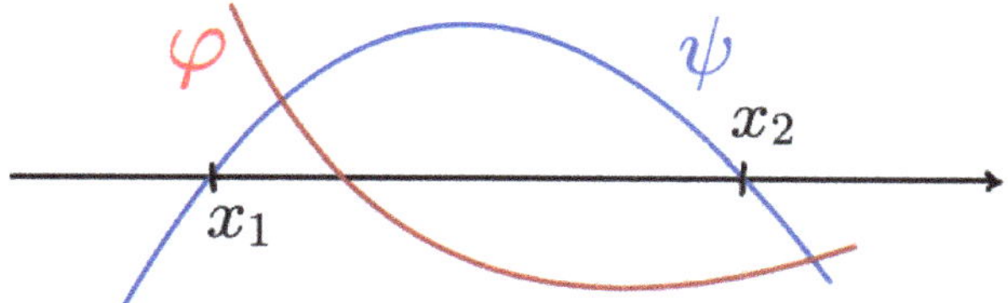

Fig. 3.2 If $\psi(x_1) = \psi(x_2) = 0$ and $E_\psi < E_\varphi$, the wave function φ must have a node in the $[x_1, x_2]$ interval

3.1.5 Nodes

We recall that the Wronskian, or Wronski determinant, of two functions is defined by

$$W(\varphi, \psi) = \det\begin{pmatrix} \varphi & \psi \\ \varphi' & \psi' \end{pmatrix} = \varphi\psi' - \varphi'\psi. \tag{3.16}$$

If φ and ψ are linearly dependent, $\psi = c\varphi$, and thus $W(\varphi, c\varphi) = \varphi c\varphi' - \varphi' c\varphi = 0$, i.e. the Wronskian vanishes.

We assume that φ and ψ are solutions of the same Hamiltonian with $E_\varphi > E_\psi$. Then we have

$$W' = \frac{\mathrm{d}}{\mathrm{d}x}(\varphi\psi' - \varphi'\psi) = \varphi\psi'' - \varphi''\psi = \frac{2m}{\hbar^2}\Delta E\, \varphi\psi, \text{ with } \Delta E = E_\varphi - E_\psi > 0. \tag{3.17}$$

If we also assume that $x_1 < x_2$ are two neighboring nodes of ψ (Fig. 3.2), i.e. $\psi(x_1) = \psi(x_2) = 0$, then, by integrating W', we find

$$W(\varphi, \psi)|_{x_1}^{x_2} = \varphi\psi'|_{x_1}^{x_2} = \varphi(x_2)\psi'(x_2) - \varphi(x_1)\psi'(x_1) = \frac{2m}{\hbar^2}\Delta E \int_{x_1}^{x_2} \mathrm{d}x\, \varphi(x)\psi(x). \tag{3.18}$$

Without loss of generality we can assume that in the $[x_1, x_2]$ interval $\psi(x) > 0$, and consequently $\psi'(x_1) > 0$ and $\psi'(x_2) < 0$. If φ in the interval $[x_1, x_2]$ does not change sign, e.g. suppose it is positive everywhere, then, the integrand in Eq. (3.18) is positive, but the expression in the middle is negative. Consequently, φ must change sign in the interval $[x_1, x_2]$. *If we have two solutions ψ and φ of the same one-dimensional Hamiltonian with $E_\psi < E_\varphi$, then between the two zeros of ψ, φ must have a node.*

Now, we consider a Dirichlet problem where the wave function should vanish on the boundaries. Let us take ψ and $\tilde{\psi}$ as two solutions with the same energy. Then, in Eq. (3.17) $\Delta E = 0$, consequently $W' = 0$. So, the Wronskian

$$W(\psi, \tilde{\psi}) = \psi\tilde{\psi}' - \psi'\tilde{\psi} = \psi\tilde{\psi}(\log\tilde{\psi} - \log\psi)' \tag{3.19}$$

is independent of x. On the boundaries however, ψ and $\tilde{\psi}$ vanish and therefore $W(\psi, \tilde{\psi}) = 0$ over the whole interval. Consequently,

$$(\log \tilde{\psi})' = (\log \psi)'. \tag{3.20}$$

So, in the neighborhood of the boundaries, where the potential is smooth, $\tilde{\psi} = c\psi$ and $\tilde{\psi}' = c\psi'$, with some constant c. Moreover, they satisfy the same linear second order differential equation. Therefore, $\tilde{\psi} = c\psi$ over the whole interval, and thus they describe the same quantum state. As a consequence, *the eigenstates in a one-dimensional potential are not degenerate.*

As far as the whole x axis is considered, a bound state wave function must vanish at the boundaries, thus having Dirichlet boundary conditions. A wave function with one node cannot be the ground state, because a no-node function would have less energy. So, *in one-dimensional Schrödinger problems with Dirichlet boundary conditions, the states are not degenerate, the ground state has no node, and the* n-th *excited state has exactly* n *nodes.*

3.2 Step Potential

The step potential is a piecewise constant potential

$$V(x) = \begin{cases} 0, & \text{if } x < 0, \\ V_0, & \text{if } 0 \le x. \end{cases} \tag{3.21}$$

The general time-dependent solution is given by

$$\begin{aligned} \psi(x) &= A \exp[-i(Et/\hbar - k(x)\,x)] + B \exp[-i(Et/\hbar + k(x)\,x)] \\ &= [A \exp(ik(x)\,x) + B \exp(-ik(x)\,x)] \exp(-iEt/\hbar), \end{aligned} \tag{3.22}$$

with $k(x) = \sqrt{2m(E - V(x))/\hbar^2}$. The first term describes a right traveling wave, while the second one a left traveling wave.

3.2.1 Scattering

Here we consider a right traveling wave coming from $x = -\infty$. After hitting the barrier the wave is partly reflected and partly transmitted (Fig. 3.3)

$$\psi(x) = \begin{cases} e^{ikx} + \mathcal{R}e^{-ikx}, & \text{if } x < 0, \\ \mathcal{T}e^{ik'x}, & \text{if } 0 \le x, \end{cases} \tag{3.23}$$

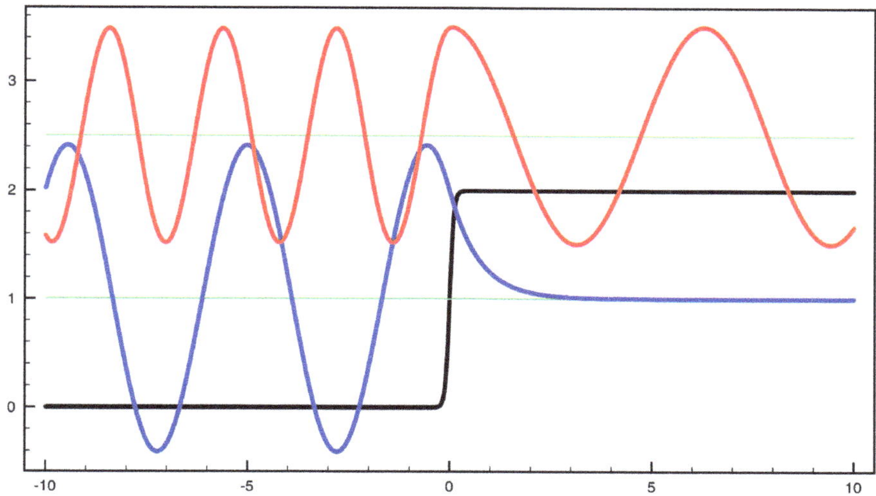

Fig. 3.3 Solutions in a potential step. If $E > V_0$ we have a scattering solution with transmitted wave, while if $E < V_0$ we observe a total reflection without transmission. The corresponding energies are denoted by horizintal lines

with $k = \sqrt{2mE/\hbar^2}$ and $k' = \sqrt{2m(E - V_0)/\hbar^2}$.

From the definition of the probability current

$$j = \frac{i\hbar}{2m}(\psi\psi^{*\prime} - \psi^*\psi') = \frac{\hbar}{m}\Im(\psi^*\psi'), \tag{3.24}$$

we find for the incoming, the reflected and the transmitted currents

$$j = \begin{cases} j_i = \dfrac{\hbar k}{m}, & \text{if } x < 0, \\ j_r = \dfrac{\hbar k}{m}R, \ \text{with } R = |\mathcal{R}|^2, & \text{if } x < 0, \\ j_t = \dfrac{\hbar k}{m}T, \ \text{with } T = \dfrac{k'}{k}\,|\mathcal{T}|^2, & \text{if } 0 \le x. \end{cases} \tag{3.25}$$

The potential has a jump at $x = 0$, but nevertheless, both ψ and ψ' should be continuous there. This results in the equations

$$1 + \mathcal{R} = \mathcal{T} \quad \text{and} \quad k(1 - \mathcal{R}) = k'\mathcal{T}, \tag{3.26}$$

which are solved by

$$\mathcal{R} = \frac{k - k'}{k + k'} \quad \text{and} \quad \mathcal{T} = \frac{2k}{k + k'}. \tag{3.27}$$

Here $E > V_0$, thus both k and k' are real and $k, k' > 0$. Therefore, we get

$$R + T = \left|\frac{k - k'}{k + k'}\right|^2 + \frac{k'}{k}\left|\frac{2k}{k + k'}\right|^2 = 1, \tag{3.28}$$

in agreement with the conservation of the probability current.

We can examine this scattering process in the wave packet formalism. A wave packet is a solution of the time-dependent Schrödinger equation. It can be constructed from the time-independent solution as

$$\psi(x,t) = \begin{cases} \int_0^\infty \mathrm{d}k f(k)\left(\exp(ikx) + \frac{k - k'}{k + k'}\exp(-ikx)\right)\exp(-iEt/\hbar), & \text{if } x < 0, \\ \int_0^\infty \mathrm{d}k f(k)\frac{2k}{k + k'}\exp(ik'x)\exp(-iEt/\hbar), & \text{if } 0 \leq x, \end{cases} \tag{3.29}$$

where $f(k)$ is sharply peaked, like a narrow Gaussian, around $k = k_0$. For a given t and x, the function in the integrand of the incoming wave,

$$f(k)\exp[i(kx - Et/\hbar)], \tag{3.30}$$

changes in k so rapidly, that contributions from outside the close vicinity of k_0 are negligible due to destructive interference. We have constructive interference only when the phase is slowly varying, or almost constant, i.e. when

$$\left.\frac{\mathrm{d}}{\mathrm{d}k}(kx - Et/\hbar)\right|_{k_0} = 0. \tag{3.31}$$

Then, from $E = \hbar^2 k^2/(2m)$, with $p_0 = \hbar k_0$, we have

$$x - \frac{p_0}{m}t = 0. \tag{3.32}$$

Here, $x < 0$ and $p_0 > 0$, so this condition is satisfied only for $t < 0$. The incoming wave vanishes for $t > 0$. Analogously, we find that for the reflected and the transmitted waves, t has to be positive.

3.2.2 Complete Reflection

In the previous section we solved the Schrödinger equation with a step potential for $E > V_0$. The results are analytic functions of the energy. Therefore, when $E < V_0$, the wave number becomes

$$k' = \sqrt{2m(E - V_0)/\hbar^2} = \sqrt{(-)2m(V_0 - E)/\hbar^2} = i\kappa', \tag{3.33}$$

with real

$$\kappa'(k) = \sqrt{2mV_0/\hbar^2 - k^2}. \tag{3.34}$$

Then, the corresponding formula for $\mathcal{R}$ can also be obtained by analytically continuing Eq. (3.27) from $k' \to i\kappa'$

$$\mathcal{R} = \frac{k - i\kappa'}{k + i\kappa'}. \tag{3.35}$$

Since the numerator and the denominator are complex conjugates to each other, we find

$$R = |\mathcal{R}|^2 = 1, \tag{3.36}$$

and consequently the incoming and reflected probability currents are the same, i.e. $j_i = j_r$.

We can also write

$$\mathcal{R} = \frac{k - i\kappa'}{k + i\kappa'} = \exp(-2i\delta(E)), \tag{3.37}$$

with

$$\tan\delta = \frac{\kappa'}{k}, \tag{3.38}$$

where δ is called the *phase shift*.

Then, the wave packet, in terms of the phase shift, reads

$$\psi(x,t) = \begin{cases} \int_0^\infty \mathrm{d}k f(k) \left[\exp(ikx) + \exp(-ikx - 2i\delta)\right] \exp(-iEt/\hbar), & \text{if } x < 0 \\ \int_0^\infty \mathrm{d}k f(k) \dfrac{2k}{k + i\kappa'} \exp(-\kappa' x) \exp(-iEt/\hbar), & \text{if } 0 \le x. \end{cases} \tag{3.39}$$

There is no transmitting wave, it turned into an exponentially decaying function. The incoming wave, as before, represents a traveling wave packet

$$x = \frac{p_0}{m} t, \quad \text{if} \quad t < 0. \tag{3.40}$$

We can see that the phase of the reflected wave is shifted with respect to the incoming wave by 2δ. For the reflected wave, the constructive interference condition gives

$$x = -\frac{p_0}{m} t - 2 \left. \frac{\mathrm{d}\delta(k)}{\mathrm{d}k} \right|_{k_0}. \tag{3.41}$$

Considering Eq. (3.34), and by taking the derivative of Eq. (3.38), we can infer that

$$\frac{d\delta}{dk} = -\frac{1}{\kappa'}, \tag{3.42}$$

and consequently we have

$$x = -\frac{p_0}{m}(t - t_0) \tag{3.43}$$

with $t_0 = 2m/(p_0\kappa'(k_0))$. We can see that the reflected wave suffers a *time delay*. To get $x < 0$ for the reflected wave, t should be bigger than t_0. This means that the wave enters the barrier, penetrates a little bit, spends some time there, and then bounces back with time delay t_0. The time delay is particularly big when $\kappa'(k)$ is small, i.e. $E \simeq V_0$.

3.3 The Box Potential

We consider a rectangular box potential

$$V(x) = \begin{cases} 0, & \text{if } x \leq -a/2, \\ V_0, & \text{if } -a/2 < x < a/2, \\ 0, & \text{if } a/2 \leq x. \end{cases} \tag{3.44}$$

We will use the notations $k = \sqrt{2m(E - V_0)/\hbar^2}$, $k' = \sqrt{2mE/\hbar^2}$, $k_0 = \sqrt{-2mV_0/\hbar^2}$ and assume that $V_0 < 0$. We emphasize here that the potential is symmetric, $V(x) = V(-x)$, so the Hamiltonian is parity invariant and the solutions should have even or odd parity symmetry.

3.3.1 Bound States

In the case of bound states, both k and k_0 are positive real numbers, but $k' = i\kappa'$ is imaginary with $\kappa' = \sqrt{2m/\hbar^2|E|}$ positive real. The even-parity wave function is given by

$$\psi_e(x) = \begin{cases} Be^{\kappa' x}, & \text{if } x < -a/2, \\ A\cos(kx), & \text{if } -a/2 \leq x < a/2, \\ Be^{-\kappa' x}, & \text{if } a/2 < x, \end{cases} \tag{3.45}$$

and the odd one is

$$\psi_o(x) = \begin{cases} Be^{\kappa' x}, & \text{if } x < -a/2, \\ A\sin(kx), & \text{if } -a/2 \leq x < a/2, \\ -Be^{-\kappa' x}, & \text{if } a/2 < x. \end{cases} \tag{3.46}$$

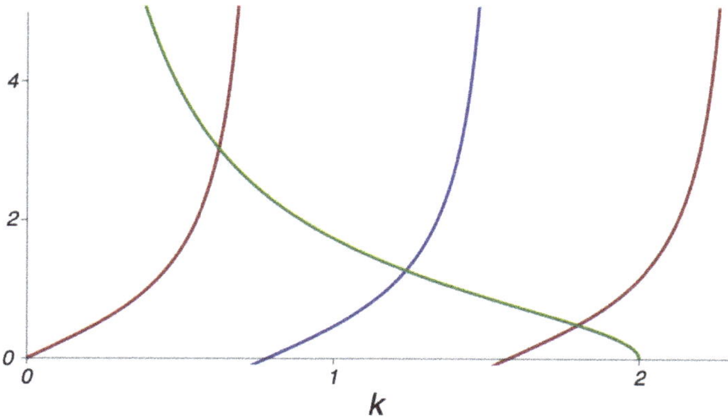

Fig. 3.4 The graphical solutions of the transcendental equation for bound states in the box potential of Eq. (3.44) with $a = 4$ and $k_0 = 2$. The solutions are the intersections of the left and right sides of Eqs. (3.49) and (3.50)

The continuity conditions at $a/2$ give

$$Be^{-\kappa' a/2} = A\cos(ka/2) \quad \text{and} \quad \kappa' Be^{-\kappa' a/2} = Ak\sin(ka/2), \quad \text{for even states,} \tag{3.47}$$

while

$$-Be^{-\kappa' a/2} = A\sin(ka/2) \quad \text{and} \quad \kappa' Be^{-\kappa' a/2} = Ak\cos(ka/2), \quad \text{for odd states.} \tag{3.48}$$

These result in transcendental equations for even states

$$\tan(ka/2) = \frac{\kappa'}{k} = \frac{1}{ka/2}\sqrt{(k_0 a/2)^2 - (ka/2)^2} = \frac{1}{k}\sqrt{k_0^2 - k^2}, \tag{3.49}$$

and for odd states

$$-\cot(ka/2) = \frac{\kappa'}{k} = \frac{1}{ka/2}\sqrt{(k_0 a/2)^2 - (ka/2)^2} = \frac{1}{k}\sqrt{k_0^2 - k^2}, \tag{3.50}$$

respectively. These transcendental equations can only be solved numerically, but a graphical plot reveals some properties of the solutions (Fig. 3.4). The right hand side behaves like $1/k$ for small k but runs to zero at $k = k_0$. This means that there are a finite number of bound states, depending on the value of k_0. On the other hand, since the tangent function starts from zero, there is at least one positive parity bound state, irrespective of how small k_0 is, i.e. how small V_0 is.

Infinite Potential Well

We can also recover the results of an infinite potential well if we take $V_0 \to -\infty$. In this case, we should measure the energy with respect to the bottom of the potential and introduce the notation $\epsilon = E - V_0$. Then $k = \sqrt{2m\epsilon/\hbar^2}$ and $k' = \sqrt{2m(V_0 + \epsilon)/\hbar^2} = i\kappa'$. In the limit of $V_0 \to -\infty$, κ' becomes infinite. Consequently, the wave function vanishes outside the $[-a/2, a/2]$ interval. This gives us the conditions for energy eigenstates:

$$\cos(ka/2) = 0 \quad \to \quad ka/2 = \pi/2 + n'\pi, \quad n' = 1, 2, \ldots, \quad \text{for the even parity states} \tag{3.51}$$

and

$$\sin(ka/2) = 0 \quad \to \quad ka/2 = n'\pi, \quad n' = 1, 2, \ldots, \quad \text{for the odd parity states.} \tag{3.52}$$

We can unify the two conditions into

$$\sin(ka) = 0 \quad \text{and} \quad ka = \sqrt{2m\epsilon/\hbar^2}\, a = n\pi, \quad n = 1, 2, \ldots, \tag{3.53}$$

which results in

$$\epsilon_n = \frac{\hbar^2}{2m}\left(\frac{n\pi}{a}\right)^2, \qquad n = 1, 2, \ldots. \tag{3.54}$$

Particle on a Circle

We observe a somewhat similar situation when a free particle of mass m moves on a circle. We denote the distance along the perimeter by x. There is no potential, therefore the solution is just a plane wave

$$\psi = Ae^{ikx}, \quad \text{with} \quad k = \sqrt{2mE/\hbar^2}, \quad E > 0. \tag{3.55}$$

Moving on a circle means that we have the condition $x \equiv x + L$, which involves

$$\psi(x + L) = \psi(x). \tag{3.56}$$

Consequently, k cannot be arbitrary, but rather such that

$$kL = 2\pi n, \quad \text{with} \quad n = 0, \pm 1, \pm 2, \ldots, \tag{3.57}$$

and the energy is discrete

$$E_n = \frac{\hbar^2}{2m}\left(\frac{2\pi}{L}\right)^2 n^2, \tag{3.58}$$

and two-fold degenerate.

However, the continuity of the probability allows a more general boundary condition

$$\psi(x+L) = e^{i\theta}\psi(x), \tag{3.59}$$

where θ is a real constant. This is a sufficient requirement since

$$|\psi(x+L)|^2 = |e^{i\theta}\psi(x)|^2 = |\psi(x)|^2. \tag{3.60}$$

The more general boundary condition requires that

$$kL = 2\pi n + \theta, \qquad \text{for} \quad n = 0, \pm 1, \pm 2, \ldots, \tag{3.61}$$

and then

$$E_n = \frac{\hbar^2}{2m}\left(\frac{2\pi}{L}\right)^2\left(n + \frac{\theta}{2\pi}\right)^2. \tag{3.62}$$

In this case, the energy is in general, not degenerate any more, and for the wave function, we have

$$\psi_n(x) = \frac{1}{\sqrt{L}}\exp\left(i\frac{2\pi n + \theta}{L}x\right). \tag{3.63}$$

If $\theta = \pi$, the energy levels become

$$E_n = \frac{\hbar^2}{2m}\left(\frac{2\pi}{L}\right)^2\left(n + \frac{1}{2}\right)^2. \tag{3.64}$$

The levels, even the lowest one, are doubly degenerate. The pairs $n = (0, -1)$, $(1, -2)$, $(2, -3)$, ... provide the same energy.

If $\theta = 2\pi$, the energy levels are

$$E_n = \frac{\hbar^2}{2m}\left(\frac{2\pi}{L}\right)^2(n+1)^2, \tag{3.65}$$

and we have the same spectrum as for $\theta = 0$. The only difference is that now the $n = -1$ is the non-degenerate ground state, while the degenerate pairs are $n = (0, -2), (1, -3), \ldots$.

3.3.2 Scattering

Now for $E > 0$, the wave numbers are positive real numbers in all regions. Consequently, the wave function is given by piecewise plane waves

$$\psi(x) = \begin{cases} e^{ik'x} + \mathcal{R}e^{-ik'x}, & \text{if } x < -a/2, \\ Ce^{ikx} + De^{-ikx}, & \text{if } -a/2 \le x < a/2, \\ \mathcal{T}e^{ik'x}, & \text{if } a/2 < x. \end{cases} \tag{3.66}$$

The continuity of ψ and ψ' at $x = \pm a/2$ leads to the solutions

$$\begin{aligned} \mathcal{R}(E) &= \frac{i}{2}\left(\frac{k}{k'} - \frac{k'}{k}\right)\sin(ka)\mathcal{T}(E), \\ C(E) &= \frac{1}{2}\left(1 + \frac{k'}{k}\right)\exp[i(k'-k)a/2]\mathcal{T}(E), \\ D(E) &= \frac{1}{2}\left(1 - \frac{k'}{k}\right)\exp[i(k'+k)a/2]\mathcal{T}(E), \end{aligned} \tag{3.67}$$

where

$$\mathcal{T}(E) = \frac{\exp(-ik'a)}{\cos(ka) - (i/2)(k'/k + k/k')\sin(ka)}. \tag{3.68}$$

For the transmission coefficient we have

$$T(E) = |\mathcal{T}(E)|^2 = \left(1 + \frac{V_0^2}{4E(E-V_0)}\sin^2(ka)\right)^{-1}. \tag{3.69}$$

Potential Barrier and Tunneling

In the case of a potential barrier $V_0 > 0$. Then, if the energy is below the barrier, $0 < E < V_0$, we find $k = i\kappa$ with $\kappa = \sqrt{2m(V_0 - E)/\hbar^2}$. The wave function in the $[-a/2, a/2]$ interval becomes a combination of exponentially rising and decaying components

$$\psi(x) = \begin{cases} e^{ik'x} + \mathcal{R}e^{-ik'x}, & \text{if } x < -a/2, \\ Ce^{-\kappa x} + De^{\kappa x}, & \text{if } -a/2 \le x < a/2, \\ \mathcal{T}e^{ik'x}, & \text{if } a/2 < x. \end{cases} \tag{3.70}$$

Scattering wave functions for energies above and below the barrier are shown in Fig. 3.5.

By performing the $k \to i\kappa$ substitution in Eqs. (3.67) and (3.69), we obtain the corresponding relations with

$$\mathcal{T} = \frac{2i\kappa k' \exp(-ik'a)}{2i\kappa k' \cosh(\kappa a) + (k'^2 - \kappa^2)\sinh(\kappa a)}, \tag{3.71}$$

which gives

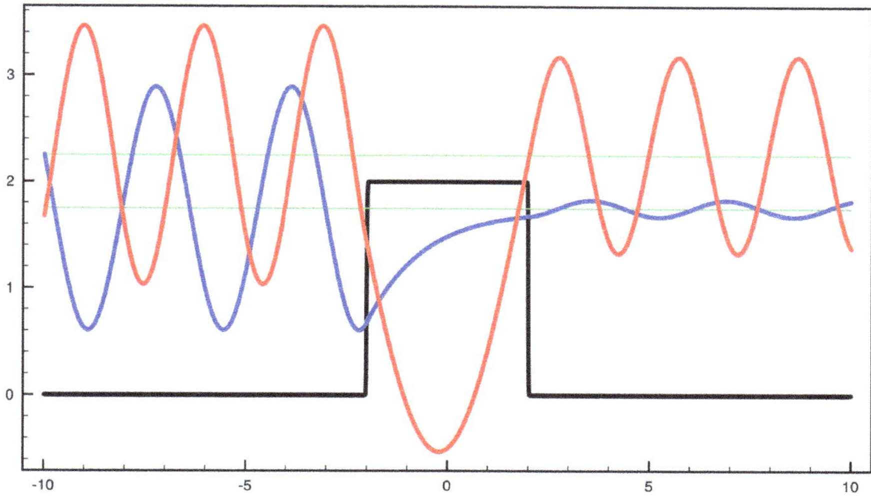

Fig. 3.5 Scattering above the barrier and tunneling below the barrier

$$T(E) = |\mathcal{T}(E)|^2 = \left(1 + \frac{V_0^2}{4E(V_0 - E)} \sinh^2(\kappa a)\right)^{-1}. \tag{3.72}$$

We can see that even for $E < V_0$ the transmission does not vanish, i.e. $T(E) \neq 0$. So, there is a tunneling through the barrier, unlike in classical mechanics. This is one of the striking phenomena of quantum mechanics.

If $\kappa a \gg 1$, i.e. the energy is well below the barrier, $\sinh(\kappa a) \to \exp(\kappa a)/2$, and the transmission coefficient simplifies

$$T(E) \approx \frac{16E(V_0 - E)}{V_0^2} \exp(-2a\sqrt{2m(V_0 - E)/\hbar^2}). \tag{3.73}$$

We can observe an interesting phenomena when the scattering energy is such that $\sin(ka) = 0$ or $\sinh(ka) = 0$. Then, the reflection coefficient becomes zero and the transmission coefficient becomes one. The system becomes fully transparent. This provides an explanation to the Ramsauer-Townsend effect. In scattering of low energy electrons on noble gas atoms, the cross section obtains a minimum value at certain energies. We note that the condition for full transparency is the same as the discrete states in an infinite box potential. In both cases the wave should "fit" in the potential.

3.4 Analytic Properties of the Scattering Amplitudes

We should notice that the transition amplitude for the box potential $\mathcal{T}$ in Eq. (3.68) has a pole if

$$\cos(ka) - \frac{i}{2}\left(\frac{k'}{k} + \frac{k}{k'}\right)\sin(ka) = 0. \tag{3.74}$$

If $E < 0$ then $k' = i\kappa'$, with $\kappa' > 0$. Then, there is no incoming or outgoing wave, i.e. the wave function outside the box decays exponentially. The pole condition of Eq. (3.74) becomes

$$\frac{1}{\tan(ka)} = \frac{i}{2}\left(\frac{i\kappa'}{k} + \frac{k}{i\kappa'}\right) = \frac{1}{2}\left(\frac{k}{\kappa'} - \frac{\kappa'}{k}\right). \tag{3.75}$$

By using the relations

$$\tan(\alpha + \beta) = \frac{\tan\alpha + \tan\beta}{1 - \tan\alpha\tan\beta} \quad \text{and} \quad \tan(\alpha) = \frac{2\tan\alpha/2}{1 - \tan^2\alpha/2} = \frac{2}{\cot\alpha/2 - \tan\alpha/2}$$

we find that

$$\cot(ka/2) - \tan(ka/2) = \frac{k}{\kappa'} - \frac{\kappa'}{k}. \tag{3.76}$$

We can see that this equation can be solved by taking either

$$\tan(ka/2) = \kappa'/k, \quad \text{or} \quad -\cot(ka/2) = \kappa'/k. \tag{3.77}$$

This means that the poles of the transition amplitude at negative energy coincide with both the even, Eq. (3.49), and the odd, Eq. (3.50), solutions of the bound-state problem.

For positive energies, $E > 0$, the wave number k' is real and positive. If $E - V_0 > 0$ then k is also positive real. While if $E - V_0 < 0$ then $k = i\kappa$ is positive imaginary and the trigonometric functions turn into hyperbolic functions. In any case, if E is positive, the left hand side of Eq. (3.74) never vanishes.

It may become zero, however, for complex energies. The transition probability $|\mathcal{T}(E)|^2$ in Eq. (3.69) is maximal when $\sin(ka) = 0$. In this region the transition amplitude reads

$$\begin{aligned} \mathcal{T}(E) &= \frac{\exp(-ik'a)}{\cos(ka) - (i/2)(k'/k + k/k')\sin(ka)} \\ &= \frac{\exp(-ik'a)}{\cos(ka)} \frac{1}{1 - (i/2)(k'/k + k/k')\tan(ka)}. \end{aligned} \tag{3.78}$$

We denote the real energy corresponding to $\sin(ka) = \tan(ka) = 0$ by E_r. In the vicinity of E_r we have

$$\left(\frac{k'}{k}+\frac{k}{k'}\right)\tan(ka)\approx\frac{\mathrm{d}}{\mathrm{d}E}\left[\left(\frac{k'}{k}+\frac{k}{k'}\right)\tan(ka)\right]_{E_r}(E-E_r)\approx\frac{4}{\Gamma}(E-E_r),\tag{3.79}$$

and the so-defined Γ is real. Then, the transmission amplitude is approximated by

$$\mathcal{T}(E)\approx\left.\frac{\exp(-ik'a)}{\cos(ka)}\right|_{E_r}\frac{1}{1-2i(E-E_r)/\Gamma}\approx\exp(-ik'a)\frac{i\Gamma/2}{E-(E_r-i\Gamma/2)}.\tag{3.80}$$

The function $\mathcal{T}(E)$ has a pole at complex energy $E=E_r-i\Gamma/2$ and for the transmission probability $T(E)$ we get

$$T(E)=|\mathcal{T}(E)|^2\approx\frac{(\Gamma/2)^2}{(E-E_r)^2+(\Gamma/2)^2}.\tag{3.81}$$

This is the *Breit-Wigner* formula. The transmission probability $T(E)$ has a maximum at real energy E_r, and the maximum value is one, signaling a perfectly transparent potential. If Γ is small, $T(E)$ changes rapidly. This is the *resonance* case, E_r is the resonance energy and Γ is the width of the resonance. Resonances are associated with the complex energy poles of the transmission or reflection coefficients and with a rapid change of the physical quantities.

We note that the resonance condition for the box potential, $\sin(ka)=0$, is the same as the condition Eq. (3.53) for the eigenvalues in an infinite box potential. Resonances happen when the scattering wave "fits" in the box. The wave that reflected from the left edge and the wave reflected from the right edge of the potential cancel each other out, resulting in a perfect transmission.

We denote the wave number corresponding to the complex resonance energy by $k=k_r-i\gamma$. Around the pole, the transmission is dominant, so the wave function at $t>0$ is almost entirely the transmitted wave function

$$\psi\sim\exp(-iEt/\hbar)\exp(ikx)=\exp(-iE_rt/\hbar)\exp(-\Gamma t/(2\hbar))\exp(ik_rx)\exp(\gamma x).\tag{3.82}$$

This state exhibits the usual stationary quantum state motion, $\exp(-iE_rt/\hbar)$, but with exponentially decaying amplitude in time, $\exp(-\Gamma t/(2\hbar))$. In the spatial coordinate, it oscillates like a scattering state $\exp(ik_rx)$, but between exponentially growing $\pm\exp(\gamma x)$ envelopes. These states are called *resonant*, or *Gamow*, states. They are the complex energy eigenstates of the Schrödinger equation, not square integrable, and do not belong to the Hilbert space. They are used to describe decaying quantum states, like the α decay, where finding a particle at a definite location is exponentially diminishing.

We can write the transmission amplitude Eq. (3.80) as

$$\begin{aligned}\mathcal{T}(E) &\approx e^{-ik'a}\frac{i\Gamma/2}{E-E_r+i\Gamma/2} = e^{-ik'a}\frac{i(E-E_r-i\Gamma/2)\Gamma/2}{(E-E_r)^2+(\Gamma/2)^2} \\ &= e^{-ik'a}\frac{(\Gamma/2)^2+i(E-E_r)\Gamma/2}{(E-E_r)^2+(\Gamma/2)^2}.\end{aligned} \tag{3.83}$$

The phase $\delta(E)$ is defined by $\mathcal{T}(E) = \exp(i\delta(E))|\mathcal{T}(E)|$, which gives

$$\delta(E) \approx -k'a + \arctan\left(\frac{2}{\Gamma}(E-E_r)\right). \tag{3.84}$$

We can see that as E goes from far below E_r to far above E_r, the phase $\delta(E)$ changes by π. If Γ is small, the change happens in a narrow energy interval. Then, the time delay

$$\tau = \frac{\mathrm{d}\delta(E)}{\mathrm{d}k} \tag{3.85}$$

is large, indicating that the scattered particle spends a long time in the potential forming a long-lived intermediate resonant state. So, resonant states can be considered as bound states with finite lifetime, while bound states are resonant states with infinite lifetime.

3.5 Dirac δ Potential

An attractive Dirac delta potential

$$V(x) = -V_0\delta(x), \tag{3.86}$$

with $V_0 > 0$, can be considered as a limit of a box potential when $a \to 0$ such that aV_0 remains constant. The Schrödinger equation takes the form

$$-\frac{\hbar^2}{2m}\psi''(x) - V_0\delta(x)\psi(x) = E\psi(x). \tag{3.87}$$

Let us denote the solutions by ψ_+ and ψ_- for positive and negative x, respectively. The wave function is continuous, therefore

$$\psi_-(0) = \psi_+(0). \tag{3.88}$$

Integrating Eq. (3.87), we find

$$-\frac{\hbar^2}{2m}\int_{-\epsilon}^{\epsilon}\mathrm{d}x\ \psi''(x) - V_0\int_{-\epsilon}^{\epsilon}\mathrm{d}x\ \delta(x)\psi(x) = E\int_{-\epsilon}^{\epsilon}\mathrm{d}x\ \psi(x). \tag{3.89}$$

Then, taking the $\epsilon \to 0$ limit, from the continuity of ψ, we obtain

$$\psi'_{+}(0) - \psi'_{-}(0) = -\frac{2mV_0}{\hbar^2}\psi(0). \tag{3.90}$$

Bound states

If $E < 0$, $\kappa = \sqrt{-2mE/\hbar^2}$ is real and the wave function becomes

$$\psi(x) = \begin{cases} A_{-}e^{\kappa x}, & \text{if } x < 0, \\ A_{+}e^{-\kappa x}, & \text{if } 0 < x. \end{cases} \tag{3.91}$$

Then,

$$\psi_{-}(0) = A_{-} = \psi_{+}(0) = A_{+} = A, \tag{3.92}$$

and, from Eq. (3.90), we have

$$-\kappa A - \kappa A = -\frac{2m}{\hbar^2}V_0 A. \tag{3.93}$$

Thus,

$$\kappa = \frac{m}{\hbar^2}V_0 = \kappa_0, \tag{3.94}$$

and then

$$E = -\frac{\hbar^2\kappa_0^2}{2m} = -\frac{mV_0^2}{2\hbar^2}. \tag{3.95}$$

For an attractive Dirac δ potential we have found only one bound state, which is symmetric (Fig. 3.6).

If the δ potential is repulsive, i.e. $V_0 < 0$, we get the solution $\kappa = -\kappa_0$. This state has negative energy, but with exponentially increasing wave function, which, of course, fails to be square integrable. These states are called *virtual* states.

Scattering states

If $0 < E$ then $k = \sqrt{2mE/\hbar^2}$ is real and the wave function is given by

$$\psi(x) = \begin{cases} e^{ikx} + \mathcal{R}e^{-ikx}, & \text{if } x < 0, \\ \mathcal{T}e^{ikx}, & \text{if } 0 < x. \end{cases} \tag{3.96}$$

The boundary conditions at $x = 0$ results in

$$1 + \mathcal{R} = \mathcal{T} \quad \text{and} \quad ik\mathcal{T} - (ik - ik\mathcal{R}) = -2\kappa_0\mathcal{T}. \tag{3.97}$$

So, finally we have

$$\mathcal{R} = \frac{i\kappa_0}{k - i\kappa_0} \quad \text{and} \quad \mathcal{T} = \frac{k}{k - i\kappa_0}, \tag{3.98}$$

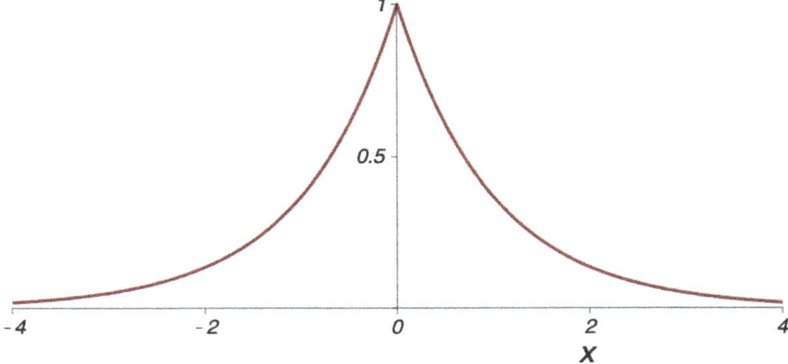

Fig. 3.6 The only bound state in an attractive Dirac delta potential

which result in the reflection and transmission coefficients

$$R = \frac{\kappa_0^2}{k^2 + \kappa_0^2} \quad \text{and} \quad T = \frac{k^2}{k^2 + \kappa_0^2}. \tag{3.99}$$

The pole appears at $k = i\kappa_0$, which gives the same bound-state energy as we found before. We can get the formulae for a repulsive δ potential by the $\kappa_0 \to -\kappa_0$ substitution, which leads to the same R and T coefficients.

3.6 Harmonic Oscillator

As we know from classical mechanics, a small amplitude motion of a bound system around the equilibrium is a harmonic oscillation. The first relevant term of a potential curve around equilibrium x_0 is a quadratic term (Fig. 3.7)

$$V(x) = V(x_0) + \frac{1}{2} \left. \frac{\mathrm{d}^2 V(x)}{\mathrm{d}x^2} \right|_{x_0} (x - x_0)^2 + \cdots . \tag{3.100}$$

So, quantum systems with one degree of freedom around equilibrium can be described as a one dimensional harmonic oscillator. Taking the coordinate system such that $x_0 = 0$ and using the notation $V''(x_0) = m\omega^2$, we find the Hamiltonian

$$\hat{H} = \frac{\hat{p}_x^2}{2m} + \frac{1}{2} m\omega^2 x^2, \tag{3.101}$$

where m is the mass and ω is the angular frequency of the motion.

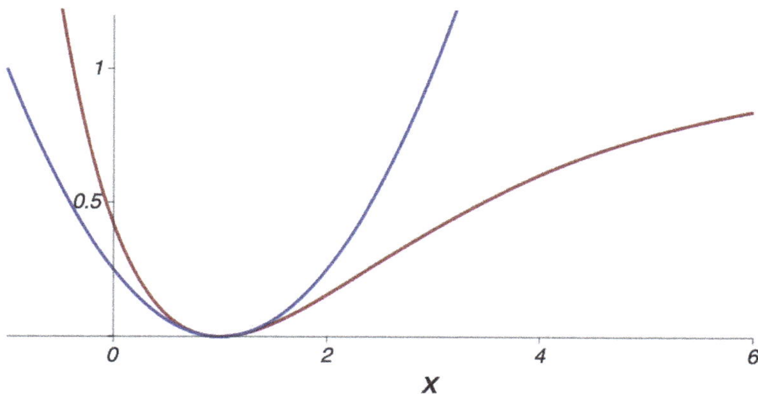

Fig. 3.7 The interaction between atoms often described by the Morse potential $V(x) = v_0(1 - \exp(-(x - x_0)/a_0))^2$. Here we plot with $v_0 = 1, x_0 = 1$ and $a_0 = 2$. Around the equilibrium it can be approximated by a quadratic potential

3.6.1 Operator Solution

In solving the harmonic oscillator problem, it would be beneficial if we could write the Hamiltonian in Eq. (3.101) as a product of first order terms. This Hamiltonian is a sum of quadratic terms, so we can try to decompose it as $u^2 + v^2 = (u + iv)(u - iv)$ with $u \sim p_x$ and $v \sim x$. The operators $\hat{x}$ and $\hat{p}_x$ in the Hamiltonian Eq. (3.101) are Hermitian operators that obey the commutation relation $[\hat{x}, \hat{p}_x] = i\hbar$. Here, it is customary to introduce the operators

$$\hat{a} = \sqrt{\frac{m\omega}{2\hbar}}\left(\hat{x} + i\frac{\hat{p}_x}{m\omega}\right) = \sqrt{\frac{m\omega}{2\hbar}}\left(\hat{x} + \frac{\hbar}{m\omega}\frac{\mathrm{d}}{\mathrm{d}x}\right) \tag{3.102}$$

and

$$\hat{a}^\dagger = \sqrt{\frac{m\omega}{2\hbar}}\left(\hat{x} - i\frac{\hat{p}_x}{m\omega}\right) = \sqrt{\frac{m\omega}{2\hbar}}\left(\hat{x} - \frac{\hbar}{m\omega}\frac{\mathrm{d}}{\mathrm{d}x}\right), \tag{3.103}$$

and express $\hat{x}$ and $\hat{p}_x$ as

$$\hat{x} = \sqrt{\frac{\hbar}{2\,m\omega}}\,(\hat{a} + \hat{a}^\dagger) \quad \text{and} \quad \hat{p}_x = -i\sqrt{\frac{\hbar m\omega}{2}}\,(\hat{a} - \hat{a}^\dagger)\,. \tag{3.104}$$

A little calculation reveals that

$$\hat{a}\hat{a}^\dagger = \frac{m\omega}{2\hbar}\left(\hat{x}^2 + \frac{\hat{p}_x^2}{m^2\omega^2} + \frac{\hbar}{m\omega}\right) \quad \text{and} \quad \hat{a}^\dagger\hat{a} = \frac{m\omega}{2\hbar}\left(\hat{x}^2 + \frac{\hat{p}_x^2}{m^2\omega^2} - \frac{\hbar}{m\omega}\right). \tag{3.105}$$

Subtracting them we obtain the commutation relation

$$\hat{a}\hat{a}^\dagger - \hat{a}^\dagger\hat{a} = [\hat{a}, \hat{a}^\dagger] = 1, \tag{3.106}$$

and by adding them we find the Hamiltonian

$$\hat{a}\hat{a}^\dagger + \hat{a}^\dagger\hat{a} = \frac{2}{\hbar\omega}\hat{H}. \tag{3.107}$$

So, we can write the Hamiltonian as

$$\hat{H} = \left(\hat{a}^\dagger\hat{a} + \frac{1}{2}\right)\hbar\omega, \tag{3.108}$$

or with the operator

$$\hat{N} = \hat{a}^\dagger\hat{a}, \tag{3.109}$$

we get

$$\hat{H} = \left(\hat{N} + \frac{1}{2}\right)\hbar\omega. \tag{3.110}$$

Obviously, $\hat{N}$ commutes with the Hamiltonian, and thus they have common eigenstates. The eigenvalue equation for $\hat{N}$ reads

$$\hat{N}|n\rangle = n|n\rangle, \tag{3.111}$$

which also solves the Hamiltonian

$$\hat{H}|n\rangle = \left(n + \frac{1}{2}\right)\hbar\omega. \tag{3.112}$$

We can immediately see that $\hat{N}$ is Hermitian,

$$\hat{N}^\dagger = (\hat{a}^\dagger\hat{a})^\dagger = \hat{a}^\dagger(\hat{a}^\dagger)^\dagger = \hat{a}^\dagger\hat{a} = \hat{N}, \tag{3.113}$$

and non-negative

$$\langle\psi|\hat{N}|\psi\rangle = \langle\psi|\hat{a}^\dagger\hat{a}|\psi\rangle = \langle\hat{a}\psi|\hat{a}\psi\rangle \geq 0. \tag{3.114}$$

The commutation relations of $\hat{N}$ with $\hat{a}^\dagger$ and $\hat{a}$ are given by

$$[\hat{N}, \hat{a}^\dagger] = \hat{a}^\dagger\hat{a}\hat{a}^\dagger - \hat{a}^\dagger\hat{a}^\dagger\hat{a} = \hat{a}^\dagger[\hat{a}, \hat{a}^\dagger] = \hat{a}^\dagger \tag{3.115}$$

and

$$[\hat{N}, \hat{a}] = \hat{a}^\dagger\hat{a}\hat{a} - \hat{a}\hat{a}^\dagger\hat{a} = [\hat{a}^\dagger, \hat{a}]\hat{a} = -\hat{a}, \tag{3.116}$$

respectively. Then, we find

$$\hat{N}(\hat{a}^\dagger|n\rangle) = (\hat{a}^\dagger\hat{N} + \hat{a}^\dagger)|n\rangle = (n+1)(\hat{a}^\dagger|n\rangle), \tag{3.117}$$

and

$$\hat{N}(\hat{a}|n\rangle) = (\hat{a}\hat{N} - \hat{a})|n\rangle = (n-1)(\hat{a}|n\rangle), \tag{3.118}$$

i.e. the states $\hat{a}^\dagger|n\rangle$ and $\hat{a}|n\rangle$ are eigenstates of $\hat{N}$ with eigenvalues $n+1$ and $n-1$, respectively. Therefore, $\hat{a}^\dagger$ is the *raising*, or *creation* operator, and $\hat{a}$ is the *lowering*, or *annihilation* operator. They are also referred to as *ladder* operators.

Let us assume that $|n\rangle \neq 0$, i.e. it is not a zero state. Then, from the relation

$$\langle n|\hat{N}|n\rangle = n\langle n|n\rangle = \langle \hat{a}n|\hat{a}n\rangle \geq 0, \tag{3.119}$$

we can see that $\hat{a}|n\rangle = 0$, only if $n = 0$. For $n \neq 0$, the state $|n-1\rangle$ is also an eigenstate with eigenvalue $n-1$, and so on with $n-2, n-3, \ldots$. This can be continued until the eigenvalue becomes negative, which is not possible, since $\hat{N}$ is a positive operator and therefore all the eigenvalues have to be non-negative. The only way to break this sequence is to have a state $|0\rangle$ such that

$$\hat{N}|0\rangle = 0, \quad \text{or} \quad \hat{a}|0\rangle = 0. \tag{3.120}$$

Then, any further application of $\hat{a}$ results in 0. However, to arrive at this $n = 0$ state, we must start from an integer n. Hence, n can only be a non-negative integer.

To determine the action of $\hat{a}$ and $\hat{a}^\dagger$, we consider

$$\hat{a}|n\rangle = \alpha|n-1\rangle \quad \text{and} \quad \hat{a}^\dagger|n\rangle = \beta|n+1\rangle, \tag{3.121}$$

and calculate

$$\langle \hat{a}n|\hat{a}n\rangle = |\alpha|^2\langle n-1|n-1\rangle = \langle n|\hat{a}^\dagger\hat{a}|n\rangle = \langle n|\hat{N}|n\rangle = n\langle n|n\rangle \tag{3.122}$$

and

$$\langle \hat{a}^\dagger n|\hat{a}^\dagger n\rangle = |\beta|^2\langle n+1|n+1\rangle = \langle n|\hat{a}\hat{a}^\dagger|n\rangle = \langle n|\hat{N}+1|n\rangle = (n+1)\langle n|n\rangle. \tag{3.123}$$

So, we can conclude that

$$\hat{a}|n\rangle = \sqrt{n}|n-1\rangle \quad \text{and} \quad \hat{a}^\dagger|n\rangle = \sqrt{n+1}|n+1\rangle, \tag{3.124}$$

or

$$|n-1\rangle = \frac{1}{\sqrt{n}}\hat{a}|n\rangle \quad \text{and} \quad |n+1\rangle = \frac{1}{\sqrt{n+1}}\hat{a}^\dagger|n\rangle, \tag{3.125}$$

respectively.

Now, starting with the state $|0\rangle$, by repeated application of $\hat{a}^\dagger$, we can build $|n\rangle$

$$|n\rangle = \frac{1}{\sqrt{n}}\hat{a}^\dagger|n-1\rangle = \frac{1}{\sqrt{n(n-1)}}(\hat{a}^\dagger)^2|n-2\rangle = \cdots = \frac{1}{\sqrt{n!}}(\hat{a}^\dagger)^n|0\rangle. \quad (3.126)$$

These states are orthogonal,

$$\langle n'|\hat{N}|n\rangle = \langle n'|\hat{N}n\rangle = n\langle n'|n\rangle = \langle \hat{N}n'|n\rangle = n'\langle n'|n\rangle, \quad (3.127)$$

i.e. $\langle n'|n\rangle = 0$ if $n' \neq n$, and form a complete set

$$\sum_{n=0}^{\infty} |n\rangle\langle n| = I. \quad (3.128)$$

3.6.2 Wave Function

Configuration Space Representation

The $|0\rangle$ state is defined by

$$\hat{a}|0\rangle = 0. \quad (3.129)$$

Taking this operator relation in coordinate representation, we have

$$\langle x|\hat{a}|0\rangle = \sqrt{\frac{m\omega}{2\hbar}}\langle x|\left(\hat{x} + i\frac{\hat{p}_x}{m\omega}\right)|0\rangle = 0, \quad (3.130)$$

and with $\psi_0(x) = \langle x|0\rangle$, we obtain the differential equation

$$\left(x + \frac{\hbar}{m\omega}\frac{\mathrm{d}}{\mathrm{d}x}\right)\psi_0(x) = 0, \quad \text{or} \quad \frac{\mathrm{d}\psi_0(x)}{\mathrm{d}x} = -\frac{m\omega}{\hbar}x\,\psi_0(x). \quad (3.131)$$

With $\beta = \sqrt{m\omega/\hbar}$, this differential equation is solved by

$$\psi_0(x) = A_0 \exp\left(-\frac{\beta^2 x^2}{2}\right). \quad (3.132)$$

With the help of the integral

$$\int_{-\infty}^{\infty} \exp(-\alpha x^2)\mathrm{d}x = \sqrt{\frac{\pi}{\alpha}}, \qquad \alpha > 0, \quad (3.133)$$

we find with

$$A_0 = \sqrt{\frac{\beta}{\sqrt{\pi}}}, \quad (3.134)$$

that the wave function ψ_0 is normalized

$$\int_{-\infty}^{\infty} |\psi_0(x)|^2 \mathrm{d}x = 1. \tag{3.135}$$

Now we can calculate $\psi_1(x) = \langle x|1\rangle$. By applying $\hat{a}^\dagger$ to $\psi_0(x)$, we find

$$\psi_1(x) = \langle x|\hat{a}^\dagger|0\rangle = \frac{\beta}{\sqrt{2}}\langle x|\left(\hat{x} - \frac{i}{m\omega}\hat{p}_x\right)|0\rangle = \frac{\beta}{\sqrt{2}}\left(x - \frac{1}{\beta^2}\frac{\mathrm{d}}{\mathrm{d}x}\right)\psi_0(x), \tag{3.136}$$

and obtain

$$\psi_1(x) = \sqrt{2}\beta x \psi_0(x). \tag{3.137}$$

In order to find a way to determine harmonic oscillator wave functions with higher n, we apply $\hat{a} + \hat{a}^\dagger = \sqrt{2}\hat{u}$, where $\hat{u} = \beta\hat{x}$, to a harmonic oscillator state

$$(\hat{a} + \hat{a}^\dagger)|n\rangle = \sqrt{n}|n-1\rangle + \sqrt{n+1}|n+1\rangle = \sqrt{2}\hat{u}|n\rangle. \tag{3.138}$$

Taking this relation in u representation we get

$$\sqrt{n}\langle u|n-1\rangle + \sqrt{n+1}\langle u|n+1\rangle = \sqrt{2}u\langle u|n\rangle, \tag{3.139}$$

which is basically a three term recursion relation

$$\sqrt{n+1}\,\psi_{n+1}(\beta x) = \sqrt{2}\beta x\,\psi_n(\beta x) - \sqrt{n}\,\psi_{n-1}(\beta x). \tag{3.140}$$

Now we can evaluate higher n harmonic oscillator eigenstates starting from ψ_0.

We may write the wave function as

$$\psi_n(u) = \sqrt{\frac{\beta}{\sqrt{\pi}2^n n!}}\exp(-u^2/2)\,H_n(u). \tag{3.141}$$

Then, from Eq. (3.140) we find

$$H_{n+1} - 2uH_n + 2nH_{n-1} = 0, \quad \text{with} \quad H_0(u) = 1. \tag{3.142}$$

This recursion relation with $H_0 \equiv 1$ is the recursion relation of the Hermite orthogonal polynomials (Fig. 3.8).

Momentum Space Representation

From the position-momentum commutation relation, we can easily find that in the momentum space representation, the position operator is given by

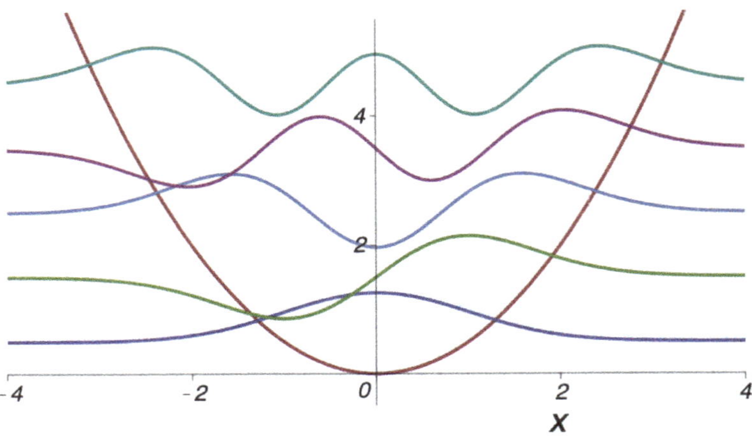

Fig. 3.8 The first few harmonic oscillator states in coordinate representation

$$\hat{x} = i\hbar \frac{\mathrm{d}}{\mathrm{d}p_x}. \tag{3.143}$$

Therefore, the relation analogous to Eq. (3.130) takes the form

$$\left(\sqrt{\frac{m\hbar\omega}{2}} \frac{\mathrm{d}}{\mathrm{d}p_x} + \frac{p_x}{\sqrt{2\,m\hbar\omega}}\right) \psi_0(p_x) = 0. \tag{3.144}$$

This equation, aside some constants, is very similar to Eq. (3.130). Therefore, the solution is analogous

$$\psi_0(p_x) = \frac{1}{(\pi m\hbar\omega)^{1/4}} \exp\left(-\frac{p_x^2}{2\,m\hbar\omega}\right). \tag{3.145}$$

From $\hat{a} - \hat{a}^\dagger = \sqrt{2}\hat{v}$, where $\hat{v} = i\,\hat{p}_x/\sqrt{m\hbar\omega}$, very much the same way as before, we arrive at the recursion relation

$$\sqrt{n+1}\psi_{n+1}(v) = -\sqrt{2}v\psi_n(v) + \sqrt{n}\psi_{n-1}(v). \tag{3.146}$$

The same way as in the coordinate representation, this recursion relation can also be linked to Hermite polynomials

$$\psi_n(v) = \frac{1}{\pi^{1/4}\sqrt{2^n n!}} \exp(-v^2/2) H_n(v). \tag{3.147}$$

3.6.3 Matrix Representation

We can use the harmonic oscillator states $\{|n\rangle\}$ as a basis. This basis is used extensively in quantum mechanical calculations and it is also very useful in treating the harmonic oscillator itself. In that basis, the operators $\hat{N}$ and $\hat{H}$ take the form of infinite diagonal matrices

$$\hat{N} = \begin{pmatrix} 0 & & & \\ & 1 & & \\ & & 2 & \\ & & & \ddots \end{pmatrix} \tag{3.148}$$

and

$$\hat{H} = \frac{\hbar\omega}{2} \begin{pmatrix} 1 & & & \\ & 3 & & \\ & & 5 & \\ & & & \ddots \end{pmatrix}, \tag{3.149}$$

respectively. The eigenstates are represented by infinite vectors

$$|0\rangle = \begin{pmatrix} 1 \\ 0 \\ 0 \\ 0 \\ \vdots \end{pmatrix}, \quad |1\rangle = \begin{pmatrix} 0 \\ 1 \\ 0 \\ 0 \\ \vdots \end{pmatrix}, \quad |2\rangle = \begin{pmatrix} 0 \\ 0 \\ 1 \\ 0 \\ \vdots \end{pmatrix}, \quad |3\rangle = \begin{pmatrix} 0 \\ 0 \\ 0 \\ 1 \\ \vdots \end{pmatrix} \quad \ldots, \tag{3.150}$$

and for the creation and annihilation operators, we find

$$\hat{a} = \begin{pmatrix} 0 & \sqrt{1} & 0 & 0 & \cdots \\ 0 & 0 & \sqrt{2} & 0 & \cdots \\ 0 & 0 & 0 & \sqrt{3} & \cdots \\ 0 & 0 & 0 & 0 & \cdots \\ \vdots & \vdots & \vdots & \vdots & \ddots \end{pmatrix} \quad \text{and} \quad \hat{a}^{\dagger} = \begin{pmatrix} 0 & 0 & 0 & 0 & \cdots \\ \sqrt{1} & 0 & 0 & 0 & \cdots \\ 0 & \sqrt{2} & 0 & 0 & \cdots \\ 0 & 0 & \sqrt{3} & 0 & \cdots \\ \vdots & \vdots & \vdots & \vdots & \ddots \end{pmatrix}. \tag{3.151}$$

We can also have

$$\hat{x} = \sqrt{\frac{\hbar}{2m\omega}}(\hat{a} + \hat{a}^{\dagger}) = \sqrt{\frac{\hbar}{2m\omega}} \begin{pmatrix} 0 & \sqrt{1} & 0 & 0 & \cdots \\ \sqrt{1} & 0 & \sqrt{2} & 0 & \cdots \\ 0 & \sqrt{2} & 0 & \sqrt{3} & \cdots \\ 0 & 0 & \sqrt{3} & 0 & \cdots \\ \vdots & \vdots & \vdots & \vdots & \ddots \end{pmatrix}, \tag{3.152}$$

and

$$\hat{p}_x = -i\sqrt{\frac{m\hbar\omega}{2}}(\hat{a} - \hat{a}^\dagger) = -i\sqrt{\frac{m\hbar\omega}{2}}\begin{pmatrix} 0 & \sqrt{1} & 0 & 0 & \cdots \\ -\sqrt{1} & 0 & \sqrt{2} & 0 & \cdots \\ 0 & -\sqrt{2} & 0 & \sqrt{3} & \cdots \\ 0 & 0 & -\sqrt{3} & 0 & \cdots \\ \vdots & \vdots & \vdots & \vdots & \ddots \end{pmatrix}, \tag{3.153}$$

which obviously satisfy the commutation relation $[\hat{x}, \hat{p}_x] = i\hbar I$.

3.6.4 Expectation Values

The representation of operators in terms of $\hat{a}$ and $\hat{a}^\dagger$ is especially handy in calculating matrix elements. For example, we can easily see that

$$\langle n|\hat{x}|n\rangle = \sqrt{\frac{\hbar}{2m\omega}}\langle n|(\hat{a} + \hat{a}^\dagger)|n\rangle = 0 \quad \text{and} \quad \langle n|\hat{p}_x|n\rangle = -i\sqrt{\frac{m\hbar\omega}{2}}\langle n|(\hat{a} - \hat{a}^\dagger)|n\rangle = 0. \tag{3.154}$$

The evaluation of the matrix elements of $\hat{x}^2$ and $\hat{p}_x^2$ is also straightforward, since

$$\hat{x}^2 = \frac{\hbar}{2m\omega}(\hat{a}^2 + (\hat{a}^\dagger)^2 + \hat{a}\hat{a}^\dagger + \hat{a}^\dagger\hat{a}) = \frac{\hbar}{2m\omega}(\hat{a}^2 + (\hat{a}^\dagger)^2 + 2\hat{a}^\dagger\hat{a} + 1), \tag{3.155}$$

and

$$\hat{p}_x^2 = -\frac{m\hbar\omega}{2}(\hat{a}^2 + (\hat{a}^\dagger)^2 - \hat{a}\hat{a}^\dagger - \hat{a}^\dagger\hat{a}) = -\frac{m\hbar\omega}{2}(\hat{a}^2 + (\hat{a}^\dagger)^2 - 2\hat{a}^\dagger\hat{a} - 1). \tag{3.156}$$

Obviously, $\langle n|\hat{a}^2|n\rangle = \langle n|(\hat{a}^\dagger)^2|n\rangle = 0$ and $\langle n|\hat{a}^\dagger\hat{a}|n\rangle = \langle n|\hat{N}|n\rangle = n$, therefore

$$\langle n|\hat{x}^2|n\rangle = \frac{\hbar}{2m\omega}(2n + 1) \quad \text{and} \quad \langle n|\hat{p}_x^2|n\rangle = \frac{m\hbar\omega}{2}(2n + 1). \tag{3.157}$$

In calculating the expectation values of the potential and kinetic energies, we obtain

$$\frac{1}{2}m\omega^2\langle n|\hat{x}^2|n\rangle = \frac{1}{2m}\langle n|\hat{p}_x^2|n\rangle = \frac{1}{2}\langle n|\hat{H}|n\rangle = \frac{\hbar\omega}{2}(n + 1/2). \tag{3.158}$$

This is the *virial theorem* for the harmonic oscillator. This result is similar to the classical mechanical one.

Uncertainty Relation

Now, we can easily calculate the uncertainties

$$\Delta x = \sqrt{\langle \hat{x}^2 \rangle - \langle \hat{x} \rangle^2} = \sqrt{\frac{\hbar}{2\,m\omega}(2n+1)} \quad \text{and} \quad \Delta p_x = \sqrt{\langle \hat{p}_x^2 \rangle - \langle \hat{p}_x \rangle^2} = \sqrt{\frac{m\hbar\omega}{2}(2n+1)}, \tag{3.159}$$

and find

$$\Delta x \cdot \Delta p_x = (n + 1/2)\hbar \longrightarrow \Delta x\, \Delta p_x \geq \hbar/2, \tag{3.160}$$

in complete agreement with the Heisenberg uncertainty relation. The $n = 0$ ground state is the minimal uncertainty harmonic oscillator state with

$$\Delta x = \sqrt{\frac{\hbar}{2\,m\omega}} \quad \text{and} \quad \Delta p_x = \sqrt{\frac{m\hbar\omega}{2}} \tag{3.161}$$

such that

$$\Delta x \cdot \Delta p_x = \frac{\hbar}{2}. \tag{3.162}$$

3.6.5 Coherent States

The harmonic oscillator states form a basis, so any state can be represented as their linear combinations. A particularly interesting state is

$$|\alpha\rangle = e^{-\frac{1}{2}|\alpha|^2} \sum_{n=0}^{\infty} \frac{\alpha^n}{\sqrt{n!}} |n\rangle, \tag{3.163}$$

where α is a complex number. It is easy to see that these states are normalized, i.e. $\langle\alpha|\alpha\rangle = 1$. Furthermore, an easy calculation gives

$$\hat{a} \sum_{n=0}^{\infty} \frac{\alpha^n}{\sqrt{n!}} |n\rangle = \sum_{n=0}^{\infty} \frac{\alpha^n}{\sqrt{n!}} \sqrt{n} |n-1\rangle = \alpha \sum_{n=1}^{\infty} \frac{\alpha^{n-1}}{\sqrt{(n-1)!}} |n-1\rangle = \alpha \sum_{n=0}^{\infty} \frac{\alpha^n}{\sqrt{n!}} |n\rangle, \tag{3.164}$$

i.e. the $|\alpha\rangle$ states are eigenstates of the annihilation operator $\hat{a}$ with eigenvalue α so that

$$\hat{a}|\alpha\rangle = \alpha|\alpha\rangle. \tag{3.165}$$

The eigenstate α can be complex, since $\hat{a}$ is not Hermitian, therefore

$$\langle\alpha|\hat{a}^\dagger = \langle\alpha|\alpha^*. \tag{3.166}$$

We can easily determine the matrix elements:

$$\begin{aligned}
\langle\alpha|\hat{x}|\alpha\rangle &= \sqrt{\frac{\hbar}{2m\omega}}\langle\alpha|(\hat{a}^\dagger+\hat{a})|\alpha\rangle = \sqrt{\frac{\hbar}{2m\omega}}(\alpha^*+\alpha),\\
\langle\alpha|\hat{p}_x|\alpha\rangle &= i\sqrt{\frac{m\omega\hbar}{2}}\langle\alpha|(\hat{a}^\dagger-\hat{a})|\alpha\rangle = i\sqrt{\frac{m\omega\hbar}{2}}(\alpha^*-\alpha),\\
\langle\alpha|\hat{x}^2|\alpha\rangle &= \frac{\hbar}{2m\omega}\langle\alpha|(\hat{a}^{\dagger 2}+\hat{a}^2+2\hat{a}^\dagger\hat{a}+1)|\alpha\rangle = \frac{\hbar}{2m\omega}(\alpha^{*2}+\alpha^2+2\alpha^*\alpha+1),\\
\langle\alpha|\hat{p}_x^2|\alpha\rangle &= -\frac{m\omega\hbar}{2}\langle\alpha|(\hat{a}^{\dagger 2}+\hat{a}^2-2\hat{a}^\dagger\hat{a}-1)|\alpha\rangle = -\frac{m\omega\hbar}{2}(\alpha^{*2}+\alpha^2-2\alpha^*\alpha-1).
\end{aligned} \tag{3.167}$$

From the above relations, it follows that

$$\Delta x = \sqrt{\langle\hat{x}^2\rangle-\langle\hat{x}\rangle^2} = \sqrt{\frac{\hbar}{2m\omega}} \quad \text{and} \quad \Delta p = \sqrt{\langle\hat{p}_x^2\rangle-\langle\hat{p}_x\rangle^2} = \sqrt{\frac{m\omega\hbar}{2}}, \tag{3.168}$$

and consequently

$$\Delta x\cdot\Delta p = \frac{\hbar}{2}. \tag{3.169}$$

So, we have found that the $|\alpha\rangle$ states are minimal uncertainty states, just like the harmonic oscillator ground state. They are called *coherent states.*

With the help of Eq. (1.69), we define the operator

$$\hat{D}(\alpha) = \exp(-\alpha^*\hat{a}+\alpha\hat{a}^\dagger) = \exp(-\alpha^*\hat{a})\exp(\alpha\hat{a}^\dagger)\exp\left(\frac{1}{2}|\alpha|^2\right). \tag{3.170}$$

We can see that $\hat{D}(\alpha)$ is unitary

$$\hat{D}^\dagger(\alpha) = \hat{D}^{-1}(\alpha) = \hat{D}(-\alpha), \tag{3.171}$$

and from Eq. (1.59) we find that

$$\hat{D}^\dagger(\alpha)\hat{a}\hat{D}(\alpha) = \hat{a}+\alpha \quad \text{and} \quad \hat{D}^\dagger(\alpha)\hat{a}^\dagger\hat{D}(\alpha) = \hat{a}^\dagger+\alpha^*. \tag{3.172}$$

Consequently,

$$\hat{D}^\dagger(\alpha)\hat{x}\hat{D}(\alpha) = \hat{x}+x_0 \quad \text{and} \quad \hat{D}^\dagger(\alpha)\hat{p}_x\hat{D}(\alpha) = \hat{p}_x+p_0, \tag{3.173}$$

with $x_0 = \sqrt{\hbar/(2m\omega)}(\alpha+\alpha^*)$ and $p_0 = -i\sqrt{\hbar m\omega/2}(\alpha-\alpha^*)$. From these relations, it follows that

$$\hat{D}^\dagger(\alpha)f(\hat{x},\hat{p}_x)\hat{D}(\alpha) = f(\hat{x}+x_0,\hat{p}_x+p_0). \tag{3.174}$$

Applying to the harmonic oscillator ground state we find

$$
\begin{aligned}
\hat{D}(\alpha)|0\rangle &= \exp(\frac{1}{2}|\alpha|^2)\exp(-\alpha^*\hat{a})\exp(\alpha\hat{a}^\dagger)|0\rangle \\
&= \exp(\frac{1}{2}|\alpha|^2)\exp(-\alpha^*\hat{a})\sum_{n=0}^{\infty}\frac{\alpha^n(\hat{a}^\dagger)^n}{n!}|0\rangle \\
&= \exp(\frac{1}{2}|\alpha|^2)\exp(-\alpha^*\hat{a})\sum_{n=0}^{\infty}\frac{\alpha^n\sqrt{n!}}{n!}|n\rangle \\
&= \exp(|\alpha|^2)\exp(-\alpha^*\hat{a})\exp(-\frac{1}{2}|\alpha|^2)\sum_{n=0}^{\infty}\frac{\alpha^n}{\sqrt{n!}}|n\rangle \\
&= \exp(|\alpha|^2)\exp(-\alpha^*\hat{a})|\alpha\rangle = \exp(|\alpha|^2)\exp(-\alpha^*\alpha)|\alpha\rangle = |\alpha\rangle .
\end{aligned}
\tag{3.175}
$$

So, a coherent state is basically a displaced harmonic oscillator ground state.

The time evolution of the coherent state is determined by the time evolution of the harmonic oscillator states

$$
\begin{aligned}
|\alpha(t)\rangle &= \exp(-\frac{1}{2}|\alpha|^2)\sum_{n=0}^{\infty}\frac{\alpha^n}{\sqrt{n!}}\exp(-i/\hbar\, E_n t)|n\rangle \\
&= \exp(-\frac{1}{2}|\alpha|^2)\sum_{n=0}^{\infty}\frac{\alpha^n}{\sqrt{n!}}\exp(-i/\hbar\,\hbar\omega(n+1/2)t)|n\rangle \\
&= \exp(-i\omega t/2)\exp(-\frac{1}{2}|\alpha|^2)\sum_{n=0}^{\infty}\frac{(\alpha\exp(-i\omega t))^n}{\sqrt{n!}}|n\rangle \\
&= \exp(-i\omega t/2)|\alpha\exp(-i\omega t)\rangle .
\end{aligned}
\tag{3.176}
$$

So we find that the coherent state, during time evolution, remains a coherent state with minimal uncertainty. The expectation values of $\hat{x}$ and $\hat{p}_x$ oscillate

$$
\begin{aligned}
\langle\alpha(t)|\hat{x}|\alpha(t)\rangle &= \sqrt{\frac{\hbar}{2m\omega}}(\alpha^*(t)+\alpha(t)) = \sqrt{\frac{\hbar}{2m\omega}}2|\alpha|\cos(\omega t-\delta), \\
\langle\alpha(t)|\hat{p}_x|\alpha(t)\rangle &= i\sqrt{\frac{m\omega\hbar}{2}}(\alpha^*(t)-\alpha(t)) = \sqrt{\frac{m\omega\hbar}{2}}2|\alpha|\sin(\omega t-\delta),
\end{aligned}
\tag{3.177}
$$

where δ is basically the phase of α, i.e. $\alpha = |\alpha|\exp(i\delta)$. A coherent state behaves like a classical mass-spring system and it traces a classical oscillator-like phase trajectory in the α-plane while maintaining the minimal uncertainty property.

Of course, we cannot beat the Heisenberg uncertainty relation, but we can make Δx smaller at the expense of making Δp bigger. These *squeezed states* are defined by the operator

$$
\hat{S}(r) = \exp\left[\frac{r}{2}((\hat{a}^\dagger)^2 - (\hat{a})^2)\right],
\tag{3.178}
$$

with r a real number. We can see that $\hat{S}$ is unitary

$$\hat{S}^\dagger(r) = \hat{S}^{-1}(r) = \hat{S}(-r). \tag{3.179}$$

It can generate the unitary transformations

$$\begin{aligned} \hat{b}(r) &= \hat{S}^\dagger(r)\hat{a}\hat{S}(r) = \cosh(r)\hat{a} + \sinh(r)\hat{a}^\dagger \\ \hat{b}^\dagger(r) &= \hat{S}^\dagger(r)\hat{a}^\dagger\hat{S}(r) = \cosh(r)\hat{a}^\dagger + \sinh(r)\hat{a}, \end{aligned} \tag{3.180}$$

which obviously does not change the commutation relation

$$[\hat{b}, \hat{b}^\dagger] = [\hat{a}, \hat{a}^\dagger] = 1. \tag{3.181}$$

Then, the functions of $\hat{a}$ and $\hat{a}^\dagger$ transform as

$$\hat{S}^\dagger(r) f(\hat{a}, \hat{a}^\dagger)\hat{S}(r) = f(\hat{b}, \hat{b}^\dagger), \tag{3.182}$$

and consequently we find

$$\hat{S}^\dagger(r)\hat{x}\hat{S}(r) = \exp(r)\hat{x} \quad \text{and} \quad \hat{S}^\dagger(r)\hat{p}_x\hat{S}(r) = \exp(-r)\hat{p}. \tag{3.183}$$

Now we can define the squeezed state as

$$|0, r\rangle = \hat{S}(r)|0\rangle. \tag{3.184}$$

This transformation does not change the expectation values

$$\langle 0, r|\hat{x}|0, r\rangle = \langle 0, r|\hat{p}_x|0, r\rangle = 0, \tag{3.185}$$

but scales the operators as

$$\langle 0, r|\hat{x}^2|0, r\rangle = \exp(2r)\frac{\hbar}{2m\omega} \quad \text{and} \quad \langle 0, r|\hat{p}_x^2|0, r\rangle = \exp(-2r)\frac{m\hbar\omega}{2}. \tag{3.186}$$

As a result, the uncertainties become squeezed

$$\Delta x = \exp(r)\sqrt{\frac{\hbar}{2m\omega}} \quad \text{and} \quad \Delta p_x = \exp(-r)\sqrt{\frac{m\hbar\omega}{2}}, \tag{3.187}$$

while the Heisenberg relation with minimal uncertainty, $\Delta x \cdot \Delta p_x = \hbar/2$, is maintained.

3.7 Periodic Potentials

In this section, we consider periodic potentials, with period d, in one dimension, such that

$$\hat{V}(x+d) \equiv \hat{V}(x), \quad -\infty \leq x \leq \infty. \tag{3.188}$$

Let us denote the operator of finite displacement

$$\hat{F}(d)|x\rangle = |x+d\rangle. \tag{3.189}$$

We should not envision this operator as an infinite product of infinitesimal displacements, but rather a finite jump by d. Nevertheless, it is a unitary operator, since it acts on the whole space and preserves the norm. Therefore,

$$\hat{F}^{\dagger}(d)\hat{F}(d) = \hat{F}(d)\hat{F}^{\dagger}(d) = I, \tag{3.190}$$

and the following relations are valid:

$$\hat{F}^{\dagger}(d)|x+d\rangle = |x\rangle, \quad \langle x+d|\hat{F}(d) = \langle x|, \quad \langle x|\hat{F}(d) = \langle x-d|. \tag{3.191}$$

Moreover,

$$\hat{F}^{\dagger}(d)\hat{x}\hat{F}(d)|x\rangle = \hat{F}^{\dagger}(d)\hat{x}|x+d\rangle = \hat{F}^{\dagger}(d)(x+d)|x+d\rangle = (x+d)|x\rangle. \tag{3.192}$$

Then, due to Eq. (1.107), for our periodic potential we have

$$\hat{F}^{\dagger}(d)\hat{V}(\hat{x})\hat{F}(d) = \hat{V}(\hat{x}+d) \equiv \hat{V}(x), \tag{3.193}$$

which means that the potential commutes with the finite displacement operator

$$[\hat{F}(d), \hat{V}(x)] = 0. \tag{3.194}$$

Since $\hat{F}(d)$ does not depend on x, it should also commute with the kinetic energy, and thus with the whole Hamiltonian:

$$[\hat{F}(d), \hat{p}_x^2/2m] = 0 \quad \text{and} \quad [\hat{F}(d), \hat{H}] = 0. \tag{3.195}$$

As a consequence, the eigenstates of the Hamiltonian are also eigenstates of $\hat{F}$:

$$\hat{H}|\psi\rangle = E|\psi\rangle \qquad \text{and} \qquad \hat{F}(d)|\psi\rangle = \lambda|\psi\rangle. \tag{3.196}$$

The displacement operator $\hat{F}(d)$ is unitary and its eigenvalues are complex numbers with modulus one, $|\lambda| = 1$. So, we can write

$$\lambda = \exp(-iKd) \text{ and } \langle x|\hat{F}(d)|\psi\rangle = \langle x-d|\psi\rangle = \psi(x-d) = \exp(-iKd)\psi(x), \tag{3.197}$$

with some real parameter K.

Let us define the function

$$u_K(x) = \exp(-iKx)\psi(x). \tag{3.198}$$

Then,

$$\begin{aligned} u_K(x-d) &= \exp(-iK(x-d))\psi(x-d) = \exp(-iK(x-d))\exp(-iKd)\psi(x) \\ &= \exp(-iKx)\psi(x) = u_K(x), \end{aligned} \tag{3.199}$$

i.e. $u_K(x)$ is periodic with period d. So, we have found that the solution of the periodic potential problem can be written as

$$\psi(x) = \exp(iKx)u_K(x), \tag{3.200}$$

where the *Bloch wave function* $u_K(x)$ is periodic with the same periodicity as the potential itself.

We can learn a lot about periodic systems by studying the Kronig-Penney model:

$$V(x) = \begin{cases} 0, & \text{if } 0 < x < a, \\ V_0, & \text{if } a < x < d, \end{cases} \tag{3.201}$$

where the whole structure is periodic, $V(x+d) \equiv V(x)$ with $d = a + b$.

The wave function is a piecewise plane wave

$$\psi(x) = \begin{cases} Ae^{ikx} + Be^{-ikx}, & \text{if } 0 < x < a, \quad k = \sqrt{2mE/\hbar^2}, \\ Ce^{ik'x} + De^{-ik'x}, & \text{if } a < x < d, \quad k' = \sqrt{2m(E-V_0)/\hbar^2}, \quad E > V_0. \end{cases} \tag{3.202}$$

Then, for the Bloch wave function and its derivative, respectively, we have

$$u_K(x) = \begin{cases} (Ae^{ikx} + Be^{-ikx})e^{-iKx}, & \text{if } 0 < x < a, \\ (Ce^{ik'x} + De^{-ik'x})e^{-iKx}, & \text{if } a < x < d, \end{cases} \tag{3.203}$$

and

$$u'_K(x) = \begin{cases} ik(Ae^{ikx} - Be^{-ikx})e^{-iKx} - iKu_K(x), & \text{if } 0 < x < a, \\ ik'(Ce^{ik'x} - De^{-ik'x})e^{-iKx} - iKu_K(x), & \text{if } a < x < d. \end{cases} \tag{3.204}$$

From the periodicity conditions $u_K(0) = u_K(a+b)$ and $u'_K(0) = u'_K(a+b)$, we find

$$(A + B) = (Ce^{ik'(a+b)} + De^{-ik'(a+b)})e^{-iKd}, \tag{3.205}$$
$$k(A - B) = k'(Ce^{ik'(a+b)} - De^{-ik'(a+b)})e^{-iKd}, \tag{3.206}$$

and from the continuity conditions for ψ and ψ' at $x = a$, it follows

$$Ae^{ika} + Be^{-ika} = Ce^{ik'a} + De^{-ik'a}, \tag{3.207}$$
$$k(Ae^{ika} - Be^{-ika}) = k'(Ce^{ik'a} - De^{-ik'a}). \tag{3.208}$$

Writing these equations in a matrix form we have

$$\begin{pmatrix} e^{iKd} & e^{iKd} & -e^{ik'(a+b)} & -e^{-ik'(a+b)} \\ ke^{iKd} & -ke^{iKd} & -k'e^{ik'(a+b)} & k'e^{-ik'(a+b)} \\ e^{ika} & e^{-ika} & -e^{ik'a} & -e^{-ik'a} \\ ke^{ika} & -ke^{-ika} & -k'e^{ik'a} & k'e^{-ik'a} \end{pmatrix} \begin{pmatrix} A \\ B \\ C \\ D \end{pmatrix} = 0, \tag{3.209}$$

which has a non-trivial solution only if the determinant is zero. After some manipulation, assuming $E > V_0$, we obtain

$$\cos(ka)\cos(k'b) - \frac{k^2 + k'^2}{2kk'}\sin(ka)\sin(k'b) = \cos(Kd), \tag{3.210}$$

where both k and k' are real. If $E < V_0$, then $k' = i\kappa'$, with $\kappa' = \sqrt{2m(V_0 - E)/\hbar^2} = \sqrt{\kappa_0^2 - k^2}$ real, and the equation becomes

$$\cos(ka)\cosh(\kappa'b) - \frac{k^2 - \kappa'^2}{2k\kappa'}\sin(ka)\sinh(\kappa'b) = \cos(Kd). \tag{3.211}$$

The right hand sides of Eqs. (3.210) and (3.211) are bounded, $-1 \leq \cos(Kd) \leq 1$. Therefore, the energy values are forbidden when the left hand sides of Eqs. (3.210) and (3.211) are outside the $[-1, 1]$ interval. These energy values are excluded from the otherwise continuous spectrum. The forbidden energy gaps appear at arbitrarily high energies, irrespective of the sign of V_0.

We can simplify the expressions in Eq. (3.211) by taking the $b \to 0$ limit while keeping $V_0 b$ constant. In this case the potential becomes a delta function. Then with $a \to d$, $\kappa'^2 b \to$ constant, $k^2 b \to 0$, and $\kappa' b \to 0$, we have $\sinh(\kappa' b) \to \kappa' b$ and $\cosh(\kappa' b) \to 1$. So, Eq. (3.211) finally becomes

$$\cos(kd) + \frac{\kappa_0 b}{2k}\sin(kd) = \cos(Kd). \tag{3.212}$$

This formula shows even more clearly the existence of gaps in the spectrum. As $\cos(kd)$ approaches its maximum or minimum values, the second term pushes beyond the $[-1, 1]$ interval and makes the corresponding energy range unacceptable (Fig. 3.9).

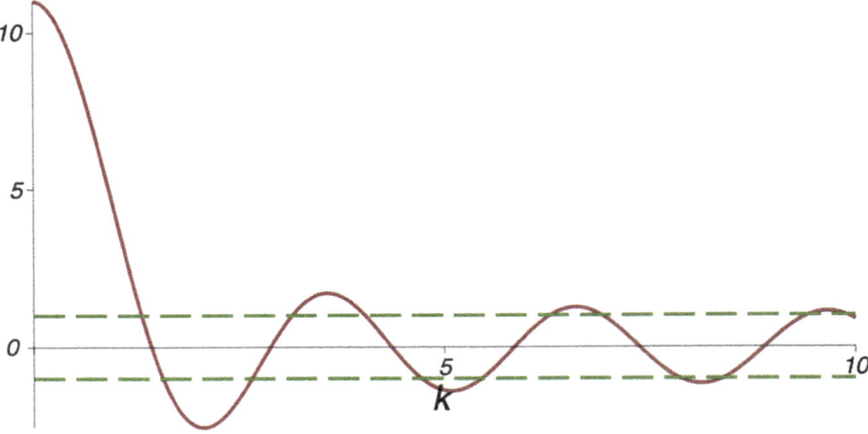

Fig. 3.9 Left hand side of Eq. (3.212) as a function of k with $d = 2$ and $\kappa_0 b = 10$. The regions where the absolute values of the left side are bigger than one are forbidden

Another interesting limit is when $a =$ constant and $b \to \infty$. Now $\sinh(\kappa' b) \sim \cosh(\kappa' b) \sim \exp(\kappa' b) \to \infty$ and we get

$$\cos(ka) - \frac{k^2 - \kappa'^2}{2k\kappa'} \sin(ka) = 0, \tag{3.213}$$

or

$$\frac{1}{\tan(ka)} = \frac{1}{2}\left(\frac{k}{\kappa'} - \frac{\kappa'}{k}\right). \tag{3.214}$$

This, as we have shown in Eq. (3.75), gives the energies of a single rectangular potential well.

We can interpret the spectrum of a periodic potential as the discrete energies in a single potential well that get broadened as we repeatedly add more and more shifted potential wells. Assume that we have a potential well with discrete bound states. If we add another identical potential well that is shifted far away, the original state becomes degenerate. A one dimensional system, however, cannot be degenerate. The proper wave function should be a symmetric and an antisymmetric combination of states localized at the potential wells. The symmetric state has fewer nodes than the antisymmetric one, consequently, it is more bound. So, the symmetric combination is shifted down a little bit and the antisymmetric one is shifted up a little bit (see Fig. 3.10). The split in energies becomes more and more prominent as the wells get closer and closer. The process is repeated if we add more and more potential wells, and if the system becomes truly periodic, the original energy level becomes a continuous band.

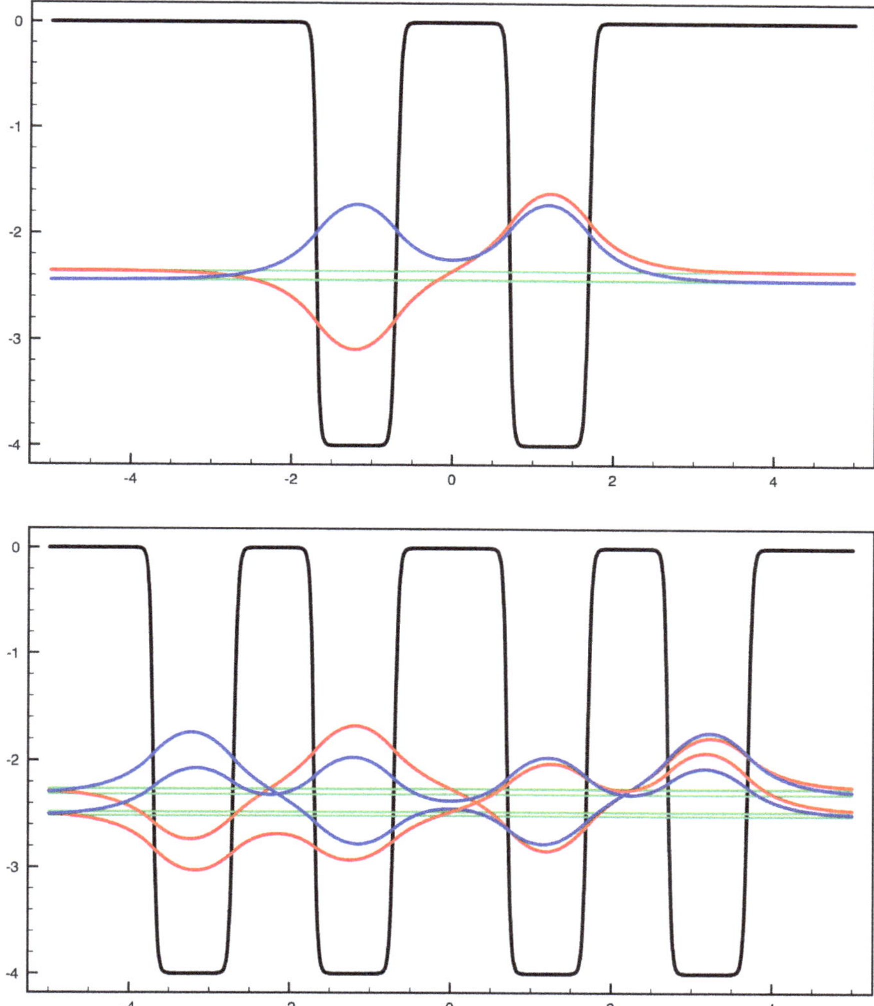

Fig. 3.10 Energy splitting of the ground state in shifted potentials

3.8 Factorization Method

We solved the harmonic oscillator problem by expressing the Hamiltonian as a product of operators that are adjoint to each other. We may try to do the same trick with more general potentials in $\hat{H}$ (see e.g. p. 875 in [1]). If we separate off the bound state energy E_0 and define

$$\hat{H}_+ = \hat{H} - E_0, \tag{3.215}$$

then $\hat{H}_+$ will have only non-negative eigenvalues

$$\hat{H}_+\psi = E\psi, \tag{3.216}$$

where $E \geq 0$ and $E = 0$ is the ground state.

We assume that

$$\hat{H}_+ = -\frac{\mathrm{d}^2}{\mathrm{d}x^2} + V_+(x), \tag{3.217}$$

and define, analogously to Eqs. (3.102) and (3.103),

$$\hat{A} = \frac{\mathrm{d}}{\mathrm{d}x} + W(x) \quad \text{and} \quad \hat{A}^\dagger = -\frac{\mathrm{d}}{\mathrm{d}x} + W(x), \tag{3.218}$$

with the requirement

$$\hat{H}_+ = \hat{A}^\dagger\hat{A} = -\frac{\mathrm{d}^2}{\mathrm{d}x^2} + W^2(x) - W'(x). \tag{3.219}$$

So, the potential problem of Eq. (3.217) can be factorized if we find the *super potential* W such that it solves the equation

$$W^2(x) - W'(x) = V_+(x). \tag{3.220}$$

By taking the substitution

$$W(x) = -\psi_0'/\psi_0 = -(\log\psi_0)', \tag{3.221}$$

Equation (3.220) becomes a Schrödinger-like equation

$$-\psi_0'' + V_+\psi_0 = \hat{H}_+\psi_0 = 0. \tag{3.222}$$

This is the same equation as Eq. (3.216) with $E = 0$. We can see that ψ_0 cannot have any node, since it is the ground state solution of a one-dimensional potential problem. So, we have found that non-negative Hamiltonians can be factorized and the potential W in $\hat{A}$ can be obtained from the solution of $\hat{H}_+\psi_0 = 0$. Since $0 = \langle\psi_0|\hat{H}_+|\psi_0\rangle = \langle\hat{A}\psi_0|\hat{A}\psi_0\rangle$, we find that

$$\hat{A}\psi_0 = \left(\frac{\mathrm{d}}{\mathrm{d}x} + W(x)\right)\psi_0 = 0. \tag{3.223}$$

This equation is a first order differential equation whose solution is given by

$$\psi_0(x) \simeq \exp\Big(-\int^x \mathrm{d}x'\, W(x')\Big), \tag{3.224}$$

which obviously satisfies Eq. (3.221).

A similar treatment goes for the corresponding adjoint operator

$$\hat{H}_- = \hat{A}\hat{A}^\dagger = -\frac{\mathrm{d}^2}{\mathrm{d}x^2} + W^2(x) + W'(x) = -\frac{\mathrm{d}^2}{\mathrm{d}x^2} + V_-(x), \tag{3.225}$$

which is just like $\hat{H}_+$, with the $W \to -W$ substitution. Therefore

$$\phi_0(x) \simeq \exp\left(\int^x \mathrm{d}x'\, W(x')\right) \tag{3.226}$$

solves the $\hat{H}_-\phi_0 = 0$ equation.

We can easily establish the intertwining relations

$$\hat{A}\hat{H}_+ = \hat{A}\hat{A}^\dagger\hat{A} = \hat{H}_-\hat{A} \quad \text{and} \quad \hat{A}^\dagger\hat{H}_- = \hat{A}^\dagger\hat{A}\hat{A}^\dagger = \hat{H}_+\hat{A}^\dagger. \tag{3.227}$$

Let us assume that ψ_+ is an eigenstate of $\hat{H}_+$ with eigenvalue E. Then we find

$$\hat{A}(E\psi_+) = \hat{A}(\hat{H}_+\psi_+) = \hat{H}_-(\hat{A}\psi_+) = E(\hat{A}\psi_+), \tag{3.228}$$

i.e. $\hat{A}\psi_+$ is an eigenstate of $\hat{H}_-$ with the same eigenvalue. Similarly, if ψ_- is an eigenstate of $\hat{H}_-$ with eigenvalue E, we have

$$\hat{A}^\dagger(E\psi_-) = \hat{A}^\dagger(\hat{H}_-\psi_-) = \hat{H}_+(\hat{A}^\dagger\psi_-) = E(\hat{A}^\dagger\psi_-), \tag{3.229}$$

i.e. $\hat{A}^\dagger\psi_-$ is an eigenstate of $\hat{H}_+$ with eigenvalue E. So, we have found that $\hat{H}_+$ and $\hat{H}_-$ have common excited state eigenvalues. These operators are called *iso-spectral*. The ground state is not conjugated, since $\hat{A}\psi_0 = \hat{A}^\dagger\phi_0 = 0$. So, if ψ_n is the normalized eigenvector of $\hat{H}_+$ with eigenvalue E_n, $1/\sqrt{E_n}\hat{A}\psi_n$ is a normalized eigenstate of $\hat{H}_-$. Likewise, if ψ_n is the normalized eigenvector of $\hat{H}_-$ with eigenvalue E_n, $1/\sqrt{E_n}\hat{A}^\dagger\psi_n$ is a normalized eigenstate of $\hat{H}_+$.

This relation between iso-spectral Hamiltonians also translates to scattering states. Assume that ψ_+ is a scattering solution of $\hat{H}_+$ with asymptotic form

$$\psi_+(x) \sim \begin{cases} e^{ikx} + \mathcal{R}_+ e^{-ikx}, & \text{for } x \to -\infty, \\ \mathcal{T}_+ e^{ik'x}, & \text{for } x \to \infty, \end{cases} \tag{3.230}$$

where k and k' are the wave numbers at $x = -\infty$ and $x = \infty$, respectively. The corresponding solution to $\hat{H}_-$ is given by

$$\psi_- \sim \hat{A}\psi_+ = \left(\frac{\mathrm{d}}{\mathrm{d}x} + W\right)\psi_+. \tag{3.231}$$

Then, for the asymptotic behavior of ψ_- we find

$$\psi_-(x) \sim \begin{cases} (W_- + ik)e^{ikx} + (W_- - ik)\mathcal{R}_+ e^{-ikx}, & \text{for } x \to -\infty, \\ (W_+ + ik')\mathcal{T}_+ e^{ik'x}, & \text{for } x \to \infty, \end{cases} \tag{3.232}$$

where $W_\pm = \lim_{x\to\pm\infty} W(x)$. From this asymptotic behavior, we can infer the reflection and transmission coefficients

$$\mathcal{R}_- = \frac{(W_- - ik)}{(W_- + ik)}\mathcal{R}_+ \quad \text{and} \quad \mathcal{T}_- = \frac{(W_+ + ik')}{(W_- + ik)}\mathcal{T}_+. \tag{3.233}$$

Further Reading

1. M. Bartelmann, B. Feuerbacher, T. Krüger, D. Lüst, A. Rebhan, A. Wipf, *Theoretische Physik*. Springer (2015)

Chapter 4
Angular Momentum

4.1 Rotation in Classical Mechanics

In classical mechanics, the rotation is a transformation

$$\boldsymbol{r}' = R(\boldsymbol{n}, \theta)\boldsymbol{r}, \tag{4.1}$$

where $\boldsymbol{r}$ is the original vector, $\boldsymbol{r}'$ is the rotated vector and $R(\boldsymbol{n}, \theta)$ denotes the rotation in the three-dimensional Euclidean space around the axis $\boldsymbol{n}$ by angle θ (Fig. 4.1). If $\boldsymbol{n} = \boldsymbol{e}_z$, i.e. we rotate about the z-axis, we have

$$\begin{pmatrix} x' \\ y' \\ z' \end{pmatrix} = \begin{pmatrix} \cos\theta & -\sin\theta & 0 \\ \sin\theta & \cos\theta & 0 \\ 0 & 0 & 1 \end{pmatrix} \begin{pmatrix} x \\ y \\ z \end{pmatrix}, \quad \text{and thus} \quad R_z(\theta) = \begin{pmatrix} \cos\theta & -\sin\theta & 0 \\ \sin\theta & \cos\theta & 0 \\ 0 & 0 & 1 \end{pmatrix}. \tag{4.2}$$

If we perform cyclic permutations of the axes $(x, y, z) \to (y, z, x)$ and $(x, y, z) \to (y, z, x) \to (z, x, y)$, the transformation of the coordinates become

$$\begin{pmatrix} y' \\ z' \\ x' \end{pmatrix} = \begin{pmatrix} \cos\theta & -\sin\theta & 0 \\ \sin\theta & \cos\theta & 0 \\ 0 & 0 & 1 \end{pmatrix} \begin{pmatrix} y \\ z \\ x \end{pmatrix} \quad \text{and} \quad \begin{pmatrix} z' \\ x' \\ y' \end{pmatrix} = \begin{pmatrix} \cos\theta & -\sin\theta & 0 \\ \sin\theta & \cos\theta & 0 \\ 0 & 0 & 1 \end{pmatrix} \begin{pmatrix} z \\ x \\ y \end{pmatrix}, \tag{4.3}$$

or by reordering in the (x, y, z) sequence we find

$$\begin{pmatrix} x' \\ y' \\ z' \end{pmatrix} = \begin{pmatrix} 1 & 0 & 0 \\ 0 & \cos\theta & -\sin\theta \\ 0 & \sin\theta & \cos\theta \end{pmatrix} \begin{pmatrix} x \\ y \\ z \end{pmatrix} \quad \text{and} \quad \begin{pmatrix} x' \\ y' \\ z' \end{pmatrix} = \begin{pmatrix} \cos\theta & 0 & \sin\theta \\ 0 & 1 & 0 \\ -\sin\theta & 0 & \cos\theta \end{pmatrix} \begin{pmatrix} x \\ y \\ z \end{pmatrix}. \tag{4.4}$$

These give us the rotation matrices around axes $\boldsymbol{e}_x$ and $\boldsymbol{e}_y$, respectively,

Z. Papp, *Mastering Quantum Mechanics*,
https://doi.org/10.1007/978-3-032-09011-9_4

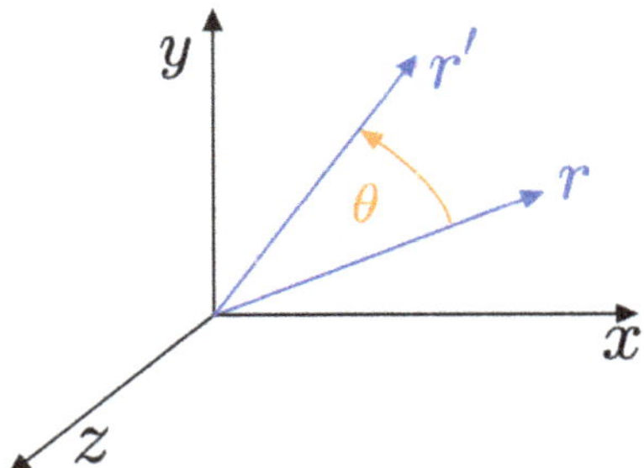

Fig. 4.1 Rotation of a vector around the z axis

$$R_x(\theta) = \begin{pmatrix} 1 & 0 & 0 \\ 0 & \cos\theta & -\sin\theta \\ 0 & \sin\theta & \cos\theta \end{pmatrix} \quad \text{and} \quad R_y(\theta) = \begin{pmatrix} \cos\theta & 0 & \sin\theta \\ 0 & 1 & 0 \\ -\sin\theta & 0 & \cos\theta \end{pmatrix}. \tag{4.5}$$

Rotation does not alter the length of the vector. Consequently, the transformation matrices are orthogonal, real and also unitary matrices, i.e. $R^{\dagger}R = 1$. These formulas represent *active* rotations, where the coordinate system is fixed and the vector is rotated. In a *passive* rotation, the vector is fixed and coordinate system is rotated. The corresponding matrices for passive rotations can be obtained by $\theta \to -\theta$ substitutions.

4.2 Rotation in Quantum Mechanics

Let us denote the operator that rotates a quantum state by $\hat{R}$, $|\psi'\rangle = \hat{R}|\psi\rangle$. Based on the correspondence principle, we assume that the quantum mechanical rotation operator $\hat{R}$ acts on a position vector $|\boldsymbol{r}\rangle$ in an analogous way that the classical mechanical rotation matrix R does on vectors, $\hat{R}|\boldsymbol{r}\rangle = |R\boldsymbol{r}\rangle$. Obviously, the normalization of the state is not changed by rotation

$$1 = \langle\psi(\boldsymbol{r})|\psi(\boldsymbol{r})\rangle = \langle\psi'(\boldsymbol{r})|\psi'(\boldsymbol{r})\rangle = \langle\hat{R}\psi(\boldsymbol{r})|\hat{R}\psi(\boldsymbol{r})\rangle = \langle\psi(\boldsymbol{r})|\hat{R}^{\dagger}\hat{R}|\psi(\boldsymbol{r})\rangle, \tag{4.6}$$

consequently

$$\hat{R}^{\dagger}\hat{R} = I, \tag{4.7}$$

i.e. the quantum mechanical rotation operator $\hat{R}$ is unitary. Then, its action on a quantum state gives

$$\psi'(\boldsymbol{r}) = (\hat{R}\psi)(\boldsymbol{r}) = \langle\boldsymbol{r}|\hat{R}\psi\rangle = \langle R^{\dagger}\boldsymbol{r}|\psi\rangle = \langle R^{-1}\boldsymbol{r}|\psi\rangle = \psi(R^{-1}\boldsymbol{r}). \tag{4.8}$$

Operators should transform according to unitary transformations

$$\hat{A}' = \hat{R}\hat{A}\hat{R}^{\dagger}, \tag{4.9}$$

consequently rotation should not change expectation values

$$\langle \hat{A} \rangle = \langle \psi(r)|\hat{A}|\psi(r)\rangle = \langle \psi(r)|\hat{R}^{\dagger}\hat{R}\hat{A}\hat{R}^{\dagger}\hat{R}|\psi(r)\rangle = \langle \psi'(r)|\hat{A}'|\psi'(r)\rangle = \langle \hat{A}' \rangle. \tag{4.10}$$

From Eq. (4.7), we have

$$\det(\hat{R}^{\dagger}\hat{R}) = \det(I) = 1, \tag{4.11}$$

and considering that infinitesimal rotations start from the unit operator, we can conclude that $\hat{R}$ must have a unit determinant

$$\det(\hat{R}) = 1, \tag{4.12}$$

just like the corresponding classical mechanical rotation matrices.

For an infinitesimal rotation $\delta\theta$ about the $\boldsymbol{e}_z$ axis, the transformation of coordinates reads

$$\hat{R}_z(\delta\theta)\boldsymbol{r} = \begin{pmatrix} x' \\ y' \\ z' \end{pmatrix} = \begin{pmatrix} 1 & -\delta\theta & 0 \\ \delta\theta & 1 & 0 \\ 0 & 0 & 1 \end{pmatrix} \begin{pmatrix} x \\ y \\ z \end{pmatrix}, \tag{4.13}$$

and

$$\hat{R}_z^{-1}(\delta\theta)\boldsymbol{r} = \begin{pmatrix} x' \\ y' \\ z' \end{pmatrix} = \begin{pmatrix} 1 & \delta\theta & 0 \\ -\delta\theta & 1 & 0 \\ 0 & 0 & 1 \end{pmatrix} \begin{pmatrix} x \\ y \\ z \end{pmatrix}. \tag{4.14}$$

Then, the wave function transforms as

$$\begin{aligned} \langle \boldsymbol{r}|\hat{R}_z(\delta\theta)\psi\rangle &= \langle \hat{R}_z^{-1}(\delta\theta)\boldsymbol{r}|\psi\rangle = \psi(\hat{R}_z^{-1}(\delta\theta)\boldsymbol{r}) = \psi(x + \delta\theta y, y - \delta\theta x, z) \\ &\approx \psi(x, y, z) + \delta\theta \left(y\frac{\partial}{\partial x} - x\frac{\partial}{\partial y} \right) \psi(x, y, z) \\ &\approx \left[1 - \frac{i}{\hbar}\delta\theta \hat{L}_z \right] \psi(x, y, z), \end{aligned} \tag{4.15}$$

where

$$\hat{L}_z = -i\hbar \left(x\frac{\partial}{\partial y} - y\frac{\partial}{\partial x} \right) = \hat{x}\hat{p}_y - \hat{y}\hat{p}_x = (\hat{\boldsymbol{r}} \times \hat{\boldsymbol{p}})_z \tag{4.16}$$

is the z-component of the orbital angular momentum operator. We can infer that

$$\hat{R}_z(\delta\theta) = \hat{R}(\boldsymbol{e}_z, \delta\theta) = \left[1 - \frac{i}{\hbar}\delta\theta \hat{L}_z \right], \tag{4.17}$$

or for arbitrary direction $\boldsymbol{n}$ we find

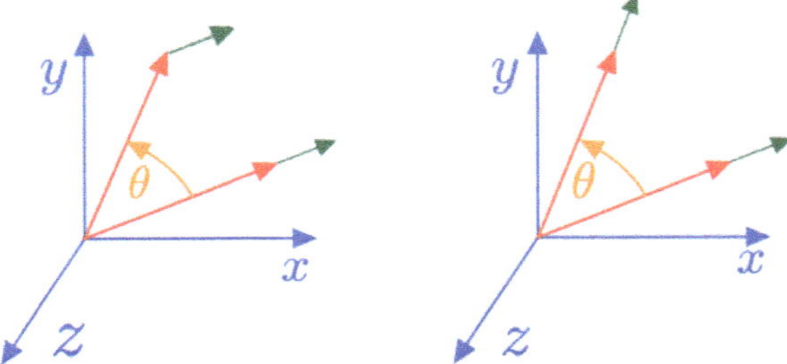

Fig. 4.2 Rotation of systems with internal structure. On the left, only the spatial coordinates are rotated by the rotation operator generated by the orbital angular momentum operator $\hat{L}$. On the right picture, the spatial and internal structure coordinates are rotated by the rotation operator generated by the total angular momentum operator $\hat{J}$

$$\hat{R}_{\boldsymbol{n}}(\delta\theta) = \hat{R}(\boldsymbol{e}_n, \delta\theta) = \left[1 - \frac{i}{\hbar}\delta\theta\, \boldsymbol{e}_n \cdot \hat{\boldsymbol{L}}\right], \tag{4.18}$$

or even further for a general angular momentum operator $\hat{\boldsymbol{J}}$ we can write

$$\hat{R}_{\boldsymbol{n}}(\delta\theta) = \hat{R}(\boldsymbol{e}_n, \delta\theta) = \left[1 - \frac{i}{\hbar}\delta\theta\, \boldsymbol{e}_n \cdot \hat{\boldsymbol{J}}\right]. \tag{4.19}$$

The angular momentum operator $\hat{\boldsymbol{L}}$ is the generator of the rotation of the space coordinates only, while the general angular momentum operator $\hat{\boldsymbol{J}}$ is the generator of the rotation of spatial and internal coordinates (see Fig. 4.2).

Rotation operators with finite θ can be constructed as an infinite number of consecutive infinitesimal transformations

$$\hat{R}_{\boldsymbol{n}}(\theta) = \lim_{k\to\infty}\left(1 - \frac{i}{\hbar}\frac{\theta}{k}\boldsymbol{e}_n \cdot \hat{\boldsymbol{J}}\right)^k = \exp\left(-\frac{i}{\hbar}\theta\, \boldsymbol{e}_n \cdot \hat{\boldsymbol{J}}\right). \tag{4.20}$$

The rotation operator $\hat{R}$ is unitary, therefore

$$\hat{R}^\dagger_{\boldsymbol{n}}(\theta) = \hat{R}^{-1}_{\boldsymbol{n}}(\theta) = \exp\left(\frac{i}{\hbar}\theta\, \boldsymbol{e}_n \cdot \hat{\boldsymbol{J}}\right), \tag{4.21}$$

and the operator $\hat{\boldsymbol{J}}$ is the generator of a unitary transformation, so it has to be Hermitian.

4.3 Angular Momentum Commutation Relations

We can derive the commutation relations of the angular momentum operators as a consequence of the geometric structure of the three-dimensional Euclidean space.

Let us consider four consecutive infinitesimal rotations:

1. Rotation about $\boldsymbol{e}_x$ by δ:

$$R_x(\delta) = \begin{pmatrix} 1 & 0 & 0 \\ 0 & 1 & -\delta \\ 0 & \delta & 1 \end{pmatrix}. \tag{4.22}$$

2. Rotation about $\boldsymbol{e}_y$ by δ:

$$R_y(\delta) = \begin{pmatrix} 1 & 0 & \delta \\ 0 & 1 & 0 \\ -\delta & 0 & 1 \end{pmatrix}. \tag{4.23}$$

3. Rotation about $\boldsymbol{e}_x$ by $-\delta$:

$$R_x(-\delta) = \begin{pmatrix} 1 & 0 & 0 \\ 0 & 1 & \delta \\ 0 & -\delta & 1 \end{pmatrix}. \tag{4.24}$$

4. Rotation about $\boldsymbol{e}_y$ by $-\delta$:

$$R_y(-\delta) = \begin{pmatrix} 1 & 0 & -\delta \\ 0 & 1 & 0 \\ \delta & 0 & 1 \end{pmatrix}. \tag{4.25}$$

By multiplying the matrices, keeping the leading terms only and discarding terms of higher order than δ^2, we obtain

$$\begin{aligned} R_y(-\delta)R_x(-\delta)R_y(\delta)R_x(\delta) &= \begin{pmatrix} 1+\delta^2 & \delta^2 & -\delta^3 \\ -\delta^2 & 1+\delta^2 & 0 \\ 0 & \delta^3 & 1+2\delta^2 \end{pmatrix} \\ &\approx \begin{pmatrix} 1 & \delta^2 & 0 \\ -\delta^2 & 1 & 0 \\ 0 & 0 & 1 \end{pmatrix} = R_z(-\delta^2). \end{aligned} \tag{4.26}$$

So, we have found that rotating infinitesimally around $\boldsymbol{e}_x$, then $\boldsymbol{e}_y$, then again around $\boldsymbol{e}_x$ and $\boldsymbol{e}_y$, but in the opposite directions, results in a rotation around the $\boldsymbol{e}_z$ axis.

Then, for the corresponding quantum mechanical rotation operators, we should have

$$\hat{R}_y(-\delta)\hat{R}_x(-\delta)\hat{R}_y(\delta)\hat{R}_x(\delta) \approx \hat{R}_z(-\delta^2), \tag{4.27}$$

which, in terms of angular momentum operators, takes the form

$$\left(1+\frac{i\delta}{\hbar}\hat{J}_y-\frac{\delta^2}{2\hbar^2}\hat{J}_y^2\right)\left(1+\frac{i\delta}{\hbar}\hat{J}_x-\frac{\delta^2}{2\hbar^2}\hat{J}_x^2\right)\left(1-\frac{i\delta}{\hbar}\hat{J}_y-\frac{\delta^2}{2\hbar^2}\hat{J}_y^2\right)\left(1-\frac{i\delta}{\hbar}\hat{J}_x-\frac{\delta^2}{2\hbar^2}\hat{J}_x^2\right)\approx\left(1+\frac{i\delta^2}{\hbar}\hat{J}_z\right). \tag{4.28}$$

Expanding the products and keeping only terms up to δ^2, we can infer that

$$\hat{J}_x\hat{J}_y - \hat{J}_y\hat{J}_x = [\hat{J}_x, \hat{J}_y] = i\hbar\hat{J}_z, \tag{4.29}$$

or in a similar fashion

$$[\hat{J}_i, \hat{J}_j] = i\hbar\hat{J}_k, \tag{4.30}$$

where (i, j, k) are cyclic permutations of $(1, 2, 3)$ or (x, y, z).

We have learned in classical mechanics that the angular momentum $\boldsymbol{J}$ is the generator of the rotation. We also learned that the components J_i follow the Poisson bracket relations

$$\{J_i, J_j\} = \epsilon_{ijk}J_k \quad \text{and} \quad \{J_i, J^2\} = 0, \tag{4.31}$$

or explicitly

$$\{J_x, J_y\} = J_z, \quad \{J_y, J_z\} = J_x, \quad \{J_z, J_x\} = J_y, \quad \text{and} \quad \{J_{x,y,z}, J^2\} = 0. \tag{4.32}$$

Then, according to the correspondence principle, the quantum mechanical angular momentum operators should follow the commutation relations

$$\frac{1}{i\hbar}[\hat{J}_x, \hat{J}_y] = \hat{J}_z, \quad \frac{1}{i\hbar}[\hat{J}_y, \hat{J}_z] = \hat{J}_x, \quad \frac{1}{i\hbar}[\hat{J}_z, \hat{J}_x] = \hat{J}_y, \tag{4.33}$$

or

$$[\hat{J}_x, \hat{J}_y] = i\hbar\hat{J}_z, \quad [\hat{J}_y, \hat{J}_z] = i\hbar\hat{J}_x, \quad [\hat{J}_z, \hat{J}_x] = i\hbar\hat{J}_y. \tag{4.34}$$

It also follows that any component of $\hat{\boldsymbol{J}}$ commutes with $\hat{\boldsymbol{J}}^2 = \hat{J}_x^2 + \hat{J}_y^2 + \hat{J}_z^2$. Indeed, for example,

$$\begin{aligned}\left[\hat{J}_x, \hat{J}_x^2 + \hat{J}_y^2 + \hat{J}_z^2\right] &= \left[\hat{J}_x, \hat{J}_y^2\right] + \left[\hat{J}_x, \hat{J}_z^2\right] \\ &= \hat{J}_y\left[\hat{J}_x, \hat{J}_y\right] + \left[\hat{J}_x, \hat{J}_y\right]\hat{J}_y + \hat{J}_z\left[\hat{J}_x, \hat{J}_z\right] + \left[\hat{J}_x, \hat{J}_z\right]\hat{J}_z \\ &= \hat{J}_y i\hbar\hat{J}_z + i\hbar\hat{J}_z\hat{J}_y + \hat{J}_z(-i\hbar\hat{J}_y) + (-i\hbar\hat{J}_y)\hat{J}_z = 0.\end{aligned} \tag{4.35}$$

These relations are in complete agreement with the definition of the orbital angular momentum operator. The orbital angular momentum is defined by

$$\hat{\boldsymbol{L}} = \hat{\boldsymbol{r}} \times \hat{\boldsymbol{p}}, \tag{4.36}$$

which, in Cartesian coordinates, reads

$$\hat{L}_x = \hat{y}\hat{p}_z - \hat{z}\hat{p}_y, \quad \hat{L}_y = \hat{z}\hat{p}_x - \hat{x}\hat{p}_z, \quad \hat{L}_z = \hat{x}\hat{p}_y - \hat{y}\hat{p}_x. \tag{4.37}$$

Indeed,

$$\begin{aligned}\left[\hat{L}_x, \hat{L}_y\right] &= \left[\hat{y}\hat{p}_z - \hat{z}\hat{p}_y, \hat{z}\hat{p}_x - \hat{x}\hat{p}_z\right] \\ &= \left[\hat{y}\hat{p}_z, \hat{z}\hat{p}_x\right] + \left[\hat{y}\hat{p}_z, -\hat{x}\hat{p}_z\right] + \left[-\hat{z}\hat{p}_y, \hat{z}\hat{p}_x\right] + \left[-\hat{z}\hat{p}_y, -\hat{x}\hat{p}_z\right] \\ &= \left[\hat{y}\hat{p}_z, \hat{z}\hat{p}_x\right] + \left[\hat{z}\hat{p}_y, \hat{x}\hat{p}_z\right] = \hat{y}\left[\hat{p}_z, \hat{z}\right]\hat{p}_x + \hat{x}\left[\hat{z}, \hat{p}_z\right]\hat{p}_y \\ &= i\hbar(\hat{x}\hat{p}_y - \hat{y}\hat{p}_x) = i\hbar\hat{L}_z,\end{aligned} \tag{4.38}$$

i.e. the orbital angular momentum operators satisfy the general angular momentum commutation relations.

4.4 Eigenvalues and Eigenstates

We have found that $\hat{\boldsymbol{J}}^2$ and any component, for example $\hat{J}_z$, commute. So, we can seek their common normalized eigenstates as

$$\hat{\boldsymbol{J}}^2|\alpha, \beta\rangle = \hbar^2\alpha|\alpha, \beta\rangle, \qquad \hat{J}_z|\alpha, \beta\rangle = \hbar\beta|\alpha, \beta\rangle \quad \text{with} \quad \langle\alpha', \beta'|\alpha, \beta\rangle = \delta_{\alpha\alpha'}\delta_{\beta\beta'}. \tag{4.39}$$

Introducing the operators

$$\hat{J}_\pm = \hat{J}_x \pm i\hat{J}_y, \tag{4.40}$$

we find that

$$\hat{J}_x = \frac{1}{2}(\hat{J}_+ + \hat{J}_-) \quad \text{and} \quad \hat{J}_y = \frac{1}{2i}(\hat{J}_+ - \hat{J}_-). \tag{4.41}$$

Obviously $\left[\hat{J}_\pm, \hat{\boldsymbol{J}}^2\right] = 0$, and simple calculations reveal that

$$\left[\hat{J}_+, \hat{J}_-\right] = 2\hbar\hat{J}_z \quad \text{and} \quad \left[\hat{J}_z, \hat{J}_\pm\right] = \pm\hbar\hat{J}_\pm. \tag{4.42}$$

We can see that

$$\begin{aligned}\hat{J}_+\hat{J}_- &= (\hat{J}_x + i\hat{J}_y)(\hat{J}_x - i\hat{J}_y) = \hat{J}_x^2 + \hat{J}_y^2 + i(\hat{J}_y\hat{J}_x - \hat{J}_x\hat{J}_y) \\ &= \hat{J}_x^2 + \hat{J}_y^2 + \hbar\hat{J}_z = \hat{\boldsymbol{J}}^2 - \hat{J}_z^2 + \hbar\hat{J}_z,\end{aligned} \tag{4.43}$$

and similarly

$$\begin{aligned}\hat{J}_-\hat{J}_+ &= (\hat{J}_x - i\hat{J}_y)(\hat{J}_x + i\hat{J}_y) = \hat{J}_x^2 + \hat{J}_y^2 - i(\hat{J}_y\hat{J}_x - \hat{J}_x\hat{J}_y) \\ &= \hat{J}_x^2 + \hat{J}_y^2 - \hbar\hat{J}_z = \hat{\boldsymbol{J}}^2 - \hat{J}_z^2 - \hbar\hat{J}_z,\end{aligned} \tag{4.44}$$

and from these relations we find

$$\hat{\boldsymbol{J}}^2 = \frac{1}{2}(\hat{J}_+\hat{J}_- + \hat{J}_-\hat{J}_+) + \hat{J}_z^2. \tag{4.45}$$

Now, let us investigate the properties of the state $\hat{J}_\pm|\alpha, \beta\rangle$. Using Eq. (4.42), we find

$$\hat{J}_z(\hat{J}_\pm|\alpha, \beta\rangle) = (\hat{J}_\pm\hat{J}_z \pm \hbar\hat{J}_\pm)|\alpha, \beta\rangle = \hbar(\beta \pm 1)\hat{J}_\pm|\alpha, \beta\rangle, \tag{4.46}$$

i.e. the state $\hat{J}_\pm|\alpha, \beta\rangle$ is an eigenstate of $\hat{J}_z$ with eigenvalue $\hbar(\beta \pm 1)$. Since $\hat{\boldsymbol{J}}^2$ and $\hat{J}_\pm$ commute, we also have

$$\hat{\boldsymbol{J}}^2(\hat{J}_\pm|\alpha, \beta\rangle) = \hat{J}_\pm\hat{\boldsymbol{J}}^2|\alpha, \beta\rangle = \hbar^2\alpha(\hat{J}_\pm|\alpha, \beta\rangle), \tag{4.47}$$

meaning that $\hat{J}_\pm|\alpha, \beta\rangle$ is still an eigenstate of $\hat{\boldsymbol{J}}^2$ with eigenvalue $\hbar^2\alpha$. So, we have found that applying $\hat{J}_\pm$ on $|\alpha, \beta\rangle$ raises or lowers the eigenvalue β by one, but does not change the value of α. Since the eigenstates of $\hat{J}_z$ are unique, up to some factor, we can write

$$\hat{J}_\pm|\alpha, \beta\rangle = C_\pm^{\alpha,\beta}|\alpha, \beta \pm 1\rangle. \tag{4.48}$$

On the other hand,

$$0 \le \langle\alpha, \beta|(\hat{J}_x^2 + \hat{J}_y^2)|\alpha, \beta\rangle = \langle\alpha, \beta|(\hat{\boldsymbol{J}}^2 - \hat{J}_z^2)|\alpha, \beta\rangle = \hbar^2(\alpha - \beta^2). \tag{4.49}$$

This means that $\alpha \ge \beta^2 \ge 0$, and consequently β is limited from below and above. Therefore, for any α there must exist a β_{max} such that it cannot be raised any further. This is possible only if $\hat{J}_+$ brings the state $|\alpha, \beta_{max}\rangle$ to a null state

$$\hat{J}_+|\alpha, \beta_{max}\rangle = 0. \tag{4.50}$$

Then applying $\hat{J}_-\hat{J}_+ = \hat{\boldsymbol{J}}^2 - \hat{J}_z^2 - \hbar\hat{J}_z$ on state $|\alpha, \beta_{max}\rangle$ we find

$$\hat{J}_-\hat{J}_+|\alpha, \beta_{max}\rangle = 0 = (\hat{\boldsymbol{J}}^2 - \hat{J}_z^2 - \hbar\hat{J}_z)|\alpha, \beta_{max}\rangle = \hbar^2(\alpha - \beta_{max}^2 - \beta_{max})|\alpha, \beta_{max}\rangle, \tag{4.51}$$

and obtain

$$\alpha = \beta_{max}(\beta_{max} + 1). \tag{4.52}$$

Similarly, there must exist a β_{min} that cannot be lowered any further

$$\hat{J}_-|\alpha, \beta_{min}\rangle = 0. \tag{4.53}$$

Then using $\hat{J}_+\hat{J}_- = \hat{\boldsymbol{J}}^2 - \hat{J}_z^2 + \hbar\hat{J}_z$ we obtain

$$\hat{J}_+\hat{J}_-|\alpha, \beta_{min}\rangle = 0 = (\hat{\boldsymbol{J}}^2 - \hat{J}_z^2 + \hbar\hat{J}_z)|\alpha, \beta_{min}\rangle = \hbar^2(\alpha - \beta_{min}^2 + \beta_{min})|\alpha, \beta_{min}\rangle, \tag{4.54}$$

and find

$$\alpha = \beta_{min}(\beta_{min} - 1). \tag{4.55}$$

Comparing with the previous result we can conclude that

$$\beta_{max} = -\beta_{min}. \tag{4.56}$$

Since $\hat{J}_\pm$ changes β by one, we must have

$$\beta_{max} = \beta_{min} + n = -\beta_{max} + n, \tag{4.57}$$

where n is integer. Consequently,

$$\beta_{max} = \frac{n}{2} \quad \text{and} \quad \beta_{min} = -\frac{n}{2}. \tag{4.58}$$

Then, with the notations $j = \beta_{max}$ and $m = \beta$, we have

$$\alpha = j(j+1) \quad \text{and} \quad -j \le m \le j, \tag{4.59}$$

and finally, we conclude that

$$\hat{\boldsymbol{J}}^2|j, m\rangle = \hbar^2\, j(j+1)|j, m\rangle \quad \text{and} \quad \hat{J}_z|j, m\rangle = \hbar m|j, m\rangle. \tag{4.60}$$

In general, the angular momentum quantum number j can take integer or half integer values, $j = 0, 1/2, 1, 3/2, 2, \ldots$ and $m = -j, -j+1, \ldots, j-1, j$.

From Eqs. (4.43) and (4.44), we derive

$$(\langle j, m|\hat{J}_-)(\hat{J}_+|j, m\rangle) = \langle j, m|\hat{J}_-\hat{J}_+|j, m\rangle = \hbar^2[j(j+1) - m(m+1)] = |C_+^{jm}|^2 \tag{4.61}$$

and

$$(\langle j, m|\hat{J}_+)(\hat{J}_-|j, m\rangle) = \langle j, m|\hat{J}_+\hat{J}_-|j, m\rangle = \hbar^2[j(j+1) - m(m-1)] = |C_-^{jm}|^2. \tag{4.62}$$

Then, following the convention by Condon and Shortley, we chose

$$C_\pm^{jm} = \hbar\sqrt{j(j+1) - m(m\pm 1)} = \hbar\sqrt{(j\mp m)(j\pm m+1)} \tag{4.63}$$

and

$$\hat{J}_\pm|j, m\rangle = \hbar\sqrt{j(j+1) - m(m\pm 1)}|j, m\pm 1\rangle = \hbar\sqrt{(j\mp m)(j\pm m+1)}|j, m\pm 1\rangle. \tag{4.64}$$

These lead to the relations

$$\hat{J}_x|j,m\rangle = \frac{1}{2}(\hat{J}_+ + \hat{J}_-)|j,m\rangle$$
$$= \frac{\hbar}{2}\left[\sqrt{j(j+1)-m(m+1)}|j,m+1\rangle + \sqrt{j(j+1)-m(m-1)}|j,m-1\rangle\right] \quad (4.65)$$

and

$$\hat{J}_y|j,m\rangle = \frac{1}{2i}(\hat{J}_+ - \hat{J}_-)|j,m\rangle$$
$$= \frac{\hbar}{2i}\left[\sqrt{j(j+1)-m(m+1)}|j,m+1\rangle - \sqrt{j(j+1)-m(m-1)}|j,m-1\rangle\right]. \quad (4.66)$$

From Eqs. (4.65) and (4.66) it is obvious that

$$\langle j,m|\hat{J}_x|j,m\rangle = \langle j,m|\hat{J}_y|j,m\rangle = 0. \quad (4.67)$$

On the other hand, we have

$$\hat{J}_x^2 = \frac{1}{4}(\hat{J}_+^2 + \hat{J}_-^2 + \hat{J}_+\hat{J}_- + \hat{J}_-\hat{J}_+), \qquad \hat{J}_y^2 = -\frac{1}{4}(\hat{J}_+^2 + \hat{J}_-^2 - \hat{J}_+\hat{J}_- - \hat{J}_-\hat{J}_+), \quad (4.68)$$

and since $\langle j,m|\hat{J}_+^2|j,m\rangle = \langle j,m|\hat{J}_-^2|j,m\rangle = 0$, we get

$$\langle j,m|\hat{J}_x^2|j,m\rangle = \langle j,m|\hat{J}_y^2|j,m\rangle = \frac{1}{4}\langle j,m|(\hat{J}_+\hat{J}_- + \hat{J}_-\hat{J}_+)|j,m\rangle$$
$$= \frac{1}{2}\langle j,m|(\hat{\boldsymbol{J}}^2 - \hat{J}_z^2)|j,m\rangle = \frac{\hbar^2}{2}[j(j+1)-m^2]. \quad (4.69)$$

Now, we can calculate the uncertainties of $\hat{J}_x$ and $\hat{J}_y$ between angular momentum eigenstates $|j,m\rangle$,

$$\Delta J_x = \Delta J_y = \sqrt{\langle\hat{J}_x^2\rangle - \langle\hat{J}_x\rangle^2} = \frac{\hbar}{\sqrt{2}}\sqrt{j(j+1)-m^2}. \quad (4.70)$$

Their product gives

$$\Delta J_x\,\Delta J_y = \frac{\hbar^2}{2}[j(j+1)-m^2] \geq \frac{\hbar^2}{2}j \geq \frac{\hbar^2}{2}|m| = \frac{1}{2}\,\langle i\hbar J_z\rangle|, \quad (4.71)$$

which is in complete agreement with the Heisenberg uncertainty relation.

4.4.1 Matrix Representation of Angular Momentum

The common eigenstates of the commuting Hermitian operators $\hat{\boldsymbol{J}}^2$ and $\hat{J}_z$ form a basis. If we restrict ourself to the given angular momentum j, the states $|j, m\rangle$ with $-j \leq m \leq j$ span a subspace with the completeness relation

$$\sum_{m=-j}^{j} |j, m\rangle\langle j, m| = I. \tag{4.72}$$

We can order the basis states as

$$\{|j, j\rangle, |j, j-1\rangle, \ldots, |j, -j+1\rangle, |j, -j\rangle\} \tag{4.73}$$

and represent the various angular momentum operators on that subspace basis. The results can be summarized as

$$\begin{aligned}
\langle j', m'|\hat{\boldsymbol{J}}^2|j, m\rangle &= \hbar^2\, j(j+1)\, \delta_{j'j}\delta_{m'm}, \\
\langle j', m'|\hat{J}_z|j, m\rangle &= \hbar m\, \delta_{j'j}\, \delta_{m'm}, \\
\langle j', m'|\hat{J}_\pm|j, m\rangle &= \hbar \sqrt{j(j+1) - m(m \pm 1)}\, \delta_{j'j}\, \delta_{m'm\pm 1}, \\
\langle j', m'|\hat{J}_x|j, m\rangle &= \frac{\hbar}{2}\Big[\sqrt{j(j+1) - m(m+1)}\, \delta_{m'm+1} \\
&\quad + \sqrt{j(j+1) - m(m-1)}\, \delta_{m'm-1}\Big]\, \delta_{j'j}, \\
\langle j', m'|\hat{J}_y|j, m\rangle &= \frac{\hbar}{2i}\Big[\sqrt{j(j+1) - m(m+1)}\, \delta_{m'm+1} \\
&\quad - \sqrt{j(j+1) - m(m-1)}\, \delta_{m'm-1}\Big]\, \delta_{j'j}.
\end{aligned} \tag{4.74}$$

The $j = 1$ Case

Let us define the states

$$|1, 1\rangle = \begin{pmatrix} 1 \\ 0 \\ 0 \end{pmatrix}, \quad |1, 0\rangle = \begin{pmatrix} 0 \\ 1 \\ 0 \end{pmatrix}, \quad |1, -1\rangle = \begin{pmatrix} 0 \\ 0 \\ 1 \end{pmatrix}. \tag{4.75}$$

We can see that they are orthonormal and complete in the $j = 1$ subspace

$$\sum_{m=-1}^{1} |1, m\rangle\langle 1, m| = \begin{pmatrix} 0 \\ 0 \\ 1 \end{pmatrix}(0\ 0\ 1) + \begin{pmatrix} 0 \\ 1 \\ 0 \end{pmatrix}(0\ 1\ 0) + \begin{pmatrix} 1 \\ 0 \\ 0 \end{pmatrix}(1\ 0\ 0) = \begin{pmatrix} 1 & 0 & 0 \\ 0 & 1 & 0 \\ 0 & 0 & 1 \end{pmatrix}, \tag{4.76}$$

and can serve as a basis. Then, we obtain the corresponding matrix representations:

$$\langle 1, m'|\hat{\boldsymbol{J}}^2|1, m\rangle = 2\hbar^2 \begin{pmatrix} 1 & 0 & 0 \\ 0 & 1 & 0 \\ 0 & 0 & 1 \end{pmatrix},$$

$$\langle 1, m'|\hat{J}_z|1, m\rangle = \hbar \begin{pmatrix} 1 & 0 & 0 \\ 0 & 0 & 0 \\ 0 & 0 & -1 \end{pmatrix},$$

$$\langle 1, m'|\hat{J}_+|1, m\rangle = \hbar\sqrt{2} \begin{pmatrix} 0 & 1 & 0 \\ 0 & 0 & 1 \\ 0 & 0 & 0 \end{pmatrix},$$

$$\langle 1, m'|\hat{J}_-|1, m\rangle = \hbar\sqrt{2} \begin{pmatrix} 0 & 0 & 0 \\ 1 & 0 & 0 \\ 0 & 1 & 0 \end{pmatrix},$$

$$\langle 1, m'|\hat{J}_x|1, m\rangle = \frac{\hbar}{\sqrt{2}} \begin{pmatrix} 0 & 1 & 0 \\ 1 & 0 & 1 \\ 0 & 1 & 0 \end{pmatrix},$$

$$\langle 1, m'|\hat{J}_y|1, m\rangle = \frac{\hbar}{\sqrt{2}} \begin{pmatrix} 0 & -i & 0 \\ i & 0 & -i \\ 0 & i & 0 \end{pmatrix}. \tag{4.77}$$

We can see that the states in Eq. (4.75) are, indeed, eigenstates of $\hat{\boldsymbol{J}}^2$ and $\hat{J}_z$.

4.5 Orbital Angular Momentum

We defined the orbital angular momentum operator before as

$$\hat{L}_x = \hat{y}\hat{p}_z - \hat{z}\hat{p}_y, \quad \hat{L}_y = \hat{z}\hat{p}_x - \hat{x}\hat{p}_z, \quad \hat{L}_z = \hat{x}\hat{p}_y - \hat{y}\hat{p}_x, \tag{4.78}$$

or

$$\hat{L}_x = -i\hbar\left(y\frac{\partial}{\partial z} - z\frac{\partial}{\partial y}\right), \quad \hat{L}_y = -i\hbar\left(z\frac{\partial}{\partial x} - x\frac{\partial}{\partial z}\right), \quad \hat{L}_z = -i\hbar\left(x\frac{\partial}{\partial y} - y\frac{\partial}{\partial x}\right). \tag{4.79}$$

Here the operators depend on all three (x, y, z) coordinates.

The spherical coordinates are defined by unit vectors $(\boldsymbol{e}_\vartheta, \boldsymbol{e}_\varphi, \boldsymbol{e}_r)$ (Fig. 4.3). If a function F is expressed in spherical coordinates, $F(\vartheta, \varphi, r)$, the derivative reads

$$\mathrm{d}F = \frac{\partial F}{\partial \vartheta}\mathrm{d}\vartheta + \frac{\partial F}{\partial \varphi}\mathrm{d}\varphi + \frac{\partial F}{\partial r}\mathrm{d}r = \nabla F \cdot \mathrm{d}\boldsymbol{r}. \tag{4.80}$$

Then, from

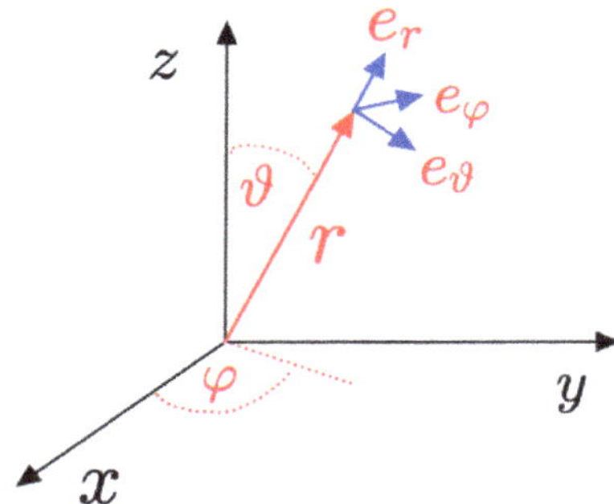

Fig. 4.3 Spherical coordinate system

$$\mathrm{d}\boldsymbol{r} = (r\mathrm{d}\vartheta, r \sin\vartheta \mathrm{d}\varphi, \mathrm{d}r), \tag{4.81}$$

we obtain for the Nabla operator

$$\nabla = \left(\frac{1}{r}\frac{\partial}{\partial\vartheta}, \frac{1}{r\sin\vartheta}\frac{\partial}{\partial\varphi}, \frac{\partial}{\partial r}\right). \tag{4.82}$$

In spherical coordinates, the vector $\boldsymbol{r}$ is given by

$$\boldsymbol{r} = (0, 0, r). \tag{4.83}$$

Consequently, for the angular momentum operator we have

$$\hat{L}'(\vartheta, \varphi, r) = -i\hbar\boldsymbol{r} \times \nabla = \left(\frac{i\hbar}{\sin\vartheta}\frac{\partial}{\partial\varphi}, -i\hbar\frac{\partial}{\partial\vartheta}, 0\right). \tag{4.84}$$

If we want $(\hat{L}_x, \hat{L}_y, \hat{L}_z)$ expressed in terms of spherical variables, we need to transform the angular momentum $\hat{L}'(\vartheta, \varphi, r)$ back to the (x, y, z) coordinate system. We can accomplish this by two consecutive Euler rotations of the coordinates $(\boldsymbol{e}_\vartheta, \boldsymbol{e}_\varphi, \boldsymbol{e}_r)$. First we rotate the coordinate system around $\boldsymbol{e}_\varphi$ by $-\vartheta$, and then rotate around the $\boldsymbol{e}_z$ axis by $-\varphi$. The transformation matrix of these two consecutive passive rotations is given by

$$\begin{aligned} R(\vartheta, \varphi) &= \begin{pmatrix} \cos\varphi & -\sin\varphi & 0 \\ \sin\varphi & \cos\varphi & 0 \\ 0 & 0 & 1 \end{pmatrix} \begin{pmatrix} \cos\vartheta & 0 & \sin\vartheta \\ 0 & 1 & 0 \\ -\sin\vartheta & 0 & \cos\vartheta \end{pmatrix} \\ &= \begin{pmatrix} \cos\vartheta\cos\varphi & -\sin\varphi & \sin\vartheta\cos\varphi \\ \cos\vartheta\sin\varphi & \cos\varphi & \sin\vartheta\sin\varphi \\ -\sin\vartheta & 0 & \cos\vartheta \end{pmatrix}. \end{aligned} \tag{4.85}$$

We can see that $R(\vartheta, \varphi)\boldsymbol{r}$ really transforms the vector $\boldsymbol{r} = (0, 0, r)$ into (x, y, z) coordinates

$$R(\vartheta,\varphi)\boldsymbol{r} = (r\sin\vartheta\cos\varphi, r\sin\vartheta\sin\varphi, r\cos\vartheta). \tag{4.86}$$

Similarly, the rotation $R(\vartheta,\varphi)\hat{L}'$ gives the orbital angular momentum in (x, y, z) coordinates

$$\begin{aligned} \hat{L}_x &= i\hbar\left(\cot\vartheta\cos\varphi\frac{\partial}{\partial\varphi} + \sin\varphi\frac{\partial}{\partial\vartheta}\right), \\ \hat{L}_y &= i\hbar\left(\cot\vartheta\sin\varphi\frac{\partial}{\partial\varphi} - \cos\varphi\frac{\partial}{\partial\vartheta}\right), \\ \hat{L}_z &= -i\hbar\frac{\partial}{\partial\varphi}. \end{aligned} \tag{4.87}$$

We can easily find that

$$\hat{L}_\pm = \hat{L}_x \pm i\hat{L}_y = e^{\pm i\varphi}\,\hbar\left(i\cot\vartheta\frac{\partial}{\partial\varphi} \pm \frac{\partial}{\partial\vartheta}\right), \tag{4.88}$$

and using Eq. (4.45), after some calculations, we obtain

$$\begin{aligned} \hat{\boldsymbol{L}}^2 &= \frac{1}{2}(\hat{L}_+\hat{L}_- + \hat{L}_-\hat{L}_+) + \hat{L}_z^2 = \hat{L}_z^2 - \hbar\hat{L}_z + \hat{L}_+\hat{L}_- \\ &= -\hbar^2\left(\frac{1}{\sin^2\vartheta}\frac{\partial^2}{\partial\varphi^2} + \frac{1}{\sin\vartheta}\frac{\partial}{\partial\vartheta}\sin\vartheta\frac{\partial}{\partial\vartheta}\right). \end{aligned} \tag{4.89}$$

We should notice that in this representation the orbital angular momentum operators do not depend on the radial variable r.

4.5.1 Kinetic Energy and Orbital Angular Momentum

In deriving the connection between $\hat{\boldsymbol{L}}^2$ and $\hat{\boldsymbol{p}}^2$ we start with the vector relations $(\boldsymbol{A}\times\boldsymbol{B})\cdot\boldsymbol{C} = \boldsymbol{A}\cdot(\boldsymbol{B}\times\boldsymbol{C})$ and $\boldsymbol{A}\times(\boldsymbol{B}\times\boldsymbol{C}) = \boldsymbol{B}(\boldsymbol{A}\cdot\boldsymbol{C}) - (\boldsymbol{A}\cdot\boldsymbol{B})\boldsymbol{C}$. Noticing from Eq. (4.78) that $(\hat{\boldsymbol{r}}\times\hat{\boldsymbol{p}}) = -(\hat{\boldsymbol{p}}\times\hat{\boldsymbol{r}})$, we find

$$\hat{\boldsymbol{L}}^2 = (\hat{\boldsymbol{r}}\times\hat{\boldsymbol{p}})\cdot(\hat{\boldsymbol{r}}\times\hat{\boldsymbol{p}}) = -\underbrace{(\hat{\boldsymbol{r}}\times\hat{\boldsymbol{p}})}_{\boldsymbol{A}\times\boldsymbol{B}}\cdot\underbrace{(\hat{\boldsymbol{p}}\times\hat{\boldsymbol{r}})}_{\boldsymbol{C}} = -\hat{\boldsymbol{r}}\cdot[\hat{\boldsymbol{p}}\times(\hat{\boldsymbol{p}}\times\hat{\boldsymbol{r}})] = -\hat{\boldsymbol{r}}\cdot[\hat{\boldsymbol{p}}(\hat{\boldsymbol{p}}\cdot\hat{\boldsymbol{r}}) - \hat{\boldsymbol{p}}^2\hat{\boldsymbol{r}}]. \tag{4.90}$$

Then from

$$[\hat{\boldsymbol{r}}, \hat{\boldsymbol{p}}] = \hat{\boldsymbol{r}}\hat{\boldsymbol{p}} - \hat{\boldsymbol{p}}\hat{\boldsymbol{r}} = [\hat{x}+\hat{y}+\hat{z}, \hat{p}_x+\hat{p}_y+\hat{p}_z] = 3i\hbar, \tag{4.91}$$

and

$$\left[\hat{\boldsymbol{r}}, \hat{\boldsymbol{p}}^2\right] = \hat{\boldsymbol{r}}\hat{\boldsymbol{p}}^2 - \hat{\boldsymbol{p}}^2\hat{\boldsymbol{r}} = \left[\hat{x} + \hat{y} + \hat{z}, \hat{p}_x^2 + \hat{p}_y^2 + \hat{p}_z^2\right] = 2i\hbar(\hat{p}_x + \hat{p}_y + \hat{p}_z) = 2i\hbar\hat{\boldsymbol{p}}, \tag{4.92}$$

we find

$$\begin{aligned}\hat{\boldsymbol{L}}^2 &= -\hat{\boldsymbol{r}} \cdot [\hat{\boldsymbol{p}} \cdot (\hat{\boldsymbol{p}} \cdot \hat{\boldsymbol{r}}) - \hat{\boldsymbol{p}}^2\hat{\boldsymbol{r}}] = -(\hat{\boldsymbol{r}}\hat{\boldsymbol{p}})(\hat{\boldsymbol{r}}\hat{\boldsymbol{p}} - 3i\hbar) + \hat{\boldsymbol{r}} \cdot (\hat{\boldsymbol{r}}\hat{\boldsymbol{p}}^2 - 2i\hbar\hat{\boldsymbol{p}}) \\ &= \hat{r}^2\hat{\boldsymbol{p}}^2 + i\hbar\hat{\boldsymbol{r}}\hat{\boldsymbol{p}} - (\hat{\boldsymbol{r}}\hat{\boldsymbol{p}})(\hat{\boldsymbol{r}}\hat{\boldsymbol{p}}).\end{aligned} \tag{4.93}$$

This relation is representation independent. In spherical coordinates, from Eqs. (4.82) and (4.83), we have

$$\hat{\boldsymbol{r}} \cdot \hat{\boldsymbol{p}} = -i\hbar r\frac{\partial}{\partial r}, \tag{4.94}$$

consequently,

$$\begin{aligned}\hat{\boldsymbol{L}}^2 &= r^2\hat{\boldsymbol{p}}^2 + \hbar^2 r\frac{\partial}{\partial r} + \hbar^2 r\frac{\partial}{\partial r}r\frac{\partial}{\partial r} = r^2\hat{\boldsymbol{p}}^2 + 2\hbar^2 r\frac{\partial}{\partial r} + \hbar^2 r^2\frac{\partial^2}{\partial r^2} \\ &= r^2\hat{\boldsymbol{p}}^2 + \hbar^2\frac{\partial}{\partial r}r^2\frac{\partial}{\partial r}.\end{aligned} \tag{4.95}$$

From the fact that $\hat{\boldsymbol{L}}$, and therefore also $\hat{\boldsymbol{L}}^2$, are functions of ϑ and φ only and they commute with any function of r, we find

$$\hat{\boldsymbol{p}}^2 = -\hbar^2\frac{1}{r^2}\frac{\partial}{\partial r}r^2\frac{\partial}{\partial r} + \frac{1}{r^2}\hat{\boldsymbol{L}}^2 = -\hbar^2\left(\frac{2}{r}\frac{\partial}{\partial r} + \frac{\partial^2}{\partial r^2}\right) + \frac{1}{r^2}\hat{\boldsymbol{L}}^2 = -\hbar^2\frac{1}{r}\frac{\partial^2}{\partial r^2}r + \frac{1}{r^2}\hat{\boldsymbol{L}}^2. \tag{4.96}$$

4.5.2 Eigenfunctions of the Orbital Angular Momentum

We are looking for the common eigenstates of $\hat{\boldsymbol{L}}^2$ and $\hat{L}_z$. We have seen before that these operators depend only on angles ϑ and φ. Therefore, their common eigenstates, the *spherical harmonics*, depend only on these two variables. With the notation

$$\langle\vartheta, \varphi|l, m\rangle = Y_{l,m}(\vartheta, \varphi), \tag{4.97}$$

the eigenvalue equations read

$$\hat{\boldsymbol{L}}^2 Y_{l,m} = \hbar^2 l(l+1)\, Y_{l,m} \quad \text{and} \quad \hat{L}_z Y_{l,m} = \hbar m\, Y_{l,m}, \tag{4.98}$$

and with the raising and lowering operators we have

$$\hat{L}_{\pm}Y_{l,m} = \hbar\sqrt{l(l+1) - m(m \pm 1)}Y_{l,m\pm 1}. \tag{4.99}$$

The $Y_{l,m}$ states are normalized

$$\langle l, m|l', m'\rangle = \int_0^{2\pi} d\varphi \int_0^{\pi} d\vartheta \sin\vartheta \, Y^*_{l,m}(\vartheta, \varphi)Y_{l',m'}(\vartheta, \varphi) = \delta_{ll'}\delta_{mm'}, \tag{4.100}$$

and complete, i.e. form a basis in the space of square integrable functions of ϑ and φ such that

$$\sum_{l=0}^{\infty}\sum_{m=-l}^{l} |l, m\rangle\langle l, m| = I, \tag{4.101}$$

or

$$\sum_{l=0}^{\infty}\sum_{m=-l}^{l} Y^*_{l,m}(\vartheta', \varphi')Y_{l,m}(\vartheta, \varphi) = \frac{1}{\sin\vartheta}\delta(\varphi' - \varphi)\delta(\vartheta' - \vartheta). \tag{4.102}$$

We can see that the φ dependence can easily be solved

$$\hat{L}_z Y_{l,m} = -i\hbar\frac{\partial}{\partial\varphi}Y_{l,m} = \hbar m \, Y_{l,m}, \tag{4.103}$$

with

$$Y_{l,m}(\vartheta, \varphi) = \frac{1}{\sqrt{2\pi}}\exp(im\varphi)\Theta_{l,m}(\vartheta). \tag{4.104}$$

The factor $1/\sqrt{2\pi}$ was introduced to satisfy the normalization condition

$$\int_0^{2\pi} d\varphi\frac{1}{\sqrt{2\pi}}\exp(-im'\varphi)\frac{1}{\sqrt{2\pi}}\exp(im\varphi) = \delta_{m'm}. \tag{4.105}$$

If we require that $Y_{l,m}(\vartheta, \varphi)$ be single valued in φ, m should take only integer values, $m = 0, \pm 1, \pm 2, \ldots$. Since m takes values $-l \leq m \leq l$, then l should also be a non-negative integer $l = 0, 1, 2, \ldots$.

Plugging Eq. (4.104) into Eqs. (4.98) and (4.89), we obtain

$$\begin{aligned} \hat{\boldsymbol{L}}^2\frac{1}{\sqrt{2\pi}}\exp(im\varphi)\Theta_{l,m}(\vartheta) &= \hbar^2 l(l+1)\frac{1}{\sqrt{2\pi}}\exp(im\varphi)\Theta_{l,m}(\vartheta) \\ &= -\hbar^2\left(\frac{1}{\sin^2\vartheta}\frac{\partial^2}{\partial\varphi^2} + \frac{1}{\sin\vartheta}\frac{\partial}{\partial\vartheta}\sin\vartheta\frac{\partial}{\partial\vartheta}\right)\frac{1}{\sqrt{2\pi}}\exp(im\varphi)\Theta_{l,m}(\vartheta), \end{aligned} \tag{4.106}$$

which results in the differential equation

$$\left[\frac{\partial}{\partial(\cos\vartheta)}\left(\sin^2\vartheta\frac{\partial}{\partial(\cos\vartheta)}\right)+\left(l(l+1)-\frac{m^2}{\sin^2\vartheta}\right)\right]\Theta_{l,m}(\vartheta)=0, \quad (4.107)$$

or in terms of the variable $x = \cos\vartheta$, we get

$$\left(\frac{\mathrm{d}}{\mathrm{d}x}(1-x^2)\frac{\mathrm{d}}{\mathrm{d}x}+l(l+1)-\frac{m^2}{1-x^2}\right)\Theta_{l,m}(x)=0. \quad (4.108)$$

The solutions of this differential equation are the associated Legendre polynomials

$$P_l^m(x)=(1-x^2)^{m/2}\left(\frac{\mathrm{d}}{\mathrm{d}x}\right)^m P_l(x) \quad \text{with} \quad P_l(x)=\frac{1}{2^l l!}\left(\frac{\mathrm{d}}{\mathrm{d}x}\right)^l (x^2-1)^l. \quad (4.109)$$

The function Θ should be normalized as

$$\int_0^\pi \mathrm{d}\vartheta \sin\vartheta\, |\Theta_{l,m}(\vartheta)|^2=\int_{-1}^1 \mathrm{d}x |\Theta_{l,m}(x)|^2=1, \quad (4.110)$$

and then the properly normalized solution is given by

$$\begin{aligned}\Theta_{l,m}&=(-)^m\sqrt{\frac{(2l+1)(l-m)!}{2(l+m)!}}P_l^m(\cos\vartheta), \quad \text{if} \quad m\geq 0,\\ \Theta_{l,-|m|}&=(-)^m\Theta_{l,|m|}(\cos\vartheta), \qquad \text{if} \quad m<0.\end{aligned} \quad (4.111)$$

Finally, we have

$$Y_{l,m}(\vartheta,\varphi)=(-)^{(m+|m|)/2}\sqrt{\frac{(2l+1)(l-|m|)!}{4\pi(l+|m|)!}}P_l^{|m|}(\cos\vartheta)\exp(im\varphi). \quad (4.112)$$

We can see that

$$Y_{l,0}(\vartheta,0)=\sqrt{\frac{2l+1}{4\pi}}P_l(\cos\vartheta) \quad \text{and} \quad Y_{l,m}^*(\vartheta,\varphi)=(-)^m Y_{l,-m}(\vartheta,\varphi). \quad (4.113)$$

Table 4.1 shows the first few $Y_{l,m}$ functions in spherical and cartesian coordinates. We can see that if $l=0$, the function is of 0-th order, if $l=1$, it is first order, and if $l=2$ it is a second order function of the variables.

4.6 Spin Angular Momentum

We have seen that the orbital angular momentum defined by $\hat{\boldsymbol{L}}=\hat{\boldsymbol{r}}\times\hat{\boldsymbol{p}}$ leads to integer quantum numbers. However, the general theory of angular momentum allows,

Table 4.1 Spherical harmonics in spherical and cartesian coordinates

$Y_{l,m}(\vartheta,\varphi)$	$Y_{l,m}(x,y,z)$
$Y_{0,0}(\vartheta,\varphi)=\sqrt{\frac{1}{4\pi}}$	$Y_{0,0}(x,y,z)=\sqrt{\frac{1}{4\pi}}$
$Y_{1,0}(\vartheta,\varphi)=\sqrt{\frac{3}{4\pi}}\cos\vartheta$	$Y_{1,0}(x,y,z)=\sqrt{\frac{3}{4\pi}}\frac{z}{r}$
$Y_{1,\pm1}(\vartheta,\varphi)=\mp\sqrt{\frac{3}{8\pi}}e^{\pm i\varphi}\sin\vartheta$	$Y_{1,\pm1}(x,y,z)=\mp\sqrt{\frac{3}{8\pi}}\frac{x\pm iy}{r}$
$Y_{2,0}(\vartheta,\varphi)=\sqrt{\frac{5}{16\pi}}(3\cos^2\vartheta-1)$	$Y_{2,0}(x,y,z)=\sqrt{\frac{5}{16\pi}}\frac{3z^2-r^2}{r^2}$
$Y_{2,\pm1}(\vartheta,\varphi)=\mp\sqrt{\frac{15}{8\pi}}e^{\pm i\varphi}\sin\vartheta\cos\vartheta$	$Y_{2,\pm1}(x,y,z)=\mp\sqrt{\frac{15}{8\pi}}\frac{(x\pm iy)z}{r^2}$
$Y_{2,\pm2}(\vartheta,\varphi)=\sqrt{\frac{15}{32\pi}}e^{\pm i2\varphi}\sin^2\vartheta$	$Y_{2,\pm2}(x,y,z)=\sqrt{\frac{15}{32\pi}}\frac{x^2-y^2\pm 2ixy}{r^2}$

and experiments like the Stern-Gerlach experiment and the fine structure of atomic spectra indicates the existence of another, non-orbital-type of angular momenta associated with the internal structure of the electron. This, non-orbital-type of angular momentum is the *spin*, and it may have integer and half-integer values. The spin is so important that we denote the spin angular momentum operator by $\hat{\mathbf{S}}$. As angular momentum operators, they obey the commutation relations

$$[\hat{S}_x,\hat{S}_y]=i\hbar\hat{S}_z,\quad [\hat{S}_y,\hat{S}_z]=i\hbar\hat{S}_x,\quad [\hat{S}_z,\hat{S}_x]=i\hbar\hat{S}_y, \tag{4.114}$$

and $\hat{\mathbf{S}}^2$ commutes with each component. For a basis, we can take the common eigenstates of $\hat{\mathbf{S}}^2$ and $\hat{S}_z$:

$$\hat{\mathbf{S}}^2|s,m_s\rangle=\hbar^2 s(s+1)|s,m_s\rangle \quad\text{and}\quad \hat{S}_z|s,m_s\rangle=\hbar m_s|s,m_s\rangle, \tag{4.115}$$

where $m_s=-s,\ldots,s$. Similar to the general case, we have $\hat{S}_\pm=\hat{S}_x\pm i\hat{S}_y$,

$$\hat{S}_\pm|s,m_s\rangle=\hbar\sqrt{s(s+1)-m_s(m_s\pm1)}|s,m_s\pm1\rangle, \tag{4.116}$$

and

$$\langle\hat{S}_x^2\rangle=\langle\hat{S}_y^2\rangle=\frac{1}{2}(\langle\hat{\mathbf{S}}^2\rangle-\langle\hat{S}_z^2\rangle)=\frac{\hbar^2}{2}(s(s+1)-m_s^2). \tag{4.117}$$

4.6.1 The Spin-1/2 Case

For $s=1/2$, m_s can take two values such that $m_s=\pm1/2$. Thus the $s=1/2$ subspace is two dimensional and the operators can be represented by 2×2 matrices:

$$\begin{aligned}\langle 1/2, m'|\hat{S}_x|1/2, m\rangle &= \frac{\hbar}{2}(\sigma_x)_{m'm},\\ \langle 1/2, m'|\hat{S}_y|1/2, m\rangle &= \frac{\hbar}{2}(\sigma_y)_{m'm},\\ \langle 1/2, m'|\hat{S}_z|1/2, m\rangle &= \frac{\hbar}{2}(\sigma_z)_{m'm},\end{aligned} \tag{4.118}$$

where

$$\sigma_x = \begin{pmatrix} 0 & 1 \\ 1 & 0 \end{pmatrix}, \quad \sigma_y = \begin{pmatrix} 0 & -i \\ i & 0 \end{pmatrix}, \quad \sigma_z = \begin{pmatrix} 1 & 0 \\ 0 & -1 \end{pmatrix}, \tag{4.119}$$

are the *Pauli matrices*. As we can see, the Pauli matrices are Hermitian, traceless and their determinant equals -1:

$$\sigma_i^\dagger = \sigma_i, \quad \mathrm{Tr}(\sigma_i) = 0, \quad \det(\sigma_i) = -1. \tag{4.120}$$

The spin operators act on the space of spin states

$$\begin{pmatrix} c_1 \\ c_2 \end{pmatrix} = c_1 \begin{pmatrix} 1 \\ 0 \end{pmatrix} + c_2 \begin{pmatrix} 0 \\ 1 \end{pmatrix}, \tag{4.121}$$

where the basis states

$$\begin{pmatrix} 1 \\ 0 \end{pmatrix} = |1/2, 1/2\rangle \equiv |\uparrow\rangle \quad \text{and} \quad \begin{pmatrix} 0 \\ 1 \end{pmatrix} = |1/2, -1/2\rangle \equiv |\downarrow\rangle, \tag{4.122}$$

represent the *spin-up* and *spin-down* states, respectively. The Pauli matrices satisfy the commutation relation

$$[\sigma_i/2, \sigma_j/2] = i\epsilon_{ijk}\sigma_k/2, \tag{4.123}$$

and possess the following properties

$$\sigma_i^2 = I, \quad \text{and} \quad \sigma_i\sigma_j + \sigma_j\sigma_i = 0 \quad \text{if} \quad i \neq j, \tag{4.124}$$

where I is the 2×2 unit matrix. We can combine the commutation and anti-commutation relations into

$$\sigma_i\sigma_j = \delta_{ij} + i\sum_k \epsilon_{ijk}\,\sigma_k. \tag{4.125}$$

We can easily verify by direct evaluation that if two vectors $\boldsymbol{A}$ and $\boldsymbol{B}$ commute with σ_i, then

$$(\boldsymbol{\sigma} \cdot \boldsymbol{A})(\boldsymbol{\sigma} \cdot \boldsymbol{B}) = (\boldsymbol{A} \cdot \boldsymbol{B})I + i\boldsymbol{\sigma} \cdot (\boldsymbol{A} \times \boldsymbol{B}). \tag{4.126}$$

We can also find that $\sigma_x \sigma_y = i\sigma_z$, and thus $\sigma_x \sigma_y \sigma_z = iI$. By utilizing the powers of σ_i, we can obtain for real α

$$\exp(i\alpha\sigma_i) = \cos\alpha\, I + i\sigma_i \sin\alpha. \tag{4.127}$$

4.7 Euler Rotation

Rotation operators form a group:

- $$\hat{R}_{n_1} \hat{R}_{n_2} = \hat{R}_{n_3}, \tag{4.128}$$

  i.e. two consecutive rotations are equivalent to a single rotation and the product of two rotation operators is another rotation operator.

- $$(\hat{R}_{n_1} \hat{R}_{n_2}) \hat{R}_{n_3} = \hat{R}_{n_1} (\hat{R}_{n_2} \hat{R}_{n_3}), \tag{4.129}$$

  i.e. rotations are associative and we can group them arbitrarily.

- $$I \hat{R}_n = \hat{R}_n I = \hat{R}_n, \tag{4.130}$$

  i.e. the identity operator I, which corresponds to no rotation, is an element of the group. From this it follows, that for each rotation there exists an inverse operator $\hat{R}_n^{-1}(\theta) = \hat{R}_{-n}(\theta) = \hat{R}_n(-\theta)$ such that

  $$\hat{R}_n^{-1}(\theta) \hat{R}_n(\theta) = \hat{R}_n(\theta) \hat{R}_n^{-1}(\theta) = I, \tag{4.131}$$

  i.e. we can always rotate back.

The generators of the rotation operator may not commute, therefore the rotation operators may not commute as well,

$$\hat{R}_{n_1} \hat{R}_{n_2} \neq \hat{R}_{n_2} \hat{R}_{n_1}, \tag{4.132}$$

in general. So, this group is a non-commutative, or non-Abelian group. However, rotations around the same axis commute and form a commutative, or Abelian, subgroup in the larger non-commutative group.

As we know from classical mechanics, an arbitrary rotation can be expressed in terms of Euler angles:

- first we rotate by $\alpha \in [0, 2\pi]$ around the z axis,
- then we rotate by $\beta \in [0, \pi]$ around the new y axis, which we denote by v,
- finally, we rotate by $\gamma \in [0, 2\pi]$ around the new z axis, denoted by z'.

So, the resulting rotation operator is

$$\hat{R}(\alpha, \beta, \gamma) = \hat{R}_{z'}(\gamma)\hat{R}_v(\beta)\hat{R}_z(\alpha) = \exp\left(-\frac{i}{\hbar}\gamma\hat{J}_{z'}\right)\exp\left(-\frac{i}{\hbar}\beta\hat{J}_v\right)\exp\left(-\frac{i}{\hbar}\alpha\hat{J}_z\right). \quad (4.133)$$

Since rotation is a unitary transformation and $\hat{J}_v$ is basically a rotated $\hat{J}_y$, we have

$$\hat{J}_v = \hat{R}_z(\alpha)\hat{J}_y\hat{R}_z^\dagger(\alpha) = \hat{R}_z(\alpha)\hat{J}_y\hat{R}_z(-\alpha), \quad (4.134)$$

and, considering that the rotation operator is a function of $\hat{J}$, we obtain

$$\hat{R}_v(\beta) = \hat{R}_z(\alpha)\hat{R}_y(\beta)\hat{R}_z(-\alpha). \quad (4.135)$$

Analogously, $\hat{J}_{z'}$ is a rotated $\hat{J}_z$, therefore

$$\hat{J}_{z'} = \hat{R}_v(\beta)\hat{R}_z(\alpha)\hat{J}_z\hat{R}_z(-\alpha)\hat{R}_v(-\beta), \quad (4.136)$$

and consequently

$$\begin{aligned}\hat{R}_{z'}(\gamma) &= \hat{R}_v(\beta)\hat{R}_z(\alpha)\hat{R}_z(\gamma)\hat{R}_z(-\alpha)\hat{R}_v(-\beta) \\ &= \hat{R}_z(\alpha)\hat{R}_y(\beta)\hat{R}_z(-\alpha)\hat{R}_z(\alpha)\hat{R}_z(\gamma)\hat{R}_z(-\alpha)\hat{R}_z(\alpha)\hat{R}_y(-\beta)\hat{R}_z(-\alpha) \\ &= \hat{R}_z(\alpha)\hat{R}_y(\beta)\hat{R}_z(\gamma)\hat{R}_y(-\beta)\hat{R}_z(-\alpha).\end{aligned} \quad (4.137)$$

Putting these results into Eq. (4.133), we obtain $\hat{R}$ in terms of the components of the $\hat{\boldsymbol{J}}$ in the original frame

$$\hat{R}(\alpha, \beta, \gamma) = \hat{R}_z(\alpha)\hat{R}_y(\beta)\hat{R}_z(\gamma) = \exp\left(-\frac{i}{\hbar}\alpha\hat{J}_z\right)\exp\left(-\frac{i}{\hbar}\beta\hat{J}_y\right)\exp\left(-\frac{i}{\hbar}\gamma\hat{J}_z\right). \quad (4.138)$$

The adjoint rotation is given by

$$\begin{aligned}\hat{R}^\dagger(\alpha, \beta, \gamma) = \hat{R}^{-1}(\alpha, \beta, \gamma) &= \hat{R}_z(-\gamma)\hat{R}_y(-\beta)\hat{R}_z(-\alpha) \\ &= \exp\left(\frac{i}{\hbar}\gamma\hat{J}_z\right)\exp\left(\frac{i}{\hbar}\beta\hat{J}_y\right)\exp\left(\frac{i}{\hbar}\alpha\hat{J}_z\right).\end{aligned} \quad (4.139)$$

4.7.1 Rotation Matrices

Let us investigate the action of $\hat{R}$ on the angular momentum state $|j, m\rangle$. Since $\hat{R}$ contains only angular momentum operators that commute with $\hat{J}^2$, $\hat{R}$ also commutes with $\hat{J}^2$. Thus,

$$\hat{J}^2(\hat{R}(\alpha, \beta, \gamma)|j, m\rangle) = \hbar^2 j(j+1)(\hat{R}(\alpha, \beta, \gamma)|j, m\rangle), \tag{4.140}$$

i.e. $\hat{R}(\alpha, \beta, \gamma)|j, m\rangle$ remains in the subspace $|j, m'\rangle$, with $-j \le m' \le j$. Therefore,

$$\hat{R}(\alpha, \beta, \gamma)|j, m\rangle = \sum_{m'=-j}^{j} |j, m'\rangle\langle j, m'|\hat{R}(\alpha, \beta, \gamma)|j, m\rangle = \sum_{m'=-j}^{j} D^{(j)}_{m',m}(\alpha, \beta, \gamma)|j, m'\rangle. \tag{4.141}$$

The coefficients of this expansion form a $(2j+1)$-dimensional representation of the $\hat{R}$ operator with

$$D^{(j)}_{m',m}(\alpha, \beta, \gamma) = \langle j, m'| \exp\left(-\frac{i}{\hbar}\alpha \hat{J}_z\right) \exp\left(-\frac{i}{\hbar}\beta \hat{J}_y\right) \exp\left(-\frac{i}{\hbar}\gamma \hat{J}_z\right)|j, m\rangle. \tag{4.142}$$

This unitary matrix is called the *Wigner D matrix*, and the matrix elements are the *Wigner D functions*. Since $|j, m\rangle$ and $|j, m'\rangle$ are eigenstates of $\hat{J}_z$, we can write

$$D^{(j)}_{m',m}(\alpha, \beta, \gamma) = e^{-im'\alpha} d^{(j)}_{m',m}(\beta) e^{-im\gamma}, \tag{4.143}$$

where

$$\begin{aligned} d^{(j)}_{m',m}(\beta) &= \langle j, m'| \exp\left(-\frac{i}{\hbar}\beta \hat{J}_y\right)|j, m\rangle \\ &= \langle j, m'| \exp\left(-\frac{\beta}{2\hbar}(\hat{J}_+ - \hat{J}_-)\right)|j, m\rangle = D^{(j)}_{m',m}(0, \beta, 0) \end{aligned} \tag{4.144}$$

are the *Wigner small d*, or *reduced rotation* matrices. We have adopted a representation that J_y is purely imaginary, which makes $d^{(j)}_{m',m}$ real. We can see that only the middle rotation $\hat{R}_y(\beta)$ mixes the states with different values of m.

The $j = 1/2$ Case

First we look at the relatively simple case of $j = 1/2$. Here, $\hat{J}_y = \hbar/2\,\sigma_y$, with $\sigma_y^2 = 1$. Then, the series expansion of the exponential function gives

$$\begin{aligned} \exp\left(-\frac{i}{\hbar}\beta \hat{J}_y\right) &= I + (-i\beta/2)\sigma_y + \frac{(-i\beta/2)^2}{2!}\sigma_y^2 + \frac{(-i\beta/2)^3}{3!}\sigma_y^3 + \cdots \\ &= \cos\left(\frac{\beta}{2}\right) - i\sigma_y \sin\left(\frac{\beta}{2}\right). \end{aligned} \tag{4.145}$$

So,

$$d^{(1/2)}_{m',m}(\beta) = \begin{pmatrix} \cos\left(\frac{\beta}{2}\right) & -\sin\left(\frac{\beta}{2}\right) \\ \sin\left(\frac{\beta}{2}\right) & \cos\left(\frac{\beta}{2}\right) \end{pmatrix}, \tag{4.146}$$

and

$$D^{(1/2)}_{m',m}(\alpha,\beta,\gamma) = \begin{pmatrix} e^{-i\alpha/2}\cos\left(\frac{\beta}{2}\right)e^{-i\gamma/2} & -e^{-i\alpha/2}\sin\left(\frac{\beta}{2}\right)e^{i\gamma/2} \\ e^{i\alpha/2}\sin\left(\frac{\beta}{2}\right)e^{-i\gamma/2} & e^{i\alpha/2}\cos\left(\frac{\beta}{2}\right)e^{i\gamma/2} \end{pmatrix}. \tag{4.147}$$

We can verify by direct calculation that the $D^{(1/2)}$ matrix is unitary and its determinant is equal to one. As expected, $D^{(1/2)}(0,0,0) = I$, the 2×2 unit matrix. It is interesting to see that, for example, a rotation by 2π around the y axis is described by

$$D^{(1/2)}(0,2\pi,0) = \begin{pmatrix} \cos\pi & -\sin\pi \\ \sin\pi & \cos\pi \end{pmatrix} = \begin{pmatrix} -1 & 0 \\ 0 & -1 \end{pmatrix}. \tag{4.148}$$

For a more general rotation $\hat{R}_{\boldsymbol{n}}(\theta)$, we can see that

$$D^{(1/2)}(\hat{R}_{\hat{n}}(2\pi k)) = (-)^n I, \quad k = 0,1,2,\ldots, \tag{4.149}$$

i.e. an even number of 2π rotations results in the identity I, but an odd number of 2π rotations results in inversion $-I$. A $1/2$-spin needs a multiple of 4π rotation to get back to the original state.

The $j = 1$ Case

In this case $m, m' = 1, 0, -1$. So,

$$J_y = \frac{1}{\sqrt{2}}\begin{pmatrix} 0 & -i & 0 \\ i & 0 & -i \\ 0 & i & 0 \end{pmatrix}, \tag{4.150}$$

$$J_y^2 = \frac{1}{2}\begin{pmatrix} 1 & 0 & -1 \\ 0 & 2 & 0 \\ -1 & 0 & 1 \end{pmatrix}, \tag{4.151}$$

and $J_y^3 = J_y$. As a consequence

$$d^{(1)}_{m',m}(\beta) = \begin{pmatrix} \frac{1}{2}(1+\cos\beta) & -\frac{1}{\sqrt{2}}\sin\beta & \frac{1}{2}(1-\cos\beta) \\ \frac{1}{\sqrt{2}}\sin\beta & \cos\beta & -\frac{1}{\sqrt{2}}\sin\beta \\ \frac{1}{2}(1-\cos\beta) & \frac{1}{\sqrt{2}}\sin\beta & \frac{1}{2}(1+\cos\beta) \end{pmatrix}. \tag{4.152}$$

The General Case

The general formula has been derived by Wigner

$$d^{(j)}_{m',m}(\beta) = \sum_s \frac{(-)^{m'-m+s}[(j+m')!(j-m')!(j+m)!(j-m)!]^{1/2}}{(j+m-s)!s!(m'-m+s)!(j-m'-s)!} \times \left(\cos\frac{\beta}{2}\right)^{2j+m-m'-2s} \left(\sin\frac{\beta}{2}\right)^{m'-m+2s}. \tag{4.153}$$

We can readily establish the following relations

$$\begin{aligned} d^{(j)}_{m,m'}(\beta) &= (-1)^{m-m'} d^{(j)}_{-m,-m'}(\beta) = (-1)^{m-m'} d^{(j)}_{m',m}(\beta) = d^{(j)}_{-m',-m}(\beta), \\ d^{(j)}_{m,m'}(-\beta) &= (-1)^{m-m'} d^{(j)}_{m,m'}(\beta) = d^{(j)}_{m',m}(\beta), \\ d^{(j)}_{m,m'}(\pi-\beta) &= (-1)^{j-m'} d^{(j)}_{-m,m'}(\beta) = (-1)^{j+m} d^{(j)}_{m,-m'}(\beta), \\ D^{(j)*}_{m',m}(\alpha,\beta,\gamma) &= D^{(j)}_{m,m'}(-\gamma,-\beta,-\alpha) = (-)^{m-m'} D^{(j)}_{-m',-m}(\alpha,\beta,\gamma). \end{aligned} \tag{4.154}$$

4.7.2 *Rotation of the Spherical Harmonics*

The rotational operator rotates the coordinate vector

$$|\boldsymbol{r}'\rangle = \hat{R}(\alpha,\beta,\gamma)|\boldsymbol{r}\rangle. \tag{4.155}$$

Taking this relation in the angular momentum basis we find

$$\langle l,m|\boldsymbol{r}'\rangle = \sum_{m'} \langle l,m|\hat{R}(\alpha,\beta,\gamma)|l,m'\rangle\langle l,m'|\boldsymbol{r}\rangle, \tag{4.156}$$

or rather

$$Y^*_{l,m}(\vartheta',\varphi') = \sum_{m'} D^{(l)}_{m,m'}(\alpha,\beta,\gamma) Y^*_{l,m'}(\vartheta,\varphi). \tag{4.157}$$

If we take a particular point along the z axis, then $\vartheta = 0$, φ becomes irrelevant, and the spherical harmonics becomes

$$Y^*_{l,m'}(0,0) = \sqrt{\frac{2l+1}{4\pi}}\delta_{0,m'}. \tag{4.158}$$

This results in a connection between the D functions and the spherical harmonics

$$D^{(l)}_{m,0}(\alpha,\beta,\gamma) = \sqrt{\frac{4\pi}{2l+1}} Y^*_{l,m}(\beta,\alpha), \tag{4.159}$$

where the (β,α) refers to the z' direction in the old coordinate system. Plugging this relation into Eq. (4.157), with $m = 0$, we obtain

$$Y_{l,0}(\vartheta', \varphi') = \sqrt{\frac{2l+1}{4\pi}} P_l(\cos\vartheta') = \sqrt{\frac{4\pi}{2l+1}} \sum_{m'} Y^*_{l,m'}(\beta, \alpha) Y_{l,m'}(\vartheta, \varphi). \quad (4.160)$$

4.8 Addition of Angular Momenta

A physical system may consist of two subsystems that have their own angular momenta $\hat{J}_1$ and $\hat{J}_2$. We assume that $\hat{J}_1$ and $\hat{J}_2$ are independent,

$$[\hat{J}_{1,i}, \hat{J}_{2,j}] = 0, \qquad \text{for all} \quad i, j = 1, 2, 3, \quad (4.161)$$

and consequently, the operators $\{\hat{J}_1^2, \hat{J}_{1,z}, \hat{J}_2^2, \hat{J}_{2,z}\}$ form a complete set of commuting observables. Their common eigenstates span a $(2j_1 + 1) \times (2j_2 + 1)$ dimensional space with a basis

$$|j_1, j_2; m_1, m_2\rangle = |j_1, m_1\rangle |j_2, m_2\rangle, \quad (4.162)$$

where $|j_1, m_1\rangle$ and $|j_2, m_2\rangle$ are common eigenstates of the pairs $\{\hat{J}_1^2, \hat{J}_{1,z}\}$ and $\{\hat{J}_2^2, \hat{J}_{2,z}\}$, respectively. Consequently,

$$\hat{J}_1^2 |j_1, j_2; m_1, m_2\rangle = \hbar^2 j_1(j_1 + 1) |j_1, j_2; m_1, m_2\rangle, \quad (4.163)$$

$$\hat{J}_{1,z} |j_1, j_2; m_1, m_2\rangle = \hbar m_1 |j_1, j_2; m_1, m_2\rangle, \quad (4.164)$$

$$\hat{J}_2^2 |j_1, j_2; m_1, m_2\rangle = \hbar^2 j_2(j_2 + 1) |j_1, j_2; m_1, m_2\rangle, \quad (4.165)$$

$$\hat{J}_{2,z} |j_1, j_2; m_1, m_2\rangle = \hbar m_2 |j_1, j_2; m_1, m_2\rangle. \quad (4.166)$$

Now, the system as a whole should possess some symmetry with respect to rotation, and this should be generated by the total angular momentum $\hat{\boldsymbol{J}}$. Let us define

$$\hat{\boldsymbol{J}} = \hat{\boldsymbol{J}}_1 + \hat{\boldsymbol{J}}_2, \quad (4.167)$$

or more rigorously

$$\hat{\boldsymbol{J}} = \hat{\boldsymbol{J}}_1 \otimes I_{j_2} + I_{j_1} \otimes \hat{\boldsymbol{J}}_2, \quad (4.168)$$

where I_{j_2} and I_{j_1} are unit operators in the subspace belonging to $\hat{\boldsymbol{J}}_2$ and $\hat{\boldsymbol{J}}_1$, respectively. We can easily see that $\hat{\boldsymbol{J}}$ satisfies the usual angular momentum commutation relations

$$[\hat{J}_i, \hat{J}_j] = i\hbar\epsilon_{ijk}\hat{J}_k. \quad (4.169)$$

For example,

$$[\hat{J}_x, \hat{J}_y] = [\hat{J}_{1,x} + \hat{J}_{2,x}, \hat{J}_{1,y} + \hat{J}_{2,y}] = [\hat{J}_{1,x}, \hat{J}_{1,y}] + [\hat{J}_{2,x}, \hat{J}_{2,y}] = i\hbar(\hat{J}_{1,z} + \hat{J}_{2,z}) = i\hbar\hat{J}_z. \quad (4.170)$$

Then,

$$\begin{aligned}\hat{J}^2 &= \hat{J}_1^2 + 2\hat{J}_1\hat{J}_2 + \hat{J}_2^2 = \hat{J}_1^2 + 2(\hat{J}_{1,x}\hat{J}_{2,x} + \hat{J}_{1,y}\hat{J}_{2,y} + \hat{J}_{1,z}\hat{J}_{2,z}) + \hat{J}_2^2 \\ &= \hat{J}_1^2 + \hat{J}_2^2 + 2\hat{J}_{1,z}\hat{J}_{2,z} + \hat{J}_{1,+}\hat{J}_{2,+} + \hat{J}_{1,-}\hat{J}_{2,-},\end{aligned} \tag{4.171}$$

should also commute with $\hat{J}_z = \hat{J}_{1,z} + \hat{J}_{2,z}$. Indeed,

$$\begin{aligned}&[\hat{J}^2, \hat{J}_{1,z} + \hat{J}_{2,z}] \\ &\quad = 2([\hat{J}_{1,x}, \hat{J}_{1,z}]\hat{J}_{2,x} + \hat{J}_{1,x}[\hat{J}_{2,x}, \hat{J}_{2,z}] + [\hat{J}_{1,y}, \hat{J}_{1,z}]\hat{J}_{2,y} + \hat{J}_{1,y}[\hat{J}_{2,y}, \hat{J}_{2,z}]) \\ &\quad = 2((-)i\hbar\hat{J}_{1,y}\hat{J}_{2,x} + \hat{J}_{1,x}(-)i\hbar\hat{J}_{2,y} + i\hbar\hat{J}_{1,x}\hat{J}_{2,y} + \hat{J}_{1,y}i\hbar J_{2,x}) = 0.\end{aligned} \tag{4.172}$$

Obviously, the operators $\hat{J}_1^2$ and $\hat{J}_2^2$ commute with $\hat{J}^2$ and $\hat{J}_z$. So, $\{\hat{J}^2, \hat{J}_z, \hat{J}_1^2, \hat{J}_2^2\}$ also form a complete set of commuting observables. We denote the common eigenstates of $\{\hat{J}^2, \hat{J}_z, \hat{J}_1^2, \hat{J}_2^2\}$ by $|j_1, j_2; j, m\rangle$ such that

$$\hat{J}_1^2|j_1, j_2; j, m\rangle = \hbar^2 j_1(j_1+1)|j_1, j_2; j, m\rangle, \tag{4.173}$$

$$\hat{J}_2^2|j_1, j_2; j, m\rangle = \hbar^2 j_2(j_2+1)|j_1, j_2; j, m\rangle, \tag{4.174}$$

$$\hat{J}^2|j_1, j_2; j, m\rangle = \hbar^2 j(j+1)|j_1, j_2; j, m\rangle, \tag{4.175}$$

$$\hat{J}_z|j_1, j_2; j, m\rangle = \hbar m|j_1, j_2; j, m\rangle. \tag{4.176}$$

Usually j_1 and j_2 are fixed, so we can drop them from the notation, $|j_1, j_2; j, m\rangle = |j, m\rangle$.

The functions $\{|j_1, j_2; m_1, m_2\rangle\}$ and $\{|j, m\rangle\}$ span the same subset of the Hilbert space and they form bases in the same Hilbert space

$$I = \sum_{m_1=-j_1}^{j_1} \sum_{m_2=-j_2}^{j_2} |j_1, j_2; m_1, m_2\rangle\langle j_1, j_2; m_1, m_2| = \sum_j \sum_{m=-j}^{j} |j, m\rangle\langle j, m|. \tag{4.177}$$

The two bases should be connected by a unitary transformation

$$\begin{aligned}|j, m\rangle &= \sum_{m_1=-j_1}^{j_1} \sum_{m_2=-j_2}^{j_2} |j_1, j_2; m_1, m_2\rangle\langle j_1, j_2; m_1, m_2|j, m\rangle \\ &= \sum_{m_1=-j_1}^{j_1} \sum_{m_2=-j_2}^{j_2} \langle j_1, j_2; m_1, m_2|j, m\rangle\, |j_1, j_2; m_1, m_2\rangle,\end{aligned} \tag{4.178}$$

and the matrix elements $\langle j_1, j_2; m_1, m_2|j, m\rangle$ are the *Clebsch-Gordan coefficients*.

The Clebsch-Gordan coefficients, by convention, are chosen to be real. Then, the unitary matrix becomes an orthogonal matrix. Consequently, the inverse Clebsch-Gordan coefficients $\langle j, m|j_1, j_2; m_1, m_2\rangle$ coincide with $\langle j_1, j_2; m_1, m_2|j, m\rangle$. From the orthogonality of the states $|j, m\rangle$, we have

$$\langle j', m'|j, m\rangle = \delta_{j,j'}\delta_{m,m'} = \sum_{m_1=-j_1}^{j_1}\sum_{m_2=-j_2}^{j_2}\langle j', m'|j_1, j_2; m_1, m_2\rangle\langle j_1, j_2; m_1, m_2|j, m\rangle$$
$$= \sum_{m_1=-j_1}^{j_1}\sum_{m_2=-j_2}^{j_2}\langle j_1, j_2; m_1, m_2|j', m'\rangle\langle j_1, j_2; m_1, m_2|j, m\rangle = \delta_{j,j'}\delta_{m,m'}. \tag{4.179}$$

A special case is when $j = j'$ and $m = m'$, then

$$\sum_{m_1=-j_1}^{j_1}\sum_{m_2=-j_2}^{j_2}\langle j_1, j_2; m_1, m_2|j, m\rangle^2 = 1. \tag{4.180}$$

Similarly, we also have

$$\sum_{j}\sum_{m=-j}^{j}\langle j_1, j_2; m_1, m_2|j, m\rangle\langle j_1, j_2; m_1', m_2'|j, m\rangle = \delta_{m_1,m_1'}\delta_{m_2,m_2'} \tag{4.181}$$

and

$$\sum_{j}\sum_{m=-j}^{j}\langle j_1, j_2; m_1, m_2|j, m\rangle^2 = 1. \tag{4.182}$$

If we calculate the matrix element of $\hat{J}_z - \hat{J}_{1,z} - \hat{J}_{2,z} = 0$, we find

$$\langle j_1, j_2; m_1, m_2|(\hat{J}_z - \hat{J}_{1,z} - \hat{J}_{2,z})|j, m\rangle = (m - m_1 - m_2)\langle j_1, j_2; m_1, m_2|j, m\rangle = 0, \tag{4.183}$$

so we can conclude that

$$\langle j_1, j_2; m_1, m_2|j, m\rangle = 0 \quad \text{if} \quad m \neq m_1 + m_2. \tag{4.184}$$

The maximal value of m_1 and m_2 are j_1 and j_2, respectively. So, the maximal value of $m_{max} = j_1 + j_2$, and this is also the maximal value of j, $j_{max} = j_1 + j_2$. The dimension of the Hilbert space equals the number of basis states in both coupling schemes. Therefore,

$$(2j_1 + 1)(2j_2 + 1) = \sum_{j_{min}}^{j_1+j_2}(2j + 1) = (j_1 + j_2 + 1 + j_{min})(j_1 + j_2 + 1 - j_{min}), \tag{4.185}$$

and thus

$$j_{min}^2 = (j_1 - j_2)^2. \tag{4.186}$$

So, we have found that the coupling of two angular momenta follows the triangular inequality of vector addition

$$|j_1 - j_2| \le j \le j_1 + j_2. \tag{4.187}$$

From this, we can see that if $j_{max} = j_1 + j_2$ is integer or half-integer, then j is also integer or half-integer, respectively. Consequently, $j_1 + j_2 + j$ has to be an integer.

A generally accepted phase convention by Condon and Shortley is that

$$\langle j_1, j_2; j_1, j - j_1 | j, j\rangle > 0. \tag{4.188}$$

Under such a phase definition the Clebsch-Gordan coefficients satisfy the relations

$$\langle j, 0; m, 0 | j, m\rangle = 1 \quad \text{and} \quad \langle j_1, j_2; j_1, j_2 | j_1 + j_2, j_1 + j_2\rangle = 1. \tag{4.189}$$

We have a systematic way of evaluating the Clebsch-Gordan coefficients by applying the lowering operator

$$\hat{J}_- = \hat{J}_{1,-} + \hat{J}_{2,-} \tag{4.190}$$

repeatedly on the state

$$|j, m\rangle = \sum_{m_1, m_2} |j_1, j_2; m_1, m_2\rangle\langle j_1, j_2; m_1, m_2 | j, m\rangle. \tag{4.191}$$

We can start with $|j_1 + j_2, j_1 + j_2\rangle$, then by applying $\hat{J}_-$, we can construct $|j_1 + j_2, j_1 + j_2 - 1\rangle$, then $|j_1 + j_2, j_1 + j_2 - 2\rangle$, and so on. The state $|j_1 + j_2 - 1, j_1 + j_2 - 1\rangle$ is determined by the phase conditions and that it is orthogonal to $|j_1 + j_2, j_1 + j_2 - 1\rangle$. Then, we apply $\hat{J}_-$ again on the states with $j = j_1 + j_2 - 1$. The procedure is illustrated below:

$$\begin{array}{lll}
|j_1 + j_2, j_1 + j_2\rangle & & \\
\downarrow \hat{J}_- & & \\
|j_1 + j_2, j_1 + j_2 - 1\rangle \longrightarrow & |j_1 + j_2 - 1, j_1 + j_2 - 1\rangle & \\
\downarrow \hat{J}_- & \downarrow \hat{J}_- & \\
|j_1 + j_2, j_1 + j_2 - 2\rangle & |j_1 + j_2 - 1, j_1 + j_2 - 2\rangle \longrightarrow & |j_1 + j_2 - 2, j_1 + j_2 - 2\rangle. \\
\downarrow \hat{J}_- & \downarrow \hat{J}_- & \downarrow \hat{J}_-
\end{array} \tag{4.192}$$

From the matrix element

$$\langle j_1, j_2; m_1, m_2 | (\hat{J}_- - \hat{J}_{1,-} - \hat{J}_{2,-}) | j, m\rangle = 0, \tag{4.193}$$

we can derive a recursion relation for the Clebsch-Gordan coefficients

$$\sqrt{(j+m)(j-m+1)}\langle j_1, j_2; m_1, m_2|j, m-1\rangle$$
$$= \sqrt{(j_1+m_1+1)(j_1-m_1)}\langle j_1, j_2; m_1+1, m_2|j, m\rangle$$
$$+ \sqrt{(j_2+m_2+1)(j_2-m_2)}\langle j_1, j_2; m_1, m_2+1|j, m\rangle. \quad (4.194)$$

A similar recursion relation can be obtained by applying the raising operator $\hat{J}_+ = \hat{J}_{1,+} + \hat{J}_{2,+}$:

$$\sqrt{(j+m+1)(j-m)}\langle j_1, j_2; m_1, m_2|j, m+1\rangle$$
$$= \sqrt{(j_1+m_1)(j_1-m_1+1)}\langle j_1, j_2; m_1-1, m_2|j, m\rangle$$
$$+ \sqrt{(j_2+m_2)(j_2-m_2+1)}\langle j_1, j_2; m_1, m_2-1|j, m\rangle. \quad (4.195)$$

We can also evaluate $\langle j_1, j_2; m_1, m_2|\hat{J}_\pm|j, m \mp 1\rangle$ to obtain

$$\sqrt{(j\mp m+1)(j\pm m)}\langle j_1, j_2; m_1, m_2|j, m\rangle$$
$$= \sqrt{(j_1\pm m_1)(j_1\mp m_1+1)}\langle j_1, j_2; m_1\mp 1, m_2|j, m\mp 1\rangle$$
$$+ \sqrt{(j_2\pm m_2)(j_2\mp m_2+1)}\langle j_1, j_2; m_1, m_2\mp 1|j, m\mp 1\rangle. \quad (4.196)$$

General Formula for Clebsch-Gordan Coefficients

There is a general, rather complicated, formula for the Clebsch-Gordan coefficients

$$\langle j_1, j_2; m_1, m_2|j, m\rangle = \delta_{m_1+m_2,m}\sqrt{2j+1}\left[\frac{(j_1+j_2-j)!(j_1-j_2+j)!(-j_1+j_2+j)!}{(j_1+j_2+j+1)!}\right]^{1/2}$$
$$\times \sum_n \frac{(-)^n\left[(j_1+m_1)!(j_1-m_1)!(j_2+m_2)!(j_2-m_2)!(j+m)!(j-m)!\right]^{1/2}}{n!(j_1+j_2-j-n)!(j_1-m_1-n)!(j_2+m_2-n)!(j-j_2+m_1+n)!(j-j_1-m_2+n)!}, \quad (4.197)$$

where the sum runs over all n integer values for which none of the factorials is negative.

We can see that in this formula, the term in front of the sum under the square root is symmetric in all three angular momenta. The sum in the numerator, except for the $(-)^n$ phase factor, is invariant under arbitrary permutations of pairs (j_1, m_1), (j_2, m_2) and (j, m). The denominator is also invariant with respect to replacing the summation index $n \to j_1 + j_2 - j - r$, where r is integer, and the interchange $(j_1, m_1) \leftrightarrow (j_2, m_2)$ introduces only a phase factor $(-)^{j_1+j_2-j}(-)^n$. As a result, we have the symmetry relation

$$\langle j_2, j_1; m_2, m_1|j, m\rangle = (-)^{j_1+j_2-j}\langle j_1, j_2; m_1, m_2|j, m\rangle. \quad (4.198)$$

Consequently, $\langle j_1, j_2; 0, 0|j, 0\rangle = 0$, if $j_1 + j_2 - j$ is odd. Therefore, j_1, j_2 and j have to be integers to have 0 projection quantum numbers. In a similar fashion, by simultaneously negating the m's and changing the summation index $n \to j_1 + j_2 - j - r$, we also obtain

$$\langle j_1, j_2; m_1, m_2|j, m\rangle = (-)^{j_1+j_2-j}\langle j_1, j_2; -m_1, -m_2|j, -m\rangle. \tag{4.199}$$

The Case of $j_1 = 1/2$ and $j_2 = 1/2$

We start from

$$|1, 1\rangle = |1/2, 1/2\rangle|1/2, 1/2\rangle, \tag{4.200}$$

and apply $\hat{J}_- = \hat{J}_{1,-} + \hat{J}_{2,-}$ to get

$$\sqrt{2}|1, 0\rangle = |1/2, 1/2\rangle|1/2, -1/2\rangle + |1/2, -1/2\rangle|1/2, 1/2\rangle, \tag{4.201}$$

which gives

$$|1, 0\rangle = \frac{1}{\sqrt{2}}|1/2, 1/2\rangle|1/2, -1/2\rangle + \frac{1}{\sqrt{2}}|1/2, -1/2\rangle|1/2, 1/2\rangle. \tag{4.202}$$

Applying $\hat{J}_-$ again on the above equation, we obtain

$$|1, -1\rangle = |1/2, -1/2\rangle|1/2, -1/2\rangle. \tag{4.203}$$

The $|0, 0\rangle$ state has the same m's as $|1, 0\rangle$, but they have to be orthogonal. The state $|1, 0\rangle$ is symmetric, therefore $|0, 0\rangle$ must be antisymmetric

$$|0, 0\rangle = \frac{1}{\sqrt{2}}|1/2, 1/2\rangle|1/2, -1/2\rangle - \frac{1}{\sqrt{2}}|1/2, -1/2\rangle|1/2, 1/2\rangle. \tag{4.204}$$

The sign of the first term is determined by the phase condition that $\langle 1/2, 1/2; 1/2, -1/2|0, 0\rangle$ has to be positive.

The Case of $j_1 = l$ Integer and $j_2 = 1/2$

This is a very important special case as it describes the coupling of the spin and orbital angular momentum, $\hat{\boldsymbol{J}} = \hat{\boldsymbol{L}} + \hat{\mathbf{S}}$. If $l > 0$, we can have $j = l \pm 1/2$.

First, we consider the $j = l + 1/2$ case. Since $m_s = \pm 1/2$, we can write

$$\begin{aligned}|l + 1/2, m\rangle =& \langle l, 1/2; m + 1/2, -1/2|l + 1/2, m\rangle|l, 1/2; m + 1/2, -1/2\rangle\\ &+ \langle l, 1/2; m - 1/2, 1/2|l + 1/2, m\rangle|l, 1/2; m - 1/2, 1/2\rangle.\end{aligned} \tag{4.205}$$

To calculate the first Clebsch-Gordan coefficient, we may start from Eq. (4.196) with $\hat{J}_+$

$$\begin{aligned}&\sqrt{(l - m + 3/2)(l + m + 1/2)}\langle l, 1/2; m + 1/2, -1/2|l + 1/2, m\rangle\\ &= \sqrt{(l + m + 1/2)(l - m + 1/2)}\langle l, 1/2; m - 1/2, -1/2|l + 1/2, m - 1\rangle,\end{aligned} \tag{4.206}$$

and then

$$\langle l, 1/2; m+1/2, -1/2|l+1/2, m\rangle = \sqrt{\frac{l-m+1/2}{l-m+3/2}}\langle l, 1/2; m-1/2, -1/2|l+1/2, m-1\rangle. \tag{4.207}$$

We can see that the Clebsch-Gordan coefficient on the right hand side of the equation is the same as the one on the left hand side with the substitution $m \to m-1$. By repeated substitutions we reach $m = -l - 1/2$, and end up with the formula

$$\langle l, 1/2; m+1/2, -1/2|l+1/2, m\rangle = \sqrt{\frac{l-m+1/2}{l-m+3/2}}\sqrt{\frac{l-m+3/2}{l-m+5/2}}\cdots\sqrt{\frac{2l}{2l+1}} \\ \times \langle l, 1/2; -l, -1/2|l+1/2, -l-1/2\rangle. \tag{4.208}$$

Due to the phase convention in Eq. (4.189), the last coefficient in the above formula equals one, and thus we obtain

$$\langle l, 1/2; m+1/2, -1/2|l+1/2, m\rangle = \sqrt{\frac{l-m+1/2}{2l+1}}. \tag{4.209}$$

Similarly as before, we can obtain the other $\langle l, 1/2; m-1/2, 1/2|l+1/2, m\rangle$ term by applying Eq. (4.196) with $\hat{J}_-$. However, the use of the orthogonality condition Eq. (4.180) offers us an easier way,

$$1 = \frac{l-m+1/2}{2l+1} + \langle l, 1/2; m-1/2, 1/2|l+1/2, m\rangle^2, \tag{4.210}$$

and consequently we find

$$\langle l, 1/2; m-1/2, 1/2|l+1/2, m\rangle = \sqrt{\frac{l+m+1/2}{2l+1}}. \tag{4.211}$$

Finally, we have

$$|l+1/2, m\rangle = \sqrt{\frac{l-m+1/2}{2l+1}}|l, 1/2; m+1/2, -1/2\rangle + \sqrt{\frac{l+m+1/2}{2l+1}}|l, 1/2; m-1/2, 1/2\rangle, \tag{4.212}$$

with $m = -l-1/2, \ldots, l+1/2$.

The $j = l - 1/2$ case goes very much the same way. Analogous calculations give

$$|l-1/2, m\rangle = \sqrt{\frac{l+m+1/2}{2l+1}}|l, 1/2; m+1/2, -1/2\rangle - \sqrt{\frac{l-m+1/2}{2l+1}}|l, 1/2; m-1/2, 1/2\rangle. \tag{4.213}$$

with $m = -l+1/2, \ldots, l-1/2$.

As an illustration, we consider the $l = 1$ case:

$$\begin{aligned}
|3/2, 3/2\rangle &= |1, 1/2; 1, 1/2\rangle, \\
|3/2, 1/2\rangle &= \sqrt{\frac{2}{3}}|1, 1/2; 0, 1/2\rangle + \sqrt{\frac{1}{3}}|1, 1/2; 1, -1/2\rangle, \\
|3/2, -1/2\rangle &= \sqrt{\frac{1}{3}}|1, 1/2; -1, 1/2\rangle + \sqrt{\frac{2}{3}}|1, 1/2; 0, -1/2\rangle, \\
|3/2, -3/2\rangle &= |1, 1/2; -1, -1/2\rangle, \\
|1/2, 1/2\rangle &= \sqrt{\frac{2}{3}}|1, 1/2; 1, -1/2\rangle - \sqrt{\frac{1}{3}}|1, 1/2; 0, 1/2\rangle, \\
|1/2, -1/2\rangle &= \sqrt{\frac{1}{3}}|1, 1/2; 0, -1/2\rangle - \sqrt{\frac{2}{3}}|1, 1/2; -1, 1/2\rangle.
\end{aligned} \tag{4.214}$$

Wigner $3-j$ Symbols

A widely used alternative form of coefficients for angular momentum couplings are the Wigner $3-j$ symbols. They are related to the Clebsch-Gordan coefficients by

$$\begin{pmatrix} j_1 & j_2 & j_3 \\ m_1 & m_2 & m_3 \end{pmatrix} = \frac{(-)^{j_1-j_2-m_3}}{\sqrt{2j_3+1}}\langle j_1, j_2; m_1, m_2|j_3, -m_3\rangle. \tag{4.215}$$

The Wigner $3-j$ symbols are invariant with respect to even permutations of their columns

$$\begin{pmatrix} j_1 & j_2 & j_3 \\ m_1 & m_2 & m_3 \end{pmatrix} = \begin{pmatrix} j_2 & j_3 & j_1 \\ m_2 & m_3 & m_1 \end{pmatrix} = \begin{pmatrix} j_3 & j_1 & j_2 \\ m_3 & m_1 & m_2 \end{pmatrix}. \tag{4.216}$$

Odd permutation, however, introduce some phase factors

$$\begin{pmatrix} j_1 & j_2 & j_3 \\ m_1 & m_2 & m_3 \end{pmatrix} = (-)^{j_1+j_2+j_3}\begin{pmatrix} j_2 & j_1 & j_3 \\ m_2 & m_1 & m_3 \end{pmatrix} = (-)^{j_1+j_2+j_3}\begin{pmatrix} j_1 & j_3 & j_2 \\ m_1 & m_3 & m_2 \end{pmatrix}. \tag{4.217}$$

Changing the sign of the m quantum numbers also introduces a similar phase change

$$\begin{pmatrix} j_1 & j_2 & j_3 \\ -m_1 & -m_2 & -m_3 \end{pmatrix} = (-)^{j_1+j_2+j_3}\begin{pmatrix} j_1 & j_2 & j_3 \\ m_1 & m_2 & m_3 \end{pmatrix}. \tag{4.218}$$

4.8.1 Coupling Rule for D Matrices

With the help of the Clebsch-Gordan coefficients, we can also calculate the rotation matrices corresponding to angular momentum coupled systems. The small d matrices are defined by

$$d^{(j)}_{m',m}(\beta) = \langle j, m'|\exp\left(-\frac{i}{\hbar}\beta\hat{J}_y\right)|j, m\rangle = \langle j, m'|\exp\left(-\frac{i}{\hbar}\beta(\hat{J}_{1,y} + \hat{J}_{2,y})\right)|j, m\rangle. \tag{4.219}$$

Expanding $|j, m\rangle$ in terms of uncoupled states we find

$$d^{(j)}_{m',m}(\beta) = \sum_{m_1,m_2,m'_1,m'_2} \langle j_1, j_2; m_1, m_2|j, m\rangle\langle j_1, j_2; m'_1, m'_2|j, m'\rangle \\ \times \langle j_1, m'_1| \exp\left(-\frac{i}{\hbar}\beta \hat{J}_{1,y}\right)|j_1, m_1\rangle\langle j_2, m'_2| \exp\left(-\frac{i}{\hbar}\beta \hat{J}_{2,y}\right)|j_2, m_2\rangle, \tag{4.220}$$

or

$$d^{(j)}_{m',m}(\beta) = \sum_{m_1,m_2,m'_1,m'_2} \langle j_1, j_2; m_1, m_2|j, m\rangle\langle j_1, j_2; m'_1, m'_2|j, m'\rangle d^{(j_1)}_{m'_1,m_1}(\beta)d^{(j_2)}_{m'_2,m_2}(\beta), \tag{4.221}$$

where $m' = m'_1 + m'_2$ and $m = m_1 + m_2$. Since

$$D^{(j)}_{m',m}(\alpha, \beta, \gamma) = e^{-i(m'\alpha+m\gamma)}d^{(j)}_{m',m}(\beta), \tag{4.222}$$

we get

$$D^{(j)}_{m',m}(\alpha, \beta, \gamma) = \sum_{m_1,m_2,m'_1,m'_2} \langle j_1, j_2; m_1, m_2|j, m\rangle\langle j_1, j_2; m'_1, m'_2|j, m'\rangle \\ \times D^{(j_1)}_{m'_1,m_1}(\alpha, \beta, \gamma)D^{(j_2)}_{m'_2,m_2}(\alpha, \beta, \gamma). \tag{4.223}$$

We can also derive the inverse relation. Starting from

$$\exp\left(-\frac{i}{\hbar}\beta \hat{J}_{1,y}\right)\exp\left(-\frac{i}{\hbar}\beta \hat{J}_{2,y}\right) = \exp\left(-\frac{i}{\hbar}\beta \hat{J}_y\right), \tag{4.224}$$

and sandwiching it with

$$|j_1, m_1\rangle|j_2, m_2\rangle = \sum_{j,m}\langle j_1, j_2; m_1, m_2|j, m\rangle|j, m\rangle, \tag{4.225}$$

we obtain

$$\langle j_1, m'_1| \exp\left(-\frac{i}{\hbar}\beta \hat{J}_{1,y}\right)|j_1, m_1\rangle\langle j_2, m'_2| \exp\left(-\frac{i}{\hbar}\beta \hat{J}_{2,y}\right)|j_2, m_2\rangle \\ = \sum_{j,m',m} \langle j_1, j_2; m_1, m_2|j, m\rangle\langle j_1, j_2; m'_1, m'_2|j, m'\rangle\langle j, m'| \exp\left(-\frac{i}{\hbar}\beta \hat{J}_y\right)|j, m\rangle, \tag{4.226}$$

or

$$d^{(j_1)}_{m'_1,m_1}(\beta)d^{(j_2)}_{m'_2,m_2}(\beta) = \sum_{j,m',m} \langle j_1, j_2; m_1, m_2|j, m\rangle\langle j_1, j_2; m'_1, m'_2|j, m'\rangle d^{(j)}_{m',m}(\beta), \tag{4.227}$$

and

$$D^{(j_1)}_{m'_1,m_1}(\alpha,\beta,\gamma)D^{(j_2)}_{m'_2,m_2}(\alpha,\beta,\gamma) = \sum_{j,m',m} \langle j_1, j_2; m_1, m_2|j, m\rangle\langle j_1, j_2; m'_1, m'_2|j, m'\rangle D^{(j)}_{m',m}(\alpha,\beta,\gamma). \tag{4.228}$$

These relations have important applications to the spherical harmonics. If $j_1 = l_1$ and $j_2 = l_2$ are integers and $m_1 = m_2 = m = 0$, we find

$$D^{(l_1)}_{m'_1,0}(\alpha,\beta,\gamma)D^{(l_2)}_{m'_2,0}(\alpha,\beta,\gamma) = \sum_{l,m'} \langle l_1, l_2; 0, 0|l, 0\rangle\langle l_1, l_2; m'_1, m'_2|l, m'\rangle D^{(l)}_{m',0}(\alpha,\beta,\gamma). \tag{4.229}$$

Using Eq. (4.159) we obtain, after removing the primes and taking the complex conjugate,

$$Y_{l_1,m_1}(\beta,\alpha)Y_{l_2,m_2}(\beta,\alpha) = \sum_{l,m} \sqrt{\frac{(2l_1+1)(2l_2+1)}{4\pi(2l+1)}} \langle l_1, l_2; 0, 0|l, 0\rangle\langle l_1, l_2; m_1, m_2|l, m\rangle Y_{l,m}(\beta,\alpha). \tag{4.230}$$

Multiplying this equation by $Y^*_{l,m}(\beta,\alpha)$ and integrating, we get

$$\begin{aligned}\int_0^{2\pi} d\alpha \int_0^{\pi} d\beta \, \sin\beta \, Y^*_{l,m}(\beta,\alpha)Y_{l_1,m_1}(\beta,\alpha)Y_{l_2,m_2}(\beta,\alpha) \\ = \sum_{l,m} \sqrt{\frac{(2l_1+1)(2l_2+1)}{4\pi(2l+1)}} \langle l_1, l_2; 0, 0|l, 0\rangle\langle l_1, l_2; m_1, m_2|l, m\rangle.\end{aligned} \tag{4.231}$$

Integrals of D-Functions

The integral over Euler angles is given by

$$\int d\Omega = \int_0^{\pi} d\beta \sin\beta \int_0^{2\pi} d\alpha \int_0^{2\pi} d\gamma. \tag{4.232}$$

Then, the integral of a single D function reads

$$\begin{aligned}\int D^{(j)}_{m',m}(\Omega)d\Omega &= \int_0^{\pi} d\beta d^{(j)}_{m',m}(\beta) \sin\beta \int_0^{2\pi} d\alpha \exp(-im'\alpha) \int_0^{2\pi} d\gamma \exp(-im\gamma) \\ &= 2\delta_{j,0}\, 2\pi\,\delta_{m',0}\, 2\pi\,\delta_{m,0}.\end{aligned} \tag{4.233}$$

With the help of this result, we can find, for the integral of the product,

$$\int D^{(j_1)}_{m_1,m'_1}(\Omega)D^{(j_2)}_{m_2,m'_2}(\Omega)\mathrm{d}\Omega = \sum_{j,m,m'} \langle j_1, j_2; m_1, m_2|j, m\rangle\langle j_1, j_2; m'_1, m'_2|j, m'\rangle \int D^{(j)}_{m,m'}(\Omega)\mathrm{d}\Omega$$
$$= \sum_{j,m,m'} \langle j_1, j_2; m_1, m_2|j, m\rangle\langle j_1, j_2; m'_1, m'_2|j, m'\rangle 8\pi^2\delta_{j,0}\delta_{m,0}\delta_{m',0}$$
$$= \langle j_1, j_2; m_1, m_2|0, 0\rangle\langle j_1, j_2; m'_1, m'_2|0, 0\rangle 8\pi^2. \tag{4.234}$$

The Clebsch-Gordan coefficients vanish, unless $m_2 = -m_1$ and $m'_2 = -m'_1$, and $j_1 = j_2$ (triangle condition). Then, by using the relation

$$\langle j, j; -m, m|0, 0\rangle = \frac{(-)^{j+m}}{\sqrt{2j+1}}, \tag{4.235}$$

and with a change of notation $j_1 \to j$, $j_2 \to j'$, $m_1 \to m$, $m'_1 \to n$, $m_2 \to m'$ and $m'_2 \to n'$, we find

$$\int D^{(j)}_{-m,-n}(\Omega)D^{(j')}_{m',n'}(\Omega)\mathrm{d}\Omega = 8\pi^2\frac{(-)^{2j+m+n}}{2j+1}\delta_{j,j'}\delta_{m,m'}\delta_{n,n'}. \tag{4.236}$$

Considering that

$$D^{(j)}_{-m,-n} = (-)^{2j+m+n}D^{(j)*}_{m,n}, \tag{4.237}$$

we obtain the orthogonality relation of the D-matrices

$$\int D^{(j)*}_{m,n}(\Omega)D^{(j')}_{m',n'}(\Omega)\mathrm{d}\Omega = \frac{8\pi^2}{2j+1}\delta_{j,j'}\delta_{m,m'}\delta_{n,n'}. \tag{4.238}$$

4.9 Scalar, Vector and Tensor Operators

4.9.1 Rotation of Operators

If we rotate the coordinate system, i.e. performing a passive rotation, the operators transform as

$$\hat{A}' = \hat{R}^\dagger\hat{A}\hat{R}. \tag{4.239}$$

If the infinitesimal rotation around $\boldsymbol{e}_n$ is generated by $\hat{\boldsymbol{J}}$,

$$\hat{R}_{\boldsymbol{n}}(\delta\theta) = 1 - \frac{i}{\hbar}\delta\theta\,\boldsymbol{e}_n\cdot\hat{\boldsymbol{J}}, \tag{4.240}$$

the operators are transformed as

$$\hat{A}' = \hat{A} - \frac{i}{\hbar}\delta\theta\left[\hat{A}, \boldsymbol{e}_n\cdot\hat{\boldsymbol{J}}\right]. \tag{4.241}$$

An operator is a *scalar* operator if it is invariant with respect to rotation. Then,

$$\hat{S}' = \hat{S} \quad \text{and} \quad \left[\hat{S}, \hat{J}_k\right] = 0 \quad (k = x, y, z). \tag{4.242}$$

The same relation holds for pseudo-scalar operators.

Vector operators transform according to Eq. (4.241). On the other hand, a vector $\hat{\boldsymbol{V}}$ transforms with respect to rotation as

$$\hat{\boldsymbol{V}}' = \hat{\boldsymbol{V}} + \delta\theta \, \boldsymbol{e}_n \times \hat{\boldsymbol{V}}. \tag{4.243}$$

So, we can infer that

$$\left[\hat{\boldsymbol{V}}, \boldsymbol{e}_n \cdot \hat{\boldsymbol{J}}\right] = i\hbar \, \boldsymbol{e}_n \times \hat{\boldsymbol{V}}, \tag{4.244}$$

or if we take $\boldsymbol{e}_n$ along the (x, y, z) axes, we have

$$\left[\hat{V}_x, \hat{J}_x\right] = \left[\hat{V}_y, \hat{J}_y\right] = \left[\hat{V}_z, \hat{J}_z\right] = 0, \tag{4.245}$$

$$\left[\hat{V}_x, \hat{J}_y\right] = i\hbar\hat{V}_z, \quad \left[\hat{V}_y, \hat{J}_z\right] = i\hbar\hat{V}_x, \quad \left[\hat{V}_z, \hat{J}_x\right] = i\hbar\hat{V}_y, \tag{4.246}$$

$$\left[\hat{V}_x, \hat{J}_z\right] = -i\hbar\hat{V}_y, \quad \left[\hat{V}_y, \hat{J}_x\right] = -i\hbar\hat{V}_z, \quad \left[\hat{V}_z, \hat{J}_y\right] = -i\hbar\hat{V}_x. \tag{4.247}$$

For $\hat{\boldsymbol{V}} = \hat{\boldsymbol{J}}$, we find

$$\left[\hat{J}_x, \hat{J}_x\right] = 0, \quad \left[\hat{J}_x, \hat{J}_y\right] = i\hbar\hat{J}_z, \quad \text{with} \quad x, y, z = \text{cyclic}. \tag{4.248}$$

Or, if we take $\hat{\boldsymbol{J}} = \hat{\boldsymbol{L}}$ and $\hat{\boldsymbol{V}} = \hat{\boldsymbol{r}}$, we obtain

$$\left[\hat{x}, \hat{L}_x\right] = 0, \quad \left[\hat{x}, \hat{L}_y\right] = i\hbar\hat{z}, \quad \left[\hat{x}, \hat{L}_z\right] = -i\hbar\hat{y}, \quad \text{with} \quad x, y, z = \text{cyclic}. \tag{4.249}$$

Similarly with $\hat{\boldsymbol{V}} = \hat{\boldsymbol{p}}$, we get

$$\left[\hat{p}_x, \hat{L}_x\right] = 0, \quad \left[\hat{p}_x, \hat{L}_y\right] = i\hbar\hat{p}_z, \quad \left[\hat{p}_x, \hat{L}_z\right] = -i\hbar\hat{p}_y, \quad \text{with} \quad x, y, z = \text{cyclic}. \tag{4.250}$$

Analogously to the $\hat{J}_\pm$ operators, we can define

$$\hat{V}_\pm = \hat{V}_x \pm i\hat{V}_y. \tag{4.251}$$

We can easily show that

$$\left[\hat{J}_x, \hat{V}_\pm\right] = \mp\hbar\hat{V}_z, \quad \left[\hat{J}_y, \hat{V}_\pm\right] = -i\hbar\hat{V}_z, \quad \left[\hat{J}_z, \hat{V}_\pm\right] = \pm\hbar\hat{V}_\pm \tag{4.252}$$

and

$$\left[\hat{J}_\pm, \hat{V}_\pm\right] = 0, \quad \left[\hat{J}_\pm, \hat{V}_\mp\right] = \pm 2\hbar \hat{V}_z. \tag{4.253}$$

4.9.2 Spherical Basis

It is customary to consider vectors in Cartesian or polar coordinates. However, to see their properties with respect to rotation, the spherical basis may be more useful. If we denote the unit vectors in Cartesian coordinates by $\boldsymbol{e}_x$, $\boldsymbol{e}_y$ and $\boldsymbol{e}_z$, the spherical basis is defined by

$$\begin{pmatrix} \boldsymbol{e}_1^1 \\ \boldsymbol{e}_1^{-1} \\ \boldsymbol{e}_1^0 \end{pmatrix} = U \begin{pmatrix} \boldsymbol{e}_x \\ \boldsymbol{e}_y \\ \boldsymbol{e}_z \end{pmatrix}, \quad \text{where} \quad U = \begin{pmatrix} -\frac{1}{\sqrt{2}} & -\frac{i}{\sqrt{2}} & 0 \\ \frac{1}{\sqrt{2}} & -\frac{i}{\sqrt{2}} & 0 \\ 0 & 0 & 1 \end{pmatrix}. \tag{4.254}$$

With the inverse relation we have

$$\begin{pmatrix} \boldsymbol{e}_x \\ \boldsymbol{e}_y \\ \boldsymbol{e}_z \end{pmatrix} = U^{-1} \begin{pmatrix} \boldsymbol{e}_1^1 \\ \boldsymbol{e}_1^{-1} \\ \boldsymbol{e}_1^0 \end{pmatrix}, \quad \text{where} \quad U^{-1} = \begin{pmatrix} -\frac{1}{\sqrt{2}} & \frac{1}{\sqrt{2}} & 0 \\ \frac{i}{\sqrt{2}} & \frac{i}{\sqrt{2}} & 0 \\ 0 & 0 & 1 \end{pmatrix}. \tag{4.255}$$

We can see that $U^{-1} = U^\dagger$, i.e. U is unitary. The basis vectors are orthogonal

$$\begin{aligned} \boldsymbol{e}_i \cdot \boldsymbol{e}_j &= \delta_{ij}, \quad \text{with} \quad i, j = x, y, z, \\ \boldsymbol{e}_1^\mu \cdot \boldsymbol{e}_1^\nu &= \delta_{\mu\nu}, \quad \text{with} \quad \mu, \nu = -1, 0, 1. \end{aligned} \tag{4.256}$$

We can represent any vector in the spherical basis

$$\hat{\boldsymbol{V}} = \hat{V}_x \boldsymbol{e}_x + \hat{V}_y \boldsymbol{e}_y + \hat{V}_z \boldsymbol{e}_z = -\hat{V}_1^{(1)} \boldsymbol{e}_1^{-1} - \hat{V}_{-1}^{(1)} \boldsymbol{e}_1^1 + \hat{V}_0^{(1)} \boldsymbol{e}_1^0 = \sum_{\mu=-1,0,1} (-)^\mu \hat{V}_\mu^{(1)} \boldsymbol{e}_1^{-\mu}, \tag{4.257}$$

and find that

$$\begin{pmatrix} \hat{V}_1^{(1)} \\ \hat{V}_{-1}^{(1)} \\ \hat{V}_0^{(1)} \end{pmatrix} = U \begin{pmatrix} \hat{V}_x \\ \hat{V}_y \\ \hat{V}_z \end{pmatrix}, \tag{4.258}$$

or

$$\hat{V}_1^{(1)} = -\frac{(\hat{V}_x + i\hat{V}_y)}{\sqrt{2}}, \quad \hat{V}_{-1}^{(1)} = \frac{(\hat{V}_x - i\hat{V}_y)}{\sqrt{2}}, \quad \hat{V}_0^{(1)} = \hat{V}_z. \tag{4.259}$$

In this respect,

$$\hat{J}_{\pm 1}^{(1)} = \mp \frac{1}{\sqrt{2}} \hat{J}_\pm \quad \text{and} \quad \hat{J}_0^{(1)} = \hat{J}_z. \tag{4.260}$$

Using Eqs. (4.252), (4.253) and (4.259), we can derive the commutators

$$\left[\hat{J}_z, \hat{V}_q^{(1)}\right] = \hbar q \hat{V}_q^{(1)}, \qquad (q = -1, 0, 1), \tag{4.261}$$

$$\left[\hat{J}_\pm, \hat{V}_q^{(1)}\right] = \hbar\sqrt{1(1+1) - q(q \pm 1)}\hat{V}_{q\pm 1}^{(1)}, \qquad (q = -1, 0, 1). \tag{4.262}$$

The scalar product of vectors $\boldsymbol{A}$ and $\boldsymbol{B}$ in the Cartesian and spherical basis can be written as

$$\boldsymbol{A} \cdot \boldsymbol{B} = A_x B_x + A_y B_y + A_z B_z = \sum_{\mu=-1,0,1} (-)^\mu A_\mu^{(1)} B_{-\mu}^{(1)} = -A_{-1}^{(1)} B_1^{(1)} + A_0^{(1)} B_0^{(1)} - A_1^{(1)} B_{-1}^{(1)}, \tag{4.263}$$

while for the cross product we have

$$\boldsymbol{A} \times \boldsymbol{B} = \epsilon_{ijk} A_i B_j = \sum_{\mu\nu} (-1)^{\mu+\nu} A_\mu^{(1)} B_\nu^{(1)} (\boldsymbol{e}_1^{-\mu} \times \boldsymbol{e}_1^{-\nu}). \tag{4.264}$$

4.9.3 Spherical Tensor Operators

The position vector in the spherical basis is given by

$$\boldsymbol{r} = -r_1^{(1)} \boldsymbol{e}_1^{-1} - r_{-1}^{(1)} \boldsymbol{e}_1^1 + r_0^{(1)} \boldsymbol{e}_1^0, \tag{4.265}$$

where

$$r_1^{(1)} = -\frac{(x + iy)}{\sqrt{2}}, \quad r_{-1}^{(1)} = \frac{(x - iy)}{\sqrt{2}}, \quad r_0^{(1)} = z. \tag{4.266}$$

In spherical coordinates, with

$$x = r \sin\vartheta \cos\varphi, \quad y = r \sin\vartheta \sin\varphi, \quad z = r \cos\vartheta, \tag{4.267}$$

we find

$$r_1^{(1)} = -\frac{r}{\sqrt{2}} \sin\vartheta\, e^{i\varphi}, \quad r_{-1}^{(1)} = \frac{r}{\sqrt{2}} \sin\vartheta\, e^{-i\varphi}, \quad r_0^{(1)} = r \cos\vartheta, \tag{4.268}$$

or

$$r_1^{(1)} = \sqrt{\frac{4\pi}{3}} r Y_{1,1}(\vartheta, \varphi), \quad r_{-1}^{(1)} = \sqrt{\frac{4\pi}{3}} r Y_{1,-1}(\vartheta, \varphi), \quad r_0^{(1)} = \sqrt{\frac{4\pi}{3}} r Y_{1,0}(\vartheta, \varphi). \tag{4.269}$$

We have expressed $\boldsymbol{r}$ in terms of $Y_{1,q}$ spherical harmonics, which are angular momentum eigenstates with $l = 1$, i.e. $\langle \boldsymbol{r} | 1, q \rangle = Y_{1,q}(\vartheta, \varphi)$. This representation is rather beneficial, since we know how they behave with respect to rotating the coordinate systems, i.e.

$$\hat{R}(\alpha,\beta,\gamma)Y_{1,q}\hat{R}^{\dagger}(\alpha,\beta,\gamma)=\sum_{q'}D^{(1)}_{q',q}(\alpha,\beta,\gamma)Y_{1,q'}. \tag{4.270}$$

This result can be generalized. A set of $2k+1$ operators $\{\hat{T}^{(k)}_q, q=-k,\dots,k\}$ are called *irreducible spherical tensors* of rank k, if they transform analogously to the angular momentum states $|k,q\rangle$

$$\hat{R}(\alpha,\beta,\gamma)\hat{T}^{(k)}_q\hat{R}^{\dagger}(\alpha,\beta,\gamma)=\sum_{q'}D^{(k)}_{q',q}(\alpha,\beta,\gamma)\hat{T}^{(k)}_{q'}. \tag{4.271}$$

This formula is also valid for infinitesimal rotations. Recalling the definition of the rotation matrix, we have

$$\hat{R}_{\boldsymbol{n}}(\delta)\hat{T}^{(k)}_q\hat{R}^{\dagger}_{\boldsymbol{n}}(\delta)=\sum_{q'}\langle k,q'|\hat{R}_{\boldsymbol{n}}(\delta)|k,q\rangle\,\hat{T}^{(k)}_{q'}, \tag{4.272}$$

where $\hat{R}_{\boldsymbol{n}}(\delta)=1-i\,\delta\,\boldsymbol{e}_n\cdot\hat{J}/\hbar$. Keeping only the leading terms in δ, we find

$$[\hat{n}\cdot\hat{J},\hat{T}^{(k)}_q]=\sum_{q'}\langle k,q'|\hat{n}\cdot\hat{J}|k,q\rangle\,\hat{T}^{(k)}_{q'}. \tag{4.273}$$

By applying this result for $\boldsymbol{e}_n$ in the directions corresponding to $\hat{J}_z$ and $\hat{J}_\pm=\hat{J}_x\pm i\,\hat{J}_y$, we obtain

$$[\hat{J}_\pm,\hat{T}^{(k)}_q]=\hbar\sqrt{(k\mp q)(k\pm q+1)}\,\hat{T}^{(k)}_{q\pm1}, \tag{4.274}$$

$$[\hat{J}_z,\hat{T}^{(k)}_q]=\hbar\,q\,\hat{T}^{(k)}_q. \tag{4.275}$$

These commutation relations can also be used as the definition for the spherical components of irreducible tensor operators.

Vectors in a spherical basis, as defined above in Eq. (4.259), are spherical tensors of rank $k=1$. The commutation relations of the spherical component are given above. The commutation relations with respect to other components, like the Cartesian one, can also be readily obtained. If the operator $\hat{S}$ is scalar, it is an irreducible tensor of rank $k=0$, and it commutes with the angular momentum operator $\hat{\boldsymbol{J}}$.

4.9.4 Wigner-Eckart Theorem

By taking the matrix elements of Eq. (4.275) between $|j,m\rangle$ states, we find

$$\langle j', m'| \left([\hat{J}_z, \hat{T}_q^{(k)}] - \hbar q \hat{T}_q^{(k)} \right) |j, m\rangle = (m' - m - q)\langle j', m'|\hat{T}_q^{(k)}|j, m\rangle = 0. \tag{4.276}$$

This may reminds us of a similar relation we have found before for the Clebsch-Gordan coefficients

$$(m' - m - q)\langle j', m'|j, k; m, q\rangle = 0. \tag{4.277}$$

Similarly, we may evaluate

$$\langle j', m'| \left([\hat{J}_\pm, \hat{T}_q^{(k)}] - \hbar\sqrt{(k \mp q)(k \pm q + 1)}\hat{T}_{q\pm 1}^{(k)} \right) |j, m\rangle = 0, \tag{4.278}$$

which results in

$$\begin{aligned} &\sqrt{(j' \pm m')(j' \mp m' + 1)}\langle j', m' \mp 1|\hat{T}_q^{(k)}|j, m\rangle \\ &\qquad = \sqrt{(j \mp m)(j \pm m + 1)}\langle j', m'|\hat{T}_q^{(k)}|j, m \pm 1\rangle \\ &\qquad + \sqrt{(k \mp q)(k \pm q + 1)}\langle j', m'|\hat{T}_{q\pm 1}^{(k)}|j, m\rangle. \end{aligned} \tag{4.279}$$

This equation has a structure similar to the recursion relations we obtained for the Clebsch-Gordan coefficients. If, in Eqs. (4.194) and (4.195), we perform the substitutions $j \to j', m \to m', j_1 \to j, m_1 \to m, j_2 \to k, m_2 \to q$, we find

$$\begin{aligned} &\sqrt{(j' \pm m')(j' \mp m' + 1)}\langle j', m' \mp 1|j, k; m, q\rangle \\ &\qquad = \sqrt{(j \mp m)(j \pm m + 1)}\langle j', m'|j, k; m \pm 1, q\rangle \\ &\qquad + \sqrt{(k \mp q)(k \pm q + 1)}\langle j', m'|j, k; m, q \pm 1\rangle. \end{aligned} \tag{4.280}$$

These relations tell us that $\langle j', m'|\hat{T}_q^{(k)}|j, m\rangle$ follows the same dependence on m', m, and q as the Clebsch-Gordan coefficients $\langle j, k; m, q|j', m'\rangle$. Therefore, they have to be linearly related

$$\langle j', m'|\hat{T}_q^{(k)}|j, m\rangle = \langle j, k; m, q|j', m'\rangle\langle j'||\hat{T}^{(k)}||j\rangle. \tag{4.281}$$

This is the *Wigner-Eckart theorem*. The factor $\langle j'||\hat{T}^{(k)}||j\rangle$ is called the *reduced matrix element*, and depends only on j', j and k. In $\langle j', m'|\hat{T}_q^{(k)}|j, m\rangle$, the dependence on m', m and q is taken care of by the Clebsch-Gordan coefficients.

Scalar Operators

For scalar operators $k = q = 0$. Therefore,

$$\langle j', m'|\hat{S}|j, m\rangle = \langle j, 0; m, 0|j', m'\rangle\langle j'||\hat{S}||j\rangle = \delta_{j,j'}\delta_{m,m'}\langle j'||\hat{S}||j\rangle. \tag{4.282}$$

Scalar operators can only connect states with the same angular momentum.

Vector Operators

As we pointed out before, the vectors in a spherical basis are spherical tensors of rank $k = 1$

$$\hat{T}_q^{(1)} = \hat{V}_q^{(1)}. \tag{4.283}$$

Therefore, according to the Wigner-Eckart theorem, we have

$$\langle j', m'|\hat{V}_q^{(1)}|j, m\rangle = \langle j, 1; m, q|j', m'\rangle\langle j'||\hat{V}||j\rangle. \tag{4.284}$$

The triangular inequality relation of the Clebsch-Gordan coefficients dictates that tensor operators with rank $k = 1$ can only connect states with $j' = j$ or $j' = j \pm 1$, but cannot connect $j' = j = 0$ states. So, a spin-0 particle with $j = 0$ cannot have a dipole moment, which is a vector operator with $k = 1$. Similarly, a spin-$1/2$ particle cannot have a quadrupole moment, which is a tensor operator of rank $k = 2$.

The angular momentum operator is itself a vector operator, therefore

$$\langle j', m'|\hat{J}_q^{(1)}|j, m\rangle = \langle j, 1; m, q|j', m'\rangle\langle j'||\hat{J}||j\rangle. \tag{4.285}$$

In particular, for $\hat{J}_0^{(1)} = \hat{J}_z$ we have

$$\langle j', m'|\hat{J}_0^{(1)}|j, m\rangle = \delta_{j,j'}\delta_{m,m'}\hbar m = \langle j, 1; m, 0|j, m\rangle\langle j||\hat{J}||j\rangle. \tag{4.286}$$

Taking $m = j$ and considering that $\langle j, 1; j, 0|j, j\rangle = \sqrt{j/(j+1)}$, we obtain for the reduced matrix element of the angular momentum operator

$$\langle j||\hat{J}||j\rangle = \hbar\sqrt{j(j+1)}. \tag{4.287}$$

Let us compare the vector $\hat{\boldsymbol{V}}$ and the angular momentum vector $\hat{\boldsymbol{J}}$. Both are vector operators of $k = 1$, therefore

$$\frac{\langle j, m'|\hat{V}_q^{(1)}|j, m\rangle}{\langle j, m'|\hat{J}_q^{(1)}|j, m\rangle} = \frac{\langle j||\hat{V}||j\rangle}{\langle j||\hat{J}||j\rangle} = \frac{\langle j||\hat{V}||j\rangle}{\hbar\sqrt{j(j+1)}}. \tag{4.288}$$

We can write the scalar product in terms of spherical components

$$\langle j, m|\hat{\boldsymbol{J}} \cdot \hat{\boldsymbol{V}}|j, m\rangle = \langle j, m|(\hat{J}_0^{(1)}\hat{V}_0^{(1)} - \hat{J}_{+1}^{(1)}\hat{V}_{-1}^{(1)} - \hat{J}_{-1}^{(1)}\hat{V}_{+1}^{(1)})|j, m\rangle. \tag{4.289}$$

Evaluating $\langle j, m|\hat{J}_q^{(1)}$, we find that $\langle j, m|\hat{\boldsymbol{J}} \cdot \hat{\boldsymbol{V}}|j, m\rangle$ is a linear combination of $\langle j, m'|\hat{V}_q^{(1)}|j, m\rangle$ with $m' = m$, or $m' = m \pm 1$, and $q = -1, 0, 1$. These, according to the Wigner-Eckart theorem, are proportional to the reduced matrix elements $\langle j||\hat{V}||j\rangle$. Consequently,

$$\langle j, m|\hat{\boldsymbol{J}} \cdot \hat{\boldsymbol{V}}|j, m\rangle = C_j\,\langle j||\hat{V}||j\rangle. \tag{4.290}$$

To evaluate C_j, let us take $\hat{\boldsymbol{V}} = \hat{\boldsymbol{J}}$

$$\langle j, m|\hat{J}^2|j, m\rangle = \hbar^2 j(j+1) = C_j \langle j||\hat{J}||j\rangle = C_j \hbar\sqrt{j(j+1)}, \tag{4.291}$$

which gives

$$C_j = \hbar\sqrt{j(j+1)}. \tag{4.292}$$

So, we have found a rather simple relation for the reduced matrix element of $\hat{\boldsymbol{V}}$

$$\langle j, m|\hat{\boldsymbol{J}} \cdot \hat{\boldsymbol{V}}|j, m\rangle = \hbar\sqrt{j(j+1)}\, \langle j||\hat{\boldsymbol{V}}||j\rangle. \tag{4.293}$$

Substituting back into Eq. (4.288), we obtain

$$\langle j, m'|\hat{V}_q^{(1)}|j, m\rangle = \frac{\langle j, m|\hat{\boldsymbol{J}} \cdot \hat{\boldsymbol{V}}|j, m\rangle}{\hbar^2 j(j+1)} \langle j, m'|\hat{J}_q^{(1)}|j, m\rangle. \tag{4.294}$$

We have expressed the matrix elements of a vector operator $\hat{\boldsymbol{V}}$ in terms of the matrix elements of the scalar operator $\hat{\boldsymbol{J}} \cdot \hat{\boldsymbol{V}}$, which can be considerably simpler.

As an application, let us evaluate the matrix elements of the spin $\hat{S}$ operator in the $|j, m\rangle$ basis. According to the formula above,

$$\langle j, m|\hat{S}_z|j, m\rangle = \frac{\langle j, m|\hat{J} \cdot \hat{S}|j, m\rangle}{\hbar^2 j(j+1)} \langle j, m|\hat{J}_z|j, m\rangle. \tag{4.295}$$

Since

$$\hat{J} \cdot \hat{S} = (\hat{L} + \hat{S}) \cdot \hat{S} = \hat{L} \cdot \hat{S} + \hat{S}^2 = \frac{(\hat{L} + \hat{S})^2 - \hat{L}^2 - \hat{S}^2}{2} + \hat{S}^2 = \frac{\hat{J}^2 - \hat{L}^2 + \hat{S}^2}{2}, \tag{4.296}$$

and $|j, m\rangle$ is a common eigenstate of $\hat{J}^2$, $\hat{L}^2$, $\hat{S}^2$ and $\hat{J}_z$, we obtain

$$\langle j, m|\hat{S}_z|j, m\rangle = \frac{j(j+1) - l(l+1) + s(s+1)}{2j(j+1)} \hbar m. \tag{4.297}$$

4.10 Discrete Symmetries and Angular Momentum

Parity

A *parity transformation* changes $\boldsymbol{r}$ to $-\boldsymbol{r}$, which, in spherical coordinates, means the transformations $\varphi \to \varphi + \pi$ and $\vartheta \to \pi - \vartheta$. From Eq. (4.109), we can see that

$$P_l^m(\cos(\pi - \vartheta)) = P_l^m(-\cos(\vartheta)) = (-)^{l+m} P_l^m(\cos(\vartheta)). \tag{4.298}$$

Hence, from Eq. (4.112), we obtain

$$\begin{aligned}\hat{\mathcal{P}}\langle \mathbf{r}|l,m\rangle &= \hat{\mathcal{P}}Y_{l,m}(\vartheta,\varphi) = Y_{l,m}(\pi-\vartheta,\varphi+\pi) = e^{im\pi}(-)^{m+l}Y_{l,m}(\vartheta,\varphi)\\ &= (-)^l Y_{l,m}(\vartheta,\varphi) = (-)^l\langle \mathbf{r}|l,m\rangle,\end{aligned} \tag{4.299}$$

i.e. the orbital angular momentum states $|l, m\rangle$ are eigenstates of the parity operator with eigenvalues $(-)^l$.

This result does not hold in general. For example, if the total angular momentum is the result of two coupled orbital angular momenta

$$|l_1,l_2;L,M\rangle = \sum_{m_1,m_2}\langle l_1,l_2;m_1,m_2|L,M\rangle|l_1,m_1\rangle|l_2,m_2\rangle, \tag{4.300}$$

then the parity transformation yields

$$\hat{\mathcal{P}}|l_1,l_2;L,M\rangle = (-)^{l_1+l_2}|l_1,l_2;L,M\rangle. \tag{4.301}$$

Since, in general, $(-)^{l_1+l_2} \neq (-)^L$, the parity of an angular momentum state is not determined by the total angular momentum. However, if the total parity $\Pi = (-)^L$, we say that the system has natural parity. If $\Pi = (-)^{L+1}$, the system has un-natural parity. For example, if $L = 1$ comes from the coupling of $l_1 = 1$ and $l_2 = 0$ or $l_1 = 0$ and $l_2 = 1$, the system has natural parity. If $L = 1$ is from the coupling of $l_1 = 1$ and $l_2 = 1$, it has un-natural parity.

Given that the parity operator $\hat{\mathcal{P}}$ commutes with the orbital angular momentum operator, it should commute with spin operator as well, i.e. $\left[\hat{\mathcal{P}}, \hat{\mathbf{S}}\right] = 0$. So, just like the orbital angular momentum, the spin states are eigenstates of $\hat{\mathcal{P}}$

$$\hat{\mathcal{P}}|s,m_s\rangle = \eta_s|s,m_s\rangle. \tag{4.302}$$

The spin is an internal quantum number, not related to any spatial coordinate, so the η_s is the intrinsic parity. Its value is determined by the internal structure of the elementary particle. By convention, the electron, the neutron, and the proton have $+1$ intrinsic parity. The intrinsic parity of other particles is assigned from parity conservation. The total parity Π of an N particle system is given by

$$\Pi = \prod_i^N (-)^{l_i}\eta_i, \tag{4.303}$$

where l_i are the relative orbital angular momenta and η_i are the internal parities of the constituents.

If the Hamiltonian is parity invariant, the eigenstates have definite parity. On the contrary, if the parity symmetry were violated, then the stationary states in an atom or in a nucleus would not possess any definite parity. Rather, they would be combinations of states with opposite parities. Experiments have shown that in electromagnetic and strong processes, transitions take place only between definite parity states and the parity is conserved. This means that if we describe the system by a Hamiltonian

$$\hat{H} = \alpha \hat{H}_s + \beta \hat{H}_p, \tag{4.304}$$

where $\hat{H}_s$ and $\hat{H}_p$ are the scalar and pseudo-scalar components, respectively, the strength of the pseudo-scalar component β is extremely small compared to the strength of the scalar component α.

However, in processes governed by the weak interactions this is not the case. Lee and Yang showed that previous experiments were sensitive only to the combination of $|\alpha|^2 + |\beta|^2$, thus they could not rule out or confirm possible parity violating pseudo-scalar terms. They proposed experiments that were sensitive to spin-momentum correlations, since these are proportional to pseudo-scalar $\hat{\boldsymbol{p}} \cdot \hat{\boldsymbol{S}}$ terms. Experimental findings confirmed that the parity is not conserved by weak interactions. Consequently in processes governed by the weak interaction, the intrinsic parity loses its meaning for leptons, like the electron, muon, tau and the corresponding neutrinos.

Time Reversal

We have found before that the time reversal, $t \to -t$, involves the transformations

$$\boldsymbol{r} \to \boldsymbol{r} \quad \text{and} \quad \boldsymbol{p} \to -\boldsymbol{p}. \tag{4.305}$$

We also found that the time reversal operator contains a complex conjugation $\hat{K}$, together with a possible unitary transformation $\hat{U}$,

$$\hat{\mathcal{T}} = \hat{U}\hat{K}. \tag{4.306}$$

It transforms $\hat{\boldsymbol{J}}$ as

$$\hat{\mathcal{T}}\hat{\boldsymbol{J}}\hat{\mathcal{T}}^{-1} = -\hat{\boldsymbol{J}}, \tag{4.307}$$

i.e.

$$\hat{J}'_x = \hat{\mathcal{T}}\hat{J}_x\hat{\mathcal{T}}^{-1} = -\hat{J}_x \quad \hat{J}'_y = \hat{\mathcal{T}}\hat{J}_y\hat{\mathcal{T}}^{-1} = -\hat{J}_y \quad \hat{J}'_z = \hat{\mathcal{T}}\hat{J}_z\hat{\mathcal{T}}^{-1} = -\hat{J}_z. \tag{4.308}$$

Consequently,

$$\hat{J}'_\pm = \hat{\mathcal{T}}\hat{J}_\pm\hat{\mathcal{T}}^{-1} = -\hat{J}_\mp, \tag{4.309}$$

where the sign reversal is due to complex conjugation. The operators $\hat{J}'_i$ are legitimate angular momentum operators and they commute with $\hat{\boldsymbol{J}}'^2 = \hat{\boldsymbol{J}}^2$. The common eigenstates of $\hat{\boldsymbol{J}}'^2$ and $\hat{J}'_z = -\hat{J}_z$ are the states $\hat{\mathcal{T}}|j, m\rangle$, so they should remain in the subspace spanned by the eigenstates of $\hat{\boldsymbol{J}}^2$ and $\hat{J}_z$. Therefore, the time reversal should involve a $m \to -m$ transformation, plus some phase factor. If we choose the phase factor

$$\hat{\mathcal{T}}|j, m\rangle = (-)^{j-m}|j, -m\rangle, \tag{4.310}$$

then if the individual components have this time reversal symmetry, the combined angular momentum will have the same symmetry.

Applying the time reversal twice on angular momentum eigenstates, we find

$$\hat{\mathcal{T}}^2|j, m\rangle = \hat{\mathcal{T}}(\hat{\mathcal{T}}|j, m\rangle) = \hat{\mathcal{T}}[(-)^{j-m}|j, -m\rangle] = (-)^{j+m}(-)^{j-m}|j, m\rangle = (-)^{2j}|j, m\rangle, \tag{4.311}$$

which gives $|j, m\rangle$, if j is an integer, and $-|j, m\rangle$, if j is a half integer.

Now, we can explicitly construct the time reversal operator. In coordinate representation the complex conjugation

$$\hat{\mathcal{T}} = \hat{K} \tag{4.312}$$

does the job. It transforms $\hat{\boldsymbol{r}} \to \hat{\boldsymbol{r}}$, $\hat{\boldsymbol{p}} \to -\hat{\boldsymbol{p}}$, and consequently the orbital angular momentum transform as $\hat{\boldsymbol{L}} \to -\hat{\boldsymbol{L}}$. In the formalism we adopted, the spin angular momentum operators $\hat{S}_z$ and $\hat{S}_x$ are represented by real matrices, while the representation of $\hat{S}_y$ is purely imaginary. Since complex conjugation acts as $\hat{K}^{-1} = \hat{K}$, we have

$$\hat{K}\hat{S}_x\hat{K} = \hat{S}_x, \quad \hat{K}\hat{S}_y\hat{K} = -\hat{S}_y, \quad \hat{K}\hat{S}_z\hat{K} = \hat{S}_z. \tag{4.313}$$

So, the definition in Eq. (4.312) has to be augmented by a linear operator,

$$\hat{\mathcal{T}} = \hat{U}\hat{K}, \tag{4.314}$$

that leaves $\hat{\boldsymbol{r}}$ and $\hat{\boldsymbol{p}}$ unchanged,

$$\hat{U}\hat{\boldsymbol{r}}\hat{U}^{-1} = \hat{\boldsymbol{r}} \quad \text{and} \quad \hat{U}\hat{\boldsymbol{p}}\hat{U}^{-1} = \hat{\boldsymbol{p}}, \tag{4.315}$$

and transforms the spin accordingly

$$\hat{U}\hat{S}_x\hat{U}^{-1} = -\hat{S}_x, \quad \hat{U}\hat{S}_y\hat{U}^{-1} = \hat{S}_y, \quad \hat{U}\hat{S}_z\hat{U}^{-1} = -\hat{S}_z. \tag{4.316}$$

Apparently, this goal can be achieved by rotating the spin around the $\boldsymbol{e}_y$ axis by π. Then, we have

$$\hat{\mathcal{T}} = \exp(-i\pi\hat{S}_y/\hbar)\hat{K}. \tag{4.317}$$

For spin-1/2 particles, $\hat{S}_y = \hbar/2\,\sigma_y$, and from Eq. (4.127) we have

$$\hat{\mathcal{T}} = -i\hat{\sigma}_y\hat{K}. \tag{4.318}$$

If we have n particles, the time reversal becomes

$$\hat{\mathcal{T}} = \exp(-i\pi \hat{S}_{1y}/\hbar)\exp(-i\pi \hat{S}_{2y}/\hbar)\cdots\exp(-i\pi \hat{S}_{ny}/\hbar)\hat{K}, \tag{4.319}$$

where $\hat{S}_{iy}$ is the y component of the i-th particle angular momentum. A repeated application of the time reversal operator gives

$$\begin{aligned}\hat{\mathcal{T}}^2 &= \exp(-i\pi \hat{S}_y/\hbar)\hat{K}\exp(-i\pi \hat{S}_y/\hbar)\hat{K} = \exp(-i\pi \hat{S}_y/\hbar)\exp(+i\pi(-\hat{S}_y)/\hbar)\\ &= \exp(-i2\pi \hat{S}_y/\hbar),\end{aligned}$$

which is equivalent to

$$\hat{\mathcal{T}}^2 = \exp(-i2\pi \hat{J}_y/\hbar), \tag{4.320}$$

since $\hat{J}_y = \hat{L}_y + \hat{S}_y$ and $\exp(-i2\pi\hat{L}_y/\hbar) = 1$. So, the $\hat{\mathcal{T}}^2$ is equivalent to a full revolution around the y axis,

$$\hat{\mathcal{T}}^2 = \exp(-i2\pi \hat{J}_y/\hbar) = R_y(2\pi). \tag{4.321}$$

Consequently, $\hat{\mathcal{T}}^2$ has eigenvalue $+1$ for integer total angular momentum, and -1 for half integer total angular momentum.

If the Hamiltonian is invariant with respect to time reversal, then it commutes with $\hat{\mathcal{T}}$ and $\hat{\mathcal{T}}\hat{H} = \hat{H}\hat{\mathcal{T}}$. Therefore, $\hat{H}\hat{\mathcal{T}}|\psi\rangle = \hat{\mathcal{T}}\hat{H}|\psi\rangle = \hat{\mathcal{T}}E|\psi\rangle = E\hat{\mathcal{T}}|\psi\rangle$. This means that if $|\psi\rangle$ is an eigenstate with eigenvalue E, so is $\hat{\mathcal{T}}|\psi\rangle$, with the same energy. Repeated application of time reversal leaves the physical situation unchanged, so $\hat{\mathcal{T}}^2|\psi\rangle = \pm|\psi\rangle$. Then, if $|\psi\rangle$ and $\hat{\mathcal{T}}|\psi\rangle$ are linearly dependent, $\hat{\mathcal{T}}|\psi\rangle = a|\psi\rangle$. A repeated application of time reversal gives $\hat{\mathcal{T}}^2|\psi\rangle = \hat{\mathcal{T}}a|\psi\rangle = a^*\hat{\mathcal{T}}|\psi\rangle = a^*a|\psi\rangle = |a|^2|\psi\rangle = |\psi\rangle$. Here we used that fact that the states are normalized and the absolute value squared of a complex number cannot be negative, $|a|^2 \neq -1$. As a result, $|\psi\rangle$ and $\mathcal{T}|\psi\rangle$ describe the same quantum state. However, if $\hat{\mathcal{T}}^2|\psi\rangle = -|\psi\rangle$, the states $|\psi\rangle$ and $\hat{\mathcal{T}}|\psi\rangle$ are linearly independent. They are different states with the same energy. So, systems with time reversal invariance and with an odd number of spin-$1/2$ particles have degenerate energy levels. This is *Kramer's theorem*. Kramer derived this relation using relativistic quantum mechanical equations, but Wigner pointed out that this is a consequence of the time reversal symmetry.

Chapter 5
Quantum Problems in Higher Dimensions

5.1 Central Potentials in Three Dimensions

Here we assume that the potential is central, i.e. it depends only on the relative distance from the center

$$\hat{V}(\boldsymbol{r}) = \hat{V}(r). \tag{5.1}$$

The Hamiltonian is given by

$$\hat{H} = \frac{1}{2m}\hat{\boldsymbol{p}}^2 + \hat{V}(r), \tag{5.2}$$

where for $\hat{\boldsymbol{p}}^2$ we have found before in Eq. (4.96)

$$\hat{\boldsymbol{p}}^2 = -\hbar^2 \frac{1}{r}\frac{\partial^2}{\partial r^2} r + \frac{1}{r^2}\hat{\boldsymbol{L}}^2. \tag{5.3}$$

We can define the first term as $\hat{p}_r^2$ and can easily establish analogous relations

$$\hat{p}_r^2 = -\hbar^2 \frac{1}{r}\frac{\partial^2}{\partial r^2} r = -\hbar^2 \left(\frac{1}{r}\frac{\partial}{\partial r} r\right)^2 = -\hbar^2 \left(\frac{\partial^2}{\partial r^2} + \frac{2}{r}\frac{\partial}{\partial r}\right) = -\hbar^2 \left(\frac{1}{r}\frac{\partial}{\partial r} r \frac{\partial}{\partial r} + \frac{1}{r}\frac{\partial}{\partial r}\right). \tag{5.4}$$

We can see that $\hat{p}_r^2$ comes as the square of the radial momentum operator

$$\hat{p}_r = -i\hbar \frac{1}{r}\frac{\partial}{\partial r} r = -i\hbar \left(\frac{\partial}{\partial r} + \frac{1}{r}\right). \tag{5.5}$$

It can easily be shown that this operator is Hermitian on the Hilbert space of square integrable functions of $r \in [0, \infty)$.

Z. Papp, *Mastering Quantum Mechanics*,
https://doi.org/10.1007/978-3-032-09011-9_5

We should notice that $\hat{\boldsymbol{L}}^2$ contains only derivatives in ϑ and φ. So $\hat{\boldsymbol{p}}^2$, appart from $\hat{\boldsymbol{L}}^2$, contains only derivatives with respect to r. Therefore, $\hat{\boldsymbol{L}}^2$ commutes with $\hat{\boldsymbol{p}}^2$ and $\hat{V}(r)$, and consequently with the total Hamiltonian as well

$$\left[\hat{H},\hat{\boldsymbol{L}}^2\right]=0,\quad \left[\hat{H},\hat{L}_z\right]=0,\quad \left[\hat{\boldsymbol{L}}^2,\hat{L}_z\right]=0. \tag{5.6}$$

So, $\{\hat{H},\hat{\boldsymbol{L}}^2,\hat{L}_z\}$ form a complete set of commuting observables and we can quantify the states as their common eigenstates. Since in the Hamiltonian the radial and the angular variables are separated, we can seek the solution in the form

$$\Psi_{Elm}(\boldsymbol{r})=\frac{u_l(r)}{r}Y_{l,m}(\vartheta,\varphi). \tag{5.7}$$

Then, the Schrödinger equation becomes

$$\left(-\frac{\hbar^2}{2m}\frac{1}{r}\frac{\partial^2}{\partial r^2}r+\frac{1}{2mr^2}\hat{\boldsymbol{L}}^2+V(r)\right)\frac{u_l(r)}{r}Y_{l,m}(\vartheta,\varphi)=E\,\frac{u_l(r)}{r}Y_{l,m}(\vartheta,\varphi), \tag{5.8}$$

which results in a radial equation for $u_l(r)$

$$\left(-\frac{\hbar^2}{2m}\frac{\mathrm{d}^2}{\mathrm{d}r^2}+\frac{\hbar^2 l(l+1)}{2mr^2}+V(r)\right)u_l(r)=Eu_l(r), \tag{5.9}$$

with normalization

$$\int_0^\infty \mathrm{d}r\,|u_l(r)|^2=1. \tag{5.10}$$

So, we turned the three-dimensional problem with spherical potential to a one-dimensional problem on the half line, $r\in[0,\infty)$.

5.1.1 Central Potentials in d Dimensions

The above results can easily be generalized to $2\le d$ dimensions. The radial Hamiltonian is given by

$$\hat{H}=-\frac{\hbar^2}{2m}\left(\frac{\partial^2}{\partial r^2}+\frac{d-1}{r}\frac{\partial}{\partial r}-\frac{l(l+d-2)}{r^2}\right)+V(r), \tag{5.11}$$

where l is the angular momentum and the normalization of the radial wave function reads

$$1=\langle\psi|\psi\rangle=\int_0^\infty \mathrm{d}r\,r^{d-1}|\psi(r)|^2. \tag{5.12}$$

Similarly to Eq. (5.7), we can write

$$\psi(r) = r^{-(d-1)/2} u(r), \tag{5.13}$$

which turns the normalization into

$$1 = \langle u|u \rangle = \int_0^\infty \mathrm{d}r\, |u(r)|^2. \tag{5.14}$$

Then, for the radial function $u(r)$ we obtain the differential equation

$$\left[-\frac{\hbar^2}{2m}\left(\frac{\partial^2}{\partial r^2} - \frac{L(L+1)}{r^2}\right) + V(r)\right] u_L(r) = E u_L(r), \tag{5.15}$$

with $L = l + (d-3)/2$.

We can see that for $d = 3$ we recover the results of the previous section. We can also see that the d-dimensional results, the energy, the radial wave function and the matrix elemets, can be obtained from the corresponding three dimensional results by performing the $l \to L$ mapping.

5.1.2 3-Dimensional Isotropic Harmonic Oscillator

The potential of an isotropic harmonic oscillator is given by

$$V(r) = \frac{1}{2} m\omega^2 r^2, \tag{5.16}$$

which leads to the differential equation for $u_l(r)$

$$\left(-\frac{\hbar^2}{2m}\frac{\mathrm{d}^2}{\mathrm{d}r^2} + \frac{\hbar^2 l(l+1)}{2mr^2} + \frac{1}{2} m\omega^2 r^2\right) u_l(r) = E u_l(r). \tag{5.17}$$

Introducing the variables

$$\rho = \sqrt{\frac{m\omega}{\hbar}}\, r \quad \text{and} \quad \lambda = \frac{2E}{\hbar\omega}, \tag{5.18}$$

the differential equation turns into

$$\left[\frac{\mathrm{d}^2}{\mathrm{d}\rho^2} - \frac{l(l+1)}{\rho^2} + \lambda - \rho^2\right] u_l(\rho) = 0. \tag{5.19}$$

We solve this differential equation using the method known as Fuchs method, or Frobenius-Sommerfeld method. In this approach, we locate the possible singular points of the differential equation and find the physically acceptable local approximate solutions. In the case of the isotropic harmonic oscillator the singular points are at $\rho = 0$ and at $\rho = \infty$.

In the $\rho \to 0$ limit, the centrifugal $1/\rho^2$ term is dominant over both the ρ^2 and the λ terms, so the differential equation takes the form

$$-\frac{\mathrm{d}^2 u_l(\rho)}{\mathrm{d}\rho^2} + \frac{l(l+1)}{\rho^2} u_l(\rho) = 0. \tag{5.20}$$

We may try to seek the solution as $u \sim \rho^\alpha$ and find

$$-\alpha(\alpha - 1)\rho^{\alpha-2} + l(l+1)\rho^{\alpha-2} = 0. \tag{5.21}$$

This gives two solutions, $\alpha = l + 1$ and $\alpha = -l$, but only the $\alpha = l + 1$ solution results in a square integrable wave function.

In the $\rho \to \infty$ limit, the potential term dominates over the centrifugal and energy terms, so we have the asymptotic form of differential equation

$$\left(\frac{\mathrm{d}^2}{\mathrm{d}\rho^2} - \rho^2\right) u_l(\rho) = 0. \tag{5.22}$$

With the ansatz

$$u_l(\rho) = e^{-\rho^2/2} \tag{5.23}$$

we find

$$\left(\frac{\mathrm{d}^2}{\mathrm{d}\rho^2} - \rho^2\right) e^{-\rho^2/2} = -e^{-\rho^2/2} \simeq 0, \quad \text{as} \quad \rho \to \infty. \tag{5.24}$$

Combining these findings, we can write

$$u_l(\rho) = f(\rho)\,\rho^{l+1} e^{-\rho^2/2}, \tag{5.25}$$

which results in the differential equation for $f(\rho)$

$$\frac{\mathrm{d}^2 f(\rho)}{\mathrm{d}\rho^2} + 2\left(\frac{l+1}{\rho} - \rho\right)\frac{\mathrm{d}f(\rho)}{\mathrm{d}\rho} + (\lambda - (2l+3))\, f(\rho) = 0. \tag{5.26}$$

The function $f(\rho)$ should be such that it does not modify the required asymptotic behavior of u_l. This means that $\lim_{\rho \to 0} f(\rho) = a_0 = \text{constant}$. Consequently, $f(\rho)$ must have a Taylor series

$$f(\rho) = \sum_{n=0} a_n \rho^n = a_0 + a_1\rho + a_2\rho^2 + \cdots + a_n\rho^n + \cdots . \tag{5.27}$$

Then, plugging this Taylor series into the differential equation of Eq. (5.26) we obtain

$$\sum_{n=0} \left\{ n(n-1)a_n\rho^{n-2} + 2\left(\frac{l+1}{\rho} - \rho\right) na_n\rho^{n-1} + (\lambda - (2l+3))\, a_n\rho^n \right\} = 0, \tag{5.28}$$

or

$$\sum_{n=0} \left\{ n(n+2l+1)a_n\rho^{n-2} + (-2n+\lambda-(2l+3))\, a_n\rho^n \right\} = 0. \tag{5.29}$$

This relation is valid for all values of ρ. Therefore, the coefficients of all powers of ρ should vanish. The lowest one is with ρ^{-2}, which corresponds to $n = 0$

$$0(0+2l+1)a_0 = 0, \tag{5.30}$$

which is satisfied for any $a_0 \neq 0$ value. The next power is ρ^{-1} with $n = 1$

$$1(1+2l+1)a_1 = 0, \tag{5.31}$$

which is satisfied only with $a_1 = 0$. From the vanishing of the coefficients of ρ^n, we find

$$(n+2)(n+2l+3)a_{n+2} = -\left(\lambda - (2n+2l+3)\right) a_n. \tag{5.32}$$

This is a two-term recursion relation. We can see that if $a_1 = 0$, so is a_3 and all the a's with odd n indices. So, the Taylor series in Eq. (5.27) has only even powers, therefore it is a power series in ρ^2.

We can see that for large n, the ratio behaves as

$$\lim_{n\to\infty} \frac{a_{n+2}}{a_n} \sim \frac{2}{n}. \tag{5.33}$$

On the other hand, from the Taylor series

$$e^{\rho^2} = \sum_{n=0}^{\infty} \frac{\rho^{2n}}{n!}, \tag{5.34}$$

we find the ratio of two consecutive coefficients,

$$\frac{\rho^{2n+2}/(n+1)!}{\rho^{2n}/n!} = \frac{\rho^2}{n+1} \sim \frac{\rho^2}{n}. \tag{5.35}$$

So, our power series in Eq. (5.27), if n tends to infinity, represents a function that behaves like $\exp(2\rho^2)$. This, however, would destroy the required asymptotic behavior of $u_l(\rho)$. The way out is that we do not let n tend to infinity. There must be an n' such that $a_{n'} \neq 0$, but $a_{n'+2} = a_{n'+4} = a_{n'+6} \ldots = 0$. So,

$$0 = \left(\lambda - (2n' + 2l + 3)\right) a_{n'}, \tag{5.36}$$

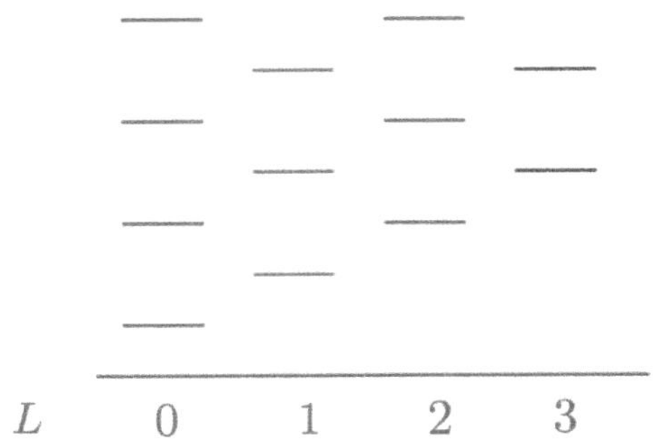

Fig. 5.1 The lowest few energy levels of the three-dimensional harmonic oscillator for $l = 0, 1, 2, 3$

and thus

$$\lambda = 2n' + 2l + 3, \tag{5.37}$$

which gives the energy eigenvalue

$$E = \hbar\omega\,(n' + l + 3/2). \tag{5.38}$$

Considering that n' is an even integer, we may write

$$E = \hbar\omega\,(2N + l + 3/2), \quad \text{with} \quad N = 0, 1, 2, \ldots . \tag{5.39}$$

Figure 5.1 shows the energy levels of the three-dimensional harmonic oscillator with different angular momenta. We can observe the equidistant level spacing and the possible degeneracies of equal parity states.

Plugging λ of Eq. (5.37) into Eq. (5.26) and introducing the new variable $\zeta = \rho^2$, we find

$$\zeta \frac{\mathrm{d}^2 f(\zeta)}{\mathrm{d}\zeta^2} + \left(l + \frac{1}{2} + 1 - \zeta\right) \frac{\mathrm{d}f(\zeta)}{\mathrm{d}\zeta} + N\, f(\zeta) = 0. \tag{5.40}$$

Comparing this result with the differential equation

$$xy'' + (\alpha + 1 - x)y' + ny = 0, \tag{5.41}$$

whose solutions are the generalized Laguerre polynomials $L_n^{(\alpha)}$, we conclude that

$$f(\rho^2) = L_N^{(l+1/2)}(\rho^2). \tag{5.42}$$

We have found that

$$u_{Nl}(\rho) \sim \rho^{l+1} e^{-\rho^2/2} L_N^{(l+1/2)}(\rho^2), \tag{5.43}$$

and the total normalized radial wave function reads

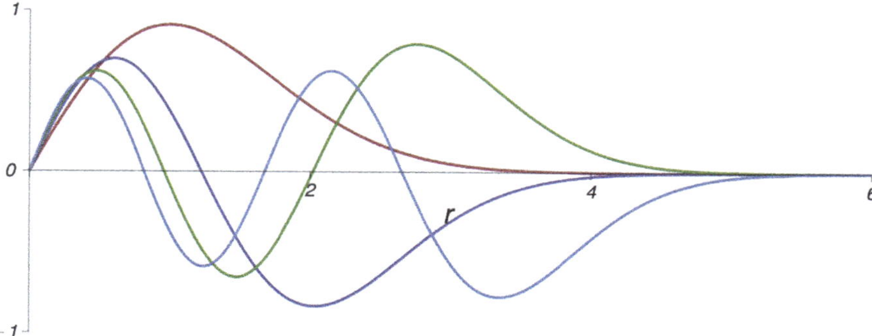

Fig. 5.2 The first four three-dimensional harmonic oscillator radial wave functions $u_{Nl}(r)$ with $\sqrt{m\omega/\hbar} = 1$, $l = 0$ and $N = 0, 1, 2, 3$. We can observe the same kind of Gaussian asymptotic behavior

$$u_{Nl}(r) = \left(\sqrt{\frac{m\omega}{\hbar}} \frac{2\Gamma(N+1)}{\Gamma(N+l+3/2)}\right)^{1/2} \rho^{l+1} \exp(-\rho^2/2)\, L_N^{(l+1/2)}(\rho^2). \tag{5.44}$$

The first few harmonic oscillator wave functions are shown in Fig. 5.2.

We note that our generalized Laguerre polynomials are the solutions of Eq. (5.41) and satisfy the three-term recursion relation

$$(n+1)L_{n+1}^{(\alpha)}(x) = (2n+1+\alpha-x)L_n^{(\alpha)}(x) - (n+\alpha)L_{n-1}^{(\alpha)}(x), \tag{5.45}$$

with $L_0^{(\alpha)}(x) = 1$ and $L_1^{(\alpha)}(x) = 1+\alpha-x$. If we take this relation with $x = \rho^2$ and $\alpha = l + 1/2$, and multiply such that the middle term makes up u_{Nl}, we can infer the recurrence relation

$$\sqrt{(N+1)(N+l+3/2)}u_{N+1,l} = (2N+l+3/2-\rho^2)u_{N,l} - \sqrt{N(N+l+1/2)}u_{N-1,l}, \tag{5.46}$$

with $u_{0,l} = \sqrt{2\sqrt{m\omega/\hbar}/\,\Gamma(l+3/2)}\exp(-\rho^2/2)\rho^{l+1}$.

2-Dimensional Harmonic Oscillator

The extension of the above results to $d = 2$ dimension is straightforward. Here the angular wave function reads

$$\Theta(\theta) = \frac{1}{2\pi}\exp(\pm il\theta), \quad \text{with} \quad l = 0, 1, 2, \ldots, \tag{5.47}$$

and we have $L = l - 1/2$. The energy eigenvalues of Eq. (5.39) turn into

$$E = \hbar\omega\,(2N+L+3/2) = \hbar\omega\,(2N+l+1), \quad \text{with} \quad N = 0, 1, 2, \ldots, \tag{5.48}$$

and the corresponding normalized radial wave function becomes

$$u_{Nl}(r) = \left(\sqrt{\frac{m\omega}{\hbar}} \frac{2\Gamma(N+1)}{\Gamma(N+l+1)}\right)^{1/2} \rho^{l+1/2} \exp(-\rho^2/2)\, L_N^{(l)}(\rho^2). \quad (5.49)$$

5.1.3 Hydrogen Atom

The Hamiltonian of the hydrogen problem in center of mass coordinates is given by

$$\hat{H} = \frac{1}{2m_e}\hat{\boldsymbol{p}}^2 + \frac{Ze^2}{r}, \quad (5.50)$$

where m_e is the reduced mass, $Z = Z_1 Z_2$, where $Z_1 e$ and $Z_2 e$ are the individual charges and e is the proton charge. For hydrogen-like systems $Z < 0$. Expanding the wave function in terms of angular momentum eigenstates, Eq. (5.7), we obtain the differential equation for the radial wave function

$$\left(-\frac{\hbar^2}{2m_e}\frac{\mathrm{d}^2}{\mathrm{d}r^2} + \frac{\hbar^2 l(l+1)}{2m_e r^2} + \frac{Ze^2}{r}\right) u_l(r) = E u_l(r). \quad (5.51)$$

Keeping in mind that $E < 0$, it is convenient to introduce the notations

$$\rho = 2\sqrt{\frac{2m_e(-E)}{\hbar^2}}\,r \quad \text{and} \quad \lambda = -\frac{Ze^2}{\hbar}\sqrt{-\frac{m_e}{2E}} = -Z\alpha\sqrt{-\frac{m_e c^2}{2E}}, \quad (5.52)$$

where $\alpha = e^2/(\hbar c) \simeq 1/137$ is the fine-structure constant. In terms of the new variables, Eq. (5.51) becomes

$$\left(\frac{\mathrm{d}^2}{\mathrm{d}\rho^2} - \frac{l(l+1)}{\rho^2} + \frac{\lambda}{\rho} - \frac{1}{4}\right) u_l(\rho) = 0. \quad (5.53)$$

In the same way as before in the harmonic oscillator case, we can see that $u_l(\rho)$ behaves like ρ^{l+1} as $\rho \to 0$. In the $\rho \to \infty$ limit, the differential equation becomes

$$\left(\frac{\mathrm{d}^2}{\mathrm{d}\rho^2} - \frac{1}{4}\right) u_l(\rho) = 0, \quad (5.54)$$

which is solved by

$$u_l(\rho) \sim e^{-\rho/2} \quad \text{as} \quad \rho \to \infty. \quad (5.55)$$

So,

$$u_l(\rho) = \rho^{l+1} e^{-\rho/2} g(\rho), \quad (5.56)$$

where

$$g(\rho) = \sum_{k=0} c_k \rho^k \quad \text{with} \quad c_0 \neq 0. \tag{5.57}$$

The function $g(\rho)$ satisfies the differential equation

$$\left(\rho \frac{\mathrm{d}^2}{\mathrm{d}\rho^2} + (2l + 2 - \rho)\frac{\mathrm{d}}{\mathrm{d}\rho} + (\lambda - l - 1)\right) g(\rho) = 0. \tag{5.58}$$

Plugging the series of Eq. (5.57) into Eq. (5.58), we find

$$\sum_k [k(k-1)c_k \rho^{k-1} + (2l + 2 - \rho)k c_k \rho^{k-1} + (\lambda - l - 1)c_k \rho^k] = 0, \tag{5.59}$$

or

$$\sum_k [(k+1)(2l + 2 + k)c_{k+1} + (\lambda - l - 1 - k)c_k]\rho^k = 0. \tag{5.60}$$

This equation is valid for all values of ρ, therefore the expression in the bracket must vanish. The coeffients c_k must satisfy the recursion relation

$$c_{k+1} = \frac{k + l + 1 - \lambda}{(k+1)(k + 2l + 2)} c_k. \tag{5.61}$$

We can see that the ratio

$$\frac{c_{k+1}}{c_k} \sim \frac{1}{k}, \quad \text{as} \quad k \to \infty, \tag{5.62}$$

matches the asymptotic behavior of the coefficients of the exponential function

$$e^\rho = \sum_{k=0}^{\infty} \frac{1}{k!}\rho^k = \sum_{k=0}^{\infty} a_k \rho^k \quad \to \quad \frac{a_{k+1}}{a_k} = \frac{1/(k+1)!}{1/k!} = \frac{1}{k+1} \sim \frac{1}{k}. \tag{5.63}$$

So, by letting $k \to \infty$, $g(\rho)$ would destroy the required asymptotic behavior of $u_l(\rho)$. The way out is to assume that the recursion relation terminates at some n_r value such that $c_{n_r} \neq 0$, but $c_{n_r+1} = c_{n_r+2} = \ldots = 0$. This is possible if

$$\lambda = n_r + l + 1 = n, \tag{5.64}$$

or

$$E_{n_r} = -\frac{m_e Z^2 e^4}{2\hbar^2} \frac{1}{(n_r + l + 1)^2} = -\frac{m_e c^2 Z^2 \alpha^2}{2} \frac{1}{n^2} = -\frac{e^2}{a_0} \frac{Z^2}{2n^2}. \tag{5.65}$$

Here, n_r and n are called the radial and principal quantum numbers, respectively, and $a_0 = \hbar^2/(m_e e^2)$ is the modified Bohr radius. Note that in the Bohr radius m_e is

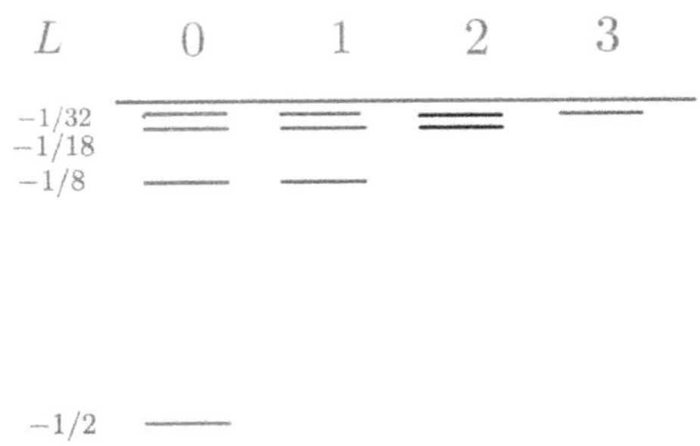

Fig. 5.3 The few lowest energy levels of the hydrogen atom for $l = 0, 1, 2, 3$. The energy levels are $2l + 1$-fold degenerate

the mass of the electron, while here, it is the reduced mass of the proton-electron system, which is slightly different. The energy levels are pictured in Fig. 5.3. We can observe the degeneracy and the accumulation of states around zero.

To get the wave function, we substitute λ into Eq. (5.58)

$$\left(\rho\frac{\mathrm{d}^2}{\mathrm{d}\rho^2} + (2l+1+1-\rho)\frac{\mathrm{d}}{\mathrm{d}\rho} + (n-l-1)\right) g(\rho) = 0. \tag{5.66}$$

Then, comparing with the differential equation of the generalized Laguerre polynomials of Eq. (5.41), we can infer that

$$g(\rho) = L_{n-l-1}^{(2l+1)}(\rho). \tag{5.67}$$

Therefore,

$$u_l(\rho) = N e^{-\rho/2} \rho^{l+1} L_{n-l-1}^{(2l+1)}(\rho), \tag{5.68}$$

with normalization constant N. Using Eqs. (5.52) and (5.65), we find that

$$\rho = 2\frac{Z}{a_0}\frac{r}{n}, \tag{5.69}$$

and the radial part of the wave function becomes

$$u_{nl}(r) = \frac{1}{n}\sqrt{\frac{Z}{a_0}}\sqrt{\frac{(n-l-1)!}{(n+l)!}}\rho^{l+1}\exp(-\rho/2)\, L_{n-l-1}^{(2l+1)}(\rho)\,, \tag{5.70}$$

or, in terms of radial quantum number n_r we have

$$u_{n_r l}(r) = \frac{1}{n_r+l+1}\sqrt{\frac{Z}{a_0}}\sqrt{\frac{n_r!}{(n_r+2l+1)!}}\rho^{l+1}\exp(-\rho/2)\, L_{n_r}^{(2l+1)}(\rho)\,. \tag{5.71}$$

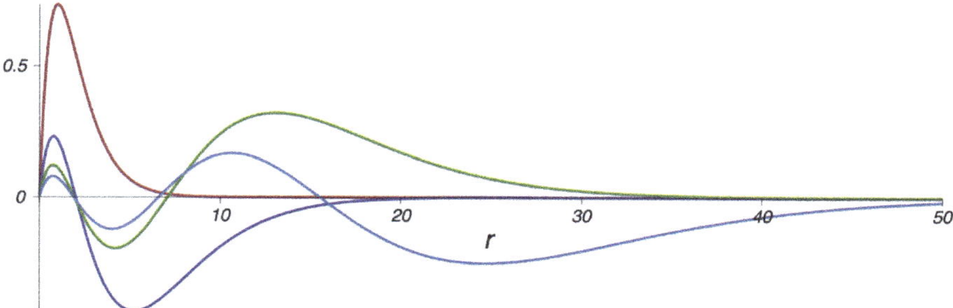

Fig. 5.4 The first four hydrogen radial wave functions $u_{n_r l}(r)$ with $Z = -1$, $a_0 = 1$, $l = 0$ and $n_r = 0, 1, 2, 3$

The first few hydrogen states are shown in Fig. 5.4. We can see that with increasing n_r, the wave functions are stretching more and more out and have more and more nodes.

5.1.4 Spherical Well Potential

Let us consider a spherical well potential

$$V(r) = \begin{cases} -V_0, & \text{if } \ 0 < r < a, \ \text{with } V_0 > 0, \\ 0, & \text{if } \ a < r. \end{cases} \tag{5.72}$$

This potential can serve as an oversimplified model for the nucleon-nucleon interaction. The radial wave function $u_l(r)$ satisfies the differential equation

$$\left(-\frac{\hbar^2}{2\,m}\frac{\mathrm{d}^2}{\mathrm{d}r^2} + \frac{\hbar^2 l(l+1)}{2\,mr^2} + V(r)\right) u_l(r) = E u_l(r) = -B u_l(r), \tag{5.73}$$

where $B > 0$ is the binding energy. We can write the equation in the form

$$\left(\frac{\mathrm{d}^2}{\mathrm{d}r^2} - \frac{l(l+1)}{r^2} - \frac{2\,m}{\hbar^2} V(r)\right) u_l(r) = \beta^2 u_l(r) \quad \text{with} \quad \beta^2 = \frac{2\,m}{\hbar^2}\,B. \tag{5.74}$$

For simplicity, we restrict ourselves to the $l = 0$ case. In the $a < r$ region, we have

$$\frac{\mathrm{d}^2}{\mathrm{d}r^2} u_0(r) = \beta^2 u_0(r), \tag{5.75}$$

and for $r < a$, we find

$$\frac{\mathrm{d}^2}{\mathrm{d}r^2}u_0(r) = -q^2 u_0(r), \quad \text{with} \quad q^2 = \frac{2m}{\hbar^2}(V_0 - E). \tag{5.76}$$

The wave function should be square integrable, thus decaying at large r, and regular at the origin

$$u_0(r) = \begin{cases} N \sin qr, & \text{if} \quad 0 < r < a, \\ N' e^{-\beta r}, & \text{if} \quad a < r. \end{cases} \tag{5.77}$$

Also, $u(r)$ and $u'(r)$ should be continuous at $r = a$

$$\begin{aligned} N e^{-\beta a} &= N' \sin qa, \\ -\beta N e^{-\beta a} &= N' q \cos qa, \end{aligned} \tag{5.78}$$

which immediately gives the transcendental equation

$$qa \cot qa = -\beta a. \tag{5.79}$$

This is the same equation we have seen before with the one-dimensional box potential with odd parity. Bound states exist only if the potential is deep enough, i.e.

$$\frac{\pi^2}{4} < \frac{2mV_0}{\hbar^2}a^2. \tag{5.80}$$

If V_0 is such that

$$\frac{\pi^2}{4} < \frac{2mV_0}{\hbar^2}a^2 < 9\frac{\pi^2}{4}, \tag{5.81}$$

there is only one bound state.

5.1.5 Features of Bound States in Three Dimensions

- For a spherical potential, after partial wave expansion, the Schrödinger equation is reduced to a one-dimensional problem on the half-line $r \in [0, \infty)$, with effective potential

$$V_{eff}(r) = V(r) + \frac{\hbar^2 l(l+1)}{2mr^2}. \tag{5.82}$$

The radial wave function $u_l(r)$ should be regular at the origin, $u_l \sim r^{l+1}$, if $V(r)$ is less singular at the origin than $1/r^2$. At large distances, the square integrability requires an asymptotically decaying solution. If the potential vanishes at large r, the differential equation takes the form

$$-\frac{\hbar^2}{2m}\frac{d^2}{dr^2}u_l(r) = Eu_l(r), \tag{5.83}$$

and the relevant solution is

$$u(r) \sim \exp(-\kappa r), \quad \text{where} \quad \kappa = \sqrt{\frac{2m}{\hbar^2}(-E)}. \tag{5.84}$$

So, the energy eigenvalue is hidden in the large r asymptotic behavior of the wave function.

- The states in a spherical potential are $(2l+1)$-fold degenerate. For a given l, however, the node rule applies. The lowest state has no node and the states with more nodes have higher energy. Since the centrifugal potential is repulsive, if two states have the same number of nodes, the one with the higher l is the least bound.
- We have learned before that in one dimension, if the potential is always negative and goes to zero as $x \to \pm\infty$, there is always a bound state, irrespective of how weak the potential may be. In three dimensions, the regularity of the wave function at $r = 0$ makes the problem similar to the one dimensional problem with odd parity. Consequently, there is no guarantee for a bound state in three dimensions, even for an everywhere attractive potential, if the potential is not strong enough.

5.2 Charged Particles in a Magnetic Field

The vector potential corresponding to a constant magnetic field $\boldsymbol{B}$ can be written as

$$\boldsymbol{A} = \frac{1}{2}(\boldsymbol{B} \times \boldsymbol{r}), \tag{5.85}$$

or

$$A_x = \frac{1}{2}(B_y z - B_z y), \quad A_y = \frac{1}{2}(B_z x - B_x z), \quad A_z = \frac{1}{2}(B_x y - B_y x), \tag{5.86}$$

with B_x, B_y and B_z constants. Calculating $\boldsymbol{B} = \nabla \times \boldsymbol{A}$, we can see that this $\boldsymbol{A}$ indeed represents a constant magnetic field.

If we add an electromagnetic field to a system, we obtain the new Hamiltonian by performing the substitutions

$$\hat{\boldsymbol{p}} \to \hat{\boldsymbol{p}} - \frac{q}{c}\boldsymbol{A} \quad \text{and} \quad \hat{H} \to \hat{H} + q\Phi \tag{5.87}$$

in the old Hamiltonian. Here, Φ is the electric potential, q is the charge and c is the speed of light. We consider cases where there is no external electric field and the potential is spherical. Then, the new Hamiltonian becomes

$$\hat{H} = \frac{1}{2m}\left(\hat{\boldsymbol{p}} - \frac{q}{c}\boldsymbol{A}\right)^2 + V(r) = \hat{H}_0 - \frac{q}{2mc}(\hat{\boldsymbol{p}}\boldsymbol{A} + \boldsymbol{A}\hat{\boldsymbol{p}}) + \frac{q^2}{2mc^2}\boldsymbol{A}^2, \quad (5.88)$$

where

$$\hat{H}_0 = \frac{\hat{\boldsymbol{p}}^2}{2m} + V(r) \quad (5.89)$$

is the original no-magnetic-field Hamiltonian.

A vector field is defined if both the curl and the divergence are set. In our case, the curl of $\boldsymbol{A}$ gives the magnetic field $\boldsymbol{B}$ and for the divergence, we adopt the Coulomb gauge

$$\nabla\boldsymbol{A} = \frac{\partial A_x}{\partial x} + \frac{\partial A_y}{\partial y} + \frac{\partial A_z}{\partial z} = 0. \quad (5.90)$$

We can see that this is true for our $\boldsymbol{A}$ field of Eq. (5.86). From a little calculation

$$(\hat{\boldsymbol{p}}\boldsymbol{A})|\varphi\rangle = -i\hbar\nabla\boldsymbol{A}|\varphi\rangle = -i\hbar(\nabla\boldsymbol{A})|\varphi\rangle + \boldsymbol{A}(-i\hbar\nabla|\varphi\rangle) = \boldsymbol{A}(-i\hbar\nabla|\varphi\rangle), \quad (5.91)$$

we can conclude that in Coulomb gauge $\hat{\boldsymbol{p}}\boldsymbol{A} = \boldsymbol{A}\hat{\boldsymbol{p}}$. Furthermore,

$$\boldsymbol{A}\hat{\boldsymbol{p}} = \frac{1}{2}(\boldsymbol{B} \times \hat{\boldsymbol{r}})\hat{\boldsymbol{p}} = \frac{1}{2}\boldsymbol{B}(\hat{\boldsymbol{r}} \times \hat{\boldsymbol{p}}) = \frac{1}{2}\boldsymbol{B}\hat{\boldsymbol{L}}. \quad (5.92)$$

Then, by using the relation $(\boldsymbol{A} \times \boldsymbol{B})(\boldsymbol{C} \times \boldsymbol{D}) = (\boldsymbol{A}\boldsymbol{C})(\boldsymbol{B}\boldsymbol{D}) - (\boldsymbol{B}\boldsymbol{C})(\boldsymbol{A}\boldsymbol{D})$, we find

$$\boldsymbol{A}^2 = \frac{1}{4}(\boldsymbol{B} \times \boldsymbol{r})(\boldsymbol{B} \times \boldsymbol{r}) = \frac{1}{4}(\boldsymbol{B}^2\boldsymbol{r}^2 - (\boldsymbol{B}\boldsymbol{r})^2). \quad (5.93)$$

So, for the Hamiltonian, we obtain

$$\hat{H} = \frac{\hat{\boldsymbol{p}}^2}{2m} + V(r) - \frac{q}{2mc}\boldsymbol{B}\hat{\boldsymbol{L}} + \frac{q^2}{8mc^2}[\boldsymbol{B}^2\boldsymbol{r}^2 - (\boldsymbol{B}\boldsymbol{r})^2], \quad (5.94)$$

which becomes

$$\hat{H} = \frac{\hat{\boldsymbol{p}}^2}{2m} + V(r) - \frac{q}{2mc}B\hat{L}_z + \frac{q^2}{8mc^2}B^2(x^2 + y^2), \quad (5.95)$$

if $\boldsymbol{B}$ is in the z direction, $\boldsymbol{B} = B\boldsymbol{e}_z$.

5.2.1 Coupling to the Spin

If the particle has spin, the coupling to the magnetic field is different. From the relativistic Dirac theory in the non-relativistic limit for the free particle, we have

$$\hat{H}_0 = \frac{1}{2\,m}(\boldsymbol{\sigma}\,\hat{\boldsymbol{p}})(\boldsymbol{\sigma}\,\hat{\boldsymbol{p}}). \tag{5.96}$$

By using Eq. (4.126), we can see that this Hamiltonian is reduced to the usual non-relativistic free Hamiltonian $\hat{H}_0 = \hat{\boldsymbol{p}}^2/(2m)$. However, when a magnetic field is present, the Hamiltonian becomes

$$\begin{aligned}\hat{H}_0 &= \frac{1}{2\,m}\left(\boldsymbol{\sigma}(\hat{\boldsymbol{p}} - \frac{q}{c}\boldsymbol{A})\right)\left(\boldsymbol{\sigma}(\hat{\boldsymbol{p}} - \frac{q}{c}\boldsymbol{A})\right)\\ &= \frac{1}{2\,m}\left[\left(\hat{\boldsymbol{p}} - \frac{q}{c}\boldsymbol{A}\right)^2 + i\boldsymbol{\sigma}\left(\hat{\boldsymbol{p}} - \frac{q}{c}\boldsymbol{A}\right)\times\left(\hat{\boldsymbol{p}} - \frac{q}{c}\boldsymbol{A}\right)\right]\\ &= \frac{1}{2\,m}\left[\left(\hat{\boldsymbol{p}} - \frac{q}{c}\boldsymbol{A}\right)^2 - \frac{iq}{c}\boldsymbol{\sigma}\left(\hat{\boldsymbol{p}}\times\boldsymbol{A} + \boldsymbol{A}\times\hat{\boldsymbol{p}}\right)\right].\end{aligned} \tag{5.97}$$

The last term can be rewritten as

$$\begin{aligned}\left(\hat{\boldsymbol{p}}\times\boldsymbol{A} + \boldsymbol{A}\times\hat{\boldsymbol{p}}\right)\psi &= -i\hbar[\nabla\times(\boldsymbol{A}\psi) + \boldsymbol{A}\times\nabla\psi]\\ &= -i\hbar[(\nabla\times\boldsymbol{A})\psi + \nabla\psi\times\boldsymbol{A} + \boldsymbol{A}\times\nabla\psi]\\ &= -i\hbar(\nabla\times\boldsymbol{A})\psi = -i\hbar\boldsymbol{B}\psi,\end{aligned} \tag{5.98}$$

which results in the Hamiltonian

$$\hat{H}_0 = \frac{1}{2\,m}\left(\hat{\boldsymbol{p}} - \frac{q}{c}\boldsymbol{A}\right)^2 - \frac{q}{mc}\boldsymbol{B}\hat{\boldsymbol{S}}, \tag{5.99}$$

where $\hat{\boldsymbol{S}} = \hbar/2\,\boldsymbol{\sigma}$. We have found that for spin, the coupling to the magnetic field is twice as strong as for the orbital angular momentum.

The non-relativistic reduction of the relativistic Dirac theory provides an additional term that represents a coupling of the magnetic moment, associated with the spin, to the orbital angular momentum. This is the spin-orbit coupling

$$\hat{H}_{so} = \frac{1}{2\,m^2c^2}\frac{1}{r}\frac{\mathrm{d}V(r)}{\mathrm{d}r}\,\hat{\boldsymbol{S}}\hat{\boldsymbol{L}}. \tag{5.100}$$

So, all together, we have

$$\hat{H} = \frac{\hat{\boldsymbol{p}}^2}{2\,m} + V(r) - \frac{q}{2mc}\hat{\boldsymbol{\mu}}\cdot\boldsymbol{B} + \frac{1}{2m^2c^2}\frac{1}{r}\frac{\mathrm{d}V(r)}{\mathrm{d}r}\,\hat{\boldsymbol{S}}\hat{\boldsymbol{L}} + \frac{q^2}{8\,mc^2}[\boldsymbol{B}^2\boldsymbol{r}^2 - (\boldsymbol{B}\boldsymbol{r})^2], \tag{5.101}$$

where

$$\hat{\boldsymbol{\mu}} = g_L\hat{\boldsymbol{L}} + g_S\hat{\boldsymbol{S}} \tag{5.102}$$

is the magnetic moment. The coefficients $g_L = 1$ and $g_S = 2$ are the gyromagnetic ratios associated with $\hat{\boldsymbol{L}}$ and $\hat{\boldsymbol{S}}$, respectively.

5.2.2 Zeeman Effect in the Hydrogen Atom

Let us consider a hydrogen atom in a homogeneous magnetic field. Here, we ignore the effect of spin and assume that the magnetic field is not very strong. Then, we can also ignore the quadratic term in B in Eq. (5.101) so that

$$\hat{H} = \frac{\hat{\boldsymbol{p}}^2}{2m_e} - \frac{e^2}{r} + \frac{e}{2m_e c} B\hat{L}_z = \hat{H}_0 + \frac{\mu_B}{\hbar} B\hat{L}_z, \tag{5.103}$$

where m_e is the reduced mass of the proton-electron system and $\mu_B = e\hbar/(2m_e c)$ is the magnetic moment of the electron.

We can see that $\{\hat{H}_0, \hat{\boldsymbol{L}}^2, \hat{L}_z\}$ form a complete set of commuting observables. Their common eigenstates are the hydrogen states $|nlm\rangle$, and they also diagonalize $\hat{H}$. Therefore,

$$\begin{aligned} E_{nlm} &= \langle nlm|\hat{H}|nlm\rangle = \langle nlm|\hat{H}_0|nlm\rangle + \frac{B\mu_B}{\hbar}\langle nlm|\hat{L}_z|nlm\rangle \\ &= -\frac{m_e e^4}{2\hbar^2}\frac{1}{n^2} + m\mu_B B. \end{aligned} \tag{5.104}$$

This is the Zeeman effect; the splitting of hydrogen spectral lines in a magnetic field. The original degeneracy in m is removed and the splitting of the energy levels is proportional to the applied magnetic field. This is the reason why the quantum number m is often called the magnetic quantum number.

5.2.3 Charged Free Particle

We consider a charged particle of mass μ and charge q moving in a homogeneous magnetic field given by the vector potential $\boldsymbol{A} = (-yB, 0, 0)$. We can see that this gives a homogeneous magnetic field in the z direction, $\boldsymbol{B} = \nabla \times \boldsymbol{A} = (0, 0, B)$. Then, we obtain

$$\hat{H}\psi = \left[\frac{1}{2\mu}\left(\hat{\boldsymbol{p}} - \frac{q}{c}\boldsymbol{A}\right)^2\right]\psi = \left[\frac{1}{2\mu}\left(\hat{p}_x + \frac{qyB}{c}\right)^2 + \frac{\hat{p}_y^2}{2\mu} + \frac{\hat{p}_z^2}{2\mu}\right]\psi = E\psi. \tag{5.105}$$

We can see that $\left[\hat{p}_x, \hat{H}\right] = \left[\hat{p}_z, \hat{H}\right] = 0$, so $\{\hat{p}_x, \hat{p}_z, \hat{H}\}$ form a complete set of commuting observables. The eigenstates of $\hat{p}_x$ and $\hat{p}_z$ are plane waves. Consequently, we can seek the solution in the form

$$\psi = \exp[i(k_x x + k_z z)]\phi(y), \tag{5.106}$$

and find that

$$\left[\frac{1}{2\mu}\left(\hbar k_x + \frac{qyB}{c}\right)^2 + \frac{\hat{p}_y^2}{2\mu} + \frac{\hbar^2 k_z^2}{2\mu}\right]\exp[i(k_x x + k_z z)]\phi(y) = E\exp[i(k_x x + k_z z)]\phi(y), \tag{5.107}$$

or

$$\left[\frac{\hat{p}_y^2}{2\mu} + \frac{1}{2}\mu\Omega^2(y-y_0)^2\right]\phi(y) = \left(E - \frac{\hbar^2 k_z^2}{2\mu}\right)\phi(y), \tag{5.108}$$

with $y_0 = -c\hbar k_x/qB$ and the cyclotron frequency $\Omega = qB/\mu c$. This is the eigenvalue problem of an one-dimensional harmonic oscillator in the y coordinate centered at y_0. Thus,

$$E_{n,k_z} = \frac{\hbar^2 k_z^2}{2\mu} + \hbar\Omega\left(n + \frac{1}{2}\right), \tag{5.109}$$

and

$$\psi_n = N_n \exp\left(-\frac{1}{2}\sqrt{\mu\Omega/\hbar}(y-y_0)^2\right)H_n\left[\sqrt{\mu\Omega/\hbar}(y-y_0)\right]\exp[i(k_x x + k_z z)], \tag{5.110}$$

where N_n is a normalization constant. We can see that the energy does not depend on k_x. The k_x dependence is buried in y_0, which just shifts the harmonic oscillator minimum point along the y coordinate. On the other hand k_x can take any value, and thus the spectrum is infinitely degenerate.

We can also address the problem by using Eq. (5.95) with $V \equiv 0$, which gives

$$\begin{aligned}\hat{H} &= \frac{\hat{p}_z^2}{2\mu} + \frac{\hat{p}_x^2}{2\mu} + \frac{\hat{p}_y^2}{2\mu} + \frac{q^2B^2}{8\mu c^2}(x^2+y^2) - \frac{q}{2\mu c}B\hat{L}_z \\ &= \frac{\hat{p}_z^2}{2\mu} + \frac{\hat{p}_x^2}{2\mu} + \frac{\hat{p}_y^2}{2\mu} + \frac{1}{8}\mu\Omega^2(x^2+y^2) - \frac{1}{2}\Omega\hat{L}_z = \frac{\hat{p}_z^2}{2\mu} + \hat{H}_\perp.\end{aligned} \tag{5.111}$$

We can recognize in $\hat{H}_\perp$ a two dimensional isotropic harmonic oscillator with frequency $\Omega/2$ and an angular momentum term. To find the eigenstates, we may introduce the operators

$$\hat{a}_1 = \frac{1}{\sqrt{\mu\Omega\hbar}}\left(i\hat{p}_x + \frac{1}{2}\mu\Omega x\right) \quad \text{and} \quad \hat{a}_2 = \frac{1}{\sqrt{\mu\Omega\hbar}}\left(i\hat{p}_y + \frac{1}{2}\mu\Omega y\right). \tag{5.112}$$

We can see that $\hat{a}_1$ and $\hat{a}_2$ commute, $\left[\hat{a}_1, \hat{a}_1^\dagger\right] = \left[\hat{a}_2, \hat{a}_2^\dagger\right] = 1$, and

$$\hat{x} = \sqrt{\frac{\hbar}{\mu\Omega}}(\hat{a}_1^\dagger + \hat{a}_1), \qquad \hat{p}_x = \frac{i}{2}\sqrt{\mu\hbar\Omega}(\hat{a}_1^\dagger - \hat{a}_1),$$
$$\hat{y} = \sqrt{\frac{\hbar}{\mu\Omega}}(\hat{a}_2^\dagger + \hat{a}_2), \qquad \hat{p}_y = \frac{i}{2}\sqrt{\mu\hbar\Omega}(\hat{a}_2^\dagger - \hat{a}_2). \tag{5.113}$$

Therefore,

$$\frac{\hat{p}_x^2}{2\mu} + \frac{\hat{p}_y^2}{2\mu} + \frac{1}{8}\mu\Omega^2(x^2 + y^2) = \frac{\hbar\Omega}{2}(\hat{a}_1^\dagger\hat{a}_1 + \hat{a}_2^\dagger\hat{a}_2 + 1), \tag{5.114}$$

and

$$\hat{L}_z = \hat{x}\hat{p}_y - \hat{y}\hat{p}_x = i\hbar(\hat{a}_2^\dagger\hat{a}_1 - \hat{a}_1^\dagger\hat{a}_2). \tag{5.115}$$

We can introduce another set of operators

$$\hat{a}_+ = -\frac{1}{\sqrt{2}}(\hat{a}_1 - i\hat{a}_2) \quad \text{and} \quad \hat{a}_- = \frac{1}{\sqrt{2}}(\hat{a}_1 + i\hat{a}_2), \tag{5.116}$$

and find the commutation relations $\left[\hat{a}_+, \hat{a}_+^\dagger\right] = \left[\hat{a}_-, \hat{a}_-^\dagger\right] = 1$. Then we can express the $\hat{H}_\perp$ and $\hat{L}_z$ in terms of $\hat{a}_+$ and $\hat{a}_-$

$$\hat{H}_\perp = \frac{\hbar\Omega}{2}(\hat{a}_+^\dagger\hat{a}_+ + \hat{a}_-^\dagger\hat{a}_- + 1) - \frac{\hbar\Omega}{2}(\hat{a}_+^\dagger\hat{a}_+ - \hat{a}_-^\dagger\hat{a}_-) = \hbar\Omega(\hat{a}_-^\dagger\hat{a}_- + 1/2), \tag{5.117}$$

and

$$\hat{L}_z = \hbar(\hat{a}_+^\dagger\hat{a}_+ - \hat{a}_-^\dagger\hat{a}_-). \tag{5.118}$$

We should notice that from their commutation relations it follows that $\hat{a}_+^\dagger\hat{a}_+$ and $\hat{a}_-^\dagger\hat{a}_-$ are number operators with integer eigenvalues n_+ and n_-, respectively. Therefore, the eigenvalues of $\hat{H}_\perp$ are $\hbar\Omega(n_- + 1/2)$ and the eigenvalues of $\hat{L}_z$ are $\hbar(n_+ - n_-)$. Their common eigenstates are the product states $|n_-, n_+\rangle = |n_-\rangle|n_+\rangle$. If we rename $n_- \to n$ and $n_+ - n_- \to m$ we obtain

$$\hat{H}_\perp = \hbar\Omega(n + 1/2) \quad \text{and} \quad \hat{L}_z = \hbar m, \tag{5.119}$$

where m takes values from $-n, -n+1, \ldots$. There is no upper limit for m, but the energy does not depend on it, so the states are infinitely degenerate. These states are called *Landau levels*.

5.3 Electromagnetic Potential and Gauge Transformation

In classical mechanics, the Hamiltonian of a charged particle moving in an electromagnetic field is given by

$$H = \frac{1}{2m}\left(\boldsymbol{p} - \frac{q}{c}\boldsymbol{A}(\boldsymbol{r})\right)^2 + q\Phi(\boldsymbol{r}), \tag{5.120}$$

where $\boldsymbol{A}$ is the vector potential and Φ is the scalar potential. These potentials determine the electric and magnetic fields

$$\boldsymbol{E} = -\nabla\Phi - \frac{1}{c}\frac{\partial}{\partial t}\boldsymbol{A} \quad \text{and} \quad \boldsymbol{B} = \nabla \times \boldsymbol{A}. \tag{5.121}$$

From the Hamilton canonical equations,

$$\dot{x}_i = \frac{\partial H}{\partial p_i} \quad \text{and} \quad \dot{p}_i = -\frac{\partial H}{\partial x_i} \tag{5.122}$$

we can readily derive the equation of motion

$$m\ddot{\boldsymbol{r}} = q\left(\boldsymbol{E} + \frac{1}{c}\boldsymbol{v} \times \boldsymbol{B}\right), \tag{5.123}$$

with the Lorentz force on the right hand side.

As we know, these potentials are not unique. If we perform the gauge transformation

$$\boldsymbol{A} \to \boldsymbol{A} + \nabla\Lambda \quad \text{and} \quad \Phi \to \Phi - \frac{1}{c}\frac{\partial\Lambda}{\partial t}, \tag{5.124}$$

the $\boldsymbol{E}$ and $\boldsymbol{B}$, and consequently the classical equations of motion, remain unchanged.

In quantum mechanics, in accordance with the correspondence principle, we should demand that the Schrödinger equation remains invariant with respect to the above gauge transformation. Let us assume a transformation that is time independent, i.e. Λ depends only on spatial coordinates and consequently

$$\boldsymbol{A} \to \boldsymbol{A}' = \boldsymbol{A} + \nabla\Lambda(\boldsymbol{r}) \quad \text{and} \quad \Phi \to \Phi' = \Phi. \tag{5.125}$$

This transformation should not change the measured quantities and should not change the probability density $|\psi|^2$. This leaves us with the possibility of a phase change only,

$$\psi'(\boldsymbol{r}) = \exp(i\alpha(\boldsymbol{r}))\,\psi(\boldsymbol{r}), \tag{5.126}$$

with a real $\alpha(\boldsymbol{r})$ phase. Calculating the action of the momentum operator, we find

$$\begin{aligned}
&\exp[-i\alpha(\boldsymbol{r})]\hat{\boldsymbol{p}}\exp[i\alpha(\boldsymbol{r})]\psi(\boldsymbol{r}) \\
&\qquad = \exp[-i\alpha(\boldsymbol{r})]\left(i(-i\hbar\nabla\alpha(\boldsymbol{r}))\exp[i\alpha(\boldsymbol{r})]\psi(\boldsymbol{r}) + \exp[i\alpha(\boldsymbol{r})]\hat{\boldsymbol{p}}\psi(\boldsymbol{r})\right) \\
&\qquad = \left(\hbar\nabla\alpha(\boldsymbol{r}) + \hat{\boldsymbol{p}}\right)\psi(\boldsymbol{r}),
\end{aligned} \tag{5.127}$$

and consequently, we have

$$\langle\psi'|\hat{\boldsymbol{p}} - \frac{q}{c}\boldsymbol{A}'|\psi'\rangle = \langle\psi|\hbar\nabla\alpha(\boldsymbol{r}) + \hat{\boldsymbol{p}} - \frac{q}{c}\boldsymbol{A} - \frac{q}{c}\nabla\Lambda|\psi\rangle. \tag{5.128}$$

So, if we take

$$\hbar\nabla\alpha(\boldsymbol{r}) = \frac{q}{c}\nabla\Lambda(\boldsymbol{r}), \quad \text{or} \quad \alpha(\boldsymbol{r}) = \frac{q}{\hbar c}\Lambda(\boldsymbol{r}), \tag{5.129}$$

the effect of the gauge transformation of the electromagnetic field is absorbed into the phase change of the wave function

$$\langle\psi'|\hat{\boldsymbol{p}} - \frac{q}{c}\boldsymbol{A}'|\psi'\rangle = \langle\psi|\hat{\boldsymbol{p}} - \frac{q}{c}\boldsymbol{A}|\psi\rangle. \tag{5.130}$$

This means that

$$\exp(-i\frac{q}{\hbar c}\Lambda(\boldsymbol{r}))\left(\hat{\boldsymbol{p}} - \frac{q}{c}\boldsymbol{A}'(\boldsymbol{r})\right)\exp(i\frac{q}{\hbar c}\Lambda(\boldsymbol{r})) = \left(\hat{\boldsymbol{p}} - \frac{q}{c}\boldsymbol{A}(\boldsymbol{r})\right), \tag{5.131}$$

and therefore,

$$\begin{aligned}
&\exp(-i\frac{q}{\hbar c}\Lambda(\boldsymbol{r}))\left(\hat{\boldsymbol{p}} - \frac{q}{c}\boldsymbol{A}'(\boldsymbol{r})\right)^2\exp(i\frac{q}{\hbar c}\Lambda(\boldsymbol{r})) \\
&= \exp(-i\frac{q}{\hbar c}\Lambda(\boldsymbol{r}))\left(\hat{\boldsymbol{p}} - \frac{q}{c}\boldsymbol{A}'(\boldsymbol{r})\right)\exp(i\frac{q}{\hbar c}\Lambda(\boldsymbol{r}))\exp(-i\frac{q}{\hbar c}\Lambda(\boldsymbol{r}))\left(\hat{\boldsymbol{p}} - \frac{q}{c}\boldsymbol{A}'(\boldsymbol{r})\right)\exp(i\frac{q}{\hbar c}\Lambda(\boldsymbol{r})) \\
&= \left(\hat{\boldsymbol{p}} - \frac{q}{c}\boldsymbol{A}(\boldsymbol{r})\right)^2.
\end{aligned} \tag{5.132}$$

As a consequence, the expectation value of the Hamiltonian is invariant with respect to the gauge transformation

$$\langle\psi'|\hat{H}'|\psi'\rangle = \langle\psi|\hat{H}|\psi\rangle. \tag{5.133}$$

We could have encountered a similar situation before when we introduced the momentum operator from the correspondence principle. We needed to satisfy the commutation relation $[x, \hat{p}_x] = i\hbar$, and we adopted the $\hat{p}_x = -i\hbar\partial/\partial x$ form. We could also have taken the operator $\hat{p}_x = -i\hbar\partial/\partial x + f(x)$, where $f(x)$ is a real function of x, and it would have had the same commutation rule with x. This choice would involve the transformation of the wave function

$$\psi(x) \to \exp(-\frac{i}{\hbar}\int^x f(x')\mathrm{d}x')\psi(x). \tag{5.134}$$

The new momentum operator $\hat{p}_x = -i\hbar\partial/\partial x + f(x)$ acting on the modified wave function result in an additional $-f(x)$, which cancels the arbitrary $f(x)$, and would result in the same physics.

Let us consider a charged particle in an environment where $\boldsymbol{A} \equiv 0$. The wave function $\psi^{(0)}(\boldsymbol{r})$ satisfies the equation

$$\left(\frac{1}{2m}\hat{\boldsymbol{p}}^2 + V(\boldsymbol{r})\right)\psi^{(0)}(\boldsymbol{r}) = E\psi^{(0)}(\boldsymbol{r}), \tag{5.135}$$

without an electromagnetic potential, but with some non-electromagnetic potential V.

Now, we generate the vector potential by a gauge transformation

$$\boldsymbol{A}'(\boldsymbol{r}) = \nabla\Lambda(\boldsymbol{r}), \tag{5.136}$$

where

$$\Lambda(\boldsymbol{r}) = \int_{\boldsymbol{r}_0}^{\boldsymbol{r}} \boldsymbol{A}'(\boldsymbol{r}')\mathrm{d}\boldsymbol{r}', \tag{5.137}$$

with arbitrary $\boldsymbol{r}_0$. Then, the Schrödinger equation becomes

$$\left[\frac{1}{2m}\left(\hat{\boldsymbol{p}} - \frac{q}{c}\boldsymbol{A}'\right)^2 + V(\boldsymbol{r})\right]\psi(\boldsymbol{r}) = E\psi(\boldsymbol{r}). \tag{5.138}$$

The wave function $\psi(\boldsymbol{r})$ is related to the no-field wave function $\psi^{(0)}(\boldsymbol{r})$ by a phase transformation

$$\psi(\boldsymbol{r}) = \exp\left(\frac{iq}{\hbar c}\int_{\boldsymbol{r}_0}^{\boldsymbol{r}} \boldsymbol{A}'(\boldsymbol{r}')\mathrm{d}\boldsymbol{r}'\right)\psi^{(0)}(\boldsymbol{r}), \tag{5.139}$$

where the path of integration can be deformed continuously with fixed end point $\boldsymbol{r}$.

5.3.1 Aharonov-Bohm Effect

In classical physics only the electric field $\boldsymbol{E}$ and magnetic field $\boldsymbol{B}$ matter. Consequently, if $\boldsymbol{B} = 0$ in a region of space, it has no effect on the motion of a charged particle. This is not necessarily true in quantum mechanics, as the Aharonov-Bohm effect shows. The motion of a charged particle can be effected in region where $\boldsymbol{B} = 0$ but the vector potential $\boldsymbol{A} \neq 0$.

Consider the experimental setup proposed by Ehrenberg and Siday [2] and Aharonov and Bohm [1]. It is basically a double slit experiment in the $x - y$ plane with a long solenoid creating a magnetic field $\boldsymbol{B}$ inside the solenoid in the z direction (Fig. 5.5). The vector potential that generates this $\boldsymbol{B}$ field, in cylindrical coordinates (r, φ, z), is given by

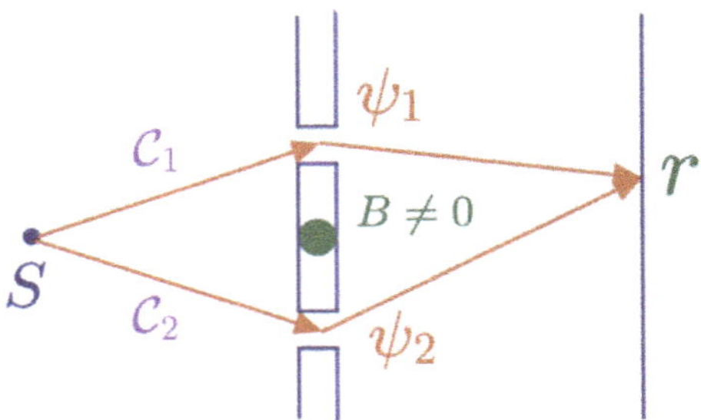

Fig. 5.5 Double-slit interference experiment in which the charged particles cannot penetrate into the region where the magnetic field $\boldsymbol{B} \neq 0$

$$\boldsymbol{A} = (0, A_\varphi, 0) = \frac{B}{2}\begin{cases} r\,\boldsymbol{e}_\varphi, & \text{if} \quad r < R \\ R^2/r\,\boldsymbol{e}_\varphi, & \text{if} \quad R < r. \end{cases} \tag{5.140}$$

Recalling the form of the curl operator in cylindrical coordinates

$$\boldsymbol{A} = \left(\frac{1}{r}\frac{\partial A_z}{\partial \varphi} - \frac{\partial A_\varphi}{\partial z}\right)\boldsymbol{e}_r + \left(\frac{\partial A_r}{\partial z} - \frac{\partial A_z}{\partial r}\right)\boldsymbol{e}_\varphi + \frac{1}{r}\left(\frac{\partial (r A_\varphi)}{\partial r} - \frac{\partial A_r}{\partial \varphi}\right)\boldsymbol{e}_z, \tag{5.141}$$

for $\boldsymbol{A}$ of Eq. (5.140), we find

$$\boldsymbol{B} = \nabla \times \boldsymbol{A} = \frac{1}{r}\frac{\partial}{\partial r}(r A_\varphi)\boldsymbol{e}_z = \begin{cases} B\,\boldsymbol{e}_z, & r < R \\ 0, & R < r. \end{cases} \tag{5.142}$$

We can also see that $\boldsymbol{A}$ satisfies the Coulomb gauge condition

$$\nabla \boldsymbol{A} = \left(\frac{1}{r}\frac{\partial}{\partial r}r, \frac{1}{r}\frac{\partial}{\partial \varphi}, \frac{\partial}{\partial z}\right)\boldsymbol{A} = 0. \tag{5.143}$$

So, this vector potential really describes a $\boldsymbol{B}$ field which is constant inside the solenoid and vanishes outside.

The magnetic field $\boldsymbol{B}$ is zero outside the solenoid, but the vector potential $\boldsymbol{A}$ does not vanish. The charged particles in the double slit experiment pass through the region where $\boldsymbol{A} \neq 0$. In the case of no $\boldsymbol{A}$ field, the wave function on the screen, due to the superposition principle, is given by $\psi^{(0)}(\boldsymbol{r}) = \psi_1^{(0)}(\boldsymbol{r}) + \psi_2^{(0)}(\boldsymbol{r})$, where $\psi_1^{(0)}$ and $\psi_2^{(0)}$ are associated with routes C_1 and C_2, respectively. Their constructive and destructive interference results in the well known observed double slit interference pattern. In the presence of the $\boldsymbol{A}$ field, according to Eq. (5.139), the wave function acquires an additional phase

$$
\begin{aligned}
\psi(\boldsymbol{r}) &= \exp(\frac{iq}{\hbar c}\int_{C_1}\boldsymbol{A}(\boldsymbol{r}')\mathrm{d}\boldsymbol{r}')\psi_1^{(0)}(\boldsymbol{r}) + \exp(\frac{iq}{\hbar c}\int_{C_2}\boldsymbol{A}(\boldsymbol{r}')\mathrm{d}\boldsymbol{r}')\psi_2^{(0)}(\boldsymbol{r}) \\
&= \exp(\frac{iq}{\hbar c}\int_{C_2}\boldsymbol{A}(\boldsymbol{r}')\mathrm{d}\boldsymbol{r}')\left[\exp(\frac{iq}{\hbar c}\int_{C_1}\boldsymbol{A}(\boldsymbol{r}')\mathrm{d}\boldsymbol{r}' - \frac{iq}{\hbar c}\int_{C_2}\boldsymbol{A}(\boldsymbol{r}')\mathrm{d}\boldsymbol{r}')\psi_1^{(0)}(\boldsymbol{r}) + \psi_2^{(0)}(\boldsymbol{r})\right] \\
&= \exp(\frac{iq}{\hbar c}\int_{C_2}\boldsymbol{A}(\boldsymbol{r}')\mathrm{d}\boldsymbol{r}')\left[\exp(\frac{iq}{\hbar c}\int_{C_1}\boldsymbol{A}(\boldsymbol{r}')\mathrm{d}\boldsymbol{r}' + \frac{iq}{\hbar c}\int_{-C_2}\boldsymbol{A}(\boldsymbol{r}')\mathrm{d}\boldsymbol{r}')\psi_1^{(0)}(\boldsymbol{r}) + \psi_2^{(0)}(\boldsymbol{r})\right] \\
&= \exp(\frac{iq}{\hbar c}\int_{C_2}\boldsymbol{A}(\boldsymbol{r}')\mathrm{d}\boldsymbol{r}')\left[\exp(\frac{iq}{\hbar c}\oint_{C_1-C_2}\boldsymbol{A}(\boldsymbol{r}')\mathrm{d}\boldsymbol{r}')\psi_1^{(0)}(\boldsymbol{r}) + \psi_2^{(0)}(\boldsymbol{r})\right],
\end{aligned}
\tag{5.144}
$$

where S is the area enclosed by going from the source to the screen along path 1 and going back along path 2. Then, by using Stokes' theorem, we obtain

$$
\begin{aligned}
\psi(\boldsymbol{r}) &= \exp(\frac{iq}{\hbar c}\int_{C_2}\boldsymbol{A}(\boldsymbol{r}')\mathrm{d}\boldsymbol{r}')\left[\exp(\frac{iq}{\hbar c}\int_S \nabla\times\boldsymbol{A}\,\mathrm{d}\boldsymbol{S})\psi_1^{(0)}(\boldsymbol{r}) + \psi_2^{(0)}(\boldsymbol{r})\right] \\
&= \exp(\frac{iq}{\hbar c}\int_{C_2}\boldsymbol{A}(\boldsymbol{r}')d\boldsymbol{r}')\left[\exp(\frac{iq}{\hbar c}\int_S \boldsymbol{B}\,\mathrm{d}\boldsymbol{S})\psi_1^{(0)}(\boldsymbol{r}) + \psi_2^{(0)}(\boldsymbol{r})\right] \\
&= \exp(\frac{iq}{\hbar c}\int_{C_2}\boldsymbol{A}(\boldsymbol{r}')\mathrm{d}\boldsymbol{r}')\left[\exp(\frac{iq}{\hbar c}\Phi_B)\psi_1^{(0)}(\boldsymbol{r}) + \psi_2^{(0)}(\boldsymbol{r})\right],
\end{aligned}
\tag{5.145}
$$

where Φ_B is the magnetic flux going through S.

Although the magnetic field $\boldsymbol{B}$ is nonzero only in the non-accessible region, it still has an effect on the diffraction pattern through the presence of the vector potential $\boldsymbol{A}$. We have found that even though the particle could pass only outside the solenoid where the magnetic field $\boldsymbol{B}$ is zero, it is the flux coming from the magnetic field inside the solenoid, where the particles cannot penetrate, that causes a relative change of the phase between ψ_1 and ψ_2. If $\psi_1^{(0)}$ and $\psi_2^{(0)}$ are such that we observe an interference minimum or maximum, we observe the same pattern if

$$
\frac{q}{\hbar c}\Phi_B = 2\pi n, \tag{5.146}
$$

with integer n. The result is gauge independent, as it depends only on the flux Φ_B. It seems that in quantum mechanics, it is the potential which is of primary importance, indicating its non-local nature. The quantum particle senses the whole region including the cross section of the solenoid though which flows the magnetic field and it is not localized to a particular place.

5.3.2 Bound-State Aharonov-Bohm Effect

In a similar fashion, we can consider a particle of mass m and charge q constrained to move along a fixed circle of radius r in the $x-y$ plane (Fig. 5.6). The momentum operator in cylindrical coordinates (r, φ, z) in the present situation has only φ component

$$\hat{p}_\varphi = -i\hbar \frac{1}{r}\frac{\mathrm{d}}{\mathrm{d}\varphi} \tag{5.147}$$

The Hamiltonian in the absence of magnetic field becomes

$$\hat{H} = \frac{1}{2m}\hat{p}_\varphi^2 = -\frac{\hbar^2}{2mr^2}\frac{\mathrm{d}^2}{\mathrm{d}\varphi^2}. \tag{5.148}$$

The operators $\hat{p}_\varphi$ and $\hat{H}$ commute and they possess common eigenstates

$$\psi^{(0)}(\varphi) = \frac{1}{\sqrt{2\pi}}\exp(il\varphi). \tag{5.149}$$

The solution should be single valued, therefore it should be periodic. Consequently, l can take only integer values, $l = 0, \pm 1, \pm 2, \ldots$. This leads to the energy

$$E_l^{(0)} = \frac{\hbar^2 l^2}{2mr^2}. \tag{5.150}$$

In the presence of a magnetic field, the Hamiltonian is modified

$$\hat{H} = \frac{1}{2m}\left(\hat{p}_\varphi - \frac{q}{c}A_\varphi\right)^2, \tag{5.151}$$

where $\hat{p}_\varphi$ is defined in Eq. (5.147) and A_φ in Eq. (5.140). The wave function acquires an additional phase calculated in Eq. (5.145),

$$\psi = \psi^{(0)}\exp\left(\frac{iq}{\hbar c}\Phi_B\right) = \frac{1}{\sqrt{2\pi}}\exp(il\varphi)\exp\left(\frac{iq}{\hbar c}\Phi_B\right). \tag{5.152}$$

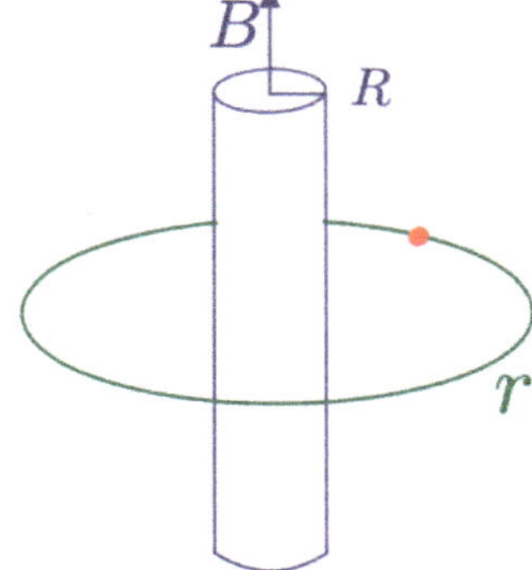

Fig. 5.6 Charged particle confined to a ring of radius r around a solenoid with radius R and magnetic field $\boldsymbol{B}$

We assume a simple circular geometry as in Eq. (5.140) with a solenoid of radius R. The vector potential outside the solenoid at distance r is given by

$$A_\varphi = \frac{BR^2\pi}{2\pi}\frac{1}{r} = \frac{\Phi_B}{2\pi r}, \tag{5.153}$$

where Φ_B is the magnetic flux. Since $\hat{p}_\varphi \psi^{(0)} = l\hbar/r\, \psi^{(0)}$, we find

$$\frac{1}{2m}\left(\hat{p}_\varphi - \frac{q}{c}A_\varphi\right)^2 \psi = \frac{1}{2m}\left(\frac{l\hbar}{r} - \frac{q}{c}\frac{\Phi_B}{2\pi r}\right)^2 \psi = E\psi. \tag{5.154}$$

Consequently,

$$E_l = \frac{\hbar^2}{2mr^2}\left(l - \frac{q\Phi_B}{2\pi\hbar c}\right)^2 = \frac{\hbar^2}{2mr^2}\left(l - \frac{q\Phi_B}{hc}\right)^2. \tag{5.155}$$

The result is gauge independent because it depends only on the flux. The charged particle feels the effect of the magnetic field even though $\boldsymbol{B}$ is zero along the ring where the particle is confined. With $\boldsymbol{B} = 0$ particles moving clockwise and counterclockwise have the same energy. Consequently the states are double degenerate, whereas with $\boldsymbol{B} \neq 0$ the degeneracy is lifted.

5.3.3 *Quantization of Magnetic Flux*

Let us consider a superconducting ring surrounding a region of magnetic field (Fig. 5.7). There is no electric field $\boldsymbol{E}$ and magnetic field $\boldsymbol{B}$ inside the superconducting tube, but there is a magnetic field $\boldsymbol{B} \neq 0$ inside the ring. As the charged particle moves along, its wave function $\psi^{(0)}$ acquires an additional phase. If it makes

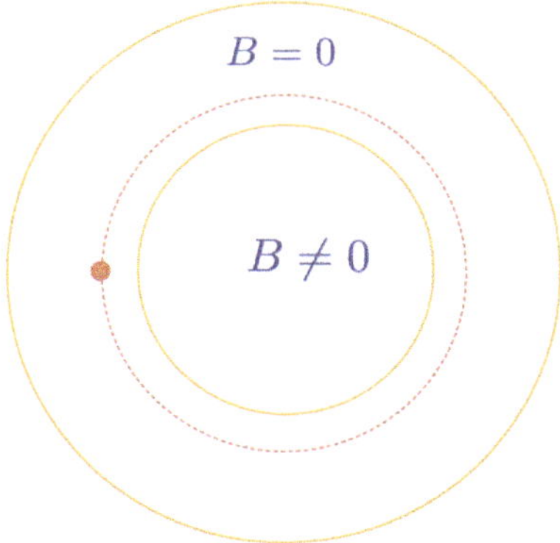

Fig. 5.7 Quantized flux in a superconducting ring

a full turn, the wave function becomes

$$\psi = \psi^{(0)} \exp\left(\frac{iq}{\hbar c}\Phi_B\right), \tag{5.156}$$

where Φ_B is the magnetic flux. The wave function is stationary, so in principle, it is allowed to turn many times. On the other hand, the wave function must be single valued at any point in space. Therefore, the acquired phase should be a multiple of 2π,

$$\frac{q}{\hbar c}\Phi_B = 2\pi\, n, \quad (n = 0, \pm 1, \pm 2, \ldots), \tag{5.157}$$

and consequently,

$$\Phi_B = n\,\frac{hc}{q}, \quad (n = 0, \pm 1, \pm 2, \ldots). \tag{5.158}$$

This means that the magnetic flux trapped inside a superconducting ring can only be an integer multiple of the flux quantum $\Phi_0 = hc/q$. It has been observed that $q = -2e$, providing experimental support for the Cooper pair concept and the BCS theory.

5.3.4 Electric Aharonov-Bohm Effect

There is an analogous effect with the electric potential. Consider again a double-slit experiment where the charged particles pass through long conducting tubes (Fig. 5.8). The time evolution of the state is governed by the Hamiltonian

$$|\psi(t)\rangle = \exp(-\frac{i}{\hbar}\int_0^t \mathrm{d}t'\, H(t'))|\psi(0)\rangle. \tag{5.159}$$

Particles passing through the tubes with different potential levels acquire different phases, which results in a phase difference at the screen

$$\phi_1 - \phi_2 = \frac{1}{\hbar}\int_0^t \mathrm{d}t'\,[V_2(t') - V_1(t')]. \tag{5.160}$$

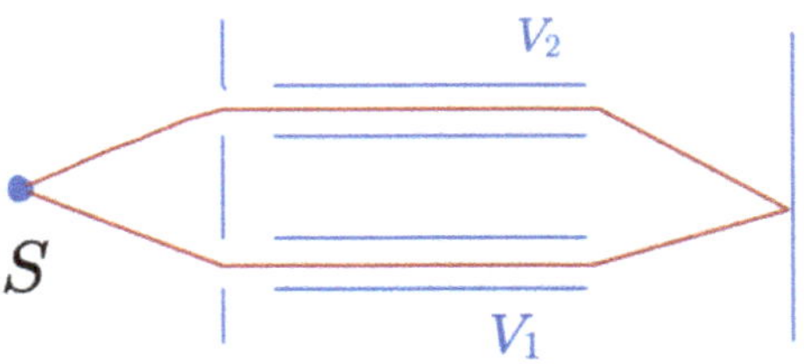

Fig. 5.8 Double-slit experiment with particles going through regions at different potential levels

This phase difference gives rise to a diffraction pattern that depends on the potential difference. This happens in spite of the fact that the particles do not experience any force inside the tubes. This effect is purely quantum mechanical. In the $\hbar \to 0$ limit the interference is washed out as the oscillation becomes infinitely rapid.

We note that this result is more general. The potential does not need to be an electric potential and the particles do not need to be charged particles. We can have the same effect for neutral particles going through regions where the gravitational potential is different.

Further Reading

1. Y. Aharonov, D. Bohm, Significance of electromagnetic potentials in the quantum theory. Phys. Rev. **115**(3), 485 (1959)
2. W. Ehrenberg, R.E. Siday, The refractive index in electron optics and the principles of dynamics. Proc. Phys. Soc. Sect. B **62**(1), 8 (1949)

Chapter 6
Identical Particles in Quantum Mechanics

6.1 Identity and Indistinguishability of Quantum Particles

Quantum mechanical systems often consist of several particles that are *identical* in some sense. The notion of *identical* means that particles cannot be distinguished from each other. The dynamical properties of the system do not change when two of the identical particles are permuted. In this respect, electrons are identical particles since they have the same mass, same charge, same spin, and thus they cannot be distinguished from each other. However, protons and neutrons are not identical paricles, because they have different charges and masses. On the other hand, in some respect, protons and neutrons can also be considered as identical particles. It has been observed that with respect to the strong nuclear force, there is no difference between them. This symmetry is broken when electromagnetic interaction is also considered. So, the notion of particle identity depends on the Hamiltonian of the model system and it is related to the indistinguishability within the applied model or performed experiment.

6.1.1 Exchange Symmetry

The Hamiltonian $\hat{H}(1, 2)$ describes a system with particle 1 in state α and particle 2 in state β. In the Hamiltonian, the first argument refers to the quantum state α and the second one to quantum state β. The corresponding wave function is determined by the Schrödinger equation

$$i\hbar\frac{\mathrm{d}}{\mathrm{d}t}\psi(1, 2) = \hat{H}(1, 2)\psi(1, 2). \tag{6.1}$$

The principle of indistinguishability requires that physical observables, in particular the Hamiltonian, is invariant, or symmetric, with respect to the exchange of the two

Z. Papp, *Mastering Quantum Mechanics*,
https://doi.org/10.1007/978-3-032-09011-9_6

identical particles

$$\hat{H}(1,2) = \hat{H}(2,1). \tag{6.2}$$

Then, $\psi(2,1)$, which describes the physical situation where particle 2 is in the physical state α and particle 1 in state β, also satisfies the same equation

$$i\hbar\frac{\mathrm{d}}{\mathrm{d}t}\psi(2,1) = \hat{H}(1,2)\psi(2,1). \tag{6.3}$$

If $\psi(1,2)$ is a possible eigenstate, so is $\psi(2,1)$, with the same energy. Then, according to the superposition principle, any linear combination of $\psi(1,2)$ and $\psi(2,1)$ is also a possible solution with the same energy.

Let $\hat{P}_{12}$ denote the operator that exchanges the particles in states α and β

$$\hat{P}_{12}\psi(1,2) = \psi(2,1). \tag{6.4}$$

Two consecutive exchanges give back the original state,

$$\hat{P}_{12}^2\psi(1,2) = \hat{P}_{12}\psi(2,1) = \psi(1,2). \tag{6.5}$$

So,

$$\hat{P}_{12}^2 = I \quad \text{and} \quad \hat{P}_{12} = \hat{P}_{12}^{-1}. \tag{6.6}$$

The operator $\hat{P}_{12}$ is an involution, i.e. it is its own inverse, and hence, it is Hermitian and unitary at the same time, just like the parity operator. Furthermore, just like the parity operator, it represents a discrete symmetry transformation.

The indistinguishability of identical particles implies that the Hamilton operator commutes with $\hat{P}_{12}$

$$[\hat{P}_{12}, \hat{H}] = 0, \qquad \text{or} \qquad \hat{H} = \hat{P}_{12}\hat{H}\hat{P}_{12}. \tag{6.7}$$

For that reason, $\hat{H}$ and $\hat{P}_{12}$ must have common eigenstates. In fact, Eq. (6.5) is an eigenvalue equation which shows that the eigenvalue of $\hat{P}_{12}^2$ is $\lambda^2 = 1$. Consequently, the eigenvalue of $\hat{P}_{12}$ is $\lambda = \pm 1$.

We have to find a combination of $\psi(1,2)$ and $\psi(2,1)$ that is, besides being an eigenstate of $\hat{H}$ with eigenvalue E, also an eigenstate of $\hat{P}_{12}$ with definite eigenvalue λ. We can introduce the operators

$$\hat{S} = \frac{1}{2}(1 + \hat{P}_{12}) \quad \text{and} \quad \hat{A} = \frac{1}{2}(1 - \hat{P}_{12}), \tag{6.8}$$

and we call them the *symmetrizing* and *antisymmetrizing* operators, respectively. These operators are projectors,

$$\hat{S}^2 = \hat{S} \quad \text{and} \quad \hat{A}^2 = \hat{A}, \tag{6.9}$$

and they project into orthogonal subspaces, since

$$\hat{S}\hat{A} = \hat{A}\hat{S} = 0. \tag{6.10}$$

We can also see that

$$\hat{P}_{12}\hat{S} = \hat{S} \quad \text{and} \quad \hat{P}_{12}\hat{A} = -\hat{A}. \tag{6.11}$$

So, acting on an arbitrary $|\psi\rangle$ gives

$$\hat{P}_{12}\hat{S}|\psi\rangle = \hat{S}|\psi\rangle \quad \text{and} \quad \hat{P}_{12}\hat{A}|\psi\rangle = -\hat{A}|\psi\rangle. \tag{6.12}$$

In other words, with

$$|\psi_s\rangle = \hat{S}|\psi\rangle \quad \text{and} \quad |\psi_a\rangle = \hat{A}|\psi\rangle, \tag{6.13}$$

we have

$$\hat{P}_{12}|\psi_s\rangle = |\psi_s\rangle \quad \text{and} \quad \hat{P}_{12}|\psi_a\rangle = -|\psi_a\rangle, \tag{6.14}$$

i.e. $|\psi_s\rangle$ and $|\psi_a\rangle$ are the symmetric and antisymmetric eigenstates of the permutation operator, respectively. They are the common eigenstates of the Hamiltonian and the $\hat{P}_{12}$ operators, so they are the physically admissible states of a system containing two identical particles. Since $\hat{P}_{12}$ commutes with $\hat{H}$, it is a constant of motion. If a system is in a particular symmetric or antisymmetric state at some time, it remains so for all time. We can see that for two identical particles, $\hat{S} + \hat{A} = I$. So, any vector $|\phi\rangle$ in the Hilbert space of two identical particles can be split into symmetric and antisymmetric components

$$|\phi\rangle = (\hat{S} + \hat{A})|\phi\rangle = \hat{S}|\phi\rangle + \hat{A}|\phi\rangle. \tag{6.15}$$

This means that the two-particle Hilbert space is separated into disjoint subspaces of symmetric and antisymmetric vectors.

From the indistinguishability of identical particles, it also follows that the dynamical states that differ only by a permutation of identical particles cannot be distinguished by other observations. So, for any observable $\hat{O}$, we must have

$$\langle\psi|\hat{O}|\psi\rangle = \langle\psi|\hat{P}_{ij}\hat{O}\hat{P}_{ij}|\psi\rangle, \tag{6.16}$$

consequently $\hat{O} = \hat{P}_{ij}\hat{O}\hat{P}_{ij}$, or

$$\hat{P}_{ij}\hat{O} = \hat{O}\hat{P}_{ij} \quad \text{and} \quad \left[\hat{O}, \hat{P}_{ij}\right] = 0. \tag{6.17}$$

The physically admissible operators must be invariant with respect to the permutation of identical particles, just like the Hamilton operator. They cannot mix symmetric and antisymmetric states

$$\langle\psi_s|\hat{O}|\psi_a\rangle = \langle\psi_s|\hat{P}_{ij}\hat{O}\hat{P}_{ij}|\psi_a\rangle = -\langle\psi_s|\hat{O}|\psi_a\rangle, \tag{6.18}$$

therefore, $\langle\psi_s|\hat{O}|\psi_a\rangle = 0$. By taking $\hat{O} = I$, the identity operator, we see that the states of different symmetry are orthogonal.

If we have more than two identical particles in the system, all two-particle exchange operators $\hat{P}_{ij}$ should commute with the Hamiltonian

$$[\hat{P}_{ij}, \hat{H}] = 0, \quad i, j = 1, \dots n, \quad i \neq j. \tag{6.19}$$

However, the exchange operators, in general, do not commute

$$\hat{P}_{ij}\hat{P}_{jk} \neq \hat{P}_{jk}\hat{P}_{ij}. \tag{6.20}$$

For example, let us consider three identical particles $(1, 2, 3)$. Then

$$\hat{P}_{12}\hat{P}_{23}\psi(1, 2, 3) = \hat{P}_{12}\psi(1, 3, 2) = \psi(3, 1, 2), \tag{6.21}$$

while

$$\hat{P}_{23}\hat{P}_{12}\psi(1, 2, 3) = \hat{P}_{23}\psi(2, 1, 3) = \psi(2, 3, 1), \tag{6.22}$$

therefore, $\left[\hat{P}_{12}, \hat{P}_{23}\right] \neq 0$. So, it is not possible to find a complete set of commuting observables that contains all the two-particle exchange operators and the Hamiltonian. On the other hand, the physical system should be a common eigenstate of $\hat{H}$ and all the $\hat{P}_{ij}$ operators

$$\hat{H}\psi = E\psi \tag{6.23a}$$

$$\hat{P}_{ij}\psi = \lambda_{ij}\psi, \quad i, j = 1, \dots n, \quad i \neq j, \tag{6.23b}$$

where $\lambda_{ij} = \pm 1$.

We can easily see that, for any i, j, k values, we have

$$\hat{P}_{ij}\hat{P}_{ik} = \hat{P}_{jk}\hat{P}_{ij} = \hat{P}_{ik}\hat{P}_{jk}. \tag{6.24}$$

By applying these exchanges on ψ we find

$$\lambda_{ij}\lambda_{ik}\psi = \lambda_{jk}\lambda_{ij}\psi = \lambda_{ik}\lambda_{jk}\psi. \tag{6.25}$$

This is possible only if

$$\lambda_{ij} = \lambda_{ik} = \lambda_{jk}, \tag{6.26}$$

i.e. all λ_{ij} eigenvalues are the same. So, the wave function of identical particles is either totally symmetric ψ_s with all $\lambda_{ij} = 1$, or totally antisymmetric ψ_a with all $\lambda_{ij} = -1$.

For three identical particles, we have permutations that contain, with respect to the original configuration $I = \hat{P}_{123}$, an even number of exchanges

$$\hat{P}_{123}, \quad \hat{P}_{312}, \quad \hat{P}_{231}, \tag{6.27}$$

and an odd number of exchanges

$$\hat{P}_{132}, \quad \hat{P}_{213}, \quad \hat{P}_{321}. \tag{6.28}$$

The symmetrizing operator is given by the sum of all configurations

$$\hat{S} = \frac{1}{6}(\hat{P}_{123} + \hat{P}_{312} + \hat{P}_{231} + \hat{P}_{132} + \hat{P}_{213} + \hat{P}_{321}). \tag{6.29}$$

On the other hand, in the anti-symmetrizing operator each exchange introduces a -1 factor

$$\hat{A} = \frac{1}{6}(\hat{P}_{123} + \hat{P}_{312} + \hat{P}_{231} - \hat{P}_{132} - \hat{P}_{213} - \hat{P}_{321}). \tag{6.30}$$

If we have more than two particles $\hat{S} + \hat{A} \neq I$. While the symmetrization procedure splits the Hilbert space into disjoint symmetric and antisymmetric subspaces, in general they do not add up to the whole multi-particle Hilbert space. So, the identity of particles and the necessary symmetrization results in a drastic reduction of the physically admissible states in the Hilbert space.

Let us consider two identical particles and express the wave function in terms of center-of-mass coordinate sytem

$$\Psi = \Phi(\boldsymbol{R})\psi(\boldsymbol{r}), \tag{6.31}$$

where $\boldsymbol{R}$ and $\boldsymbol{r}$ are the center-of-mass and the relative coordinates, respectively. In the angular basis we can have $\psi(\boldsymbol{r})$ expanded in terms of $Y_{l,m}$ functions, and the two spins $s_1 = s_2 = s$ are coupled to the total spin S

$$|S, M\rangle = \sum_{m_1, m_2} \langle s, s; m_1, m_2|S, M\rangle |s, m_1\rangle |s, m_2\rangle. \tag{6.32}$$

The exchange of two particles implies a transformation $\boldsymbol{r} \to -\boldsymbol{r}$ for the spatial coordinate. Consequently, the $Y_{l,m}$ function transforms as Eq. (4.299)

$$\hat{P}_{12} Y_{l,m} = (-)^l Y_{l,m}. \tag{6.33}$$

On the other hand, the spin function in Eq. (6.32) transforms, due to Eq. (4.198), as

$$\hat{P}_{12}|S, M\rangle = (-)^{2s-S}|S, M\rangle. \tag{6.34}$$

So, by exchanging two identical particles with spin, the wave function in relative coordinates acquires the phase $(-)^{l+2s-S}$.

We can again see that the identity of particles results in a reduction of the Hilbert space. For example, if two $s = 1/2$ spins are coupled to the $S = 1$ triplet state, which is symmetric in spin coordinates, the relative angular momentum l should be even for a symmetric, and should be odd for an antisymmetric total wave function. However, if the spins are coupled to the $S = 0$ singlet state, the situation is the opposite. In this case the system is antisymmetric in spin coordinates and therefore l should be odd for a symmetric and even for an antisymmetric total wave function.

6.1.2 Spin and Statistics

We have found that the exchange of two identical particles results in a phase-change of the wave function. The spin-statistic theorem relates the spin of the particle to its exchange properties. By exchanging two integer-spin particles, the wave function experiences a $+1$ phase change, and the many particle system follows Bose-Einstein statistics. On the other hand, by exchanging two half-integer spin particles, the wave function experiences a -1 phase change, and the many particle system follows Fermi-Dirac statistics. So, an integer spin particle is a *boson*, while a half-integer spin particle is a *fermion*.

We can treat the exchange of two identical particles with spin s as a rotation of each particle by π around their center of mass. This is, however, equivalent to rotation of one particle by 2π around the other particle. This rotation introduces a phase $(-)^{2s}$, which give rise to $+1$ phase for integer spin, and -1 phase for half-integer spin particles.

As an example, we consider two identical $s = 1/2$ particles. Assuming that we rotate in the $x - y$ plane, the rotation by π is given by

$$\hat{R}_z(\pi) = \exp(i\pi\sigma_z/2) = \cos\pi/2 + i\sin\pi/2\,\sigma_z = i\sigma_z = \begin{pmatrix} i & 0 \\ 0 & -i \end{pmatrix}. \tag{6.35}$$

The rotation of single states yield

$$\hat{R}_z(\pi)|\psi_1\rangle = \begin{pmatrix} i & 0 \\ 0 & -i \end{pmatrix}|\psi_1\rangle \quad \text{and} \quad \hat{R}_z(\pi)|\psi_2\rangle = \begin{pmatrix} i & 0 \\ 0 & -i \end{pmatrix}|\psi_2\rangle, \tag{6.36}$$

and for a two-particle system we find

$$\begin{aligned} \hat{P}_{12}|\psi_1\rangle|\psi_2\rangle &= \hat{R}_z(\pi)(|\psi_1\rangle|\psi_2\rangle) = (\hat{R}_z(\pi)|\psi_1\rangle)(\hat{R}_z(\pi)|\psi_2\rangle) \\ &= \begin{pmatrix} i & 0 \\ 0 & -i \end{pmatrix}^2 |\psi_1\rangle|\psi_2\rangle = -|\psi_1\rangle|\psi_2\rangle. \end{aligned} \tag{6.37}$$

So, the wave function of a system consisting two identical spin-1/2 particles is antisymmetric with respect to exchange of the particles.

We should note that this exchange symmetry of identical particles is the consequence of the topological structure of the configuration space. We can perform the rotation in any plane, and we should get the same result. We can even take particle 2 around in a plane, which does not contain particle 1, and then shrink the rotation to just a point. The three-dimensional space minus the origin, $\mathbf{R}^3/\{0\}$, is simply connected. The wave function cannot make a jump, it is single-valued, therefore $\lambda^2 = 1$ and $\lambda = \pm 1$.

This argument does not hold in two dimensions. The two-dimensional space minus a point singularity leads to multiply connected space. A circle around the singularity cannot shrink to a point without crossing the singularity. Taking a particle around a circle should result in the same physics, i.e. the same $|\psi|^2$. Therefore $|\lambda|^2 = 1$, which gives $\lambda = e^{i\theta}$. If $\theta = 0$ and $e^{i0} = 1$, we recover the bosons. While, if $\theta = \pi$ and $e^{i\pi} = -1$, then we have fermions. In two dimensions, however, any values of θ are possible, they are the *anyons*, and they follow anyon, or fractional, statistics.

6.2 Independent Particle Approximation

In order to describe a quantum state of N identical non-interacting particles, it is useful to consider the problem in a single-particle basis. Let $\mathcal{H}^{(N)}$ be the direct product of single-particle Hilbert spaces

$$\mathcal{H}^{(N)} = \mathcal{H}_1 \otimes \mathcal{H}_2 \otimes \ldots \otimes \mathcal{H}_N, \tag{6.38}$$

with single particle Hamiltonians $\hat{H}_i$ acting in each $\mathcal{H}_i$. Since the particles are identical, each $\hat{H}_i$ has the same functional form. We denote the single particle state with quantum number n_i and energy ε_{n_i}, by $\psi_{n_i}(\xi_i)$. The energy of the N-particle system is given by $E = \sum_{i=1}^{N} \varepsilon_{n_i}$. It may happen that not all the n_i quantum numbers are different. We denote by N_i how many times the quantum number n_i shows up.

The symmetric wave function for identical bosons is given as a sum of all permutations of all particles in all possible quantum states

$$\Psi_s(\xi_1, \xi_2, \ldots, \xi_N) = \left(\frac{N_1!N_2!\cdots N_N!}{N!}\right)^{1/2} \sum_P \hat{P}\, \psi_{n_1}(\xi_1)\psi_{n_2}(\xi_2)\cdots\psi_{n_N}(\xi_N), \tag{6.39}$$

where $\hat{P}$ is the permutation operator $\hat{P} : (1, 2, \ldots, N) \to (i_1, i_2, \ldots, i_N)$. For example, if we have three bosons in two states, $n_1 = n_2 \neq n_3$, we find

$$\Psi_s(\xi_1, \xi_2, \xi_3) = \left(\frac{2!1!}{3!}\right)^{1/2} \sum_P \hat{P}\, \psi_{n_1}(\xi_1)\psi_{n_1}(\xi_2)\psi_{n_3}(\xi_3)$$

$$= \sqrt{\frac{1}{3}}\Big[\psi_{n_1}(\xi_1)\psi_{n_1}(\xi_2)\psi_{n_3}(\xi_3) + \psi_{n_1}(\xi_1)\psi_{n_3}(\xi_2)\psi_{n_1}(\xi_3) + \psi_{n_3}(\xi_1)\psi_{n_1}(\xi_2)\psi_{n_1}(\xi_3)\Big]. \tag{6.40}$$

In the case of fermions, the N_i numbers can take only values 0 and 1. We either have a particle in a quantum state n_i, or not, but we cannot have more than one particle there. If we had two identical fermions in the same quantum state, their exchange would not change the physical system. On the other hand, their exchange would change the sign of the wave function. So, in this case we would have $\psi = -\psi$, and thus the wave function has to vanish. Therefore, *no two identical fermions can occupy the same quantum state*. This is the *Pauli principle*, or the *Pauli exclusion principle*. The Pauli principle is a consequence of the exchange symmetry of identical fermions.

The wave function of an N-fermion system that reflects the Pauli principle is given by

$$\Psi_a(\xi_1, \xi_2, \ldots, \xi_N) = \left(\frac{1}{N!}\right)^{1/2} \sum_P (-)^P \hat{P}\, \psi_{n_1}(\xi_1)\psi_{n_2}(\xi_2)\cdots\psi_{n_N}(\xi_N), \tag{6.41}$$

where $(-)^P = \pm 1$, depending on whether the sequence is an even or odd permutation of the original $(1, 2, \ldots, N)$ sequence. The determinant wave function of single particle states, the *Slater determinant*, automatically provides the required antisymmetry property

$$\Psi_a(\xi_1, \xi_2, \ldots, \xi_N) = \frac{1}{\sqrt{N!}} \begin{vmatrix} \psi_{n_1}(\xi_1) & \psi_{n_1}(\xi_2) & \cdots & \psi_{n_1}(\xi_N) \\ \psi_{n_2}(\xi_1) & \psi_{n_2}(\xi_2) & \cdots & \psi_{n_2}(\xi_N) \\ \vdots & \vdots & \ddots & \vdots \\ \psi_{n_N}(\xi_1) & \psi_{n_N}(\xi_2) & \cdots & \psi_{n_N}(\xi_N) \end{vmatrix}. \tag{6.42}$$

6.2.1 Occupation Number Representation

We described the quantum states in terms of the eigenstates $|n\rangle$ of the one-particle Hamiltonian. Let us consider an N-boson state, where particle 1 is in the $|n_1\rangle$ quantum state, particle 2 is in the $|n_2\rangle$ state, and so on. A completely equivalent picture is obtained if we say that there are n_0 particles in quantum state $|0\rangle$, n_1 particles in state in state $|1\rangle$, and so on, i.e. we describe the state by the infinite vector

$$|n_0, n_1, n_2, \ldots\rangle, \tag{6.43}$$

with the constraint

$$\sum_{k=0}^{\infty} n_k = N. \tag{6.44}$$

For example, the 4-particle system

$$|1, 0, 2, 0, 1, 0, 0, \ldots\rangle \tag{6.45}$$

has 1 particle in the ground state, 2 particles in the second excited state and 1 particle in the fourth excited state. Of course, this state is a boson state. An N-fermion system would have only zeros or ones in each quantum level.

If the Hamilton operator is written as a sum of one-particle Hamiltonians

$$\hat{H} = \sum_{i=1}^{N} \hat{H}_i(\xi_i), \quad \text{with} \quad \hat{H}_i(\xi_i)|k\rangle = \varepsilon_k|k\rangle, \tag{6.46}$$

then,

$$\hat{H}|n_0, n_1, \ldots\rangle = \left(\sum_k n_k \varepsilon_k\right)|n_0, n_1, \ldots\rangle. \tag{6.47}$$

We may introduce the number operator

$$\hat{N}_k|n_0, n_1, \ldots\rangle = n_k|n_0, n_1, \ldots\rangle, \tag{6.48}$$

and, analogously to the one-dimensional harmonic oscillator case, we can define the annihilation and creation operators

$$\hat{a}_k|n_0, n_1, \ldots, n_k, \ldots\rangle = \sqrt{n_k}|n_0, n_1, \ldots, n_k - 1, \ldots\rangle, \tag{6.49}$$

and

$$\hat{a}_k^\dagger|n_0, n_1, \ldots, n_k, \ldots\rangle = \sqrt{n_k + 1}|n_0, n_1, \ldots, n_k + 1, \ldots\rangle, \tag{6.50}$$

with commutation relations

$$\left[\hat{a}_k, \hat{a}_j^\dagger\right] = \delta_{kj}, \quad \left[\hat{a}_k, \hat{a}_j\right] = \left[\hat{a}_k^\dagger, \hat{a}_j^\dagger\right] = 0. \tag{6.51}$$

These lead us to the Hamiltonian

$$\hat{H} = \sum_k \varepsilon_k \hat{N}_k = \sum_k \varepsilon_k \hat{a}_k^\dagger \hat{a}_k. \tag{6.52}$$

The N-boson state is created from the ground, or *vacuum*, state by repeated applications of creation operators

$$|n_0, n_1, \ldots\rangle = \frac{(\hat{a}_0^\dagger)^{n_0}(\hat{a}_1^\dagger)^{n_1}\cdots}{\sqrt{n_0!}\sqrt{n_1!}\cdots}|0, 0, \ldots\rangle. \tag{6.53}$$

These states are orthonormal and form a basis in the N-particle boson space

$$\sum_{\sum n_k=N} |n_0, n_1, \ldots\rangle\langle n_0, n_1, \ldots| = 1. \tag{6.54}$$

In the case of fermions, the state

$$|n_0, n_1, \ldots, n_k, \ldots\rangle, \tag{6.55}$$

is such that n_k can take only values of 0 and 1. As a consequence, the number operator $\hat{N}_k = \hat{a}_k^\dagger \hat{a}_k$ must have the property

$$\hat{N}_k^2 = \hat{N}_k. \tag{6.56}$$

Indeed, from its eigenvalue equation we can see that $N_k^2 = N_k$, which has only two solutions, $n_k = (0, 1)$. The N-fermion state is also created from the vacuum

$$|n_0, n_1, \ldots\rangle = \left((\hat{a}_0^\dagger)^{n_0}(\hat{a}_1^\dagger)^{n_1}\cdots\right)|0, 0, \ldots\rangle. \tag{6.57}$$

From the fact that $n_k = (0, 1)$, it follows that $(\hat{a}_k^\dagger)^2 = (\hat{a}_k)^2 = 0$.

To find the commutation relation between the operators $\hat{a}$ and $\hat{a}^\dagger$, we put fermions in the k-th and in the j-th states

$$|0, 0, \ldots, 1, \ldots, 1, \ldots\rangle = \hat{a}_k^\dagger \hat{a}_j^\dagger\, |0, 0, \ldots, 0, \ldots, 0, \ldots\rangle. \tag{6.58}$$

This two-particle state is antisymmetric with respect to the exchange of the two fermions, therefore

$$\hat{a}_k^\dagger \hat{a}_j^\dagger = -\hat{a}_j^\dagger \hat{a}_k^\dagger, \tag{6.59}$$

or

$$\hat{a}_k^\dagger \hat{a}_j^\dagger + \hat{a}_j^\dagger \hat{a}_k^\dagger = 0. \tag{6.60}$$

The Hermitian adjoint gives

$$\hat{a}_k \hat{a}_j + \hat{a}_j \hat{a}_k = 0, \tag{6.61}$$

i.e. these operators anti-commute

$$\left[\hat{a}_k^\dagger, \hat{a}_j^\dagger\right]_+ = \left[\hat{a}_k, \hat{a}_j\right]_+ = 0. \tag{6.62}$$

In order to satisfy the $\hat{N}_k^2 = \hat{N}_k$ relation, we need to postulate the anti-commutation relation

$$[\hat{a}_j, \hat{a}_k^\dagger]_+ = \hat{a}_j\hat{a}_k^\dagger + \hat{a}_k^\dagger\hat{a}_j = \delta_{jk}. \tag{6.63}$$

This choice guaranties Eq. (6.56), since

$$\hat{N}_k^2 = \hat{a}_k^\dagger\hat{a}_k\hat{a}_k^\dagger\hat{a}_k = \hat{a}_k^\dagger(1 - \hat{a}_k^\dagger\hat{a}_k)\hat{a}_k = \hat{a}_k^\dagger\hat{a}_k - (\hat{a}_k^\dagger)^2(\hat{a}_k)^2 = \hat{a}_k^\dagger\hat{a}_k = \hat{N}_k. \tag{6.64}$$

Similarly to the N-boson system, we find analogous relations for fermions

$$\hat{N} = \sum_k \hat{N}_k = \sum_k \hat{a}_k^\dagger\hat{a}_k, \tag{6.65}$$

and

$$\hat{H} = \sum_k \varepsilon_k\hat{a}_k^\dagger\hat{a}_k. \tag{6.66}$$

Furthermore,

$$|\Psi\rangle = \sum_{\sum n_k = N} |n_0, n_1, \ldots\rangle\langle n_0, n_1, \ldots|\Psi\rangle. \tag{6.67}$$

6.3 Isospin Formalism

It is relatively easy to see the identity of particles if they are identical in all respects. Electrons are identical since all electrons have the same properties. It is not so easy to realize the underlying identity of particles when they are different in many respects. For example, protons and neutrons are different particles as their masses and charges are different, although their masses are rather close, $m_p = 938.272$ MeV, while $m_n = 939.565$ MeV. An atomic nucleus has N number of neutrons and Z number of protons. The corresponding mirror partner has N number of protons and Z number of neutrons. It has been observed that the mirror nuclei, like ^{3}H-^{3}He, ^{9}B-^{9}Be, ^{14}C-^{14}O, ^{15}N-^{15}O, ^{24}Na-^{24}Al, ^{73}Rb-^{73}Kr, etc., have rather similar properties and rather similar energy spectra.

So, protons and neutrons behave rather similarly in the nucleus. Maybe we can consider them identical, just like a degenerate quantum state, which splits if we add some weak perturbation. This is similar to how the energy levels of hydrogen atoms split if we add a magnetic interaction. This physical picture is formulated in the *isospin* formalism.

In this picture, the proton and neutron, called nucleons, are isospin $t = 1/2$ particles. The proton is considered a quantum state with isospin up, $t_3 = +1/2$, and the neutron with isospin down, $t_3 = -1/2$. So, we can write the proton and neutron in two-component form

$$|\eta_p\rangle = |t = 1/2, t_3 = 1/2\rangle = \begin{pmatrix} 1 \\ 0 \end{pmatrix} \quad \text{and} \quad |\eta_n\rangle = |t = 1/2, t_3 = -1/2\rangle = \begin{pmatrix} 0 \\ 1 \end{pmatrix}. \tag{6.68}$$

In analogy with spin operators, we introduce the isospin operators

$$\hat{\boldsymbol{t}}_i = \frac{1}{2}\boldsymbol{\tau}_i, \qquad i = 1, 2, 3, \tag{6.69}$$

where

$$\tau_1 = \begin{pmatrix} 0 & 1 \\ 1 & 0 \end{pmatrix}, \quad \tau_2 = \begin{pmatrix} 0 & -i \\ i & 0 \end{pmatrix}, \quad \tau_3 = \begin{pmatrix} 1 & 0 \\ 0 & -1 \end{pmatrix}, \tag{6.70}$$

are the isospin Pauli matrices. Formally, they are just like the spin-1/2 Pauli operators, but they act in the isospin space. We can also introduce

$$\hat{t}^2 = \hat{t}_1^2 + \hat{t}_2^2 + \hat{t}_3^2, \tag{6.71}$$

which commutes with the components t_i. Thus, $\hat{t}^2$ and $\hat{t}_3$ form a complete set of commuting observables in the isospin space. The states $|\eta_p\rangle$ and $|\eta_n\rangle$ are eigenstates of $\hat{t}_3$ with eigenvalues $t_3 = \pm 1/2$, so

$$\hat{t}_3|\eta_p\rangle = \frac{1}{2}|\eta_p\rangle \quad \text{and} \quad \hat{t}_3|\eta_n\rangle = -\frac{1}{2}|\eta_n\rangle, \tag{6.72}$$

and also eigenstates of $\hat{\boldsymbol{t}}^2$

$$\hat{\boldsymbol{t}}^2|t, t_3\rangle = t(t+1)|t, t_3\rangle, \tag{6.73}$$

with $t = 1/2$.

We can also construct

$$\hat{t}_+ = \hat{t}_1 + i\hat{t}_2 = \frac{1}{2}(\tau_1 + i\tau_2) = \begin{pmatrix} 0 & 1 \\ 0 & 0 \end{pmatrix} \quad \text{and} \quad \hat{t}_- = \hat{t}_1 - i\hat{t}_2 = \frac{1}{2}(\tau_1 - i\tau_2) = \begin{pmatrix} 0 & 0 \\ 1 & 0 \end{pmatrix}, \tag{6.74}$$

which act as ladder operators in the isospin space

$$\hat{t}_\pm|t, t_3\rangle = \sqrt{t(t+1) - t_3(t_3 \pm 1)}\,|t, t_3 \pm 1\rangle, \tag{6.75}$$

or

$$\hat{t}_+|\eta_p\rangle = 0, \quad \hat{t}_-|\eta_p\rangle = |\eta_n\rangle, \quad \hat{t}_+|\eta_n\rangle = |\eta_p\rangle, \quad \hat{t}_-|\eta_n\rangle = 0. \tag{6.76}$$

We can see that $\hat{t}_+$ converts a neutron state into a proton state, while $\hat{t}_-$ converts a proton state into a neutron state.

6.3.1 Two-Nucleon Systems

Since the isospin operators follow the same algebra as the corresponding angular momentum operators, it is reasonable to assume that we can push the analogy a little further. Just like in the case of coupling two spin-1/2 particles, we can also couple two isospin-1/2 particles. Then, we obtain the following combinations

$$\begin{aligned}
|t = 1, t_3 = 1\rangle &= |\eta_p(1)\rangle|\eta_p(2)\rangle, \\
|t = 1, t_3 = -1\rangle &= |\eta_n(1)\rangle|\eta_n(2)\rangle, \\
|t = 1, t_3 = 0\rangle &= \frac{1}{\sqrt{2}}\left[|\eta_p(1)\rangle|\eta_n(2)\rangle + |\eta_n(1)\rangle|\eta_p(2)\rangle\right], \\
|t = 0, t_3 = 0\rangle &= \frac{1}{\sqrt{2}}\left[|\eta_p(1)\rangle|\eta_n(2)\rangle - |\eta_n(1)\rangle|\eta_p(2)\rangle\right].
\end{aligned} \tag{6.77}$$

The $t = 1$ combinations are symmetric with respect to isospin exchange, while the $t = 0$ combination is antisymmetric. The nucleons are fermions, so the two-nucleon total wave function should be antisymmetric with respect to spatial, spin and isospin exchanges. It is an experimental fact that we have only one bound pn system, the deuteron. It must be the $t = 0$ isospin state. From the experimental observation that the magnetic moment of the deuteron is almost the sum of the proton and neutron magnetic moments, it follows that the relative angular momentum should be $l = 0$. Since the deuteron quadruple moment is not zero, it must have another angular momentum component, which can only be $l = 2$, as the strong nuclear interaction preserves the parity. Anyway, its orbital angular momentum is even; it is predominantly $l = 0$, or S-state, with a little $l = 2$, D-state, admixture. The internal parity of the nucleons is even, the relative angular momentum is even, so, the parity of the deuteron is even. The spatial part of the wave function is symmetric, the isospin part is antisymmetric, and since the deuteron is made of two identical nucleons, the spin part of the wave function has to be symmetric. Indeed, the deuteron is a spin-1 nucleus, it is a ${}^3S_1 -{}^3 D_1$ coupled state. This also shows the validity of the isospin concept. If the proton and neutron were not identical particles, we would not have been able to say any definite statement about the spin of the deuteron.

In two-nucleon systems, the deuteron state, $|t = 0, t_3 = 0\rangle$, is bound, but the $t = 1$ states have higher energy and they are unbound. There are no stable di-neutrons; the $t = 1$ two-nucleon systems have been observed only as short lived correlations in nuclear scattering experiments. This isospin pattern, with two-nucleon correlation, is repeated in several systems. For example, in the $\alpha + 2N$ systems, the ${}^6Li = \alpha + np$ with $t = 0$, is more bound than the $t = 1$ states which are like the ${}^6He = \alpha + nn$, the excited ${}^6Li = \alpha + np$, and the ${}^6Be = \alpha + pp$ nuclei. These $t = 1$ systems have very similar binding energies, only slightly different due to the Coulomb interaction.

6.3.2 Pions and Pion-Nucleon Systems

Pions appear in three charge states as π^-, π^0, π^+. Their masses are very close, $m_{\pi^0} = 135$ MeV and $m_{\pi^\pm} = 139.59$ MeV, respectively. So, we may identify them as $t = 1$ particles with $t_3 = -1, 0, 1$:

$$|t=1, t_3=1\rangle = -|\pi^+\rangle, \quad |t=1, t_3=0\rangle = |\pi^0\rangle, \quad |t=1, t_3=-1\rangle = |\pi^-\rangle, \tag{6.78}$$

where the phases are arbitrary, with the adopted conventions. The states

$$|\pi_1\rangle = \frac{1}{\sqrt{2}}\left(|\pi^+\rangle + |\pi^-\rangle\right), \quad |\pi_2\rangle = -\frac{i}{\sqrt{2}}\left(|\pi^+\rangle - |\pi^-\rangle\right), \quad |\pi_3\rangle = |\pi^0\rangle \tag{6.79}$$

transform in iso-space like rank-1 spherical tensors.

If we consider a pion-nucleon system, we can couple the $t_\pi = 1$ pion and the $t_N = 1/2$ nucleon to a $t = 3/2$ and a $t = 1/2$ isospin states. With the help of Clebsch-Gordan coefficients, we obtain the physical sates as linear combinations of various isospin components

$$\begin{aligned}
\pi^+ + p &: |3/2, 3/2\rangle, \\
\pi^- + p &: \langle 1, 1/2; -1, 1/2|3/2, -1/2\rangle|3/2, -1/2\rangle \\
&\quad + \langle 1, 1/2; -1, 1/2|1/2, -1/2\rangle|1/2, -1/2\rangle \\
&= \sqrt{1/3}|3/2, -1/2\rangle - \sqrt{2/3}|1/2, -1/2\rangle, \\
\pi^0 + p &: \langle 1, 1/2; 0, 1/2|3/2, 1/2\rangle|3/2, 1/2\rangle \\
&\quad + \langle 1, 1/2; 0, 1/2|1/2, 1/2\rangle|1/2, 1/2\rangle \\
&= \sqrt{2/3}|3/2, 1/2\rangle - \sqrt{1/3}|1/2, 1/2\rangle, \\
\pi^0 + n &: \langle 1, 1/2; 0, -1/2|3/2, -1/2\rangle|3/2, -1/2\rangle \\
&\quad + \langle 1, 1/2; 0, -1/2|1/2, -1/2\rangle|1/2, -1/2\rangle \\
&= \sqrt{2/3}|3/2, -1/2\rangle + \sqrt{1/3}|1/2, -1/2\rangle.
\end{aligned} \tag{6.80}$$

Let us consider the following pion-nucleon reactions:

(a) $\pi^+ + p \to \pi^+ + p$ direct scattering,
(b) $\pi^- + p \to \pi^- + p$ direct scattering,
(c) $\pi^- + p \to \pi^0 + n$ charge exchange reaction.

We assume that these reactions are governed primarily by the isospin dependent strong interaction and we take the interaction potential as

$$H_{int} = V(3/2)\,\delta_{t,3/2} + V(1/2)\,\delta_{t,1/2}. \tag{6.81}$$

Then, the amplitudes of the reactions are given by the matrix elements

$$\begin{aligned}
\langle \pi^+ + p | H_{int} | \pi^+ + p \rangle &= \langle 3/2, 3/2 | H_{int} | 3/2, 3/2 \rangle = V(3/2), \\
\langle \pi^- + p | H_{int} | \pi^- + p \rangle &= \frac{1}{3} V(3/2) + \frac{2}{3} V(1/2), \\
\langle \pi^0 + n | H_{int} | \pi^- + p \rangle &= \frac{\sqrt{2}}{3} V(3/2) - \frac{\sqrt{2}}{3} V(1/2).
\end{aligned} \tag{6.82}$$

An experimental observation is that the ratios of cross sections, which are proportional to the absolute value square of the amplitudes, are

$$\sigma(a) : \sigma(b) : \sigma(c) \approx 1 : \frac{1}{9} : \frac{2}{9}. \tag{6.83}$$

This suggests that $V(1/2) \approx 0$ and the reactions go predominantly through the $t = 3/2$ channel.

We can encounter a similar situation in nucleon-proton reactions

$$\begin{aligned}
&\text{(a):} \quad p + p \to d + \pi^+, \\
&\text{(b):} \quad n + p \to d + \pi^0.
\end{aligned} \tag{6.84}$$

Since the deuteron is an isospin-0 nucleus, $t_d = 0$ in both reactions, and the final state is in the $t = 1$ channel. In reaction (a) the initial system can exist only in the $t = 1$ channel. So,

$$p + p \to d + \pi^+ : |1, 1\rangle \to |1, 1\rangle. \tag{6.85}$$

In reaction (b), the initial system can exist either in $t = 1$ or in $t = 0$ isospin state

$$\begin{aligned}
n + p : &\langle 1/2, 1/2; -1/2, 1/2 | 1, 0 \rangle |1, 0\rangle + \langle 1/2, 1/2; -1/2, 1/2 | 0, 0 \rangle |0, 0\rangle \\
&= \sqrt{1/2} |1, 0\rangle + \sqrt{1/2} |0, 0\rangle.
\end{aligned} \tag{6.86}$$

As a result,

$$n + p \to d + \pi^0 : \sqrt{1/2} |1, 0\rangle + \sqrt{1/2} |0, 0\rangle \to |1, 1\rangle. \tag{6.87}$$

Assuming again an isospin dependent interaction

$$H_{int} = V(1)\delta_{t,1} + V(0)\delta_{t,0}, \tag{6.88}$$

the relevant matrix elements are

$$\begin{aligned}
\langle p + p | H_{int} | \pi^+ + d \rangle &= \langle 1, 1 | H_{int} | 1, 1 \rangle = V(1), \\
\langle n + p | H_{int} | \pi^0 + d \rangle &= \frac{1}{\sqrt{2}} \langle 1, 0 | H_{int} | 1, 0 \rangle + \frac{1}{\sqrt{2}} \langle 0, 0 | H_{int} | 1, 0 \rangle = \frac{1}{\sqrt{2}} V(1).
\end{aligned} \tag{6.89}$$

These predict the ratio of cross sections

$$\sigma(a) : \sigma(b) = 1 : (1/\sqrt{2})^2 = 2, \tag{6.90}$$

which is in perfect agreement with experimental results. This shows that there is no isospin mixing interaction term in these processes and proves again the validity of isospin formalism.

Another interesting consequence of the isospin formalism is that the reaction

$$d + d \rightarrow \alpha + \pi^0, \tag{6.91}$$

is forbidden since two $t_d = 0$ deuterons cannot couple to an α particle with $t_\alpha = 0$ plus a pion with $t_\pi = 1$.

Let us consider again reaction (a) of Eq. (6.84). We know that the pions are spin-0 particles and the deuteron has $S = 1$ spin, so the total spin of the $\pi^+ + d$ system is one. The nucleons are spin-1/2 particles, so the $p + p$ system should couple to $S = 1$ spin as well. Consequently, the $p + p$ system is symmetric in isospin and spin coordinates, but as they are fermions, the spatial part should be antisymmetric. As a result, their relative angular momentum should be $L = 1$. Since the nucleons have positive internal parity, the parity of the $p + p$ system is odd. The parity is conserved by the strong interaction and the deuteron has even parity. As a consequence, the pions are odd parity particles.

6.3.3 Charge Conjugation

The charge conjugation operator $\hat{C}$ transforms the particle state into its counterpart with opposite charge. If the system is a bound state of a particle-antiparticle pair, the $\hat{C}$ transformation inverts the relative position vector, which is basically a parity operation, and introduces a $(-)^L$ phase, where L is the relative orbital angular momentum. For a fermion-antifermion system of individual spins s and total spin S, the $\hat{C}$ transformation introduces a spin exchange as well. This involves a $(-)^{2s-S} = (-)^{2s+S}$ factor and an additional $(-)$ factor for exchanging fermions

$$\hat{C}|f\bar{f}\rangle = (-)^L(-)^{2s+S}(-)|f\bar{f}\rangle, \tag{6.92}$$

which becomes

$$\hat{C}|f\bar{f}\rangle = (-)^{L+S}|f\bar{f}\rangle \tag{6.93}$$

for spin-1/2 particles. So, a fermion-antifermion spin-1/2 system has $\eta_C = (-)^{L+S}$ C-parity.

For example, the electron-positron pair with $L = 0$ can couple to a $S = 0$ singlet or the $S = 1$ triplet state. The singlet state has $\eta_C(^1S_0) = (-)^{0+0} = 1$, and it can decay into two photons, since $(-1) \cdot (-1) = 1$. On the other hand, the triplet state has $\eta_C(^3S_1) = (-)^{0+1} = -1$, and it can decay into three photons, since $(-1) \cdot (-1) \cdot (-1) = -1$. This means that we should assign $\eta_C = -1$ C-parity for photons to make the picture consistent. Similarly, the π^0 is a bound state of a quark-antiquark pair with $L = S = 0$ and $\eta_C(\pi^0) = 1$. Therefore, it can decay into 2γ, but the 3γ decay is forbidden.

We should note however, that the use of C-parity is rather limited since most of the particles are not C-parity eigenstates. For example, $\hat{C}$ converts π^+ to π^-, or in the isospin picture, it converts the $|1, 1\rangle$ state to the $|1, -1\rangle$ state. On the other hand, a rotation by π in the isospin space turns these states into each other, $|1, 1\rangle = \exp(i\pi\hat{t}_2)|1, -1\rangle$. So, the $\mathcal{G}$-parity transformation

$$\hat{\mathcal{G}} = \hat{C}\exp(i\pi\hat{t}_2), \tag{6.94}$$

a combination of the charge conjugation and a rotation by π in the isospin space, leaves the pion states unchanged. This means that pions are $\mathcal{G}$-parity eigenstates. In general, a rotation by π introduces a phase $(-)^t$, which gives

$$\eta_G = \eta_C(-)^t. \tag{6.95}$$

This rule, for particle-antiparticle bound systems amounts to

$$\eta_G = (-)^{L+S+t} = (-)^{J+t}. \tag{6.96}$$

Since, for pions, $J = 0$ and $t = 1$, then $\eta_G(\pi) = -1$. The ρ meson, with $J = 1$ and $t = 1$, has $\eta_G(\rho) = 1$, and consequently it can only decay into two pions. While the ω meson, with $J = 1$ and $t = 0$, has $\eta_G(\omega) = -1$, and it decays into three pions.

6.3.4 $\mathcal{CP}$ Symmetry

Nucleons are spin-$1/2$ particles and pions are spin-0 particles. Particles, similar to nucleons are called *baryons*, while particles similar to pions are called *mesons*. The baryons are made up from three quarks, while mesons are made up from quark-antiquark pairs. It has been observed that in nuclear reactions, the number of baryons are conserved, while the number of mesons are not conserved. So, we may introduce the *baryon number* B as a conserved quantity. It has also been observed that a certain kind of conservation occurs involving particles like Λ, Σ, Ξ baryons and K mesons, which also indicate the conservation of another type of quantum number, the *strangeness* S. Examining the charges of baryons and mesons, we can infer that

$$q = t_3 + \frac{B+S}{2} = t_3 + \frac{Y}{2}, \tag{6.97}$$

where $Y = B + S$ is the *hypercharge*. This relation is known as the Gell-Mann—Nishijima formula.

The kaons, or K mesons, are the lightest strange mesons. There are four kinds of kaons: K^-, K^+, K^0 and $\bar{K}^0$. The K^+ and K^0 have strangeness $S = 1$ and K^- and $\bar{K}^0$ have strangeness $S = -1$. The $\mathcal{CP}$ transformation, the combination of spatial inversion and charge conjugation, transforms the neutral K mesons

$$\hat{C}\hat{\mathcal{P}}|K^0\rangle = |\bar{K}^0\rangle \quad \text{and} \quad \hat{C}\hat{\mathcal{P}}|\bar{K}^0\rangle = |K^0\rangle. \tag{6.98}$$

As we can see, K^0 and $\bar{K}^0$ are not eigenstates of $\mathcal{CP}$. We can construct, however, linear combinations

$$|K_1^0\rangle = \frac{1}{\sqrt{2}}(|K^0\rangle + |\bar{K}^0\rangle) \quad \text{and} \quad |K_2^0\rangle = \frac{1}{\sqrt{2}}(|K^0\rangle - |\bar{K}^0\rangle), \tag{6.99}$$

which are $\mathcal{CP}$ eigenstates

$$\hat{C}\hat{\mathcal{P}}|K_1^0\rangle = |K_1^0\rangle \quad \text{and} \quad \hat{C}\hat{\mathcal{P}}|K_2^0\rangle = -|K_2^0\rangle. \tag{6.100}$$

As a consequence, the neutral K mesons, with definite strangeness, are linear combinations of $\mathcal{CP}$ eigenstates

$$|K^0\rangle = \frac{1}{\sqrt{2}}(|K_1^0\rangle + |K_2^0\rangle) \quad \text{and} \quad |\bar{K}^0\rangle = \frac{1}{\sqrt{2}}(|K_1^0\rangle - |K_2^0\rangle). \tag{6.101}$$

In processes governed by the strong interaction, the strangeness is conserved, while in processes governed by weak interactions, $\mathcal{CP}$ is conserved. We have found before that pions have odd parity. Therefore, $|K_1^0\rangle$, with $\mathcal{CP}$ symmetry $+1$, can decay into two pions only

$$K_1^0 \to \begin{cases} \pi^+ + \pi^-, \\ \pi^0 + \pi^0, \end{cases} \tag{6.102}$$

while $|K_2^0\rangle$, with $\mathcal{CP}$ symmetry -1, decays into three pions

$$K_2^0 \to \begin{cases} \pi^+ + \pi^- + \pi^0, \\ \pi^0 + \pi^0 + \pi^0. \end{cases} \tag{6.103}$$

The total mass of three pions are very close to the mass of the kaon. Therefore, the energy release is very small compared to the two pion decay. Consequently, the lifetime of K_2^0 is much longer than the lifetime of K_1^0, 5.2×10^{-8}s versus 0.89×10^{-10}s.

In collision experiments, when neutral K^0 mesons are produced, the K_1^0 component decays into two pions within a distance of a few centimeters away from the source, while the K_2^0 component travels a few meters before it decays into three pions. However, it has been observed that about 0.2% of K_2^0 decays goes into the two-pion channel, indicating a slight violation of the $\mathcal{CP}$ symmetry in weak processes.

The K_2^0 component can decay into semi-leptonic channels as well

$$\begin{aligned} K_2^0 &\to \pi^+ + e^- + \bar{\nu}_e, \\ K_2^0 &\to \pi^- + e^+ + \nu_e. \end{aligned} \tag{6.104}$$

These two processes are $\mathcal{CP}$ symmetric

$$\hat{\mathcal{C}}\hat{\mathcal{P}}(\pi^+ + e^- + \bar{\nu}_e) = \pi^- + e^+ + \nu_e. \tag{6.105}$$

It has been observed, however, that nature prefers the decay into the positron channel, in about 3.3×10^{-3} excess, indicating again a slight violation of the $\mathcal{CP}$ symmetry.

It is strongly believed that nature is invariant with respect to the $\mathcal{CPT}$ transformation, the combination of charge conjugation, spatial mirroring and time reversal. The $\mathcal{CPT}$ theorem states that a Lorentz invariant local field theory with Hermitian Hamiltonian must have $\mathcal{CPT}$ symmetry.

Chapter 7
Approximation Methods for Time-Independent Problems

In practice, we can only find analytic solutions for a few problems in quantum mechanics. We can solve some one-dimensional rectangular potential problems, the Dirac δ potential, the harmonic oscillator and the hydrogen atom problem. Maybe we could list a few more cases, but not too many. Obviously, we need numerical and approximation methods to explore the beauty and power of quantum mechanics.

7.1 Hellmann-Feynman Theorem

The Hellmann-Feynman theorem relates the parameter dependence of the energy to the expectation value of the parameter dependence of the Hamiltonian.

Let us assume that the Hamiltonian depends on a parameter, $\hat{H} = \hat{H}(\lambda)$. Then, with a normalized $\psi(\lambda)$, we have

$$\langle\psi(\lambda)|\psi(\lambda)\rangle = 1, \quad \frac{\mathrm{d}}{\mathrm{d}\lambda}\langle\psi(\lambda)|\psi(\lambda)\rangle = 0, \quad \text{and} \quad E(\lambda) = \langle\psi(\lambda)|\hat{H}(\lambda)|\psi(\lambda)\rangle. \tag{7.1}$$

By taking the derivative of $E(\lambda)$ and considering that $\hat{H}(\lambda)|\psi(\lambda)\rangle = E(\lambda)|\psi(\lambda)\rangle$, we obtain

$$\begin{aligned}
\frac{\mathrm{d}E(\lambda)}{\mathrm{d}\lambda} &= \langle\frac{\partial\psi(\lambda)}{\partial\lambda}|\hat{H}(\lambda)|\psi(\lambda)\rangle + \langle\psi(\lambda)|\frac{\partial\hat{H}(\lambda)}{\partial\lambda}|\psi(\lambda)\rangle + \langle\psi(\lambda)|\hat{H}(\lambda)|\frac{\partial\psi(\lambda)}{\partial\lambda}\rangle \\
&= E(\lambda)\langle\frac{\partial\psi(\lambda)}{\partial\lambda}|\psi(\lambda)\rangle + \langle\psi(\lambda)|\frac{\partial\hat{H}(\lambda)}{\partial\lambda}|\psi(\lambda)\rangle + E(\lambda)\langle\psi(\lambda)|\frac{\partial\psi(\lambda)}{\partial\lambda}\rangle \\
&= E(\lambda)\frac{\partial}{\partial\lambda}\langle\psi(\lambda)|\psi(\lambda)\rangle + \langle\psi(\lambda)|\frac{\partial\hat{H}(\lambda)}{\partial\lambda}|\psi(\lambda)\rangle \\
&= \langle\psi(\lambda)|\frac{\partial\hat{H}(\lambda)}{\partial\lambda}|\psi(\lambda)\rangle.
\end{aligned} \tag{7.2}$$

Z. Papp, *Mastering Quantum Mechanics*,
https://doi.org/10.1007/978-3-032-09011-9_7

A possible benefit of this theorem is that $\partial\hat{H}(\lambda)/\partial\lambda$ could be much simpler than the original Hamiltonian.

If the wave function is time-dependent, satisfying

$$i\hbar\frac{\partial\psi(\lambda,t)}{\partial t}=\hat{H}(\lambda,t)\psi(\lambda,t), \tag{7.3}$$

the Hellmann-Feynman theorem is not valid. However, assuming that the derivatives with respect to λ and t can be interchanged, we find the relation

$$\begin{aligned}\langle\psi|\frac{\partial H(\lambda,t)}{\partial\lambda}|\psi\rangle &= \frac{\partial}{\partial\lambda}\langle\psi|\hat{H}(\lambda,t)|\psi\rangle-\langle\frac{\partial\psi}{\partial\lambda}|\hat{H}(\lambda,t)|\psi\rangle-\langle\psi|\hat{H}(\lambda,t)|\frac{\partial\psi}{\partial\lambda}\rangle\\ &= i\hbar\frac{\partial}{\partial\lambda}\langle\psi|\frac{\partial\psi}{\partial t}\rangle-i\hbar\langle\frac{\partial\psi}{\partial\lambda}|\frac{\partial\psi}{\partial t}\rangle+i\hbar\langle\frac{\partial\psi}{\partial t}|\frac{\partial\psi}{\partial\lambda}\rangle\\ &= i\hbar\langle\psi|\frac{\partial^2\psi}{\partial\lambda\,\partial t}\rangle+i\hbar\langle\frac{\partial\psi}{\partial t}|\frac{\partial\psi}{\partial\lambda}\rangle=i\hbar\frac{\partial}{\partial t}\langle\psi|\frac{\partial\psi}{\partial\lambda}\rangle.\end{aligned} \tag{7.4}$$

As an application of the Hellmann-Feynman theorem, we calculate some matrix elements involving hydrogen wave functions. The Hamiltonian of the hydrogen atom problem is given by

$$\hat{H}=\frac{\hat{\boldsymbol{p}}^2}{2m}-\frac{e^2}{r}=-\frac{\hbar^2}{2m_e}\frac{\mathrm{d}^2}{\mathrm{d}r^2}+\frac{\hbar^2 l(l+1)}{2m_e r^2}-\frac{e^2}{r}. \tag{7.5}$$

The energy eigenvalues are given in Eq. (5.65)

$$E_n=-\frac{m_e e^4}{2\hbar^2}\frac{1}{(n_r+l+1)^2}=-\frac{e^2}{a_0}\frac{1}{2n^2}, \tag{7.6}$$

where $n=n_r+l+1$, with n_r radial quantum number, and $a_0=\hbar^2/(m_e e^2)$ is the Bohr radius.

Taking e^2 as a parameter we have

$$\langle nlm|\frac{1}{r}|nlm\rangle=-\langle nlm|\frac{\partial\hat{H}}{\partial e^2}|nlm\rangle=\frac{\mathrm{d}}{\mathrm{d}e^2}(-E_{n_r})=\frac{1}{a_0}\frac{1}{(n_r+l+1)^2}. \tag{7.7}$$

We can take l as a parameter as well. Then, from

$$\frac{\partial\hat{H}}{\partial l}=\frac{\hbar^2(2l+1)}{2m_e r^2}, \tag{7.8}$$

we obtain

$$\langle nlm|\frac{\partial \hat{H}}{\partial l}|nlm\rangle = \frac{\hbar^2(2l+1)}{2m_e}\langle nlm|\frac{1}{r^2}|nlm\rangle = \frac{\partial}{\partial l}\left[-\frac{e^2}{a_0}\frac{1}{2(n_r+l+1)^2}\right] = \frac{e^2}{a_0 n^3}, \tag{7.9}$$

which yields

$$\langle nlm|\frac{1}{r^2}|nlm\rangle = \frac{2}{a_0^2(2l+1)n^3}. \tag{7.10}$$

We can also take r as a parameter. Then, the Hellmann-Feynman theorem gives

$$\langle nlm|\frac{\partial \hat{H}}{\partial r}|nlm\rangle = \langle nlm|\left[-\frac{\hbar^2 l(l+1)}{m_e r^3} + \frac{e^2}{r^2}\right]|nlm\rangle = \frac{\partial E_n}{\partial r} = 0, \tag{7.11}$$

and consequently, we get

$$\langle nlm|\frac{1}{r^3}|nlm\rangle = \frac{2}{a_0^3 l(l+1)(2l+1)n^3}. \tag{7.12}$$

7.2 Virial Theorem

The virial theorem was introduced originally in classical mechanics, but valid also in quantum mechanics. We consider a Hamiltonian

$$\hat{H} = \hat{H}(\hat{\boldsymbol{p}}, \hat{\boldsymbol{r}}), \tag{7.13}$$

and scale the variable $\hat{\boldsymbol{r}}$ as $\hat{\boldsymbol{r}}' = \lambda\hat{\boldsymbol{r}}$. Since

$$\hat{p}_x' = -i\hbar\frac{\partial}{\partial x'} = -i\hbar\frac{1}{\lambda}\frac{\partial}{\partial x} = \hat{p}_x/\lambda, \tag{7.14}$$

this scaling does not change the commutator

$$[\hat{x}', \hat{p}_x'] = [\hat{x}/\lambda, \lambda\hat{p}_x] = [\hat{x}, \hat{p}_x] = i\hbar. \tag{7.15}$$

Scaling is just using different units for $\boldsymbol{r}$ and $\boldsymbol{p}$, which obviously does not affect the energy. Observing that E_n does not depend on λ and taking $\lambda = 1$, due to the Hellmann-Feynman theorem, we obtain

$$\left.\frac{\partial E_n(\lambda)}{\partial \lambda}\right|_{\lambda=1} = \left.\left\langle \psi_n|\frac{\partial \hat{H}(\lambda)}{\partial \lambda}|\psi_n\right\rangle\right|_{\lambda=1} = 0. \tag{7.16}$$

We consider the Hamiltonian

$$\hat{H} = \frac{\hat{\boldsymbol{p}}^2}{2m} + V(\boldsymbol{r}), \tag{7.17}$$

and the scaling results in

$$\hat{H}(\lambda) = \hat{H}(\hat{\boldsymbol{p}}/\lambda, \lambda\boldsymbol{r}) = \frac{\hat{\boldsymbol{p}}^2}{\lambda^2 2m} + V(\lambda\boldsymbol{r}). \tag{7.18}$$

Then from Eq. (7.16) we have

$$0 = -2\left\langle \frac{\hat{\boldsymbol{p}}^2}{2m} \right\rangle + \langle \boldsymbol{r} \cdot \nabla V \rangle. \tag{7.19}$$

This is the virial theorem; a relation between the expectation value of the kinetic energy and the expectation value of the product of the distance and the gradient of the potential

$$2\langle T \rangle = \langle \boldsymbol{r} \cdot \nabla V \rangle. \tag{7.20}$$

For the hydrogen atom $V(r) \sim 1/r$, therefore we find

$$\langle \boldsymbol{r} \cdot \nabla V \rangle = \langle r\, \partial_r V \rangle = -\langle V \rangle. \tag{7.21}$$

For the matrix elements between hydrogen states of Eq. (5.70) we have

$$2\langle T \rangle_n + \langle V \rangle_n = 0. \tag{7.22}$$

From

$$\langle T \rangle_n + \langle V \rangle_n = E_n, \tag{7.23}$$

we can conclude that

$$\langle T \rangle_n = -E_n \quad \text{and} \quad \langle V \rangle_n = 2E_n, \tag{7.24}$$

in agreement with what we found in Eq. (7.7).

Following a similar analysis for the harmonic oscillator with $V \sim r^2$, we obtain

$$\langle \boldsymbol{r} \cdot \nabla V \rangle = 2\langle V \rangle. \tag{7.25}$$

Consequently,

$$2\langle T \rangle_n = 2\langle V \rangle_n \quad \text{and} \quad \langle T \rangle_n + \langle V \rangle_n = E_n, \tag{7.26}$$

and therefore,

$$\langle T \rangle_n = \langle V \rangle_n = E_n/2. \tag{7.27}$$

7.3 Perturbation Theory for Stationary States

7.3.1 *The Basic Principle of Perturbation Methods*

Perturbation methods represent a large variety of approximation methods in mathematical physics. The general idea is that somehow we introduce a continuous parameter λ into the problem. By taking $\lambda \in [0, 1]$, we switch some part of the problem on and off. If we can solve the $\lambda = 0$ version of the problem, we may try to solve the $\lambda = 1$ problem as a power series in λ.

As an illustrative example (p. 319 in [1]), let us consider the cubic equation

$$x^3 - 4.1x + 0.2 = 0. \tag{7.28}$$

This problem does not look promising since we do not have any small parameter here. We can rewrite as

$$x^3 - (4 + \lambda)x + 2\lambda = 0, \tag{7.29}$$

which with $\lambda = 0.1$ gives back the original equation. For $\lambda = 0$, we have

$$x^3 - 4x = 0, \tag{7.30}$$

and we find three solutions $x_1^{(0)} = -2$, $x_2^{(0)} = 0$ and $x_3^{(0)} = 2$.

Now, we may seek the first root of Eq. (7.28) as a power series

$$x_1(\lambda) = -2 + a_1\lambda + a_2\lambda^2 + \cdots . \tag{7.31}$$

Plugging back into Eq. (7.29) and requiring that the coefficients of each power of λ separately vanish, we find

$$8a_1 + 4 = 0, \qquad 8a_2 - a_1 - 6a_1^2 = 0, \tag{7.32}$$

and so on. The solutions of these equations are $a_1 = -1/2$ and $a_2 = 1/8$, which gives

$$x_1(\lambda) = -2 - \frac{1}{2}\lambda + \frac{1}{8}\lambda^2 + \cdots . \tag{7.33}$$

A similar procedure results in

$$x_2(\lambda) = 0 + \frac{1}{2}\lambda - \frac{1}{8}\lambda^2 + \cdots \quad \text{and} \quad x_3(\lambda) = 2 + 0\lambda + 0\lambda^2 + \cdots . \tag{7.34}$$

The exact numerical solutions of Eq. (7.28) are $x_1 = -2.048808$, $x_2 = 0.048808$ and $x_3 = 2$, while the perturbation method up to second order in λ gives $x_1 = -2.048750$, $x_2 = 0.048750$ and $x_3 = 2$, which is a very reasonable agreement.

7.3.2 Rayleigh-Schrödinger Perturbation Theory

This version of perturbation theory has been proposed by Schrödinger, but he utilized the method worked out previously by Rayleigh to investigate perturbed harmonic vibrations. It is assumed that we can split the Hamiltonian as

$$\hat{H} = \hat{H}_0 + \hat{W}, \tag{7.35}$$

where we know the solution of $\hat{H}_0$. This means that we know

$$\hat{H}_0|\psi_n^{(0)}\rangle = E_n^{(0)}|\psi_n^{(0)}\rangle \tag{7.36}$$

for all n. We also assume that the spectrum $E_n^{(0)}$ is discrete and non-degenerate.

Now we write the Hamiltonian as

$$\hat{H} = \hat{H}_0 + \lambda\hat{W}, \tag{7.37}$$

with continuous parameter $\lambda \in [0, 1]$. With $\lambda = 0$, the system is described by $\hat{H}_0$ and with $\lambda = 1$, we recover the original Hamiltonian $\hat{H}$ in Eq. (7.35).

The eigenvalue problem takes the form

$$(\hat{H}_0 + \lambda\hat{W})|\psi_n(\lambda)\rangle = E_n(\lambda)|\psi_n(\lambda)\rangle, \tag{7.38}$$

where $E_n(\lambda)$ and $|\psi_n(\lambda)\rangle$ are functions of the continuous parameter λ. We can perform a Taylor expansion around $\lambda = 0$

$$E_n(\lambda) = E_n^{(0)} + \lambda E_n^{(1)} + \lambda^2 E_n^{(2)} + \cdots, \tag{7.39}$$

and

$$|\psi_n(\lambda)\rangle = |\psi_n^{(0)}\rangle + \lambda|\psi_n^{(1)}\rangle + \lambda^2|\psi_n^{(2)}\rangle + \cdots. \tag{7.40}$$

Substituting back into Eq. (7.38), we find

$$\begin{aligned}&(\hat{H}_0 + \lambda\hat{W})\left(|\psi_n^{(0)}\rangle + \lambda|\psi_n^{(1)}\rangle + \lambda^2|\psi_n^{(2)}\rangle \ldots\right)\\ &= (E_n^{(0)} + \lambda E_n^{(1)} + \lambda^2 E_n^{(2)} + \ldots)\left(|\psi_n^{(0)}\rangle + \lambda|\psi_n^{(1)}\rangle + \lambda^2|\psi_n^{(2)}\rangle \ldots\right).\end{aligned} \tag{7.41}$$

This equation is valid for any value of λ. This is only possible, if the coefficients match at each order in λ. For the first few orders we find:

- λ^0 order:

  $$\hat{H}_0|\psi_n^{(0)}\rangle = E_n^{(0)}|\psi_n^{(0)}\rangle, \tag{7.42}$$

  which is just the unperturbed Hamiltonian.

- λ^1 order:

$$\hat{H}_0|\psi_n^{(1)}\rangle + \hat{W}|\psi_n^{(0)}\rangle = E_n^{(0)}|\psi_n^{(1)}\rangle + E_n^{(1)}|\psi_n^{(0)}\rangle. \quad (7.43)$$

- λ^2 order:

$$\hat{H}_0|\psi_n^{(2)}\rangle + \hat{W}|\psi_n^{(1)}\rangle = E_n^{(0)}|\psi_n^{(2)}\rangle + E_n^{(1)}|\psi_n^{(1)}\rangle + E_n^{(2)}|\psi_n^{(0)}\rangle. \quad (7.44)$$

- λ^k order:

$$\hat{H}_0|\psi_n^{(k)}\rangle + \hat{W}|\psi_n^{(k-1)}\rangle = E_n^{(0)}|\psi_n^{(k)}\rangle + E_n^{(1)}|\psi_n^{(k-1)}\rangle + \ldots + E_n^{(k)}|\psi_n^{(0)}\rangle. \quad (7.45)$$

We assume normalization for the unperturbed states, $\langle\psi_n^{(0)}|\psi_n^{(0)}\rangle = 1$. Switching on the perturbation introduces a variation of the wave function $\delta\psi_n = \psi_n - \psi_n^{(0)}$. So,

$$\delta(\langle\psi_n^{(0)}|\psi_n^{(0)}\rangle) = \delta(1) = 0 = \langle\delta\psi_n|\psi_n^{(0)}\rangle + \langle\psi_n^{(0)}|\delta\psi_n\rangle. \quad (7.46)$$

The most straightforward way to satisfy this condition is to assume that $|\delta\psi_n\rangle$ is orthogonal to $|\psi_n^{(0)}\rangle$. In other words, the change caused by the perturbation should be orthogonal to the unperturbed solution

$$\langle\delta\psi_n|\psi_n^{(0)}\rangle = \langle\psi_n - \psi_n^{(0)}|\psi_n^{(0)}\rangle = \lambda^1\langle\psi_n^{(1)}|\psi_n^{(0)}\rangle + \lambda^2\langle\psi_n^{(2)}|\psi_n^{(0)}\rangle + \lambda^3\langle\psi_n^{(3)}|\psi_n^{(0)}\rangle + \cdots = 0. \quad (7.47)$$

This orthogonality relation is valid for any value of λ, and therefore the coefficients of all powers of λ should vanish. This means that the perturbation corrections in all orders are orthogonal to the unperturbed state

$$\langle\psi_n^{(i)}|\psi_n^{(0)}\rangle = 0, \quad \text{for} \quad i = 1, 2, \ldots. \quad (7.48)$$

To get the *first order correction*, we multiply Eq. (7.43) by $\langle\psi_n^{(0)}|$ and find

$$\langle\psi_n^{(0)}|\hat{H}_0|\psi_n^{(1)}\rangle + \langle\psi_n^{(0)}|\hat{W}|\psi_n^{(0)}\rangle = E_n^{(0)}\langle\psi_n^{(0)}|\psi_n^{(1)}\rangle + E_n^{(1)}\langle\psi_n^{(0)}|\psi_n^{(0)}\rangle, \quad (7.49)$$

which immediately gives

$$E_n^{(1)} = \langle\psi_n^{(0)}|\hat{W}|\psi_n^{(0)}\rangle. \quad (7.50)$$

To determine the first order correction to the wave function, we use the completeness of the $\{|\psi_n^{(0)}\rangle\}$ states

$$|\psi_n^{(1)}\rangle = \left(\sum_m |\psi_m^{(0)}\rangle\langle\psi_m^{(0)}|\right)|\psi_n^{(1)}\rangle = \left(\sum_{m\neq n} |\psi_m^{(0)}\rangle\langle\psi_m^{(0)}|\right)|\psi_n^{(1)}\rangle. \quad (7.51)$$

So, to determine $|\psi_n^{(1)}\rangle$, we need the overlap $\langle\psi_m^{(0)}|\psi_n^{(1)}\rangle$. We can obtain this by multiplying Eq. (7.43) by $\langle\psi_m^{(0)}|$,

$$\langle\psi_m^{(0)}|\hat{H}_0|\psi_n^{(1)}\rangle + \langle\psi_m^{(0)}|\hat{W}|\psi_n^{(0)}\rangle = E_n^{(0)}\langle\psi_m^{(0)}|\psi_n^{(1)}\rangle + E_n^{(1)}\langle\psi_m^{(0)}|\psi_n^{(0)}\rangle, \tag{7.52}$$

and get

$$|\psi_n^{(1)}\rangle = \sum_{m\neq n} \frac{\langle\psi_m^{(0)}|\hat{W}|\psi_n^{(0)}\rangle}{E_n^{(0)} - E_m^{(0)}} |\psi_m^{(0)}\rangle. \tag{7.53}$$

We can see that this procedure works only if the unperturbed $E_n^{(0)}$ states are not degenerate. The first order correction to the wave function is applicable and sufficient if the perturbative correction is small compared to the energy difference of the unperturbed spectrum

$$|\langle\psi_m^{(0)}|\hat{W}|\psi_n^{(0)}\rangle| \ll |E_n^{(0)} - E_m^{(0)}|. \tag{7.54}$$

The contributions in Eq. (7.53) may vanish if the respective matrix elements vanish due to some symmetry, and become less important if n and m, consequently $E_n^{(0)}$ and $E_m^{(0)}$, differ considerably.

In order to find the *second order correction*, we need to proceed analogously. Multiplying Eq. (7.44) by $\langle\psi_n^{(0)}|$, we obtain

$$E_n^{(2)} = \langle\psi_n^{(0)}|\hat{W}|\psi_n^{(1)}\rangle = \sum_{m\neq n} \frac{\langle\psi_n^{(0)}|\hat{W}|\psi_m^{(0)}\rangle\langle\psi_m^{(0)}|\hat{W}|\psi_n^{(0)}\rangle}{E_n^{(0)} - E_m^{(0)}} = \sum_{m\neq n} \frac{|\langle\psi_n^{(0)}|\hat{W}|\psi_m^{(0)}\rangle|^2}{E_n^{(0)} - E_m^{(0)}}. \tag{7.55}$$

Then, in the same way as before, multiplying Eq. (7.44) by $\langle\psi_m^{(0)}|$ we find

$$\langle\psi_m^{(0)}|\hat{H}_0|\psi_n^{(2)}\rangle + \langle\psi_m^{(0)}|\hat{W}|\psi_n^{(1)}\rangle = E_n^{(0)}\langle\psi_m^{(0)}|\psi_n^{(2)}\rangle + E_n^{(1)}\langle\psi_m^{(0)}|\psi_n^{(1)}\rangle + E_n^{(2)}\langle\psi_m^{(0)}|\psi_n^{(0)}\rangle, \tag{7.56}$$

which results in

$$|\psi_n^{(2)}\rangle = \left(\sum_{m\neq n} |\psi_m^{(0)}\rangle\langle\psi_m^{(0)}|\right)|\psi_n^{(2)}\rangle = \sum_{m\neq n} \frac{\langle\psi_m^{(0)}|\hat{W}|\psi_n^{(1)}\rangle - E_n^{(1)}\langle\psi_m^{(0)}|\psi_n^{(1)}\rangle}{E_n^{(0)} - E_m^{(0)}}|\psi_m^{(0)}\rangle. \tag{7.57}$$

We can puzzle out an iterative procedure for the higher terms. Multiplying Eq. (7.45) by $\langle\psi_n^{(0)}|$ we find

$$E_n^{(k)} = \langle\psi_n^{(0)}|W|\psi_n^{(k-1)}\rangle, \tag{7.58}$$

and multiplying by $\langle\psi_m^{(0)}|$ with $m \neq n$ we obtain

$$\langle\psi_m^{(0)}|\psi_n^{(k)}\rangle = \frac{1}{E_n^{(0)} - E_m^{(0)}} \left\langle\psi_m^{(0)}|W|\psi_n^{(k-1)}\right\rangle - \frac{1}{E_n^{(0)} - E_m^{(0)}} \sum_{j=0}^{k-1} E_n^{(j)}\langle\psi_m^{(0)}|\psi_n^{(k-j)}\rangle \tag{7.59}$$

for $|\psi_n^{(k)}\rangle = \sum_{m\neq n} |\psi_m^{(0)}\rangle\langle\psi_m^{(0)}|\psi_n^{(k)}\rangle$.

The second order perturbative correction to the energy has a nice physical interpretation. We can consider the term

$$\langle \psi_n^{(0)} | \hat{W} | \psi_m^{(0)} \rangle \langle \psi_m^{(0)} | \hat{W} | \psi_n^{(0)} \rangle \tag{7.60}$$

as the transition from $\psi_n^{(0)}$ to the intermediate virtual state $\psi_m^{(0)}$ and return to the state $\psi_n^{(0)}$. This transition is virtual, because it may violate the energy conservation principle. In principle, it is possible within the time interval allowed by the time-energy uncertainty relation. The quantum system "borrows" some energy for a short period of time to jump into another quantum state and then jumps back returning the "borrowed" energy. This correction is negligible if $E_n^{(0)}$ and $E_m^{(0)}$ differ considerably. It is interesting to see that if $n = 0$, i.e. for the ground state, the second order contribution is negative, so the ground state energy is lowered.

7.3.3 *Degenerate Perturbation Theory*

The above procedure is applicable only if the spectrum of $\hat{H}_0$ is non-degenerate. Otherwise, in Eq. (7.53) for $|\psi_n^{(1)}\rangle$, some denominators might become zero and the whole expression becomes singular. In the degenerate perturbation theory, we assume that

$$\hat{H}_0 | \psi_{n,\alpha}^{(0)} \rangle = E_n^{(0)} | \psi_{n,\alpha}^{(0)} \rangle, \qquad \alpha = 1, \ldots, f, \tag{7.61}$$

i.e. the eigenstate $E_n^{(0)}$ is f-fold degenerate and the states are chosen such that $\langle \psi_{n,\alpha}^{(0)} | \psi_{n,\beta}^{(0)} \rangle = \delta_{\alpha,\beta}$. A linear combination

$$|\psi_n\rangle = \sum_{\alpha=1}^{f} a_\alpha | \psi_{n,\alpha}^{(0)} \rangle \tag{7.62}$$

is also an eigenstate of $\hat{H}_0$ with eigenvalue $E_n^{(0)}$. We assume normalization, so

$$1 = \langle \psi_n | \psi_n \rangle = \sum_{\alpha,\beta}^{f} a_\alpha^* a_\beta \delta_{\alpha\beta} = \sum_{\alpha}^{f} |a_\alpha|^2. \tag{7.63}$$

The state $|\psi_n\rangle$ is in the subspace spanned by the degenerate eigenstates of $\hat{H}_0$. We seek the solution of the total Hamiltonian in that subspace

$$(\hat{H}_0 + \hat{W}) | \psi_n \rangle = E_n | \psi_n \rangle, \tag{7.64}$$

or, with Eq. (7.62), we have

$$\sum_{\alpha=1}^{f}(\hat{H}_0|\psi_{n,\alpha}^{(0)}\rangle + \hat{W}|\psi_{n,\alpha}^{(0)}\rangle)a_\alpha = E_n \sum_{\alpha=1}^{f} a_\alpha|\psi_{n,\alpha}^{(0)}\rangle. \tag{7.65}$$

Multiplying from the left by $\langle\psi_{n,\beta}^{(0)}|$, we find

$$\sum_{\alpha=1}^{f}(E_n^{(0)}\delta_{\alpha\beta} + \langle\psi_{n,\beta}^{(0)}|\hat{W}|\psi_{n,\alpha}^{(0)}\rangle)a_\alpha = E_n \sum_{\alpha=1}^{f} a_\alpha\delta_{\alpha\beta}, \tag{7.66}$$

or, by introducing the notations $E_n^{(1)} = E_n - E_n^{(0)}$ and $W_{\beta\alpha} = \langle\psi_{n,\beta}|\hat{W}|\psi_{n,\alpha}\rangle$, we obtain the eigenvalue equation

$$\underline{W}\,\underline{a} = E_n^{(1)}\underline{a}. \tag{7.67}$$

The eigenvalues $E_{n,\beta}^{(1)}$ give the first order corrections to the degenerate energies and the corresponding eigenvectors $\underline{a}^\beta$ determine the wave function

$$|\psi_n^\beta\rangle = \sum_{\alpha=1}^{f} a_\alpha^\beta|\psi_{n,\alpha}^{(0)}\rangle. \tag{7.68}$$

7.3.4 Applications of Perturbation Theory

Charged Oscillator in an Electric Field

We consider a charged particle subject to a quadratic potential and an electric field

$$\hat{H} = -\frac{\hbar^2}{2m}\frac{d^2}{dx^2} + \frac{1}{2}m\omega^2x^2 + q\mathcal{E}x, \tag{7.69}$$

where q is the charge and $\mathcal{E}$ is the x-component of the electric field. With $x = \sqrt{\hbar/(2m\omega)}(\hat{a} + \hat{a}^\dagger)$ and $[\hat{a}, \hat{a}^\dagger] = 1$, we can rewrite the Hamiltonian as

$$\hat{H} = \hbar\omega\left(\hat{a}^\dagger\hat{a} + \frac{1}{2}\right) + v_0(\hat{a} + \hat{a}^\dagger), \tag{7.70}$$

where $v_0 = q\mathcal{E}\sqrt{\hbar/(2m\omega)}$.

We can easily find the exact solution of this problem. Introducing the shifted operators

$$\hat{b} = \hat{a} + \alpha \quad \text{and} \quad \hat{b}^\dagger = \hat{a}^\dagger + \alpha^*, \tag{7.71}$$

we find

$$[\hat{b}, \hat{b}^\dagger] = [\hat{a}, \hat{a}^\dagger] = 1. \tag{7.72}$$

Expressing $\hat{H}$ in terms of $\hat{b}$, we obtain

$$\hat{H} = \hbar\omega\left((\hat{b}^\dagger - \alpha^*)(\hat{b} - \alpha) + \frac{1}{2}\right) + v_0(\hat{b} - \alpha + \hat{b}^\dagger - \alpha^*)$$
$$= \hbar\omega\left(\hat{b}^\dagger\hat{b} + \frac{1}{2}\right) + \hat{b}(v_0 - \hbar\omega\alpha^*) + \hat{b}^\dagger(v_0 - \hbar\omega\alpha) + \hbar\omega|\alpha|^2 - v_0(\alpha + \alpha^*). \tag{7.73}$$

By choosing the auxiliary constant $\alpha = v_0/(\hbar\omega)$, we eliminate the terms linear in $\hat{b}$ and $\hat{b}^\dagger$ and arrive at

$$\hat{H} = \hbar\omega\left(\hat{b}^\dagger\hat{b} + \frac{1}{2}\right) - \frac{v_0^2}{\hbar\omega}. \tag{7.74}$$

This is just another harmonic oscillator with shifted energy eigenvalues

$$E_n = \hbar\omega\left(n + \frac{1}{2}\right) - \frac{v_0^2}{\hbar\omega}. \tag{7.75}$$

The eigenstates are the harmonic oscillator wave functions constructed from $\hat{b}$ and $\hat{b}^\dagger$. In this case $x' = \sqrt{\hbar/(2m\omega)}(\hat{b} + \hat{b}^\dagger) = \sqrt{\hbar/(2m\omega)}(\hat{a} + \hat{a}^\dagger + 2v_0/(\hbar\omega)) = x + x_0$, with $x_0 = q\mathcal{E}/(m\omega^2)$. The eigenstates are harmonic oscillator states in terms of $x + x_0$

$$\langle x|\psi_n'\rangle = \langle x + x_0|\psi_n\rangle = \psi_n(x + x_0), \tag{7.76}$$

with the center of the harmonic oscillator states shifted.

We can also treat the problem in the framework of perturbation theory. Here,

$$\hat{H}_0 = \hbar\omega\left(\hat{a}^\dagger\hat{a} + \frac{1}{2}\right) \quad \text{and} \quad \hat{W} = v_0(\hat{a} + \hat{a}^\dagger). \tag{7.77}$$

The solutions of the $\hat{H}_0$ Hamiltonian problem are the energies $E_n^0 = \hbar\omega(n + 1/2)$ and the one dimensional harmonic oscillator eigenstates $\{|n\rangle\}$. We can see that the first order corrections vanish

$$E_n^{(1)} \sim v_0\langle n|(\hat{a} + \hat{a}^\dagger)|n\rangle = 0, \tag{7.78}$$

so we need to consider the second order corrections, Eq. (7.55),

$$E_n^{(2)} = \sum_{m\neq n} \frac{|v_0\langle m|(\hat{a} + \hat{a}^\dagger)|n\rangle|^2}{E_n^{(0)} - E_m^{(0)}}. \tag{7.79}$$

Since the non-vanishing terms are

$$\langle n-1|\hat{a}|n\rangle = \sqrt{n} \quad \text{and} \quad \langle n+1|\hat{a}^\dagger|n\rangle = \sqrt{n+1}, \tag{7.80}$$

we obtain

$$E_n^{(2)} = v_0^2 \left[\frac{|\langle n-1|\hat{a}|n\rangle|^2}{E_n^{(0)} - E_{n-1}^{(0)}} + \frac{|\langle n+1|\hat{a}^\dagger|n\rangle|^2}{E_n^{(0)} - E_{n+1}^{(0)}} \right] = v_0^2 \left[\frac{n}{\hbar\omega} + \frac{n+1}{-\hbar\omega} \right] = -\frac{v_0^2}{\hbar\omega}. \tag{7.81}$$

So, the solution for the energy is

$$E_n = E_n^{(0)} + E_n^{(1)} + E_n^{(2)} = \hbar\omega \left(n + \frac{1}{2} \right) - \frac{v_0^2}{\hbar\omega}, \tag{7.82}$$

i.e. the perturbation theory up to second order recovers the exact results.

The first order correction to the wave function couples only the two neighboring terms

$$|\psi_n\rangle = |n\rangle + \frac{v_0}{\hbar\omega} \left(\sqrt{n}|n-1\rangle - \sqrt{n+1}|n+1\rangle \right). \tag{7.83}$$

Hydrogen Atom in an Electric Field: The $n = 1$ Ground State

An effect, called the Stark effect, occurs when the energy states of hydrogen atoms are shifted and split in the presence of an electric field (Johannes Stark - Antonio Lo Surdo).

The Hamiltonian of the problem is given by

$$\hat{H} = \hat{H}_0 + \hat{W} = \frac{\hat{p}^2}{2m} - \frac{e^2}{r} - e\mathcal{E}z, \tag{7.84}$$

where $\hat{H}_0$ is the hydrogen Hamiltonian, m_e is the mass of the electron, e is the electric charge and we assume that the electric field is in the z direction. The hydrogen states are the common eigenstates of $\{\hat{H}_0, \hat{L}^2, \hat{L}_z\}$,

$$\hat{H}_0|nlm\rangle = E_n^{(0)}|nlm\rangle, \tag{7.85}$$

where n is the principal quantum number, $E_n^{(0)} = -e^2/(2a_0n^2)$, and a_0 is the Bohr radius. The first order perturbative correction to the ground state vanishes

$$E_{100}^{(1)} \sim e\mathcal{E}\langle 100|z|100\rangle = 0. \tag{7.86}$$

The states in the matrix element have definite parity, so their combined parity is positive. On he other hand, z has negative parity, therefore, the integral over the whole space cancels out.

The second order correction reads

$$E_{100}^{(2)} = e^2\mathcal{E}^2 \sum_{nlm \neq 100} \frac{|\langle 100|z|nlm\rangle|^2}{E_1^{(0)} - E_n^{(0)}}, \tag{7.87}$$

i.e. the Stark effect on the ground state of hydrogen is quadratic in the strength of the electric field. We note that $z = r\cos\theta = r\sqrt{4\pi/3}Y_{1,0}$, and thus the matrix elements become

$$\langle 100|z|nlm\rangle = \int \mathrm{d}r\mathrm{d}\theta \sin\theta \,\mathrm{d}\phi \, Y^*_{0,0}(\theta,\phi)\, r\sqrt{4\pi/3}Y_{1,0}(\theta,\phi)Y_{l,m}(\theta,\phi)\, u_{10}(r)\, u_{nl}(r), \quad (7.88)$$

where u_{nl} are the hydrogen radial wave functions of Eq. (5.70). The angular integrals vanish unless $m = 0$ and $l = 1$ yielding

$$\langle 100|z|nlm\rangle = \sqrt{\frac{1}{3}}\delta_{l1}\delta_{m0}\int_0^\infty \mathrm{d}r\, r\, u_{10}(r)\, u_{n1}(r). \quad (7.89)$$

So, the second order correction becomes

$$E^{(2)}_{100} = e^2\mathcal{E}^2 \sum_{n=2}^{\infty} \frac{|\sqrt{\frac{1}{3}}\int_0^\infty \mathrm{d}r\, r\, u_{10}(r)\, u_{n1}(r)|^2}{-e^2/(2a_0) + e^2/(2a_0 n^2)}. \quad (7.90)$$

Evaluating the radial integrals numerically, we observe a rather slow convergence in n. The first 6 terms give

$$E^{(2)}_{100} = -a_0^3\mathcal{E}^2(1.4798 + 0.2002 + 0.0660 + 0.0303 + 0.0165 + 0.0100 + \ldots) \approx -1.8a_0^3\mathcal{E}^2, \quad (7.91)$$

which is in an acceptable agreement with the exact second order correction $E^{(2)}_{100} = -2.25a_0^3\mathcal{E}^2$. We have shown before that the second order correction is always negative and quadratic in terms of the coupling constant.

Hydrogen Atom in an Electric Field: The $n = 2$ Excited State

The first excited state of the hydrogen atom is four-fold degenerate (if we ignore the spin). Those 4 states are $|nlm\rangle = |200\rangle, |210\rangle, |21-1\rangle, |211\rangle$. Since $z = r\cos\theta = r\sqrt{4\pi/3}Y_{1,0}$ and there is no dependence on ϕ, we find again that the first order perturbation matrix elements $\langle 2lm|r\sqrt{4\pi/3}Y_{1,0}|2l'm'\rangle$ vanish unless $m = m'$. So, only the states $|200\rangle$ and $|210\rangle$ are affected in the first order of degenerate perturbation theory, because z has odd parity, and the bra and the ket states should have opposite parities. Therefore the only non-vanishing matrix elements are

$$\begin{aligned}&\langle 200|z|210\rangle = \langle 210|z|200\rangle \\ &= \int \mathrm{d}r\, \mathrm{d}\theta \, \sin\theta\, \mathrm{d}\phi\, u_{2p}(r)u_{2s}(r)\, r\sqrt{\frac{4\pi}{3}}Y_{1,0}(\theta,\phi)\, Y^*_{1,0}(\theta,\phi)Y_{0,0}(\theta,\phi) = \xi = -3a_0.\end{aligned} \quad (7.92)$$

The secular equation $|\underline{H}' - \lambda \underline{I}| = 0$ gives $\lambda^2 - \xi^2 = 0$. So, we have

$$\begin{aligned}\psi_s^{(1)} &= \frac{1}{\sqrt{2}}(|200\rangle + |210\rangle), \quad \text{for eigenvalue} \quad +\xi, \\ \psi_a^{(1)} &= \frac{1}{\sqrt{2}}(|200\rangle - |210\rangle), \quad \text{for eigenvalue} \quad -\xi.\end{aligned} \tag{7.93}$$

We have found that the states $|21-1\rangle$ and $|211\rangle$ are not affected in the first order of degenerate perturbation theory, while the antisymmetric combination of $|200\rangle$ and $|210\rangle$ is shifted up and the symmetric one is shifted down by $3e\mathcal{E}a_0$. The energy shift is proportional to the electric field.

Spin-Orbit Interaction

The electron with mass m_e, moving in a central potential $V(r)$ generated by the nucleus and the other electrons, experiences an interaction between its spin $\hat{S}$ and its orbital angular momentum $\hat{L}$. The actual form of the interaction can be derived as a non-relativistic limit of the relativistic Dirac equation. The spin-orbit interaction reads

$$\hat{H}_{so} = \frac{1}{2m_e^2c^2}\frac{1}{r}\frac{\mathrm{d}V(r)}{\mathrm{d}r}\hat{L}\hat{S}. \tag{7.94}$$

For a hydrogen atom

$$V(r) = -\frac{e^2}{r}, \qquad \text{and} \qquad \frac{\mathrm{d}V(r)}{\mathrm{d}r} = \frac{e^2}{r^2}, \tag{7.95}$$

therefore,

$$\hat{H}_{so} = \frac{e^2}{2m_e^2c^2}\frac{1}{r^3}\hat{L}\hat{S}. \tag{7.96}$$

The first order perturbative matrix elements can easily be calculated if we take the $|nlsjm\rangle$ basis. Since $\hat{J}^2 = \hat{L}^2 + \hat{S}^2 + 2\hat{L}\hat{S}$

$$\begin{aligned}\langle nlsjm|\hat{L}\hat{S}|nlsjm\rangle &= \langle nlsjm|\frac{\hat{J}^2 - \hat{L}^2 - \hat{S}^2}{2}|nlsjm\rangle \\ &= \frac{1}{2}\hbar^2[j(j+1) - l(l+1) - s(s+1)],\end{aligned} \tag{7.97}$$

where $s = 1/2$ and $j = l \pm 1/2$ and the expectation value of $1/r^3$ between hydrogen-like functions has been found in Eq. (7.12). So, the first order perturbative correction, due to the spin-orbit interaction, reads

$$E_{so}^{(1)} = \frac{e^2\hbar^2}{2m_e^2c^2}\frac{j(j+1)-l(l+1)-3/4}{n^3l(l+1)(2l+1)a_0^3}. \tag{7.98}$$

Expressed with the help of the *fine structure, or Sommerfeld constant*

$$\alpha = \frac{e^2}{\hbar c} = \frac{1}{137.035999084} \simeq \frac{1}{137}, \tag{7.99}$$

we have

$$E_{so}^{(1)} = \frac{\alpha^4 m_e c^2}{2n^3}\frac{j(j+1)-l(l+1)-3/4}{l(l+1)(2l+1)}. \tag{7.100}$$

Sommerfeld introduced the constant α in the relativistic extension of the Bohr model of atoms. In his work α appeared as the ratio of the velocity of the orbiting electron with respect to the speed of light.

Relativistic Correction to the Kinetic Energy

The relativistic correction to the kinetic energy gives a term, which is the same order in c as the spin-orbit interaction. From the relativistic energy-momentum relation we find

$$\sqrt{p^2c^2 + m_e^2c^4} \simeq m_ec^2 + \frac{p^2}{2m_e} - \frac{p^4}{8m_e^3c^2} + \cdots. \tag{7.101}$$

The second term is the non-relativistic kinetic energy and the third one is the leading relativistic correction

$$\hat{H}_r = -\frac{\hat{\boldsymbol{p}}^4}{8m_e^3c^2}, \tag{7.102}$$

which, like the spin-orbit interaction, is of order c^{-2}. From Eq. (7.5) we can infer that

$$\frac{\hat{\boldsymbol{p}}^2}{2m}|nlm\rangle = \frac{e^2}{r}|nlm\rangle - \frac{e^2}{a_0}\frac{1}{2n^2}|nlm\rangle, \tag{7.103}$$

and using Eqs. (7.7) and (7.10), the first order perturbation correction easily follows

$$E_r^{(1)} = \langle nljm|\hat{H}_r|nljm\rangle = -\frac{\alpha^4 m_e c^2}{8n^4}\left(\frac{8n}{2l+1} - 3\right). \tag{7.104}$$

The spin-orbit and the relativistic corrections together make up the fine-structure correction

$$\hat{H}_{fs} = \hat{H}_{so} + \hat{H}_r. \tag{7.105}$$

Considering that $j = l \pm 1/2$ and thus $l = j \mp 1/2$, we obtain

$$E_{fs}^{(1)} = E_{so}^{(1)} + E_r^{(1)} = \frac{\alpha^4 m_e c^2}{8n^4}\left(3 - \frac{4n}{j+1/2}\right). \tag{7.106}$$

The Hydrogen Atom in a Magnetic Field: The Zeeman Effect

Both the orbital and spin angular momenta have associated dipole moments, and putting them in a magnetic field gives

$$\hat{H}_Z = -\hat{\boldsymbol{\mu}}_l \boldsymbol{B} - \hat{\boldsymbol{\mu}}_s \boldsymbol{B}, \tag{7.107}$$

where $\hat{\boldsymbol{\mu}}_l = -e\hat{\boldsymbol{L}}/(2m_e c)$ and $\hat{\boldsymbol{\mu}}_s = -e\hat{\boldsymbol{S}}/(m_e c)$. Note that the orbital and spin angular momenta couple to the magnetic field with different coupling strengths. This is an experimental observation, but it comes out nicely from the relativistic Dirac theory. So,

$$\hat{H}_Z = \frac{e}{2m_e c}(\hat{\boldsymbol{L}} + 2\hat{\boldsymbol{S}})\boldsymbol{B} = \frac{eB}{2m_e c}(\hat{L}_z + 2\hat{S}_z), \tag{7.108}$$

if we take $\boldsymbol{B} = B\boldsymbol{e}_z$, i.e. the magnetic field in the z direction.

The total Hamiltonian is given by

$$\hat{H} = \hat{H}_0 + \hat{H}_{fs} + \hat{H}_Z. \tag{7.109}$$

Now, we have two scenarios. If the magnetic field is strong and the Zeeman term is stronger than the fine structure term, we can neglect the fine structure and have

$$\hat{H} \simeq \hat{H}_0 + \hat{H}_Z. \tag{7.110}$$

We notice that this Hamiltonian is diagonal in the $|nlm_l m_s\rangle$ basis. Consequently,

$$\begin{aligned} E_{nlm_l m_s} &= \langle nlm_l m_s|\hat{H}|nlm_l m_s\rangle = \langle nlm_l m_s|\hat{H}_0 + \frac{eB}{2m_e c}(\hat{L}_z + 2\hat{S}_z)|nlm_l m_s\rangle \\ &= E_n^{(0)} + \frac{eB\hbar}{2m_e c}(m_l + 2m_s) = -\frac{e^2}{2a_0 n^2} + \frac{eB\hbar}{2m_e c}(m_l + 2m_s). \end{aligned} \tag{7.111}$$

If the magnetic field is weak and the fine structure term is not negligible, we should take the $|nljm_j\rangle$ basis and put the Hamiltonian in the form

$$\hat{H} = \hat{H}_0 + \hat{H}_{fs} + \hat{H}_Z = \hat{H}_0 + \hat{H}_{fs} + \frac{eB}{2m_e c}(\hat{J}_z + \hat{S}_z). \tag{7.112}$$

Then, additional to the fine-structure result of Eq. (7.106), we treat $\hat{H}_Z$ as a perturbation. For the matrix elements of $\hat{S}_z$ we found in Eq. (4.297)

$$\begin{aligned}\langle nljm_j|\hat{S}_z|nljm_j\rangle &= \frac{\langle nljm_j|\hat{J}\hat{S}|nljm_j\rangle}{\hbar^2 j(j+1)}\langle nljm_j|\hat{J}_z|nljm_j\rangle \\ &= \frac{j(j+1)-l(l+1)+s(s+1)}{2j(j+1)}\hbar m_j,\end{aligned} \tag{7.113}$$

which gives

$$E_Z^{(1)} = \frac{eB\hbar}{2m_e c}\left[1+\frac{j(j+1)-l(l+1)+s(s+1)}{2j(j+1)}\right]m_j. \tag{7.114}$$

So, in a weak magnetic field we can observe unequal spacing of energy levels, unlike in Eq. (7.111). This is called the *anomalous Zeeman effect.*

7.3.5 General Theory of Perturbations

We have seen that perturbation methods provide valuable insights to quantum mechanical problems with relatively low cost. However, it can easily become rather complicated if we need higher order terms or if the unperturbed spectrum is degenerate or quasi-degenerate. Now, we present a method which offers broader flexibility (see p. 242 in [2] or p. 425 in [3]).

Consider again the eigenvalue problem

$$\hat{H}|\lambda\rangle = E_\lambda|\lambda\rangle, \tag{7.115}$$

with

$$\hat{H} = \hat{H}_0 + \hat{W}. \tag{7.116}$$

We assume that $\hat{W}$ is "small" in some sense and the spectrum of $\hat{H}_0$ is known

$$\hat{H}_0|a\rangle = \epsilon_a|a\rangle. \tag{7.117}$$

With the help of the Green's operator

$$\hat{G}(E) = (E-\hat{H})^{-1}, \tag{7.118}$$

we define the operators

$$\hat{F}(E) = \hat{G}(E)(\hat{g}(E))^{-1} \quad \text{and} \quad \hat{R}(E) = \hat{W}\hat{F}(E), \tag{7.119}$$

where

$$\hat{g}(E) = \sum_a |a\rangle G_a(E)\langle a|, \quad \text{with} \quad G_a(E) = \langle a|\hat{G}(E)|a\rangle. \tag{7.120}$$

From the defining equation of the Green's operator

$$(E - \hat{H})\hat{G}(E) = I, \tag{7.121}$$

we find

$$(E - \hat{H})\hat{F}(E)|a\rangle = (E - \hat{H})\hat{G}(E)[\hat{g}(E)]^{-1}|a\rangle = (\hat{g}(E))^{-1}|a\rangle = \frac{1}{G_a(E)}|a\rangle. \tag{7.122}$$

For the diagonal matrix element of $\hat{R}$ we get

$$\begin{aligned} R_a(E) &= \langle a|\hat{R}(E)|a\rangle = \langle a|\hat{W}\hat{F}(E)|a\rangle = \langle a|(\hat{H} - \hat{H}_0)\hat{F}(E)|a\rangle \\ &= \langle a|(E - \hat{H}_0)\hat{F}(E)|a\rangle - \langle a|(E - \hat{H})\hat{F}(E)|a\rangle \\ &= (E - \epsilon_a)\langle a|\hat{F}(E)|a\rangle - \frac{1}{G_a(E)} = (E - \epsilon_a) - \frac{1}{G_a(E)}, \end{aligned} \tag{7.123}$$

since $\langle a|\hat{F}(E)|a\rangle = \langle a|[\hat{G}(E)\sum_{a'}|a'\rangle G_{a'}^{-1}(E)\langle a'|]|a\rangle = 1$. Reorganizing, we obtain the implicit equation

$$E = \epsilon_a + R_a(E) + \frac{1}{G_a(E)}. \tag{7.124}$$

If the parameter E approaches the eigenvalue E_λ of Eq. (7.115), $G_a(E)$ becomes singular and the last term vanishes. Consequently, we have

$$E_\lambda = \epsilon_a + R_a(E_\lambda) \tag{7.125}$$

and

$$(E_\lambda - \hat{H})\hat{F}(E_\lambda)|a\rangle = 0. \tag{7.126}$$

Here R_a represents the *level shift* from the unperturbed ϵ_a to E_λ and the level shift operator $\hat{F}(E_\lambda)$ transforms the eigenstate $|a\rangle$ of $\hat{H}_0$ to eigenstate $|\lambda_{E_\lambda}\rangle = \hat{F}(E_\lambda)|a\rangle$ of $\hat{H}$.

So far we have not made any approximation. The operator $\hat{F}$ is rather complicated since it contains the Green's operator of the whole Hamiltonian. In order to introduce some approximations, we derive an equation for $\hat{F}$. For any invertible operator $\hat{A}$ and $\hat{B}$, we have

$$\hat{A}^{-1} = \hat{B}^{-1} + \hat{B}^{-1}(\hat{B} - \hat{A})\hat{A}^{-1}. \tag{7.127}$$

If we take $\hat{A} = \hat{g}(E)(E - \hat{H})$ and $\hat{B} = \hat{g}(E)(E - \hat{H}_0 - \hat{O})$, we find $\hat{A}^{-1} = (E - \hat{H})^{-1}[\hat{g}(E)]^{-1} = \hat{G}(E)[\hat{g}(E)]^{-1} = \hat{F}(E)$ and $\hat{B}^{-1} = (E - \hat{H}_0 - \hat{O})^{-1}[\hat{g}(E)]^{-1}$. Consequently,

$$\hat{F}(E) = (E - \hat{H}_0 - \hat{O})^{-1}[\hat{g}(E)]^{-1} + (E - \hat{H}_0 - \hat{O})^{-1}(\hat{W} - \hat{O})\hat{F}(E). \tag{7.128}$$

Here, $\hat{O}$ is an arbitrary operator that we can choose for our convenience. We consider the following form

$$\hat{O} = \sum_{a'} O_{a'} |a'\rangle\langle a'|, \tag{7.129}$$

where $O_{a'}$ are numbers. This choice ensures that $\hat{O}$ and $\hat{H}_0$ commute. Inserting Eq. (7.124) into Eq. (7.120), we have

$$[\hat{g}(E)]^{-1} = \sum_a |a\rangle \frac{1}{G_a(E)} \langle a| = \sum_a |a\rangle (E - \epsilon_a - R_a(E)) \langle a|. \tag{7.130}$$

Then, applying Eq. (7.128) on the state $|a\rangle$, we find

$$\hat{F}(E)|a\rangle = (E - \hat{H}_0 - \hat{O})^{-1}(E - \epsilon_a - R_a(E))|a\rangle + (E - \hat{H}_0 - \hat{O})^{-1}(\hat{W} - \hat{O})\hat{F}(E)|a\rangle. \tag{7.131}$$

The choice of

$$\hat{O} = R_a(E)|a\rangle\langle a| \tag{7.132}$$

leads us to the representation of the inverse operator

$$(E - \hat{H}_0 - \hat{O})^{-1} = \frac{1}{E - \epsilon_a - R_a(E)} |a\rangle\langle a| + \sum_{a' \neq a} \frac{1}{E - \epsilon_{a'}} |a'\rangle\langle a'|. \tag{7.133}$$

Since $\hat{R} = \hat{W}\hat{F}$ and $\langle a|\hat{F}|a\rangle = 1$, we obtain

$$\langle a|(\hat{W} - \hat{O})\hat{F}(E)|a\rangle = \langle a|\hat{W}\hat{F}(E)|a\rangle - R_a\langle a|\hat{F}(E)|a\rangle = \langle a|\hat{W}\hat{F}(E)|a\rangle - R_a = 0. \tag{7.134}$$

Inserting Eq. (7.133) into Eq. (7.131), we can see that in the first term, only the diagonal element remains, while in the second term, due to Eq. (7.134), the diagonal element vanishes. As a result, Eq. (7.131) turns into

$$\hat{F}(E)|a\rangle = |a\rangle + \sum_{a' \neq a} \frac{1}{E - \epsilon_{a'}} |a'\rangle\langle a'|\hat{W}\hat{F}(E)|a\rangle. \tag{7.135}$$

This is basically an implicit equation for $\hat{F}(E)|a\rangle$. The solution could a challenging problem, but it can be cast into a recursion relation. Assume we have the solution up to rank k, $\hat{F}^{(k)}(E)|a\rangle$, then the rank $k+1$ solution can be obtained by the formula

$$\hat{F}^{(k+1)}(E)|a\rangle = |a\rangle + \sum_{a' \neq a} \frac{1}{E - \epsilon_{a'}} |a'\rangle\langle a'|\hat{W}\,\hat{F}^{(k)}(E)|a\rangle. \tag{7.136}$$

Starting with $\hat{F}^{(0)}(E)|a\rangle = |a\rangle$, by successive substitutions we obtain the series representation

$$\hat{F}(E)|a\rangle = |a\rangle + \sum_{a'\neq a} \frac{1}{E-\epsilon_{a'}}|a'\rangle\langle a'|\hat{W}|a\rangle + \sum_{a'\neq a}\sum_{a''\neq a} \frac{1}{E-\epsilon_{a'}}\frac{1}{E-\epsilon_{a''}}|a'\rangle\langle a'|\hat{W}|a''\rangle\langle a''|\hat{W}|a\rangle + \cdots . \tag{7.137}$$

The eigenvalues of $\hat{H}$ are given by

$$\begin{aligned} E_{\lambda(a)} &= \epsilon_a + \langle a|\hat{W}\hat{F}(E_{\lambda(a)})|a\rangle \\ &= \epsilon_a + \langle a|\hat{W}|a\rangle + \sum_{a'\neq a} \frac{1}{E_{\lambda(a)}-\epsilon_{a'}}\langle a|\hat{W}|a'\rangle\langle a'|\hat{W}|a\rangle \\ &\quad + \sum_{a'\neq a}\sum_{a''\neq a} \frac{1}{E_{\lambda(a)}-\epsilon_{a'}}\frac{1}{E_{\lambda(a)}-\epsilon_{a''}}\langle a|\hat{W}|a'\rangle\langle a'|\hat{W}|a''\rangle\langle a''|\hat{W}|a\rangle + \cdots . \end{aligned} \tag{7.138}$$

This is an implicit equation for the eigenvalue $E_{\lambda(a)}$ in terms of unperturbed energies ϵ_a and the matrix elements of $\hat{W}$ between the unperturbed states $|a\rangle$. The state $\hat{F}(E_{\lambda(a)})|a\rangle$ of Eq. (7.137) provides the corresponding state of the total Hamiltonian $\hat{H}$. These formulas were derived by a special choice of $\hat{O}$ in Eq. (7.132) and they constitute the *Brillouin-Wigner* perturbation theory.

We may make another choice for $\hat{O}$

$$\hat{O} = R_a(E)|a\rangle\langle a| + \sum_{a'\neq a} |a'\rangle(E-\epsilon_a)\langle a' , \tag{7.139}$$

and we find

$$(E - \hat{H}_0 - \hat{O})^{-1} = \frac{1}{E-\epsilon_a - R_a(E)}|a\rangle\langle a| + \sum_{a'\neq a} \frac{1}{\epsilon_a - \epsilon_{a'}}|a'\rangle\langle a'| . \tag{7.140}$$

Then, from Eq. (7.131), with $E = E_{\lambda(a)}$, we obtain

$$\hat{F}(E_{\lambda(a)})|a\rangle = |a\rangle + \sum_{a'\neq a} \frac{1}{\epsilon_a - \epsilon_{a'}}|a'\rangle\langle a'|[\hat{W} - (E_{\lambda(a)} - \epsilon_a))\hat{F}(E_{\lambda(a)})]|a\rangle, \tag{7.141}$$

and for the level shift we get

$$E_{\lambda(a)} - \epsilon_a = R_a(E_{\lambda(a)}) = \langle a|\hat{W}\hat{F}(E_{\lambda(a)})|a\rangle . \tag{7.142}$$

These relations are also implicit relations that can be solved by iteration. In the 0-th order we have

$$R_a^{(0)} = 0 \quad \text{and} \quad \hat{F}^{(0)}|a\rangle = |a\rangle, \tag{7.143}$$

while in the first order we find

$$R_a^{(1)} = \langle a|\hat{W}\hat{F}^{(0)}|a\rangle = \langle a|\hat{W}|a\rangle \tag{7.144}$$

and

$$\begin{aligned}\hat{F}^{(1)}|a\rangle &= |a\rangle + \sum_{a' \neq a} \frac{1}{\epsilon_a - \epsilon_{a'}}|a'\rangle\langle a'|(\hat{W} - R_{a'}^{(0)})\hat{F}^{(0)}|a\rangle \\ &= |a\rangle + \sum_{a' \neq a} \frac{1}{\epsilon_a - \epsilon_{a'}}|a'\rangle\langle a'|\hat{W}|a\rangle.\end{aligned} \tag{7.145}$$

Iterating again, we get in the second order

$$R_a^{(2)} = \langle a|\hat{W}|a\rangle + \sum_{a' \neq a} \frac{1}{\epsilon_a - \epsilon_{a'}}\langle a|\hat{W}|a'\rangle\langle a'|\hat{W}|a\rangle \tag{7.146}$$

and

$$\begin{aligned}\hat{F}^{(2)}|a\rangle &= |a\rangle + \sum_{a' \neq a} \frac{1}{\epsilon_a - \epsilon_{a'}}|a'\rangle\langle a'|\hat{W}|a\rangle \\ &+ \sum_{a' \neq a}\sum_{a'' \neq a} \frac{1}{\epsilon_a - \epsilon_{a'}}\frac{1}{\epsilon_a - \epsilon_{a''}}|a'\rangle[\langle a'|\hat{W}|a''\rangle - \delta_{a'a''}\langle a|\hat{W}|a\rangle]\langle a''|\hat{W}|a\rangle.\end{aligned} \tag{7.147}$$

We just recovered the Rayleigh-Schrödinger perturbation method.

With a clever choice of $\hat{O}$, we can create various methods for perturbative calculations. The Brillouin-Wigner method offers several advantages compared to the Rayleigh-Schrödinger approach. The Brillouin-Wigner method works for degenerate cases. In most cases, the rate of convergence is faster. A little drawback is that in Eq. (7.138), the $E_{\lambda(a)}$ is implicit and cannot be given in closed form. It can be solved numerically, for example, by iteration.

7.4 Variational Methods

Variational methods are the other main approximation methods in quantum mechanics. They can be more robust than methods based on perturbation theory as they do not require an exact knowledge of some underlying part of the problem. In variational methods, we make an educated guess about the wave function with some adjustable parameters, called *variational parameters*. Those parameters are chosen such that the *trial function* becomes *optimal* in some sense.

7.4.1 The Variational Principle

Our aim is to find the solution of the eigenvalue problem

$$\hat{H}|\psi_n\rangle = E_n|\psi_n\rangle, \tag{7.148}$$

with the constraint

$$\langle\psi_n|\psi_m\rangle = \delta_{nm}. \tag{7.149}$$

We assume that $|\psi\rangle$ is a small variation of the true eigenstate

$$|\psi\rangle = |\psi_n\rangle + |\delta\psi\rangle. \tag{7.150}$$

Calculating the variation of the energy, we obtain

$$\begin{aligned}\delta(E) &= \langle\psi|\hat{H}|\psi\rangle - \langle\psi_n|\hat{H}|\psi_n\rangle \\ &= \langle\psi_n|\hat{H}|\delta\psi\rangle + \langle\delta\psi|\hat{H}|\psi_n\rangle + \langle\delta\psi|\hat{H}|\delta\psi\rangle \\ &\approx E_n(\langle\psi_n|\delta\psi\rangle + \langle\delta\psi|\psi_n\rangle).\end{aligned} \tag{7.151}$$

Since ψ is normalized,

$$1 = \langle\psi|\psi\rangle, \tag{7.152}$$

its variation vanishes, hence

$$\delta(1) = 0 = \langle\psi|\delta\psi\rangle + \langle\delta\psi|\psi\rangle \approx \langle\psi_n|\delta\psi\rangle + \langle\delta\psi|\psi_n\rangle. \tag{7.153}$$

Consequently, the variation of the energy also vanishes

$$\delta(E) = 0. \tag{7.154}$$

We have found that $\langle\psi|\hat{H}|\psi\rangle$ is stationary at $\psi = \psi_n$. So, if we parametrize the wave function $|\psi\rangle$ with parameter vector $\underline{\alpha}$, the energy

$$E(\underline{\alpha}) = \langle\psi(\underline{\alpha})|\hat{H}|\psi(\underline{\alpha})\rangle, \tag{7.155}$$

as a function of $\underline{\alpha}$, is stationary at E_n. Moreover, we can see that

$$\langle\psi|\hat{H}|\psi\rangle = \sum_n\langle\psi|\hat{H}|\psi_n\rangle\langle\psi_n|\psi\rangle = \sum_n E_n\langle\psi|\psi_n\rangle\langle\psi_n|\psi\rangle \geq E_0\sum_n\langle\psi|\psi_n\rangle\langle\psi_n|\psi\rangle \geq E_0, \tag{7.156}$$

i.e. the ground state is a true minimum.

In order to extend the variational method to excited states, we need to select trial functions that are orthogonal to the lower states. This can be achieved by taking trial functions with different symmetry, or explicitly orthogonalizing them to the lower states. If $|\psi_0\rangle$ is the ground state and $|\phi_1\rangle$ is a trial function, then

$$|\phi_1'\rangle = |\phi_1\rangle - |\psi_0\rangle\langle\psi_0|\phi_1\rangle, \tag{7.157}$$

is orthogonal to $|\psi_0\rangle$

$$\langle\psi_0|\phi_1'\rangle = \langle\psi_0|\phi_1\rangle - \langle\psi_0|\phi_1\rangle\langle\psi_0|\psi_0\rangle = 0. \tag{7.158}$$

Therefore,

$$\begin{aligned}\langle\phi_1'|\hat{H}|\phi_1'\rangle &= \sum_n \langle\phi_1'|\hat{H}|\psi_n\rangle\langle\psi_n|\phi_1'\rangle \\ &= \sum_n E_n\langle\phi_1'|\psi_n\rangle\langle\psi_n|\phi_1'\rangle \geq E_1\sum_n\langle\phi_1'|\psi_n\rangle\langle\psi_n|\phi_1'\rangle \geq E_1.\end{aligned} \tag{7.159}$$

The generalization of the procedure gives

$$|\phi_i'\rangle = |\phi_i\rangle - \sum_{k=0}^{i-1}\langle\psi_k|\phi_i\rangle|\psi_k\rangle, \tag{7.160}$$

and

$$\langle\phi_i'|\hat{H}|\phi_i'\rangle \geq E_i. \tag{7.161}$$

7.4.2 Rayleigh-Ritz Method

An interesting and very important special case of variational methods is when $|\psi\rangle$ is taken as a linear combinations of some basis states $\{|\phi_n\rangle\}$

$$|\psi(\underline{a})\rangle = \sum_{n=1}^{N} a_n|\phi_n\rangle, \tag{7.162}$$

with a_n variational parameters. Then we find that

$$E(\underline{a}) = \frac{\langle\psi(\underline{a})|\hat{H}|\psi(\underline{a})\rangle}{\langle\psi(\underline{a})|\psi(\underline{a})\rangle} = \frac{\sum_{mn} a_m^* a_n\langle\phi_m|\hat{H}|\phi_n\rangle}{\sum_n a_n^* a_n}, \tag{7.163}$$

or with $H_{mn} = \langle\phi_m|\hat{H}|\phi_n\rangle$, we have

$$\sum_{mn} a_m^* a_n H_{mn} = E\sum_n a_n^* a_n. \tag{7.164}$$

We can consider a_n^* and a_n as independent variables. The condition

$$\frac{\partial E}{\partial a_n} = 0, \tag{7.165}$$

gives

$$\sum_m a_m^* H_{mn} = E a_n^*, \tag{7.166}$$

while the condition

$$\frac{\partial E}{\partial a_n^*} = 0, \tag{7.167}$$

provides us with

$$\sum_m a_m H_{mn} = E a_n. \tag{7.168}$$

In either way, we get the matrix eigenvalue equation

$$\underline{H}\,\underline{a} = E\,\underline{a}. \tag{7.169}$$

In the $N \to \infty$ limit the basis states $|\phi_n\rangle$ span the whole Hilbert space and we recover the Schrödinger eigenvalue problem in matrix representation

$$\underline{H}\,\underline{\psi} = E\,\underline{\psi}. \tag{7.170}$$

So, the truncation of the Hilbert space to a finite subset is basically a variational method.

7.4.3 Lower Limits

Variational methods always approach the true eigenvalues from above and provide upper limits to the true solutions. However, the Rayleigh-Ritz method provides us with a tool for estimating the lower limits as well (see p. 247 in [4]).

Let us consider

$$\sigma = \langle \hat{H}\psi | \hat{H}\psi \rangle - \langle \psi | \hat{H} | \psi \rangle^2 = D_\psi - E_\psi^2, \tag{7.171}$$

with an arbitrary $|\psi\rangle$. If the $|\psi_k\rangle$ states are true normalized eigenstates of $\hat{H}$, then

$$\hat{H}|\psi_k\rangle = E_k|\psi_k\rangle, \qquad \text{and also} \quad |\psi\rangle = \sum_k a_k|\psi_k\rangle, \tag{7.172}$$

since they form a basis. So,

$$E_\psi = \langle\psi|\hat{H}|\psi\rangle = \sum_k E_k|a_k|^2 \quad \text{and} \quad D_\psi = \langle\hat{H}\psi|\hat{H}\psi\rangle = \sum_k E_k^2|a_k|^2. \tag{7.173}$$

Assuming a normalized $|\psi\rangle$, we have $\sum_k |a_k|^2 = 1$, and therefore

$$\begin{aligned}\sigma = D_\psi - 2E_\psi E_\psi + E_\psi^2 &= \sum_k E_k^2|a_k|^2 - 2E_\psi \sum_k E_k|a_k|^2 + E_\psi^2 \sum_k |a_k|^2 \\ &= \sum_k |a_k|^2(E_k - E_\psi)^2 \geq 0.\end{aligned} \tag{7.174}$$

Assume that E_j is the closest to E_ψ. Then

$$(E_k - E_\psi)^2 \geq (E_j - E_\psi)^2, \quad \text{for all } k. \tag{7.175}$$

Therefore,

$$\sigma \geq (E_j - E_\psi)^2 \sum_k |a_k|^2 = (E_j - E_\psi)^2, \tag{7.176}$$

and consequently,

$$E_\psi - \sqrt{\sigma} \leq E_j \leq E_\psi + \sqrt{\sigma}, \tag{7.177}$$

i.e. $E_j \in [E_\psi - \sqrt{\sigma}, E_\psi + \sqrt{\sigma}]$ interval. So, $E_\psi - \sqrt{\sigma}$ provides a lower limit and $E_\psi + \sqrt{\sigma}$ an upper limit to E_j.

7.4.4 Applications of Variational Methods

One-Dimensional Harmonic Oscillator

To determine the ground state of the one dimensional harmonic oscillator

$$\hat{H} = \frac{\hat{p}_x^2}{2m} + \frac{1}{2}m\omega^2x^2, \tag{7.178}$$

we take the anzatz

$$\psi_0(x) = A\exp(-\alpha x^2). \tag{7.179}$$

By using

$$\hat{p}_x\psi_0(x) = -i\hbar\frac{\mathrm{d}}{\mathrm{d}x}A\exp(-\alpha x^2) = 2i\hbar A\alpha x\exp(-\alpha x^2) \tag{7.180}$$

and the relation $\int_{-\infty}^{\infty} dx\, x^{2n} \exp(-\alpha x^2) = \sqrt{\pi/\alpha}(2n-1)!!/(2\alpha)^n = \sqrt{\pi/\alpha} \cdot 1 \cdot 3 \cdots (2n-1)/(2\alpha)^n$, we obtain $A = (2\alpha/\pi)^{1/4}$. Here, $n!! = \prod_{k=0}^{[n/2]-1}(n-2k)$ is the double factorial. Then, we find

$$E(\alpha) = A^2 \int_{-\infty}^{\infty} dx \exp(-\alpha x^2) \left(\frac{\hat{p}_x^2}{2m} + \frac{1}{2} m\omega^2 x^2 \right) \exp(-\alpha x^2) = \frac{\hbar^2}{2m}\alpha + \frac{m\omega^2}{8\alpha}. \tag{7.181}$$

The stationary condition gives

$$\frac{\partial E(\alpha)}{\partial \alpha} = \frac{\hbar^2}{2m} - \frac{m\omega^2}{8\alpha^2} = 0, \tag{7.182}$$

which leads to

$$\alpha_0 = \frac{m\omega}{2\hbar} \quad \text{and} \quad E_0(\alpha_0) = \frac{\hbar\omega}{2}, \quad \text{with} \quad \psi_0(x, \alpha_0) = \left(\frac{m\omega}{\pi\hbar}\right)^{1/4} \exp(-m\omega x^2/2\hbar^2). \tag{7.183}$$

So, the variational method with this choice of trial function recovers the exact ground state solution.

To calculate the first excited state, we take another ansatz

$$\psi_1(x) = Bx \exp(-\alpha x^2), \tag{7.184}$$

with normalization constant $B = \sqrt{8\alpha^{3/2}/\sqrt{2\pi}}$. We can immediately notice that $\langle \psi_1 | \psi_0 \rangle = 0$, since ψ_0 is even and ψ_1 is odd function of the variable x. In the same manner as before, we find

$$E_1(\alpha) = B^2 \int_{-\infty}^{\infty} dx\, x^2 \exp(-\alpha x^2) \left(-\frac{\hbar^2}{2m}\frac{d^2}{dx^2} + \frac{1}{2} m\omega^2 x^2 \right) \exp(-\alpha x^2) = \frac{3\hbar^2}{2m}\alpha + \frac{3m\omega^2}{8\alpha}, \tag{7.185}$$

which results in

$$\alpha_0 = \frac{m\omega}{2\hbar} \quad \text{and} \quad E_1(\alpha_0) = \frac{3\hbar\omega}{2} \quad \text{with} \quad \psi_1(x, \alpha_0) = \left(\frac{4m^3\omega^3}{\pi\hbar^3}\right)^{1/4} x \exp(-m\omega x^2/2\hbar^2), \tag{7.186}$$

i.e. the variational method with this particular trial function recovers the first excited state wave function.

Variational methods are very powerful. If, with a good sense of physics, we parametrize the trial function such that it catches the character of an exact solution, we may get the exact result or come very close to it.

Ground State of the Helium Atom

The helium atom, an α particle plus two electrons, looks like two hydrogen-like atoms plus an e^2/r_{12} repulsive potential, where r_{12} is the relative coordinate of the two electrons. The exact solution for the two hydrogen-like atoms is just the product of two ground state wave functions

$$\phi = \phi_1(r_1)\phi_2(r_2) = \frac{Z^3}{\pi a_0^3} \exp\left(-\frac{Z}{a_0}(r_1 + r_2)\right), \tag{7.187}$$

where r_1 and r_2 are the electron coordinates.

A reasonable physical picture is that in the helium atom, the electron feels some kind of screened charge due to the other electron. Therefore, we take the charge in the hydrogen-like atom as a variational parameter and replace Z in the Eq. (7.187) by Z'. Now, the Hamiltonian of the helium atom takes the form

$$\hat{H} = \hat{H}_1^0(r_1, Z') + \hat{H}_2^0(r_2, Z') + (Z' - Z)e^2\left(\frac{1}{r_1} + \frac{1}{r_2}\right) + \frac{e^2}{r_{12}}, \tag{7.188}$$

where $\hat{H}^0$ is the hydrogen Hamiltonian with charge Z'. The expectation value of this Hamiltonian with Z' in the trial function (7.187) reads

$$\begin{aligned} \langle\phi_1\phi_2|\hat{H}|\phi_1\phi_2\rangle &= \langle\phi_1|\hat{H}_1^0(Z')|\phi_1\rangle + \langle\phi_2|\hat{H}_2^0(Z')|\phi_2\rangle \\ &+ (Z' - Z)e^2\langle\phi_1|\frac{1}{r_1}|\phi_1\rangle + (Z' - Z)e^2\langle\phi_2|\frac{1}{r_2}|\phi_2\rangle + \langle\phi_1\phi_2|\frac{e^2}{r_{12}}|\phi_1\phi_2\rangle. \end{aligned} \tag{7.189}$$

The first and the second terms are

$$\langle\phi_1|\hat{H}_1^0(Z')|\phi_1\rangle = \langle\phi_2|\hat{H}_2^0(Z')|\phi_2\rangle = -Z'^2 E_0, \tag{7.190}$$

where $E_0 = m_e e^4/(2\hbar^2)$. Then, from a direct integration, or from the Hellmann-Feynman theorem Eq. (7.7), we get

$$\begin{aligned} \langle\phi_1|\frac{e^2}{r_1}|\phi_1\rangle &= \langle\phi_2|\frac{e^2}{r_2}|\phi_2\rangle \\ &= \frac{Z'^3 e^2}{\pi a_0^3}\int_0^{2\pi} d\phi \int_0^{\pi} d\theta \sin\theta \int_0^{\infty} dr\, r \exp\left(-\frac{2Z'r}{a_0}\right) \\ &= \frac{Z'^3 e^2}{\pi a_0^3} 4\pi \frac{1}{(2Z'/a_0)^2} = \frac{Z'e^2}{a_0} = 2Z'E_0. \end{aligned} \tag{7.191}$$

The last term is slightly more complicated. From the expansion

$$\frac{1}{r_{12}} = \sum_{l=0}^{\infty}\sum_{m=-l}^{l} \frac{4\pi}{2l+1}\frac{r_<^l}{r_>^{l+1}} Y_{l,m}^*(\theta_1,\phi_1)Y_{l,m}(\theta_2,\phi_2), \tag{7.192}$$

where $r_< = \min(r_1, r_2)$ and $r_> = \max(r_1, r_2)$, and using the fact that $4\pi Y_{0,0}Y_{0,0}^* = 1$, we get

$$\begin{aligned}\langle\phi_1\phi_2|\frac{e^2}{r_{12}}|\phi_1\phi_2\rangle &= \frac{16Z'^6e^2}{a_0^6}\sum_{l=0}^{\infty}\sum_{m=-l}^{l}\int_0^{\infty}\int_0^{\infty} \mathrm{d}r_1\mathrm{d}r_2 \exp\left(-\frac{2Z'(r_1+r_2)}{a_0}\right)\frac{r_<^l}{r_>^{l+1}}r_1^2r_2^2 \\ &\times \int_0^{2\pi}\mathrm{d}\phi_1\int_0^{\pi}\mathrm{d}\theta_1 \sin\theta_1 Y_{l,m}^*(\theta_1,\phi_1)Y_{0,0}(\theta_1,\phi_1) \\ &\times \int_0^{2\pi}\mathrm{d}\phi_2\int_0^{\pi}\mathrm{d}\theta_2 \sin\theta_2 Y_{l,m}(\theta_2,\phi_2)Y_{0,0}^*(\theta_2,\phi_2).\end{aligned} \tag{7.193}$$

The $Y_{l,m}$ functions are orthogonal and therefore the angular integrals vanish unless $l = m = 0$. So, we can perform the integrations as

$$\begin{aligned}\langle\phi_1\phi_2|\frac{e^2}{r_{12}}|\phi_1\phi_2\rangle &= \frac{16Z'^6e^2}{a_0^6}\int_0^{\infty}\int_0^{\infty}\mathrm{d}r_1\mathrm{d}r_2 \exp\left(-\frac{2Z'(r_1+r_2)}{a_0}\right)\frac{1}{r_>}r_1^2r_2^2 \\ &= \frac{16Z'^6e^2}{a_0^6}\int_0^{\infty}\mathrm{d}r_2 \exp\left(-\frac{2Z'r_2}{a_0}\right)r_2\left[\int_0^{r_2}\mathrm{d}r_1 \exp\left(-\frac{2Z'r_1}{a_0}\right)r_1^2\right] \\ &+ \frac{16Z'^6e^2}{a_0^6}\int_0^{\infty}\mathrm{d}r_2 \exp\left(-\frac{2Z'r_2}{a_0}\right)r_2^2\left[\int_{r_2}^{\infty}\mathrm{d}r_1 \exp(-\frac{2Z'r_1}{a_0})r_1\right] \\ &= \frac{5}{4}Z'E_0.\end{aligned} \tag{7.194}$$

Putting all these terms together, we find

$$E(Z') = -2Z'^2E_0 + 4(Z'-Z)Z'E_0 + \frac{5}{4}Z'E_0. \tag{7.195}$$

The stationary condition

$$\frac{\partial E(Z')}{\partial Z'} = -4Z'E_0 + 8Z'E_0 - 4ZE_0 + \frac{5}{4}E_0 = 0 \tag{7.196}$$

gives

$$Z' = Z - \frac{5}{16} \quad \text{and} \quad E(Z') = -2\left(Z - \frac{5}{16}\right)^2 E_0. \tag{7.197}$$

With $E_0 = 13.6$ eV, we get for the ground state of helium $E(Z') = -77.46$ eV. The experimental value is $E_{He} = -78.975$ eV. A slightly more complicated trial function, introduced by Hylleraas,

$$\phi = A(1 + Cr_{12}) \exp\left(-\frac{Z'}{a_0}(r_1 + r_2)\right), \tag{7.198}$$

with Z' and C as variational parameters, gives $E = -78.7$ eV. We can see also in this example that sometimes, with very simple functions, if we manage to grasp the essential character of the physical system, we can obtain remarkably good results.

7.5 Semiclassical Approximation

This method is called the WKB-method by Wentzel, Kramers and Brillouin, but it has been known before by Jeffreys (JWKB), and even before by Rayleigh, and even earlier by Green and Liouville.

We consider the one-dimensional Schrödinger equation

$$\psi'' + \frac{2m}{\hbar^2}(E - V(x))\psi = 0, \tag{7.199}$$

or

$$\psi'' + k^2(x)\psi = 0, \tag{7.200}$$

where $k(x) = \sqrt{2m(E - V(x))/\hbar^2}$ is the local wave number. In the WKB approach, we seek the solution in the form

$$\psi(x) = \vartheta(x) \exp\left(i\phi(x)/\hbar\right). \tag{7.201}$$

We take the derivatives

$$\begin{aligned} \psi' &= \left(\vartheta' + i\vartheta\phi'/\hbar\right) \exp\left(i\phi(x)/\hbar\right), \\ \psi'' &= \left(\vartheta'' + i\vartheta'\phi'/\hbar + i\vartheta\phi''/\hbar\right) \exp\left(i\phi(x)/\hbar\right) + \left(\vartheta' + i\vartheta\phi'/\hbar\right) \exp\left(i\phi(x)/\hbar\right) i\phi'/\hbar, \end{aligned} \tag{7.202}$$

and insert them into Eq. (7.199). Then, we equate both the real and imaginary parts to zero, and obtain

$$\phi'^2 - 2m(E - V) = \hbar^2 \frac{\vartheta''}{\vartheta}, \tag{7.203}$$

and

$$2\vartheta'\phi' + \vartheta\phi'' = 0. \tag{7.204}$$

We can see that Eq. (7.204) can be easily solved by

$$\vartheta = \frac{C}{\sqrt{\phi'}}, \tag{7.205}$$

where C is a constant. Inserting back into Eq. (7.203), we find

$$\phi'^2 = 2m(E - V) + \hbar^2 \left(\frac{3}{4} \left(\frac{\phi''}{\phi'} \right)^2 - \frac{1}{2} \frac{\phi'''}{\phi'} \right). \tag{7.206}$$

So far, we have not made any approximation. In the WKB approximation we express ϕ in powers of $\hbar^2$

$$\phi = \phi_0 + \hbar^2 \phi_1 + \ldots, \tag{7.207}$$

and keep the lowest term only,

$$\phi'^2 \approx \phi_0'^2 = 2m(E - V(x)). \tag{7.208}$$

The solution is rather straightforward. If $E > V(x)$,

$$\psi(x) \simeq A \frac{1}{\sqrt{k(x)}} \exp\left(\pm i \int^x k(x')\mathrm{d}x' \right). \tag{7.209}$$

In this region, the physical wave function is a linear combination of oscillating solutions. If $E < V(x)$, we simply replace $k(x)$ by $i\kappa(x)$, where $\kappa(x) = \sqrt{2m(V(x) - E)/\hbar^2}$. Here, we have two different kinds of solutions, an exponentially growing one and an exponentially decaying one. Obviously, this procedure is not applicable around classical turning points, where $E \simeq V(x)$. Here, $k(x) \simeq 0$ and the solutions become singular. The regions where we have solutions are separated by a singularity. Nevertheless, these different kinds of solutions should be matched over the singularity.

We assume that x_0 is a classical turning point, $E = V(x_0)$, and the potential is rising around x_0. Then, in the region $x_0 \ll x$ the physically acceptable solution is exponentially decaying

$$\psi_>(x) \sim \frac{B}{\sqrt{\kappa(x)}} \exp\left(- \int_{x_0}^x \mathrm{d}x' \, \kappa(x') \right), \tag{7.210}$$

while in the region $x \ll x_0$, the solution is a combination of oscillating waves

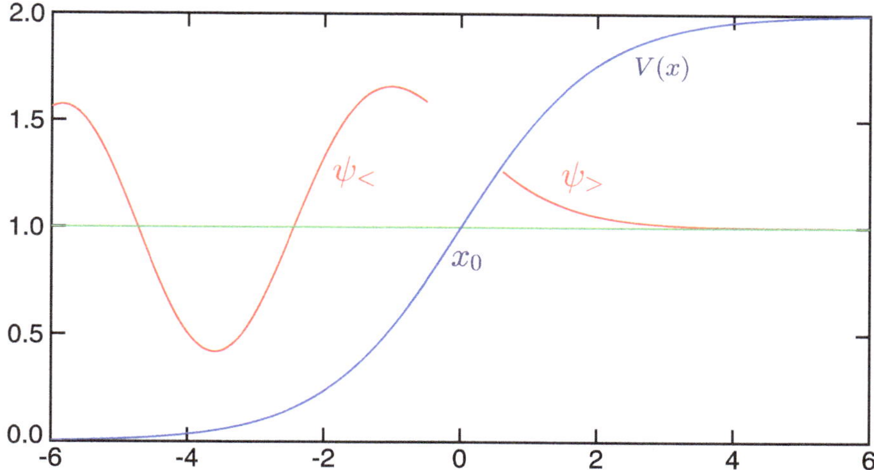

Fig. 7.1 WKB solutions around the classical turning point when the energy equals to the potential. The $\psi_>$ and the $\psi_<$ solutions should be matched across the singularity

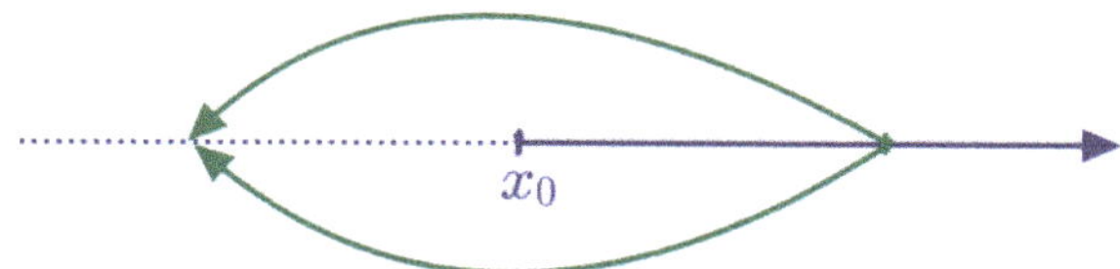

Fig. 7.2 Analytic continuation $k(x) \to \kappa(x)$ across the branch-cut singularity at the classical turning point

$$\psi_<(x) \sim \frac{A_1}{\sqrt{k(x)}} \exp\left(i \int^x k(x')\mathrm{d}x'\right) + \frac{A_2}{\sqrt{k(x)}} \exp\left(-i \int^x k(x')\mathrm{d}x'\right). \quad (7.211)$$

However, the solutions on both sides of x_0 are solutions of the same differential equation. They differ only in the value of a parameter x, which is a continuous analytic parameter of the differential equation. Therefore, the solutions in different regions should be an analytic continuation of each other. Furthermore, the differential equation is real and the boundary conditions are real, so the solutions can be taken to be real (Fig. 7.1).

In the $x_0 < x$ region, $\kappa(x) = \sqrt{2m(V(x) - E)/\hbar^2} \sim (x - x_0)^{1/2}$. This function has a branch-cut singularity at $x = x_0$ (Fig. 7.2). If we continue from $x_0 < x$ to $x < x_0$, we find that $\kappa(x) \to e^{i\pi/2}k(x)$, if we use the upper half complex x plane, and $\kappa(x) \to e^{-i\pi/2}k(x)$, if we use the lower half x plane. To keep the final result real, we split the $x_0 < x$ solution in half, and continue the first part on the upper half complex x-plane and the second part on the lower half of the complex x-plane

$$
\begin{aligned}
\psi_>(x) &\sim \frac{B}{2\sqrt{\kappa(x)}}\exp\left(-\int_{x_0}^{x} dx'\,\kappa(x')\right) + \frac{B}{2\sqrt{\kappa(x)}}\exp\left(-\int_{x_0}^{x} dx'\,\kappa(x')\right) \\
&\to \frac{B}{2\sqrt{e^{i\pi/2}k(x)}}\exp\left(-i\int_{x_0}^{x} dx'\,k(x')\right) + \frac{B}{2\sqrt{e^{-i\pi/2}k(x)}}\exp\left(i\int_{x_0}^{x} dx'\,k(x')\right) \\
&= \frac{B}{2\sqrt{k(x)}}\left[e^{-i\pi/4}\exp\left(-i\int_{x_0}^{x} dx'\,k(x')) + e^{i\pi/4}\exp(i\int_{x_0}^{x} dx'\,k(x')\right)\right] \\
&= \frac{B}{\sqrt{k(x)}}\cos\left(\int_{x_0}^{x} dx'\,k(x') + \frac{\pi}{4}\right) \\
&= \frac{B}{\sqrt{k(x)}}\cos\left(-\int_{x}^{x_0} dx'\,k(x') + \frac{\pi}{4}\right).
\end{aligned}
\tag{7.212}
$$

Using the identity $\cos(\theta) = \sin(\pi/2 - \theta)$, we find the properly connected solution

$$
\psi_< \simeq \frac{B}{\sqrt{k(x)}}\sin\left(\int_{x}^{x_0} dx'\,k(x') + \frac{\pi}{4}\right). \tag{7.213}
$$

7.5.1 Semiclassical Quantization

We consider a particle of mass m moving in a one-dimensional potential $V(x)$, with energy E. We assume that there are two classical turning points $x_1 < x_2$, where $V(x_1) = V(x_2) = E$, (Fig. 7.3).

In region III, where $x_2 < x$ and $E < V(x)$, the solution is given by

$$
\psi_{III} \simeq \frac{B}{\sqrt{\kappa(x)}}\exp\left(-\int_{x_2}^{x} dx'\kappa(x')\right), \tag{7.214}
$$

with $\kappa = \sqrt{2m(V(x) - E)/\hbar^2}$. This can be continued analytically to region II, where $x_1 < x < x_2$ and $V(x) < E$,

$$
\psi_{II} \simeq \frac{B}{\sqrt{k(x)}}\sin\left(\int_{x}^{x_2} dx'\,k(x') + \frac{\pi}{4}\right). \tag{7.215}
$$

Very much in the same way, the solution in region I is given by

$$
\psi_{I} \simeq \frac{B'}{\sqrt{\kappa(x)}}\exp\left(-\int_{x}^{x_1} dx'\kappa(x')\right), \tag{7.216}
$$

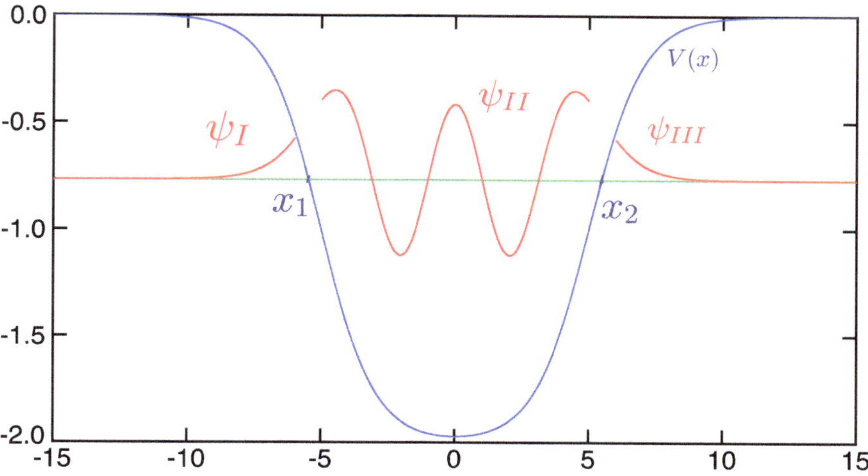

Fig. 7.3 WKB solution in a a potential well (blue line). The solutions should be matched across the singularities at the classical turning points

which can again be continued to the region II

$$\psi_{II} \simeq \frac{B'}{\sqrt{k(x)}} \sin\left(\int_{x_1}^{x} \mathrm{d}x'\, k(x') + \frac{\pi}{4}\right). \tag{7.217}$$

These two analytic continuations in Eqs. (7.215) and (7.217) must represent exactly the same function. So, we should be able to transform one solution to the another

$$\begin{aligned} B' \sin\left(\int_{x_1}^{x} \mathrm{d}x'\, k(x') + \frac{\pi}{4}\right) &= -B' \sin\left(-\int_{x_1}^{x} \mathrm{d}x'\, k(x') - \frac{\pi}{4}\right) \\ &= -B' \sin\left(\int_{x}^{x_1} \mathrm{d}x'\, k(x') - \frac{\pi}{4}\right) \\ &= -B' \sin\left(\int_{x}^{x_2} \mathrm{d}x'\, k(x') + \int_{x_2}^{x_1} \mathrm{d}x'\, k(x') - \frac{\pi}{4}\right) \\ &= -B' \sin\left(\left\{\int_{x}^{x_2} \mathrm{d}x'\, k(x') + \frac{\pi}{4}\right\} - \left\{\int_{x_1}^{x_2} \mathrm{d}x'\, k(x') + \frac{\pi}{2}\right\}\right). \end{aligned} \tag{7.218}$$

We can see that if we assign

$$\theta = \int_{x}^{x_2} \mathrm{d}x'\, k(x') + \frac{\pi}{4}, \tag{7.219}$$

and

$$\int_{x_1}^{x_2} \mathrm{d}x'\, k(x') + \frac{\pi}{2} = (n+1)\pi, \tag{7.220}$$

then by using the relation $\sin(\theta - (n+1)\pi) = \sin(\theta)\cos[(n+1)\pi] = \sin(\theta)(-1)^{n+1}$ we find that with $B = -B'(-)^{n+1}$, the two solutions coincide. Reformulating Eq. (7.220), we obtain

$$\int_{x_1}^{x_2} \mathrm{d}x'\, k(x') = \left(n + \frac{1}{2}\right)\pi\,, \qquad n = 0, 1, \ldots. \tag{7.221}$$

So, if the energy is such that this condition is satisfied, we can continue analytically our solutions over the entire range of x, and these pieces make up the physically acceptable WKB solution.

Recalling that $p = \hbar k$, we obtain

$$\int_{x_1}^{x_2} \mathrm{d}x'\, p(x') = \left(n + \frac{1}{2}\right)\frac{h}{2}\,, \tag{7.222}$$

and by extending the integral over a cycle between classical turning points, i.e. from x_1 to x_2 and back, we find

$$\oint \mathrm{d}x'\, p(x') = \left(n + \frac{1}{2}\right) h. \tag{7.223}$$

This rule corresponds to the Bohr-Sommerfeld quantization condition of old quantum mechanics. The $1/2$ term is absent in the original Bohr-Sommerfeld formula. This WKB formula, however, gives better agreements with exact results.

As an application, we may consider the one dimensional harmonic oscillator with $V(x) = 1/2\, m\omega^2 x^2$. The classical turning points are given by the solutions of $1/2\, m\omega^2 x^2 = E$, or by $x_{1,2} = \pm\sqrt{2E/(m\omega^2)}$. Then the quantization condition reads

$$\int_{-\sqrt{2E/(m\omega^2)}}^{\sqrt{2E/(m\omega^2)}} \mathrm{d}x \sqrt{\frac{2m}{\hbar^2}\left(E - \frac{1}{2}m\omega^2 x^2\right)} = \left(n + \frac{1}{2}\right)\pi. \tag{7.224}$$

By making the substitution $\sqrt{m\omega^2/(2E)}x = \xi$ and using the relation $\int_{-1}^{1} \sqrt{1-\xi^2}\,\mathrm{d}\xi = \pi/2$ we find

$$E = \hbar\omega(n + 1/2), \tag{7.225}$$

i.e. we recover the exact result.

Semiclassical Quantization in Three Dimensions

We assume that the potential is spherical, attractive and we have only one classical turning point at r_0, with $V(r_0) = E$. For simplicity, we consider only the $l = 0$ case. Then the three-dimensional problem becomes a one-dimensional problem on the half-line

$$\frac{\mathrm{d}^2 u}{\mathrm{d}r^2} = -\frac{2m}{\hbar^2}[V(r) - E]u(r), \quad \text{with} \quad r \in [0, \infty). \tag{7.226}$$

Sufficiently far beyond the classical turning point, $r_0 \ll r$, the WKB solution is given by

$$\frac{B}{\sqrt{\kappa(r)}} \exp\left(-\int_{r_0}^{r} \kappa(r')\mathrm{d}r'\right), \quad \text{with} \quad \kappa(r) = \sqrt{2m(V(r) - E)/\hbar^2}, \tag{7.227}$$

and represents a solution that tends to zero as $r \to \infty$. If we continue back to the classically allowed region, we obtain

$$u(r) \sim \frac{B}{\sqrt{k(r)}} \sin\left[\int_{r}^{r_0} k(r')\mathrm{d}r' + \frac{\pi}{4}\right]. \tag{7.228}$$

However, $u(r)$ should vanish at $r = 0$. This provides us with the quantization condition

$$\int_{0}^{r_0} k(r')\mathrm{d}r' + \frac{\pi}{4} = n\pi, \quad \text{with} \quad n = 1, 2, \ldots. \tag{7.229}$$

Here, $n = 0$ is not possible because the integral $\int_0^{r_0} k(r')\mathrm{d}r'$ is always positive.

For an oscillator potential, $V(r) = 1/2\, m\omega^2 r^2$, the ground state condition

$$\int_0^{r_0} k(r')\mathrm{d}r' = \frac{3\pi}{4}, \tag{7.230}$$

gives

$$E = \frac{3}{2}\hbar\omega, \tag{7.231}$$

which again coincides with the exact result.

The quantization condition was the consequence of the fact that the wave function vanishes at $r = 0$. The same rule should apply when the wave function vanishes for any other reason, like the infinite potential. So, if the potential is infinite at the left and right boundaries, at $a < b$, respectively, the quantization condition becomes

$$\int_a^b k(r')\mathrm{d}r' = n\pi, \quad \text{with} \quad n = 1, 2, \ldots. \tag{7.232}$$

7.5.2 Tunneling Through a Potential Barrier

Let us consider tunneling through a barrier (Fig. 7.4). Here, the energy E is less than the height of the barrier. The motion is scattering type in regions $(-\infty, a]$ and $[b, \infty)$ and bound type in the region $[a, b]$. The region $a < x < b$ is classically forbidden since $E < V$. The WKB solution, far away from a or b, is given by

$$\frac{B}{\sqrt{\kappa(x)}} \exp\left(-\int_a^x \kappa(x')\mathrm{d}x'\right), \quad \text{with} \quad \kappa(x) = \sqrt{2m(V(x) - E)/\hbar^2}. \tag{7.233}$$

This is a good approximation if $E \ll V$. Now, if we continue analytically to the left, far away from the turning point a, we find

$$\frac{B}{2\sqrt{k(x)}} \left[e^{-i\pi/4} \exp\left(-i\int_a^x k(x')\mathrm{d}x'\right) + e^{i\pi/4} \exp\left(i\int_a^x k(x')\mathrm{d}x'\right)\right], \tag{7.234}$$

which is a combination of incident and reflected waves with identical amplitudes. So, this approximation is acceptable if the tunneling through the barrier is very small.

To get the tunneling probability we need to continue Eq. (7.233) analytically to the right, well beyond the classical turning point b

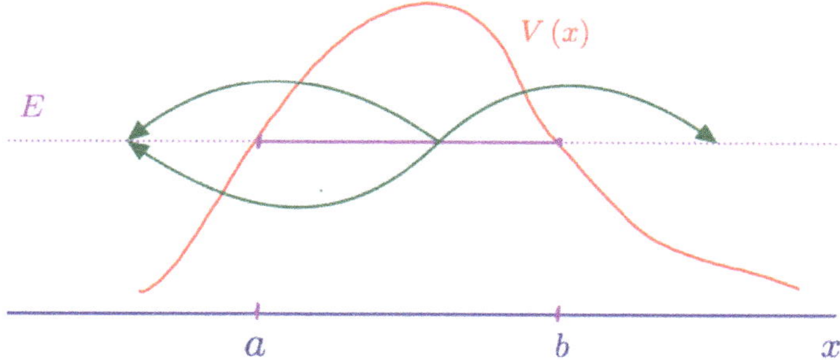

Fig. 7.4 Tunneling trough barrier and the analytic continuation of $\kappa(x) \to k(x)$ into the classically allowed regions

$$\frac{B}{\sqrt{\kappa(x)}} \exp\left(-\int_a^b \kappa(x')\mathrm{d}x'\right) \exp\left(-\int_b^x \kappa(x')\mathrm{d}x'\right)$$
$$\to \frac{B}{\sqrt{k(x)}} \exp\left(-\int_a^b \kappa(x')\mathrm{d}x'\right) e^{i\pi/4} \exp\left(i\int_b^x k(x')\mathrm{d}x'\right). \tag{7.235}$$

This is a right-traveling wave. By comparing the amplitudes of the incident and transmitted waves, we can extract the penetration probability

$$T \sim \exp\left(-2\int_a^b \kappa(x')\mathrm{d}x'\right). \tag{7.236}$$

This is called the *Gamow factor*. The transition probability is proportional to the area under the barrier.

In Eq. (7.233), we were concerned only about the exponentially decaying solution. However, in a finite potential barrier we cannot rule out the exponentially rising solution. The magnitude of this solution should be such that its analytic continuation to the right would eliminate the left-traveling wave. Then, its analytic continuation to the left would break the balance of the incoming and reflected waves. This more complicated treatment, however, for a broad and high barrier, would lead to the same result of Eq. (7.236).

Alpha Decay

An interesting early application of the WKB tunneling solution was the interpretation of α decay of certain radioactive nuclei. An α particle, consisting of two protons and two neutrons, formed randomly inside the nucleus, penetrate through the potential

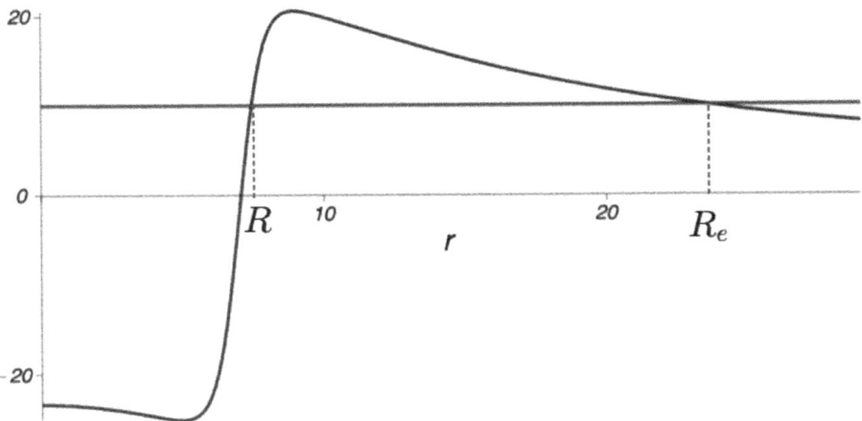

Fig. 7.5 α particle tunneling through a potential barrier. The classical turning points are defined by the crossings of the energy of the α particle and the potential

barrier and escape from the nucleus. The α particle experiences a deep finite-range nuclear and a repulsive Coulomb potential (Fig. 7.5).

The first classical turning point, denoted by R, is about the same as the range of the strong nuclear potential and it is roughly equal to the radius of the nucleus. The other turning point is R_e, the escape radius, which is defined by the condition $2Ze^2/R_e = E$, where E is the energy of the α decay. Then, the WKB approximation for the penetration probability is

$$T \simeq \exp\left(-2\int_R^{R_e} \sqrt{2m(V(r)-E)/\hbar^2}\,\mathrm{d}r\right). \tag{7.237}$$

The integral in the exponent, with the help of the substitution $r = R_e\cos^2\theta$ and with the integral $\int 2\sin^2\theta\,\mathrm{d}\theta = \theta - \sin\theta\cos\theta$, can be calculated as

$$\begin{aligned} I &\simeq \sqrt{\frac{2m}{\hbar^2}}\int_R^{R_e}\mathrm{d}r\sqrt{\frac{2Ze^2}{r}-E} = \sqrt{\frac{2m}{\hbar^2}E}\int_R^{R_e}\mathrm{d}r\sqrt{\frac{R_e}{r}-1} \\ &= \sqrt{\frac{2m}{\hbar^2}ER_e}\int_0^{\arccos\sqrt{R/R_e}}\mathrm{d}\theta\,2\sin^2\theta \\ &= R_e\sqrt{\frac{2m}{\hbar^2}E}\left[\arccos\sqrt{\frac{R}{R_e}} - \sqrt{\frac{R}{R_e}\left(1-\frac{R}{R_e}\right)}\right]. \end{aligned} \tag{7.238}$$

Now, if the energy is low and the barrier is high, then $R/R_e \ll 1$ and $\arccos\sqrt{R/R_e} \simeq \pi/2$. So, with $R_e = 2Ze^2/E$, the integral becomes

$$I \simeq \frac{2Ze^2}{E}\sqrt{\frac{2m}{\hbar^2}E}\left(\pi/2 - \sqrt{\frac{R}{2Ze^2}E}\right) \simeq e^2\pi\sqrt{\frac{2m}{\hbar^2}}\frac{Z}{\sqrt{E}} - 2\sqrt{\frac{m}{e^2\hbar^2}}\sqrt{Z}\sqrt{R}. \tag{7.239}$$

The decay rate λ is proportional to the transition probability, therefore

$$\log\lambda \simeq c_1\frac{Z}{\sqrt{E}} - c_2\sqrt{Z}\sqrt{R}, \tag{7.240}$$

with c_1 and c_2 constants. Consequently, $\log\lambda$ versus $Z/\sqrt{E}$ gives a straight line. This relation has been observed empirically over a wide range of nuclei and decay lifetimes (Geiger-Nuttal law). This law even allows extraction of information about R, the size of the nuclei.

Further Reading

1. C.M. Bender, S.A. Orszag, *Advanced Mathematical Methods for Scientists and Engineers I: Asymptotic Methods and Perturbation Theory* (Springer Science & Business Media, 2013)
2. A. Böhm, *Quantum Mechanics: Foundations and Applications* (Springer Science & Business Media, 2013)
3. M.L. Goldberger, K.M. Watson, *Collision Theory* (Courier Corporation, 2004)
4. K. Konishi, G. Paffuti, *Quantum Mechanics: A New Introduction* (OUP Oxford, 2009)

Chapter 8
Time-Dependent Quantum Mechanics

In the previous chapters, we were mainly concerned with stationary solutions of time-independent problems. However, quite a lot of problems and processes involve time and time evolution of quantum systems. In this chapter we are going to study methods that enable us to approach the time-evolution of quantum systems.

8.1 Time-Dependent Perturbation Theory

We consider a time-independent Hamiltonian $\hat{H}_0$ with a time-dependent potential

$$\hat{H} = \hat{H}_0 + \hat{V}(t). \tag{8.1}$$

We assume that we have a complete knowledge of $\hat{H}_0$:

$$\hat{H}_0|\psi_n\rangle = E_n|\psi_n\rangle \quad \text{and} \quad |\psi_n(t)\rangle = \exp(-i\hat{H}_0 t/\hbar)|\psi_n\rangle = \exp(-iE_n t/\hbar)|\psi_n\rangle. \tag{8.2}$$

We also assume that the perturbation acts only for a finite time interval (Fig. 8.1)

$$\hat{V}(t) = \begin{cases} \hat{V}(t), & \text{if} \quad 0 \le t \le \tau, \\ 0, & \text{if} \quad t < 0 \quad \text{or} \quad \tau < t. \end{cases} \tag{8.3}$$

The Standard Procedure

In the standard solution method, we solve the time-dependent Schrödinger equation

Z. Papp, *Mastering Quantum Mechanics*,
https://doi.org/10.1007/978-3-032-09011-9_8

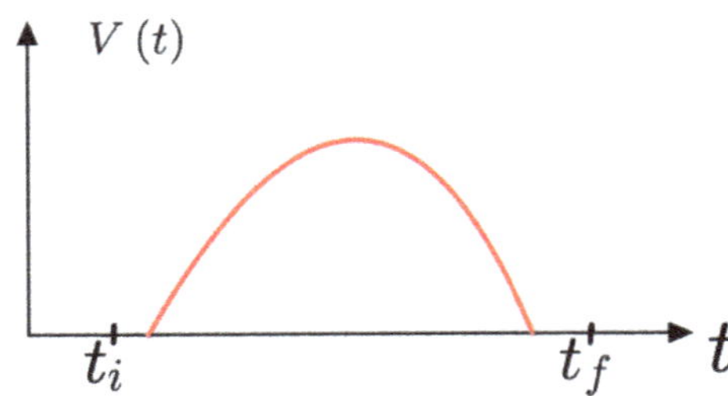

Fig. 8.1 The time-dependent potential acts only for a finite time interval

$$i\hbar\frac{\mathrm{d}|\psi(t)\rangle}{\mathrm{d}t} = (\hat{H}_0 + \hat{V}(t))|\psi(t)\rangle, \tag{8.4}$$

with the ansatz

$$|\psi(t)\rangle = \sum_n c_n(t)\exp(-iE_nt/\hbar)|\psi_n\rangle. \tag{8.5}$$

Substituting into Eq. (8.4), we get

$$\begin{aligned} i\hbar\sum_n \dot{c}_n(t)\exp(-iE_nt/\hbar)|\psi_n\rangle + i\hbar\sum_n c_n(t)\,(-iE_n/\hbar)\exp(-iE_nt/\hbar)|\psi_n\rangle = \\ \sum_n c_n(t)\exp(-iE_nt/\hbar)E_n|\psi_n\rangle + \sum_n c_n(t)\exp(-iE_nt/\hbar)\hat{V}(t)|\psi_n\rangle, \end{aligned} \tag{8.6}$$

and acting by $\langle\psi_k|$, we obtain a set of coupled differential equations

$$i\hbar\dot{c}_k(t) = \sum_n c_n(t)\exp(i\omega_{kn}t)\langle\psi_k|\hat{V}(t)|\psi_n\rangle, \tag{8.7}$$

where $\omega_{kn} = (E_k - E_n)/\hbar$. This differential equation is first order in the time derivative. It can be solved if $|\psi(0)\rangle$, or equivalently $c_n(0)$ for all n, is given.

8.1.1 Interaction Picture Solution

In an alternative procedure we solve the Schrödinger equation in the interaction picture representation. We start with the Schrödinger time-evolution operator

$$|\psi(t)\rangle = \hat{U}(t, t_i)|\psi(t_i)\rangle, \tag{8.8}$$

with $t_i < t$, and transform this relation into the interaction picture (see Sect. 2.2.2)

$$
\begin{aligned}
|\psi(t)\rangle_I &= \exp(i\hat{H}_0 t/\hbar)|\psi(t)\rangle = \exp(i\hat{H}_0 t/\hbar)\hat{U}(t,t_i)|\psi(t_i)\rangle \\
&= \underbrace{\exp(i\hat{H}_0 t/\hbar)\hat{U}(t,t_i)\exp(-i\hat{H}_0 t_i/\hbar)}_{\hat{U}_I(t,t_i)}\underbrace{\exp(i\hat{H}_0 t_i/\hbar)|\psi(t_i)\rangle}_{|\psi(t_i)\rangle_I} = \hat{U}_I(t,t_i)|\psi(t_i)\rangle_I .
\end{aligned}
\tag{8.9}
$$

Taking the time derivative, we obtain

$$
i\hbar\frac{\mathrm{d}|\psi(t)\rangle_I}{\mathrm{d}t} = i\hbar\frac{\mathrm{d}\hat{U}_I(t,t_i)}{\mathrm{d}t}|\psi(t_i)\rangle_I . \tag{8.10}
$$

On the other hand, the interaction picture Schrödinger equation of Eq. (2.27) reads

$$
i\hbar\frac{\mathrm{d}|\psi(t)\rangle_I}{\mathrm{d}t} = \hat{V}_I(t)|\psi(t)\rangle_I = \hat{V}_I(t)\hat{U}_I(t,t_i)|\psi(t_i)\rangle_I . \tag{8.11}
$$

Combining these relations, we can infer a differential equation for $\hat{U}_I$

$$
i\hbar\frac{\mathrm{d}\hat{U}_I(t,t_i)}{\mathrm{d}t} = \hat{V}_I(t)\hat{U}_I(t,t_i), \tag{8.12}
$$

with the initial condition

$$
\hat{U}_I(t_i,t_i) = I. \tag{8.13}
$$

This differential equation with the initial condition can be unified into an integral equation

$$
\hat{U}_I(t,t_i) = I - \frac{i}{\hbar}\int_{t_i}^{t}\mathrm{d}t'\,\hat{V}_I(t')\hat{U}_I(t',t_i). \tag{8.14}
$$

The $\hat{U}_I(t,t_i)$ solution of Eq. (8.14) obviously satisfies the differential equation Eq. (8.12) and the initial condition Eq. (8.13). Integral equations of this type are Volterra-type integral equations of the second kind.

We can solve this integral equation by adopting an iterative procedure. If we have an approximate solution in the n-th step, $\hat{U}_I^{(n)}$, we can insert it back into the right hand side of Eq. (8.14) to get the next approximation

$$
\hat{U}_I^{(n+1)}(t,t_i) = I - \frac{i}{\hbar}\int_{t_i}^{t}\mathrm{d}t'\,\hat{V}_I(t')\hat{U}_I^{(n)}(t',t_i). \tag{8.15}
$$

We repeat the procedure until we find convergence in some sense, i.e.

$$
||\hat{U}_I^{(n+1)} - \hat{U}_I^{(n)}|| < \varepsilon, \tag{8.16}
$$

with ε an arbitrary small number. The procedure starts with the unit operator and generates the sequence

$$\begin{aligned}\hat{U}_I^{(0)}(t,t_i) &= I,\\ \hat{U}_I^{(1)}(t,t_i) &= I - \frac{i}{\hbar}\int_{t_i}^{t} \mathrm{d}t' \hat{V}_I(t'),\\ \hat{U}_I^{(2)}(t,t_i) &= I - \frac{i}{\hbar}\int_{t_i}^{t} \mathrm{d}t' \hat{V}_I(t') + \left(-\frac{i}{\hbar}\right)^2 \int_{t_i}^{t} \mathrm{d}t_1 \int_{t_i}^{t_1} \mathrm{d}t_2 \hat{V}_I(t_1)\hat{V}_I(t_2),\end{aligned} \tag{8.17}$$

and so on, resulting in the Dyson series

$$\hat{U}_I = \sum_{n=0}^{\infty}\left(-\frac{i}{\hbar}\right)^n \int_{t_i}^{t} \mathrm{d}t_1 \int_{t_i}^{t_1} \mathrm{d}t_2 \cdots \int_{t_i}^{t_{n-1}} \mathrm{d}t_n \hat{V}_I(t_1)\hat{V}_I(t_2)\cdots \hat{V}_I(t_n), \tag{8.18}$$

where the time arguments in $\hat{V}_I$ should be ordered as

$$t_i \le t_n \le t_{n-1} \le \cdots \le t_1 \le t. \tag{8.19}$$

8.1.2 Transition Probability

As we mentioned before, we assume that $\hat{V}(t) \neq 0$ only for a finite time interval. At t around t_i, the system is in the initial state $|\psi_i\rangle$ and governed by the Hamiltonian $\hat{H}_0$ with the time evolution

$$|\psi_i(t)\rangle = \exp(-i\hat{H}_0 t/\hbar)|\psi_i\rangle. \tag{8.20}$$

Later, at $t \sim t_f$, the system is again governed by the Hamiltonian $\hat{H}_0$ and is in the final state $|\psi_f\rangle$

$$|\psi_f(t)\rangle = \exp(-i\hat{H}_0 t/\hbar)|\psi_f\rangle. \tag{8.21}$$

As the potential $\hat{V}(t)$ starts acting, the time evolution operator of the full Hamiltonian evolves the system from the initial state

$$|\psi_i(t)\rangle = \hat{U}(t,t_i)|\psi_i(t_i)\rangle. \tag{8.22}$$

The transition probability from ψ_i to ψ_f is given by the overlap

$$\begin{aligned}P_{i\to f}(t) = \left|\langle\psi_f(t)|\psi_i(t)\rangle\right|^2 &= \left|\langle\psi_f|\exp(i\hat{H}_0 t/\hbar)\hat{U}(t,t_i)\exp(-i\hat{H}_0 t_i/\hbar)|\psi_i\rangle\right|^2\\ &= \left|\langle\psi_f|\hat{U}_I(t,t_i)|\psi_i\rangle\right|^2.\end{aligned} \tag{8.23}$$

Taking the iterative solution for $\hat{U}_I(t,t_i)$ of Eq. (8.18) and assuming $t_i = 0$, we obtain the time dependent perturbation series

$$
\begin{aligned}
P_{i\to f}(t) =& \Bigg| \langle\psi_f|\psi_i\rangle - \frac{i}{\hbar}\int_0^t \mathrm{d}t' \exp(i\omega_{fi}\, t')\langle\psi_f|\hat{V}(t')|\psi_i\rangle \\
& + \left(\frac{-i}{\hbar}\right)^2 \sum_n \int_0^t \mathrm{d}t_1 \exp(i\omega_{fn}\, t_1)\langle\psi_f|\hat{V}(t_1)|\psi_n\rangle \\
& \times \int_0^{t_1} \mathrm{d}t_2 \exp(i\omega_{ni}\, t_2)\langle\psi_n|\hat{V}(t_2)|\psi_i\rangle + \ldots \Bigg|^2 \\
=& \left| c_f^{(0)} + c_f^{(1)} + c_f^{(2)} + \ldots \right|^2,
\end{aligned}
\tag{8.24}
$$

where this relation defines the different orders of perturbation $c_f^{(i)}$. Here, we have used the fact that

$$
\begin{aligned}
\langle\psi_f|\hat{V}_I(t)|\psi_i\rangle &= \langle\psi_f|\exp(i\hat{H}_0 t/\hbar)\hat{V}(t)\exp(-i\hat{H}_0 t/\hbar)|\psi_i\rangle \\
&= \exp(i(E_f - E_i)t/\hbar)\langle\psi_f|\hat{V}(t)|\psi_i\rangle = \exp(i\omega_{fi}t)\langle\psi_f|\hat{V}(t)|\psi_i\rangle,
\end{aligned}
\tag{8.25}
$$

with $\omega_{fi} = (E_f - E_i)/\hbar$.

In general, we should evaluate this series term by term, but in practice, we can explain a lot of physics by considering only terms up to first order

$$
\begin{aligned}
c_f^{(0)} &= \langle\psi_f|\psi_i\rangle = \delta_{fi}, \\
c_f^{(1)} &= -\frac{i}{\hbar}\int_0^t \mathrm{d}t' \exp(i\omega_{fi}t')\langle\psi_f|\hat{V}(t')|\psi_i\rangle.
\end{aligned}
\tag{8.26}
$$

Then, for the transition probability, we have

$$
P_{i\to i}(t) = \left| 1 - \frac{i}{\hbar}\int_0^t \mathrm{d}t' \langle\psi_i|\hat{V}(t')|\psi_i\rangle \right|^2, \qquad \text{if } f = i, \tag{8.27}
$$

$$
P_{i\to f}(t) = \left| -\frac{i}{\hbar}\int_0^t \mathrm{d}t' \exp(i\omega_{fi}t')\langle\psi_f|\hat{V}(t')|\psi_i\rangle \right|^2, \qquad \text{if } f \neq i. \tag{8.28}
$$

8.1.3 Constant Perturbation

First, we consider a perturbation constant in time

$$
\hat{V}(t') = \begin{cases} 0, & \text{if} \quad t' \le 0, \\ \hat{V}, & \text{if} \quad 0 \le t' \le t, \\ 0, & \text{if} \quad t < t', \end{cases} \tag{8.29}
$$

which may not be constant in spatial coordinates. Then, for the transition probability, in the first order of perturbation theory, we have

$$
\begin{aligned}
P_{i\to f}(t) &\simeq \frac{1}{\hbar^2}\left|\langle\psi_f|\hat{V}|\psi_i\rangle\int_0^t \mathrm{d}t'\exp(i\omega_{fi}t')\right|^2 = \frac{1}{\hbar^2}\left|\langle\psi_f|\hat{V}|\psi_i\rangle\right|^2\left|\frac{\exp(i\omega_{fi}t)-1}{i\omega_{fi}}\right|^2 \\
&= \frac{\left|\langle\psi_f|\hat{V}|\psi_i\rangle\right|^2}{\hbar^2\omega_{fi}^2}\left|2\exp(i\omega_{fi}t/2)\frac{\exp(i\omega_{fi}t/2)-\exp(-i\omega_{fi}t/2)}{2i}\right|^2 \\
&= \frac{4\left|\langle\psi_f|\hat{V}|\psi_i\rangle\right|^2}{\hbar^2\omega_{fi}^2}\sin^2\left(\frac{\omega_{fi}t}{2}\right).
\end{aligned}
\tag{8.30}
$$

The transition probability oscillates with time and it is the strongest with $\omega_{fi}=0$.

By using the l'Hospital rule, we can easily find the limit

$$
\begin{aligned}
\lim_{t\to\infty}\frac{\sin^2(\omega t)}{\pi t\omega^2} &= \lim_{t\to\infty}\frac{2\sin(\omega t)\cos(\omega t)\omega}{\pi\omega^2} = \lim_{t\to\infty}\frac{\sin(2\omega t)}{\pi\omega} \\
&= \lim_{\epsilon\to 0}\frac{2\sin(2\omega/\epsilon)}{\pi 2\omega} = 2\delta(2\omega) = \delta(\omega),
\end{aligned}
\tag{8.31}
$$

and then obtain at large times

$$
P_{i\to f}(t) \simeq \frac{2\pi}{\hbar}t\left|\langle\psi_f|\hat{V}|\psi_i\rangle\right|^2\delta(E_f-E_i),
\tag{8.32}
$$

which gives for the transition rate

$$
\Gamma_{i\to f} = \lim_{t\to\infty}\frac{P_{i\to f}(t)}{t} = \frac{2\pi}{\hbar}\left|\langle\psi_f|\hat{V}|\psi_i\rangle\right|^2\delta(E_f-E_i).
\tag{8.33}
$$

We can see that with constant perturbation, in the $t\to\infty$ limit, transition can happen only between states with the same energy.

Transition to the Continuum

If the final state is in the continuum, we can write the energy interval as $[E_f, E_f + dE_f] = \rho(E_f)\,\mathrm{d}E_f$, where $\rho(E_f)$ is the density of states. Then the transition rate becomes

$$
\begin{aligned}
\Gamma_{i\to f} &= \lim_{t\to\infty}\int\frac{P_{i\to f}(t)}{t}\rho(E_f)\,\mathrm{d}E_f = \frac{2\pi}{\hbar}\left|\langle\psi_f|\hat{V}|\psi_i\rangle\right|^2\int\delta(E_f-E_i)\rho(E_f)\mathrm{d}E_f \\
&= \frac{2\pi}{\hbar}\left|\langle\psi_f|\hat{V}|\psi_i\rangle\right|^2\rho(E_i).
\end{aligned}
\tag{8.34}
$$

This relation is known as Fermi's golden rule (1950), but it had been known well before by Dirac around 1927.

8.1.4 Harmonic Perturbations

Harmonic perturbations are periodic in time with frequency ω

$$\hat{V}(t) = \hat{V}\exp(i\omega t) + \hat{V}^{\dagger}\exp(-i\omega t). \tag{8.35}$$

The first order perturbation to the transition amplitude reads

$$P_{i\to f}(t) \simeq \frac{1}{\hbar^2}\left|\langle\psi_f|\hat{V}|\psi_i\rangle\int_0^t \mathrm{d}t'\exp(i(\omega_{fi}+\omega)t') + \langle\psi_f|\hat{V}^{\dagger}|\psi_i\rangle\int_0^t \mathrm{d}t'\exp(i(\omega_{fi}-\omega)t')\right|^2. \tag{8.36}$$

After neglecting the cross terms, since they vanish for large t due to destructive interference, we find

$$\begin{aligned} P_{i\to f}(t) &\simeq \frac{1}{\hbar^2}\left|\langle\psi_f|\hat{V}|\psi_i\rangle\right|^2\left|\frac{\exp(i(\omega_{fi}+\omega)t)-1}{\omega_{fi}+\omega}\right|^2 + \frac{1}{\hbar^2}\left|\langle\psi_f|\hat{V}^{\dagger}|\psi_i\rangle\right|^2\left|\frac{\exp(i(\omega_{fi}-\omega)t)-1}{\omega_{fi}-\omega}\right|^2 \\ &\simeq \frac{4}{\hbar^2}\left(\left|\langle\psi_f|\hat{V}|\psi_i\rangle\right|^2\frac{\sin^2((\omega_{fi}+\omega)t/2)}{(\omega_{fi}+\omega)^2} + \left|\langle\psi_f|\hat{V}^{\dagger}|\psi_i\rangle\right|^2\frac{\sin^2((\omega_{fi}-\omega)t/2)}{(\omega_{fi}-\omega)^2}\right), \end{aligned} \tag{8.37}$$

and for the transition rate we have

$$\begin{aligned} \Gamma_{i\to f} &= \lim_{t\to\infty}\frac{P_{i\to f}(t)}{t} \\ &= \frac{2\pi}{\hbar}\left(\left|\langle\psi_f|\hat{V}|\psi_i\rangle\right|^2\delta(E_f-E_i+\hbar\omega) + \left|\langle\psi_f|\hat{V}^{\dagger}|\psi_i\rangle\right|^2\delta(E_f-E_i-\hbar\omega)\right). \end{aligned} \tag{8.38}$$

We can see that for harmonic perturbations in the $t\to\infty$ limit, transitions are possible only to final states with $E_f = E_i - \hbar\omega$ and $E_f = E_i + \hbar\omega$. These transitions are called emission and absorption, respectively (Fig. 8.2). In absorption, the quantum system absorbs $\hbar\omega$ energy from the field, and its energy rises to $E_f = E_i + \hbar\omega$. In the case of emission, the system emits $\hbar\omega$ energy to the surroundings, and its energy becomes $E_f = E_i - \hbar\omega$.

8.2 Sudden and Adiabatic Approximations

8.2.1 Sudden Approximation

In this section, we assume that the system, initially at $t = t_0$, is described by the Hamiltonian $\hat{H}_0$ which changes continuously. At a later time, $t = t_0 + T$, the Hamiltonian turns into $\hat{H}$. We parametrize the change by introducing $\tau = (t - t_0)/T$, and then $t = t_0 + \tau T$ with $\tau \in [0, 1]$. The corresponding time evolution operator in the Schrödinger picture is defined by the integral equation

$$\hat{U}(t) = 1 - \frac{i}{\hbar}\int_{t_0}^{t} \mathrm{d}t' \hat{H}(t')\hat{U}(t') = 1 - \frac{i}{\hbar}T\int_{0}^{1} \mathrm{d}\tau \hat{H}(\tau)\hat{U}(\tau), \tag{8.39}$$

and the time evolution of the system is given by

$$|\psi(t)\rangle = \hat{U}(t, t_0)|\psi(t_0)\rangle. \tag{8.40}$$

In the *sudden approximation*, we assume that the change happens very rapidly, almost instantaneously, in the $T \to 0$ limit. Then, the second term in Eq. (8.39) vanishes and

$$\lim_{T\to 0} \hat{U}(T) = 1. \tag{8.41}$$

This means that the system remains unchanged, even if the change in the Hamiltonian is substantial. The change is so rapid that the system has no time to adjust and the wave function remains the same. The transition from an eigenstate $|\phi_i\rangle$ of $\hat{H}_0$ to the eigenstate $|\psi_j\rangle$ of $\hat{H}$ is given by

$$P_{i\to j} = \lim_{T\to 0}\left|\langle\psi_j(T)|\hat{U}(T)\phi_i\rangle\right|^2 = \left|\langle\psi_j|\phi_i\rangle\right|^2. \tag{8.42}$$

We can estimate the accuracy of the sudden approximation. Assume that the system at $t = t_0$ is in a state $|\phi\rangle$, that evolves to $\hat{U}(T)|\phi\rangle$. The probability amplitude

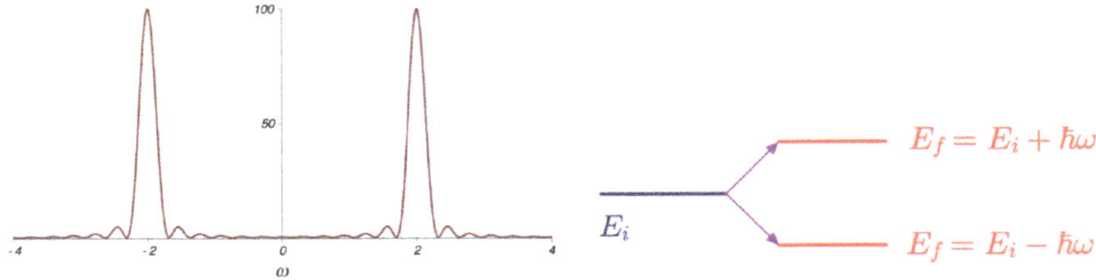

Fig. 8.2 Illustrative plot of Eq. (8.37) with $\omega_{fi} = 2$ and $t = 20$. In the $t \to \infty$ limit the peaks become δ functions resulting in emission, with $E_f = E_i - \hbar\omega$, and absorption, with $E_f = E_i + \hbar\omega$

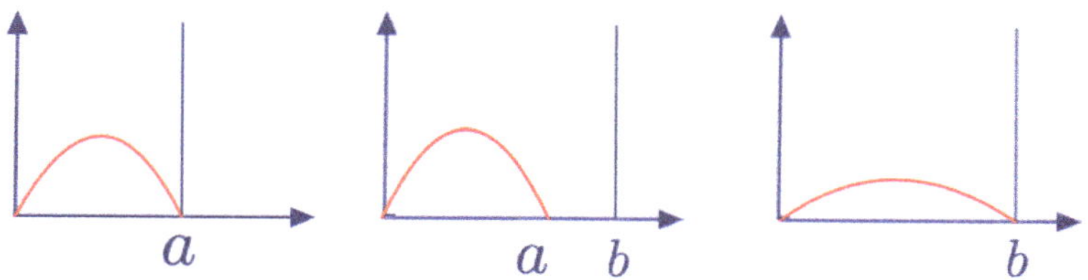

Fig. 8.3 Wave function by sudden and adiabatic change of the Hamiltonian

of not finding the system at $t = t_0 + T$ in the state $|\phi\rangle$ is given by

$$\bar{\omega} = \langle\phi|\hat{U}^\dagger(T)\hat{Q}\hat{U}(T)|\phi\rangle, \tag{8.43}$$

where $\hat{Q} = 1 - |\phi\rangle\ \langle\phi|$ is the projection onto the orthogonal subspace. We may approximate $\hat{U}$ up to first order in $\hat{H}$

$$\hat{U}(t, t_0) \simeq 1 - \frac{i}{\hbar}T\int_0^1 \mathrm{d}\tau \hat{H}(\tau) \simeq 1 - \frac{i}{\hbar}T\hat{\bar{H}}, \tag{8.44}$$

where $\hat{\bar{H}}$ is the time average of the Hamiltonian. Considering that $\hat{Q}|\phi\rangle = 0$, we find in leading order

$$\bar{\omega} \simeq \langle\phi|(1 + \frac{i}{\hbar}T\hat{\bar{H}})\hat{Q}(1 - \frac{i}{\hbar}T\hat{\bar{H}})|\phi\rangle \simeq \frac{T^2}{\hbar^2}\langle\phi|\hat{\bar{H}}\hat{Q}\hat{\bar{H}}|\phi\rangle. \tag{8.45}$$

Evaluating further, we obtain

$$\bar{\omega} \simeq \frac{T^2}{\hbar^2}\langle\phi|\hat{\bar{H}}(1 - |\phi\rangle\ \langle\phi|)\hat{\bar{H}}|\phi\rangle = \frac{T^2}{\hbar^2}\left[\langle\phi|\hat{\bar{H}}^2|\phi\rangle - \langle\phi|\hat{\bar{H}}|\phi\rangle^2\right] = \frac{T^2}{\hbar^2}(\Delta\bar{H})^2. \tag{8.46}$$

The sudden approximation is valid if $\bar{\omega} \ll 1$, which implies that

$$T \ll \frac{\hbar}{\Delta\bar{H}} = \Delta T, \tag{8.47}$$

where ΔT is the characteristic time period of the system defined by the time-energy uncertainty relation $\Delta T \cdot \Delta E \simeq \hbar$ (Fig. 8.3).

8.2.2 Adiabatic Approximation

Here, we consider a Hamiltonian that changes slowly in time. We assume that the change in the Hamiltonian happens through a parameter λ with $\hat{H}(t) = \hat{H}(\lambda(t))$. The time evolution of the state is governed by the time-dependent Schrödinger equation

$$i\hbar\frac{\mathrm{d}}{\mathrm{d}t}|\psi(\lambda(t))\rangle = \hat{H}(\lambda(t))|\psi(\lambda(t))\rangle. \tag{8.48}$$

As $\lambda(t)$ varies slowly, the system at all moments can be considered as static and we can have stationary solutions

$$\hat{H}(\lambda(t))|n,\lambda(t)\rangle = E_n(\lambda(t))|n,\lambda(t)\rangle, \tag{8.49}$$

which form a basis $\{|n,\lambda(t)\rangle\}$. Now, for the sake of easy notation, we drop the $\lambda(t)$ dependence from the formulae, but we still assume implicitly that the time dependence is through the λ dependence.

We can expand $|\psi(t)\rangle$ in terms of the basis states and seek the general solution in the form

$$|\psi(t)\rangle = \sum_n a_n(t)\exp(-i\int_0^t \mathrm{d}t' E_n(t')/\hbar)|n(t)\rangle, \tag{8.50}$$

with time dependent coefficients $a_n(t)$. If $\hat{H}$ is time-independent, so is E_n, and the exponential terms fall back to the usual stationary time dependence of the eigenstates and a_n remain time-independent.

Inserting Eqs. (8.50) into (8.48), we find

$$i\hbar\sum_n \dot{a}_n \exp\left(-i\int_0^t \mathrm{d}t' E_n(t')/\hbar\right)|n\rangle + \sum_n a_n E_n \exp\left(-i\int_0^t \mathrm{d}t' E_n(t')/\hbar\right)|n\rangle +$$
$$\sum_n a_n \exp\left(-i\int_0^t \mathrm{d}t' E_n(t')/\hbar\right) i\hbar|\dot{n}\rangle = \sum_n a_n \exp\left(-i\int_0^t \mathrm{d}t' E_n(t')/\hbar\right) E_n|n\rangle. \tag{8.51}$$

Some terms cancel out, and if we multiply from the left by $\langle k|\exp(i\int_0^t \mathrm{d}t' E_k(t')/\hbar)$, we obtain a set of first order time dependent differential equations

$$\dot{a}_k(t) = -\sum_n a_n(t)\langle k|\dot{n}\rangle \exp\left(i\int_0^t \mathrm{d}t'\omega_{kn}(t')\right), \tag{8.52}$$

where $\omega_{kn} = (E_k - E_n)/\hbar$.

In order to determine $\langle k|\dot{n}\rangle$, we take the time derivative of Eq. (8.49)

$$\frac{\partial \hat{H}}{\partial t}|n\rangle + \hat{H}|\dot{n}\rangle = \dot{E}_n|n\rangle + E_n(t)|\dot{n}\rangle, \tag{8.53}$$

and multiplying by $\langle k|$, we find

$$\langle k|\frac{\partial \hat{H}}{\partial t}|n\rangle + E_k\langle k|\dot{n}\rangle = \dot{E}_n\langle k|n\rangle + E_n\langle k|\dot{n}\rangle. \tag{8.54}$$

If $k = n$, $\langle n|\dot{n}\rangle$ drops out and we find

$$\dot{E}_n = \langle n|\frac{\partial \hat{H}}{\partial t}|n\rangle, \tag{8.55}$$

which is basically a special case of the Hellmann-Feynman theorem. For $k \neq n$, assuming that the eigenstates are not degenerate, we obtain

$$\langle k|\dot{n}\rangle = -\frac{1}{E_k - E_n}\langle k|\frac{\partial \hat{H}}{\partial t}|n\rangle. \tag{8.56}$$

To determine $\langle n|\dot{n}\rangle$ we take the derivative of $\langle n|n\rangle = 1$ and obtain

$$\langle n|\dot{n}\rangle + \langle \dot{n}|n\rangle = 0. \tag{8.57}$$

These two terms are complex conjugates of each other and add up to zero. Therefore, they must be purely imaginary, and we can write that

$$\langle n|\dot{n}\rangle = -i\gamma_n, \tag{8.58}$$

with γ_n real.

Then Eq. (8.52) becomes

$$\dot{a}_k = -a_k\langle k|\dot{k}\rangle + \sum_{n\neq k}\frac{a_n}{\hbar\omega_{kn}}\langle k|\frac{\partial \hat{H}}{\partial t}|n\rangle \exp\left(i\int_0^t \mathrm{d}t'\omega_{kn}(t')\right). \tag{8.59}$$

If we assume a slow change in the Hamiltonian, then we should have a slow variation in a_n, ω_{kn} and $|n\rangle$ as well. Furthermore, if we assume that at $t = 0$ the system is in the quantum state $|n\rangle$, then $a_n(0) = 1$ and $a_k(0) = 0$, for $k \neq n$. Then, at $t \sim 0$, in the summation term of Eq. (8.59), the main contribution comes from the single $|n\rangle$ state

$$\dot{a}_k(t) \simeq \frac{1}{\hbar\omega_{kn}}\langle k|\frac{\partial \hat{H}}{\partial t}|n\rangle \exp(i\omega_{kn}t), \qquad k \neq n. \tag{8.60}$$

As the change is slow in $\lambda(t)$, all terms are almost constant, and thus we can integrate out with the initial condition $a_k(0) = 0$

$$\begin{aligned}
a_k(t) &\simeq \frac{1}{\hbar\omega_{kn}}\langle k|\frac{\partial \hat{H}}{\partial t'}|n\rangle \frac{1}{i\omega_{kn}} \exp(i\omega_{kn}t')\Big|_0^t \\
&\simeq \frac{2}{\hbar\omega_{kn}^2}\langle k|\frac{\partial \hat{H}}{\partial t}|n\rangle \frac{\exp(i\omega_{kn}t)-1}{2i} \\
&\simeq \frac{2}{\hbar\omega_{kn}^2}\langle k|\frac{\partial \hat{H}}{\partial t}|n\rangle \exp(i\omega_{kn}t/2)\frac{\exp(i\omega_{kn}t/2)-\exp(-i\omega_{kn}t/2)}{2i} \\
&\simeq \frac{2}{\hbar\omega_{kn}^2}\langle k|\frac{\partial \hat{H}}{\partial t}|n\rangle \exp(i\omega_{kn}t/2)\sin(\frac{\omega_{kn}t}{2}).
\end{aligned} \tag{8.61}$$

The transition probability $|a_k(t)|^2$ is oscillating with time. However, if the change in the Hamiltonian is slow, $\langle k|\partial\hat{H}/\partial t|n\rangle$ is small and the transition $n \to k$ is negligible, even if the overall change of the Hamiltonian is substantial.

For $k = n$, we have

$$\dot{a}_n(t) \simeq -\langle n|\dot{n}\rangle a_n(t) \simeq i\gamma_n(t)a_n(t), \tag{8.62}$$

with slow variation in γ_n. Integrating, we find

$$a_n(t) \simeq a_n(0)\exp\left(i\int_0^t \mathrm{d}t'\gamma_n\right). \tag{8.63}$$

So, for the time evolution of the state, we obtain

$$|\psi(t)\rangle \simeq a_n(0)\exp\left(i\int_0^t \mathrm{d}t'\gamma_n\right)\exp\left(-i\int_0^t \mathrm{d}t' E_n(\lambda(t'))/\hbar\right)|n\rangle. \tag{8.64}$$

This means that the system remains in the nth eigenstate $|n, \lambda(t)\rangle$, which may change considerably with time. This is the *adiabatic theorem*, derived by Born and Fock in 1928.

The adiabatic theorem is valid if the change is slow, i.e. in Eq. (8.60) the right hand side is small

$$\frac{1}{\hbar\omega_{kn}^2}\left|\langle k|\frac{\partial \hat{H}}{\partial t}|n\rangle\right| \ll 1, \tag{8.65}$$

for all t. This condition is obviously not satisfied if the states, during the time evolution, become degenerate or quasi degenerate and thus some of the ω_{kn} become small. The eigenvalues cannot be degenerate, cannot cross, and the level ordering should be maintained at all time.

8.2.3 Berry Phase

We have found that with an adiabatic change of the Hamiltonian, if the states do not become degenerate or quasi-degenerate, the system remains in the same quantum state. The history of the time evolution shows up in the overall phase of the wave function. The total phase has two components. The term

$$\delta_k = -\frac{1}{\hbar}\int_0^t E_n(\lambda(t'))\mathrm{d}t' \tag{8.66}$$

has dynamical origin and it is called the *dynamical* phase. The other term,

$$\Gamma_n(t) = \int_0^t \gamma_n(t')\mathrm{d}t', \tag{8.67}$$

has geometric origin and it is called the *geometric* or *Berry* phase. In the past, it was usually absorbed into the definition of $|n, \lambda(t)\rangle$, and thus was omitted. This, however, may not be the correct procedure, as has been shown by Berry.

We assume a slow time dependence and that the parameters are periodic in time, $\boldsymbol{\lambda}(T) = \boldsymbol{\lambda}(0)$ and $\hat{H}(T) = \hat{H}(0)$. So, in the $[0, T]$ time interval $\boldsymbol{\lambda}$ makes a closed contour

$$\mathcal{C} = \{\boldsymbol{\lambda}(t) : \ 0 \leq t \leq T; \boldsymbol{\lambda}(T) = \boldsymbol{\lambda}(0)\}\,. \tag{8.68}$$

The role of the Berry phase becomes clear if we calculate it for a period. Evaluating γ_n we find

$$\gamma_n = i\langle n, \lambda(t)|\frac{\mathrm{d}}{\mathrm{d}t}|n, \lambda(t)\rangle = i\langle n, \lambda(t)|\frac{\mathrm{d}}{\mathrm{d}\lambda}|n, \lambda(t)\rangle\frac{\mathrm{d}\lambda}{\mathrm{d}t} = \mathbf{A}_n(\lambda)\frac{\mathrm{d}\lambda}{\mathrm{d}t}, \tag{8.69}$$

where $\mathbf{A}_n = i\langle n, \lambda|\mathrm{d}/\mathrm{d}\lambda|n, \lambda\rangle$, called the Berry connection, is a vector in the parameter space of λ. Then, for the geometric phase, we obtain

$$\Gamma_n(T) = \int_0^T \gamma_n(t)\,\mathrm{d}t = \int_0^T \mathbf{A}_n(\lambda)\frac{\mathrm{d}\lambda}{\mathrm{d}t}\,\mathrm{d}t = \oint \mathbf{A}_n(\lambda)\,\mathrm{d}\lambda = \oint \mathrm{d}\mathbf{S}\,\boldsymbol{\nabla}\times\mathbf{A}_n. \tag{8.70}$$

The quantum state of a time-periodic Hamiltonian, over the period T, acquires a phase equal to the integral of the *vector potential* $\mathbf{A}_n$ along a closed path in the parameter space λ. This, by Stokes' theorem, is the same as the surface integral of the Berry curvature $\mathbf{B}_n = \nabla \times \mathbf{A}_n$. This phase may lead to interference effects.

8.2.4 Aharonov-Bohm Effect as Geometric Phase

In Sects. 5.3.1–5.3.2 we introduced the Aharonov-Bohm effect of a quantum particle in a magnetic field using gauge transformation. Here we show that this effect can be described in terms of the geometric phase of the wave function of the particle.

We assume that we have a narrow magnetic flux tube and we put a particle of charge q in a small box that is $\mathbf{R}$ distance away. The vector $\mathbf{r}$ denotes the coordinate of the particle with respect to an outside reference frame and $\mathbf{r}'$ denotes its coordinate with respect to the box, $\mathbf{r} = \mathbf{R} + \mathbf{r}'$. Here $\mathbf{R}$ is the adiabatic variable, and $r' \ll R$ (Fig. 8.4).

In the no-magnetic-field case, the particle is in quantum state $|n\rangle$ described by the wave function $\psi_n(\mathbf{r}')$. If the magnetic field is turned on, the vector potential $\mathbf{A}$ is not zero any more, and the state acquires a phase

$$\begin{aligned}\langle\mathbf{r}|n, \mathbf{R}\rangle &= \exp\left(\frac{iq}{\hbar c}\int_{\mathbf{R}}^{\mathbf{R}+\mathbf{r}'} d\mathbf{r}''\mathbf{A}(\mathbf{r}'')\right)\psi_n(\mathbf{r}') = \exp\left(\frac{iq}{\hbar c}\int_{\mathbf{R}}^{\mathbf{r}} d\mathbf{r}''\mathbf{A}(\mathbf{r}'')\right)\psi_n(\mathbf{r}-\mathbf{R}) \\ &= \exp\big(i\Theta(\mathbf{r})\big)\,\psi_n(\mathbf{r}-\mathbf{R}).\end{aligned} \tag{8.71}$$

If we take the box adiabatically around the flux tube without rotation, during one turn the particle acquires an additional geometric phase

$$\begin{aligned}\langle n, \mathbf{R}|\nabla_{\mathbf{R}}|n, \mathbf{R}\rangle &= \int d\mathbf{r}\langle n, \mathbf{R}|\mathbf{r}\rangle\langle\mathbf{r}|\nabla_{\mathbf{R}}|n, \mathbf{R}\rangle \\ &= \int d\mathbf{r}\exp\big(-i\Theta(\mathbf{r})\big)\psi_n^*(\mathbf{r}-\mathbf{R})\nabla_{\mathbf{R}}\left[\exp\big(i\Theta(\mathbf{r})\big)\psi_n(\mathbf{r}-\mathbf{R})\right] \\ &= \int d\mathbf{r}\exp\big(-i\Theta(\mathbf{r})\big)\psi_n^*(\mathbf{r}-\mathbf{R}) \\ &\times\left[-\frac{iq}{\hbar c}\mathbf{A}(\mathbf{R})\exp\big(i\Theta(\mathbf{r})\big)\psi_n(\mathbf{r}-\mathbf{R}) + \exp\big(i\Theta(\mathbf{r})\big)\nabla_{\mathbf{R}}\psi_n(\mathbf{r}-\mathbf{R})\right] \\ &= -\frac{iq}{\hbar c}\mathbf{A}(\mathbf{R}) + \int d\mathbf{r}\,\psi_n^*(\mathbf{r}-\mathbf{R})\nabla_{\mathbf{R}}\psi_n(\mathbf{r}-\mathbf{R}).\end{aligned} \tag{8.72}$$

If, in the last term, we perform the correspondence $\mathbf{r} \to \mathbf{r} + \mathbf{R}$, which we can do since we integrate over the whole space, we find

$$\int d\mathbf{r}\langle n|\mathbf{r}\rangle\nabla_{\mathbf{R}}\langle\mathbf{r}|n\rangle = \nabla_{\mathbf{R}}\langle n|n\rangle = 0. \tag{8.73}$$

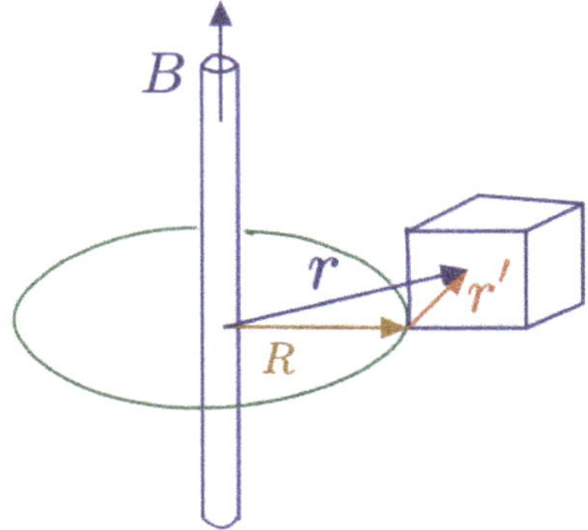

Fig. 8.4 As the box moves around adiabatically, without rotation, the quantum state acquires an Aharonov-Bohm phase

Then, performing the integration around the magnetic flux, we immediately obtain the Aharonov-Bohm phase

$$\Gamma_C = \frac{q}{\hbar c} \oint_C \mathrm{d}\mathbf{R}\, \mathbf{A}(\mathbf{R}) = \frac{q}{\hbar c} \Phi, \tag{8.74}$$

where Φ is the magnetic flux.

8.2.5 *Adiabatic and Sudden Approximation as Time-Dependent Perturbation*

We derived previously in Eq. (8.26) the transition amplitude in the first order perturbation theory

$$c_f^{(1)}(t) = -\frac{i}{\hbar} \int_0^t \mathrm{d}t' \exp(i\omega_{fi}t') \langle \psi_f | \hat{V}(t') | \psi_i \rangle = \frac{1}{i\hbar} \int_0^t \mathrm{d}t' \exp(i\omega_{fi}t') V_{fi}(t'). \tag{8.75}$$

Since $\exp(i\omega_{fi}t) = \big(1/(i\omega_{fi})\big)\partial \exp(i\omega_{fi}t)/\partial t$, integration by parts gives

$$\begin{aligned} c_f^{(1)}(t) &= -\frac{1}{\hbar\omega_{fi}} \int_0^t \mathrm{d}t' V_{fi}(t') \frac{\partial \exp(i\omega_{fi}t')}{\partial t'} \\ &= -\frac{1}{\hbar\omega_{fi}}\, V_{fi}(t') \exp(i\omega_{fi}t')\Big|_{t'=0}^{t'=t} + \frac{1}{\hbar\omega_{fi}} \int_0^t \mathrm{d}t' \frac{\mathrm{d}V_{fi}(t')}{\mathrm{d}t'} \exp(i\omega_{fi}t'). \end{aligned} \tag{8.76}$$

The perturbation $\hat{V}$ acts for a finite time interval and $\hat{V}(0) = \hat{V}(t) = 0$. So, the first part vanishes and we obtain an alternative expression for the transition amplitude in the first order of perturbation theory

$$c_f^{(1)}(t) = \frac{1}{\hbar\omega_{fi}} \int_0^t \mathrm{d}t' \frac{\mathrm{d}V_{fi}(t')}{\mathrm{d}t'} \exp(i\omega_{fi}t'). \tag{8.77}$$

This form is very advantageous for studying transitions in the adiabatic and the sudden limits.

Adiabatic Approximation

If the perturbation $\hat{V}(t)$ varies slowly with time, then $\mathrm{d}V_{fi}(t)/\mathrm{d}t$ is small, it varies slowly, and it is almost constant. Consequently, it can be taken outside of the integral and can be approximated by an averaged value

$$c_f^{(1)}(t) = \frac{1}{\hbar\omega_{fi}} \overline{\frac{\mathrm{d}V_{fi}}{\mathrm{d}t}} \int_0^t \mathrm{d}t' \exp(i\omega_{fi}t') = \frac{1}{\hbar\omega_{fi}} \overline{\frac{\mathrm{d}V_{fi}}{\mathrm{d}t}} \frac{\exp(i\omega_{fi}t) - 1}{i\omega_{fi}}. \tag{8.78}$$

The approximation is valid if

$$\frac{1}{\hbar\omega_{fi}^2} \overline{\frac{\mathrm{d}V_{fi}}{\mathrm{d}t}} \ll 1. \tag{8.79}$$

Then, the transition amplitude becomes

$$P_{i\to f}(t) = \left|c_f^{(1)}(t)\right|^2 = \frac{1}{\hbar^2\omega_{fi}^4} \left|\overline{\frac{\mathrm{d}V_{fi}}{\mathrm{d}t}}\right|^2 4\sin^2\left(\frac{1}{2}\omega_{fi}t\right). \tag{8.80}$$

We can draw the same conclusion as before. If the change is slow, i.e. $\mathrm{d}V_{fi}(t)/\mathrm{d}t$ is small and almost constant, the transition to other final states are negligible. As a result, the system remains in the initial state even if the change itself is substantial.

Sudden Approximation

If the perturbation $\hat{V}(t)$ varies rapidly over a short time interval, then during this time, the exponential term does not change much. Consequently, it can be pulled out of the integral

$$c_f^{(1)}(t) = \frac{1}{\hbar\omega_{fi}} \exp(i\omega_{fi}t) \int_0^t \mathrm{d}t' \frac{\mathrm{d}V_{fi}(t')}{\mathrm{d}t'} = \frac{1}{\hbar\omega_{fi}} \exp(i\omega_{fi}t) V_{fi}(t). \tag{8.81}$$

So, for the transition amplitude, we have

$$P_{i\to f}(t) = \left|c_f^{(1)}(t)\right|^2 = \frac{1}{\hbar^2\omega_{fi}^2} \left|V_{fi}(t)\right|^2. \tag{8.82}$$

The sudden approximation is applicable if $\left|V_{fi}\right| \ll \hbar\omega_{fi}$, or equivalently if $\left|V_{fi}\right| \ll \Delta E$, where $\Delta E = \left|E_f - E_i\right|$ is the energy difference between the initial and final states.

8.3 Time-Dependent Two-State Systems

We consider a two-state system with time-dependent Hamiltonian

$$\hat{H}(t) = \Delta(t)\big(|1\rangle\langle 1| - |2\rangle\langle 2|\big) + Z(t)|2\rangle\langle 1| + Z^*(t)|1\rangle\langle 2|, \tag{8.83}$$

where states $|1\rangle$ and $|2\rangle$ are orthonormal and the level splitting Δ and coupling term Z may explicitly be time-dependent. Remember that Pauli matrices together with the 2×2 unit matrix form a basis in a two-dimensional Hilbert space. So, we can write Eq. (8.83) in the form

$$\hat{H}(t) = \Delta(t)\sigma_z + Z(t)\sigma_- + Z^*(t)\sigma_+, \tag{8.84}$$

with

$$\sigma_z = \begin{pmatrix} 1 & 0 \\ 0 & -1 \end{pmatrix}, \quad \sigma_- = \begin{pmatrix} 0 & 0 \\ 1 & 0 \end{pmatrix}, \quad \sigma_+ = \begin{pmatrix} 0 & 1 \\ 0 & 0 \end{pmatrix}. \tag{8.85}$$

Since $\sigma_\pm = 1/2\,(\sigma_x \pm i\sigma_y)$, we may rewrite the Hamiltonian in Eq. (8.83)

$$\begin{aligned} \hat{H}(t) &= \Delta(t)\sigma_z + \frac{1}{2}(Z(t) + Z^*(t))\sigma_x - \frac{i}{2}(Z(t) - Z^*(t))\sigma_y \\ &= g\big(\sigma_x B_x(t) + \sigma_y B_y(t) + \sigma_z B_z(t)\big) = g\boldsymbol{\sigma} \cdot \mathbf{B}(t). \end{aligned} \tag{8.86}$$

Then, the correspondence

$$\Delta(t) = gB_z(t) \quad \text{and} \quad Z(t) = g[B_x(t) + iB_y(t)] \tag{8.87}$$

shows that a general two-state system is mathematically equivalent to a spin-1/2 particle moving in a time-dependent magnetic field.

8.3.1 Two-State System with Harmonic Potential

Let us consider a two-state system with a harmonic potential

$$\begin{aligned} \hat{H} &= E_1|1\rangle\langle 1| + E_2|2\rangle\langle 2| + \frac{\lambda}{2}\exp(i\omega t)|1\rangle\langle 2| + \frac{\lambda}{2}\exp(-i\omega t)|2\rangle\langle 1| \\ &= \begin{pmatrix} E_1 & \lambda/2\exp(i\omega t) \\ \lambda/2\exp(-i\omega t) & E_2 \end{pmatrix}. \end{aligned} \tag{8.88}$$

We solve for the two-component wave function

$$|\Psi(t)\rangle = \begin{pmatrix} \psi_1(t) \\ \psi_2(t) \end{pmatrix}, \quad \text{with} \quad |\Psi(0)\rangle = \begin{pmatrix} 1 \\ 0 \end{pmatrix}. \tag{8.89}$$

Inserting into the time-dependent Schrödinger equation we obtain

$$
\begin{aligned}
i\hbar\dot{\psi}_1 &= E_1\psi_1 + \frac{\lambda}{2}\exp(i\omega t)\psi_2, \\
i\hbar\dot{\psi}_2 &= E_2\psi_2 + \frac{\lambda}{2}\exp(-i\omega t)\psi_1.
\end{aligned} \tag{8.90}
$$

We may separate off the stationary time dependence, $\psi_1 = \tilde{\psi}_1\exp(-i\omega_1 t)$ and $\psi_2 = \tilde{\psi}_2\exp(-i\omega_2 t)$, where $\omega_1 = E_1/\hbar$ and $\omega_2 = E_2/\hbar$, respectively. Then, for $\tilde{\psi}_1$ and $\tilde{\psi}_2$ we find

$$
\begin{aligned}
\dot{\tilde{\psi}}_1 &= -i\frac{\tilde{\lambda}}{2}\exp\big(i(\omega-\tilde{\omega})t\big)\tilde{\psi}_2, \\
\dot{\tilde{\psi}}_2 &= -i\frac{\tilde{\lambda}}{2}\exp\big(-i(\omega-\tilde{\omega})t\big)\tilde{\psi}_1,
\end{aligned} \tag{8.91}
$$

where $\tilde{\lambda} = \lambda/\hbar$ and $\tilde{\omega} = (\omega_2 - \omega_1)$.

We can rearrange the second equation, take the derivative

$$
\frac{\tilde{\lambda}}{2}\tilde{\psi}_1 = i\dot{\tilde{\psi}}_2\exp\big(i(\omega-\tilde{\omega})t\big), \quad \frac{\tilde{\lambda}}{2}\dot{\tilde{\psi}}_1\exp\big(-i(\omega-\tilde{\omega})t\big) = i\ddot{\tilde{\psi}}_2 - \dot{\tilde{\psi}}_2(\omega-\tilde{\omega}), \tag{8.92}
$$

and insert into the first one to obtain

$$
\ddot{\tilde{\psi}}_2 + i(\omega-\tilde{\omega})\dot{\tilde{\psi}}_2 + \frac{\tilde{\lambda}^2}{4}\tilde{\psi}_2 = 0. \tag{8.93}
$$

Using the ansatz $\tilde{\psi}_2 \sim \exp(i\Omega t)$, we find

$$
[-\Omega^2 - (\omega-\tilde{\omega})\Omega + \tilde{\lambda}^2/4]\exp(i\Omega t) = 0, \tag{8.94}
$$

which gives

$$
\Omega_\pm = \frac{1}{2}\left[-(\omega-\tilde{\omega}) \pm \Omega_0\right], \tag{8.95}
$$

where $\Omega_0 = \sqrt{(\omega-\tilde{\omega})^2 + \tilde{\lambda}^2}$. Thus, for the wave function, we get

$$
\begin{aligned}
\tilde{\psi}_2(t) &= A_+\exp(i\Omega_+ t) + A_-\exp(i\Omega_- t), \\
\tilde{\psi}_1(t) &= \frac{2i}{\tilde{\lambda}}\dot{\psi}_2(t)\exp\big(i(\omega-\tilde{\omega})t\big) \\
&= -\frac{2}{\tilde{\lambda}}\exp\big(i(\omega-\tilde{\omega})t\big)\big(A_+\Omega_+\exp(i\Omega_+ t) + A_-\Omega_-\exp(i\Omega_- t)\big).
\end{aligned} \tag{8.96}
$$

From the initial conditions $\tilde{\psi}_2(0) = 0$ and $\tilde{\psi}_1(0) = 1$, we can conclude that

$$
A_+ + A_- = 0 \quad \text{and} \quad -\frac{2}{\tilde{\lambda}}(A_+\Omega_+ + A_-\Omega_-) = 1, \tag{8.97}
$$

which results in

$$A_+ = -A_- = -\frac{\tilde{\lambda}/2}{\Omega_+ - \Omega_-} = -\frac{\tilde{\lambda}/2}{\Omega_0}. \tag{8.98}$$

Finally we have

$$\begin{aligned}\psi_2(t) &= \frac{-i\tilde{\lambda}}{\Omega_0}\exp(-i\omega_2 t)\exp\bigl(-i(\omega-\tilde{\omega})t/2\bigr)\sin(\Omega_0 t/2),\\ \psi_1(t) &= \exp\bigl(-i\omega_1 t\bigr)\exp\bigl(i(\omega-\tilde{\omega})t/2\bigr)\left(\cos(\Omega_0 t/2) - i\frac{\omega-\tilde{\omega}}{\Omega_0}\sin(\Omega_0 t/2)\right).\end{aligned} \tag{8.99}$$

The transition probability from state $|1\rangle$ to state $|2\rangle$ is given by

$$|\psi_2(t)|^2 = \frac{\tilde{\lambda}^2}{(\omega-\tilde{\omega})^2+\tilde{\lambda}^2}\sin^2\left(\frac{1}{2}\sqrt{(\omega-\tilde{\omega})^2+\tilde{\lambda}^2\,t}\right). \tag{8.100}$$

We can see that the probability oscillates with time. If $\tilde{\lambda}$ is small, the probability of the transition from $|1\rangle$ to $|2\rangle$ is also small. However, if $\omega \simeq \tilde{\omega} = \omega_2 - \omega_1$, i.e. when the driving frequency matches the energy difference, the probability oscillates between zero and one. This is the resonant case. Here,

$$\psi_1(t) \simeq \cos\tilde{\lambda}t \quad \text{and} \quad \psi_2(t) \simeq -i\sin\tilde{\lambda}t, \tag{8.101}$$

i.e. ψ_1 and ψ_2 oscillate with opposite phases. If $\omega - \tilde{\omega} = \tilde{\lambda}$, the maximum value of $|\psi_2(t)|^2$ drops to 1/2, meaning that $\tilde{\lambda}$ is the width of the resonance.

If the system is in the Ω_+ energy eigenstate, then $A_- = 0$, while if it is in the Ω_- energy eigenstate, $A_+ = 0$. The corresponding normalized eigenstates are given by

$$\Psi_\pm(t) = \frac{1}{\sqrt{4\Omega_\pm^2+\tilde{\lambda}^2}}\begin{pmatrix}-2\Omega_\pm\exp(i\omega t/2)\\ \tilde{\lambda}\exp(-i\omega t/2))\exp(i\Omega_0 t/2)\exp(-i(\omega_2+\omega_1)t/2\end{pmatrix}. \tag{8.102}$$

Nuclear Magnetic Resonance

The above analysis provides us with the physical basis for nuclear magnetic resonance (NMR) phenomena. Let us consider a nuclear spin in a strong magnetic field oriented in the z direction. This aligns the spins along the z axis. Then, the Hamiltonian is given by

$$\hat{H}_0 = -\boldsymbol{\mu}\cdot\mathbf{B} = -\hbar\mu\sigma_z B_z = -\omega_0\begin{pmatrix}1 & 0\\ 0 & -1\end{pmatrix}. \tag{8.103}$$

In an NMR experiment, a small magnetic field is applied along the x direction for a short period of time

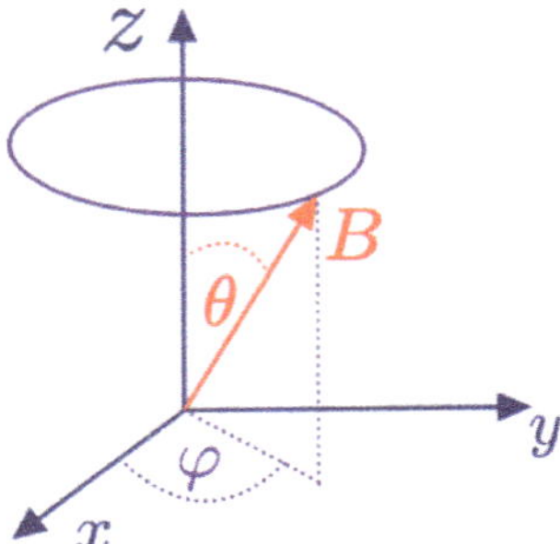

Fig. 8.5 The magnetic field rotates around the z axis

$$\hat{H}' = -\hbar\mu\sigma_x B_x \cos\omega t \sim \lambda/2 \begin{pmatrix} 0 & \exp(i\omega t) \\ \exp(-i\omega t) & 0 \end{pmatrix}. \tag{8.104}$$

Since B_x is small, the coupling term λ is also small, so the transition is negligible, unless ω meets the resonant condition. The duration of the field along x is chosen such that the spins flip, and as they return to the energetically preferred state, they emit radiation, which can be detected. If the B_z field, and consequently ω_0, has some slight gradient, the resonance condition is satisfied only for a small region of space. This way we can map the local magnetic field, which provides information on the structure of the material.

Spin in Rotating Magnetic Field

In another example, we consider a spin-1/2 particle with magnetic moment μ interacting with an external time dependent magnetic field that rotates around the z axis (Fig. 8.5)

$$\mathbf{B}(t) = B(\sin\theta\cos\omega t, \sin\theta\sin\omega t, \cos\theta). \tag{8.105}$$

The Hamiltonian is given by

$$\hat{H} = -\boldsymbol{\mu}\cdot\mathbf{B} = -\hbar\mu\boldsymbol{\sigma}\cdot\mathbf{B} = -\hbar\mu B \begin{pmatrix} \cos\theta & \sin\theta\exp(-i\omega t) \\ \sin\theta\exp(i\omega t) & -\cos\theta \end{pmatrix}. \tag{8.106}$$

We can see, that with correspondences $\omega_1 \to -\mu B\cos\theta, \omega_2 \to \mu B\cos\theta, \omega \to -\omega$, $\tilde{\omega} \to 2\mu B\cos\theta$, and $\tilde{\lambda} \to -2\mu B\sin\theta$ the Hamiltonian of Eq. (8.88) coincides with the above Hamiltonian. In terms of these variables, we have

$$\begin{aligned} \Omega_\pm &= \frac{1}{2}\left[(\omega + 2\mu B\cos\theta) \pm \sqrt{(\omega + 2\mu B\cos\theta)^2 + (2\mu B\sin\theta)^2}\right] \\ &= \frac{1}{2}\left[(\omega + 2\mu B\cos\theta) \pm \sqrt{\omega^2 + 4\omega\mu B\cos\theta + 4\mu^2 B^2}\right]. \end{aligned} \tag{8.107}$$

We assume that the system is in energy eigenstate Ω_+ with $A_- = 0$. Then, the eigenstate corresponding to Eq. (8.102) becomes

$$\Psi_+(t) = \frac{1}{\sqrt{4\Omega_+^2 + \tilde{\lambda}^2}} \begin{pmatrix} -\left[(\omega + 2\mu B\cos\theta) + \sqrt{\omega^2 + 4\omega\mu B\cos\theta + 4\mu^2 B^2}\right]\exp(-i\omega t/2) \\ -2\mu B\sin\theta\exp(i\omega t/2) \end{pmatrix} \times \exp(i\sqrt{\omega^2 + 4\omega\mu B\cos\theta + 4\mu^2 B^2}\, t/2). \tag{8.108}$$

In the sudden approximation the rotation is very fast, $\omega \gg \mu B$. Consequently, the upper component dominates over the lower one. Additionally, the time dependence drops out and the state becomes

$$\Psi_+(t) \simeq \begin{pmatrix} 1 \\ 0 \end{pmatrix}, \tag{8.109}$$

i.e. the state remains about the same.

In the adiabatic limit, the rotation is slow, $\omega \ll 2\mu B\cos\theta$ and $\sqrt{\omega^2 + 4\omega\mu B\cos\theta + 4\mu^2 B^2} \to 2\mu B + \omega\cos\theta$. Then, for the wave function, we find

$$\Psi_+(t) \simeq \frac{-1}{\sqrt{2(1+\cos\theta)}} \begin{pmatrix} 1+\cos\theta \\ \sin\theta\,\exp(i\omega t) \end{pmatrix} \exp(i\mu Bt)\exp\big(-i\omega(1-\cos\theta)t/2\big). \tag{8.110}$$

This wave function describes a spin-1/2 particle in a slowly rotating magnetic field. The system is in the $E = -\hbar\mu B$ energy eigenstate of the instantaneous Hamiltonian and has the corresponding dynamical phase $\exp(i\mu Bt)$. The last term is the Berry phase. When the magnetic field turns around the z-axis over a time T, we have $\omega T = 2\pi$. Then, the system acquires a phase $\exp(i\pi(1-\cos\theta)) = \exp(im\Omega)$. Here $m = 1/2$ for the spin and Ω is the area swept by the magnetic field vector

$$\Omega = \int_0^\theta d\theta' \sin\theta' \int_0^{2\pi} d\phi = 2\pi(1-\cos\theta). \tag{8.111}$$

8.3.2 Landau-Zener Transition

We should notice that in our treatment of two-level systems presented so far, $\tilde{\omega}$ should not vanish, i.e. the states should not be degenerate. Now, we consider the case where, during time evolution, the states become degenerate. The two-state Hamiltonian is given by

$$\hat{H} = \begin{pmatrix} E_1(t) & W \\ W^* & E_2(t) \end{pmatrix}, \tag{8.112}$$

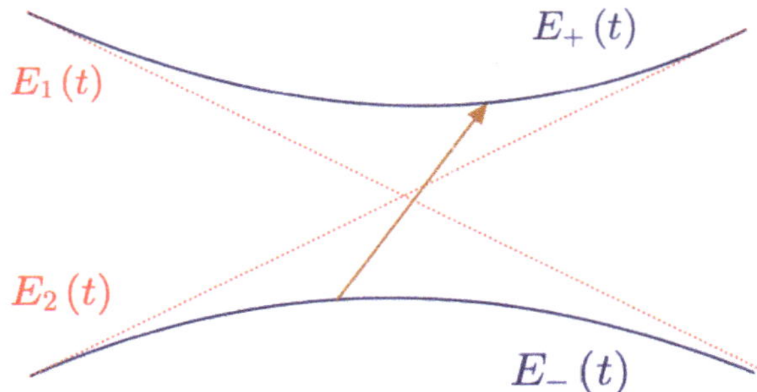

Fig. 8.6 Avoided crossing of energy levels. The Landau-Zener process represents transition between adiabatic states $E_-(t) \to E_+(t)$ as $t \to \infty$

where $E_1(-\infty) > E_2(-\infty)$, $E_2(\infty) > E_1(\infty)$, $E_1(0) = E_2(0)$ and W is constant. So, the energies $E_1(t)$ and $E_2(t)$ cross each other at $t = 0$.

Solving for the energy eigenvalues we find

$$E_\pm(t) = \frac{1}{2}\left[E_1(t) + E_2(t) \pm \sqrt{(E_1(t) - E_2(t))^2 + 4|W|^2}\right]. \tag{8.113}$$

We can see that

$$\lim_{t\to-\infty} E_-(t) = E_2(t) \quad \text{and} \quad \lim_{t\to\infty} E_-(t) = E_1(t), \tag{8.114}$$

and

$$\lim_{t\to-\infty} E_+(t) = E_1(t) \quad \text{and} \quad \lim_{t\to\infty} E_+(t) = E_2(t). \tag{8.115}$$

Due to the non-vanishing coupling W, there is a finite gap between E_- and E_+. This is the phenomenon of avoided crossing (Fig. 8.6). In the adiabatic limit, if at $t \to -\infty$ the systems is in the E_2 state, it will follow the $E_-(t)$ trajectory, and if it is in the E_1 state, it will follow the $E_+(t)$ trajectory.

However, there is a finite probability that the system jumps from E_- state to E_+ state or vice versa. This is the Landau-Zener (-Stückelberg-Majorana) transition.

To analyze the process we rewrite the Hamiltonian in the vicinity of $t = 0$ [2, 3]. We assume for simplicity that around $t = 0$ we have

$$E_2(t) - E_1(t) = vt, \tag{8.116}$$

with $v > 0$. Then, the Hamiltonian can be written in the form

$$\hat{H} = \hat{H}_0 + \hat{V} = \hbar\begin{pmatrix} vt/2 & 0 \\ 0 & -vt/2 \end{pmatrix} + \hbar\begin{pmatrix} 0 & \Delta/2 \\ \Delta/2 & 0 \end{pmatrix}. \tag{8.117}$$

We write the wave function as

$$\psi(t) = c_1(t)|\uparrow\rangle + c_2(t)|\downarrow\rangle, \tag{8.118}$$

where $|\uparrow\rangle$ and $|\downarrow\rangle$ are the rising and declining energy eigenstates of $\hat{H}_0$, with eigenvalues $\hbar vt/2$ and $-\hbar vt/2$, respectively.

The time-dependent Schrödinger equation results in the coupled equations

$$\begin{aligned}\dot{c}_1 &= -\frac{ivt}{2}c_1 - \frac{i\Delta}{2}c_2,\\ \dot{c}_2 &= -\frac{i\Delta}{2}c_1 + \frac{ivt}{2}c_2,\end{aligned} \tag{8.119}$$

and, after some simple manipulations, we have

$$\ddot{c}_1 + \left(\frac{iv}{2} + \frac{\Delta^2}{4} + \frac{v^2t^2}{4}\right)c_1 = 0. \tag{8.120}$$

In an adiabatic process, the system starts with $c_1(-\infty) = 1$ and remains in the lowest energy state, i.e. in the $t \to \infty$ limit we find that $c_2(\infty) = 1$ and $c_1(\infty) = 0$. In the Landau-Zener process the system jumps to the higher energy level and remains on the $|\uparrow\rangle$ trajectory. This transition probability is described by the $c_1(\infty)$ value.

We can get this value by evaluating the ratio $\dot{c}_1/c_1$ in the $t \to \infty$ limit. The modulus of c_1 has a probabilistic physical interpretation and it tends to a definite value as $t \to \infty$. Therefore, in the asymptotic region, we can write $c_1(t) = |c_1|\exp(-i\phi(t))$. Then, from Eq. (8.120), we obtain

$$\left(-\dot{\phi}^2 - i\ddot{\phi} + \frac{iv}{2} + \frac{\Delta^2}{4} + \frac{v^2t^2}{4}\right)c_1 = 0. \tag{8.121}$$

Separating out the real part, we find

$$\begin{aligned}\dot{\phi}(t) &= \pm\frac{1}{2}\sqrt{\Delta^2 + v^2t^2} = \pm\frac{1}{2}v\,|t|\sqrt{1 + \frac{\Delta^2}{v^2t^2}}\\ &\approx \frac{1}{2}vt\left(1 + \frac{1}{2}\frac{\Delta^2}{v^2t^2}\right) \approx \frac{1}{2}vt + \frac{1}{4}\frac{\Delta^2}{vt} \quad \text{as} \quad t \to \pm\infty.\end{aligned} \tag{8.122}$$

Therefore,

$$\left.\frac{\dot{c}_1}{c_1}\right|_{t\pm\infty} = -i\dot{\phi}_{t\to\pm\infty} = -i\left(\frac{1}{2}vt + \frac{1}{4}\frac{\Delta^2}{vt}\right)_{t\to\pm\infty}. \tag{8.123}$$

We can assume that $\dot{c}_1/c_1$ is well-behaved and can be analytically continued to the complex plane. Since c_1 has definite physical meaning, it never vanishes. Consequently $\dot{c}_1/c_1$, as the function of a complex time variable, does not have any pole. Then, due to the Cauchy integral theorem, we have

$$\int_{-\infty}^{\infty}\frac{\dot{c}_1}{c_1}\mathrm{d}t = -\int_C \frac{\dot{c}_1}{c_1}(z)\mathrm{d}z, \tag{8.124}$$

where $\mathcal{C}$ is a contour taken in the complex plane on the infinite semicircle above or below the real axis (Fig. 8.7). For the left-hand side of Eq. (8.124), we have

$$\int_{-\infty}^{\infty} \frac{\dot{c}_1(t)}{c_1(t)} \mathrm{d}t = \int_{-\infty}^{\infty} \frac{\mathrm{d}}{\mathrm{d}t} (\log c_1(t)) \mathrm{d}t = \log c_1(t)|_{-\infty}^{\infty} = \log \frac{c_1(\infty)}{c_1(-\infty)}. \tag{8.125}$$

To calculate the right hand side, we parametrize the semicircle $\mathcal{C}: z = Re^{i\theta}$, $\mathrm{d}z = iRe^{i\theta}\mathrm{d}\theta$ and we obtain

$$\begin{aligned} -\int_{\mathcal{C}} \frac{\dot{c}_1(t)}{c_1(t)} \mathrm{d}z &= i \int_{\mathcal{C}} \left(\frac{1}{2} vz + \frac{1}{4} \frac{\Delta^2}{vz} \right) \mathrm{d}z = i \lim_{R \to \infty} \int_0^{\pm\pi} \left(\frac{i}{2} vR^2 e^{2i\theta} + \frac{i}{4} \frac{\Delta^2}{v} \right) \mathrm{d}\theta \\ &= i \left(\frac{1}{4} vR^2 e^{2i\theta} + \frac{i}{4} \frac{\Delta^2}{v} \theta \right) \Bigg|_0^{\pm\pi} = \mp \frac{\pi \Delta^2}{4v}, \end{aligned} \tag{8.126}$$

where the $\pm\pi$ in the upper limit of integration refers to the upper and lower half circles, respectively.

So, we have found that

$$c_1(\infty) = c_1(-\infty) \exp\left(\mp \frac{\pi \Delta^2}{4v} \right), \tag{8.127}$$

or

$$|c_1(\infty)|^2 = |c_1(-\infty)|^2 \exp\left(\mp \frac{\pi \Delta^2}{2v} \right) = \exp\left(\mp \frac{\pi \Delta^2}{2v} \right), \tag{8.128}$$

since $|c_1(-\infty)|^2 = 1$. Here, we are considering a transition amplitude that cannot be bigger than one, therefore only the upper half circle makes any sense. This gives the Landau-Zener transition probability that describes the transition probability through the avoided level crossing region

$$P_{LZ} = \exp\left(-\frac{\pi \Delta^2}{2v} \right). \tag{8.129}$$

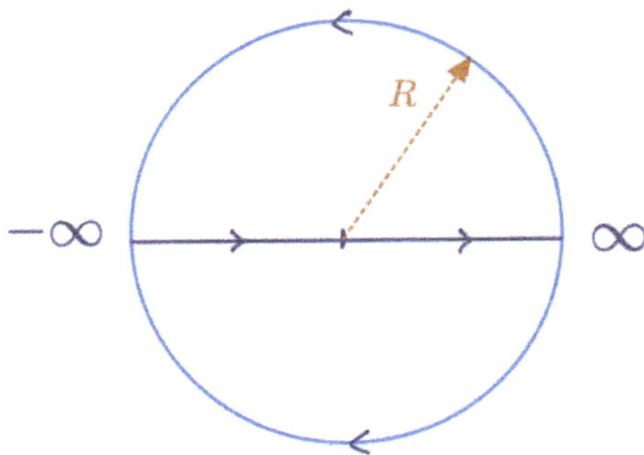

Fig. 8.7 Integration path with upper and lower semicircles

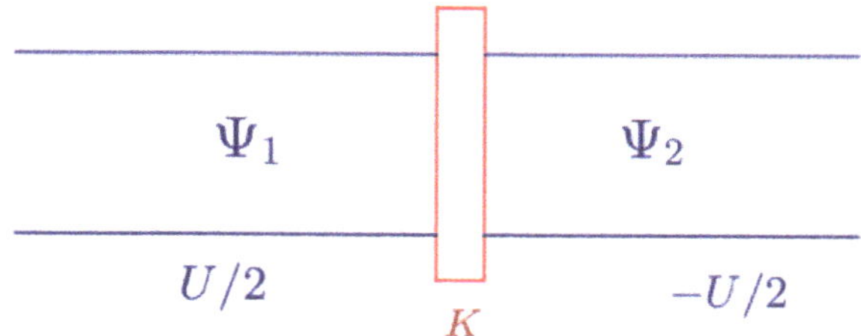

Fig. 8.8 Two superconductors separated by a thin insulator

We can see that the Landau-Zener transition can be substantial if the coupling between crossing levels, the parameter Δ, is small, and consequently the gap between E_- and E_+ is small as well.

8.3.3 Josephson Effect

In the BCS (Bardeen-Cooper-Shrieffer) theory of superconductivity an effective attractive interaction between electrons in the vicinity of Fermi level leads to an instability of the Fermi sea and the binding of electrons into Cooper pairs. These pairs are are spin-0, or maybe spin-1, systems, behaving like bosons, therefore we may describe their collective behavior by one wave function for all Cooper pairs [1]

$$\psi(\mathbf{r}, t) = |\psi(\mathbf{r}, t)| \exp[i\theta(\mathbf{r}, t)], \tag{8.130}$$

where $\mathbf{r}$ is the center-of-mass coordinate of the Cooper pairs and θ is some phase. As Cooper pairs are moving coherently, their density can be described by $|\psi(\mathbf{r}, t)| = \sqrt{\rho(\mathbf{r}, t)}$. Then, the continuity equation $\mathrm{d}\rho/\mathrm{d}t + \nabla\mathbf{J} = 0$ delivers the superconducting current density

$$\mathbf{J} = \frac{\hbar}{M}\rho\left(2\nabla\theta - \frac{2e}{\hbar c}\mathbf{A}\right), \tag{8.131}$$

where M and $-2e$ are the mass and the charge of the Cooper pair, respectively, and $\mathbf{A}$ is the external magnetic vector potential.

Now, we consider two superconductors in states ψ_1 and ψ_2 separated by a thin insulator. We assume that a potential U is applied to the junction. The quantum state in superconductor 1 is at potential level $U/2$ and in superconductor 2 is at $-U/2$. The states are coupled by the potential term K, representing the thin insulator (Fig. 8.8).

Basically, we have a two-state system

$$i\hbar\frac{\mathrm{d}}{\mathrm{d}t}\begin{pmatrix}\psi_1\\ \psi_2\end{pmatrix} = \begin{pmatrix}U/2 & K\\ K & -U/2\end{pmatrix}\begin{pmatrix}\psi_1\\ \psi_2\end{pmatrix}, \tag{8.132}$$

with

$$\psi_1 = \sqrt{\rho_1}\exp(i\theta_1) \quad \text{and} \quad \psi_2 = \sqrt{\rho_2}\exp(i\theta_2), \tag{8.133}$$

where ρ is the density of Cooper-pairs and θ_1 and θ_2 are the phases of the superconducting wave function in 1 and 2. Substituting into Eq. (8.132) and separating the real and imaginary parts, we obtain

$$\begin{aligned}
\dot{\rho}_1 &= +\frac{2K}{\hbar}\sqrt{\rho_1\rho_2}\sin\delta, \\
\dot{\rho}_2 &= -\frac{2K}{\hbar}\sqrt{\rho_1\rho_2}\sin\delta, \\
\dot{\theta}_1 &= -\frac{K}{\hbar}\sqrt{\frac{\rho_1}{\rho_2}}\cos\delta - \frac{U}{2\hbar}, \\
\dot{\theta}_2 &= -\frac{K}{\hbar}\sqrt{\frac{\rho_1}{\rho_2}}\cos\delta + \frac{U}{2\hbar},
\end{aligned} \tag{8.134}$$

with $\delta = \theta_2 - \theta_1$. We can see that $\dot{\rho}_1 = -\dot{\rho}_2$, indicating charge exchange between regions 1 and 2. Therefore, if we assume that the superconductors are of the same type and $\rho_1 = \rho_2 = \rho_0$, the currents through the insulating barrier is directly given by $\dot{\rho}_1$ or $\dot{\rho}_2$:

$$J = J_0 \sin\delta, \tag{8.135}$$

with $J_0 = 2K\rho_0/\hbar$.

From the last two equations of Eqs. (8.134), we also find that

$$\dot{\delta} = \dot{\theta}_2 - \dot{\theta}_1 = U/\hbar, \tag{8.136}$$

and integrating out, we obtain

$$\delta(t) = \delta_0 + \frac{1}{\hbar}\int^t \mathrm{d}t' U(t'). \tag{8.137}$$

Let us first assume that $U = U_0$ is constant. Then,

$$J = J_0 \sin(\delta_0 + U_0 t/\hbar). \tag{8.138}$$

We can observe some current even in absence of an applied potential, when $U_0 = 0$. The current can be anywhere between $-J_0$ to J_0, depending on the phase difference δ_0. If we apply an external voltage U_0, the current becomes sinusoidal with frequency $\omega_0 = U_0/\hbar$.

We can observe some interesting phenomena if we add a time-dependent high frequency harmonic potential in addition to a constant one

$$U(t) = U_0 + U_1 \cos\omega t, \tag{8.139}$$

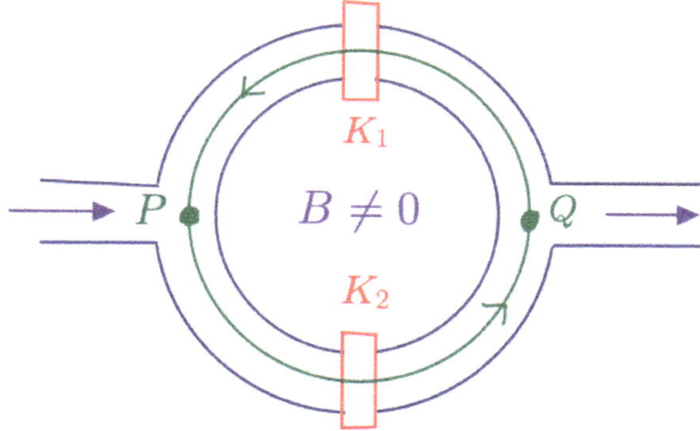

Fig. 8.9 Superconducting ring with two Josephson junctions

with $U_1 \ll U_0$. Integrating, we obtain from Eq. (8.137) for the phase

$$\delta(t) = \delta_0 + \omega_0 t + \frac{U_1}{\hbar\omega} \sin \omega t. \tag{8.140}$$

Assuming that the third term is much smaller than the two others, with the help of the approximation $\sin(x + \Delta x) \approx \sin x + \Delta x \cos x$, we find for the current

$$J = J_0 \left[\sin(\delta_0 + \omega_0 t) + \frac{U_1}{\hbar\omega} \sin \omega t \, \cos(\delta_0 + \omega_0 t) \right]. \tag{8.141}$$

The time-average of the first term vanishes. However, if the parameters meet the resonant condition $\omega = \omega_0$, the second term does not vanish.

Coupled Josephson Junctions

The physics becomes even more interesting if we couple two Josephson junctions in parallel (Fig. 8.9). If there is an additional magnetic field, the phase δ_0 of the no-field system gains an additional contribution from the vector potential, and the current becomes

$$J = J_0 \sin\left(\delta_0 + \frac{2q_e}{\hbar} \int \mathbf{A} d\mathbf{s} \right). \tag{8.142}$$

The phases gained along routes $P \to K_1 \to Q$ and $P \to K_2 \to Q$ are given by

$$\Delta\delta_{PK_1Q} = \delta_{K_1} + \frac{2q_e}{\hbar} \int_{PK_1Q} \mathbf{A} d\mathbf{s} \quad \text{and} \quad \Delta\delta_{PK_2Q} = \delta_{K_2} + \frac{2q_e}{\hbar} \int_{PK_2Q} \mathbf{A} d\mathbf{s}, \tag{8.143}$$

respectively, with δ_{K_1} and δ_{K_2} being the phase differences through insulators K_1 and K_2. The phase is a well-defined function at locations P and Q, therefore $\Delta\delta_{PK_1Q} =$

$\Delta\delta_{PK_2Q}$. Consequently, we have

$$\delta_{K_2} - \delta_{K_1} = \frac{2q_e}{\hbar} \oint_{\Gamma} \mathbf{A} \mathrm{d}\mathbf{s}, \tag{8.144}$$

where Γ is a closed loop through PK_2QK_1P. The contour integral of $\mathbf{A}$ gives rise to the magnetic flux Φ through the loop

$$\delta_{K_2} - \delta_{K_1} = \frac{2q_e}{\hbar}\Phi. \tag{8.145}$$

Assuming that junctions K_1 and K_2 are of the same kind with phase δ_0 and the arrangement is symmetric, we can conclude that

$$\delta_{K_1} = \delta_0 - \frac{q_e}{\hbar}\Phi \quad \text{and} \quad \delta_{K_2} = \delta_0 + \frac{q_e}{\hbar}\Phi. \tag{8.146}$$

The total current is the sum of currents J_{K_1} and J_{K_2},

$$J = J_0\left[\sin\left(\delta_0 - \frac{q_e}{\hbar}\Phi\right) + \sin\left(\delta_0 + \frac{q_e}{\hbar}\Phi\right)\right] = 2J_0 \sin\delta_0 \cos\frac{q_e\Phi}{\hbar}. \tag{8.147}$$

So, we have found that the transmitting current is modulated by the magnetic flux Φ. The current takes the maximum value whenever

$$\Phi_n = n\frac{\pi\hbar}{q_e}, \quad \text{with} \quad n = 0, \pm1, \pm2, \ldots. \tag{8.148}$$

These are the same values for Φ we found before in quantization of the magnetic flux in superconducting rings. Figure 8.9 is the basic configuration of the superconducting quantum interference device (SQID) used as very sensitive magnetometer.

8.4 Floquet Approach to Time-Periodic Hamiltonians

In this section, we consider Hamiltonians that are periodic in time, with period T. Our aim is to solve the time-dependent Schrödinger equation

$$i\hbar\frac{\mathrm{d}}{\mathrm{d}t}|\psi(x,t)\rangle = \hat{H}(t)|\psi(x,t)\rangle, \tag{8.149}$$

with the periodicity condition on the Hamiltonian

$$\hat{H}(x,t) = \hat{H}(x,t+T). \tag{8.150}$$

We have encountered an analogous problem before. When the Hamiltonian was periodic in the spatial variable x, with period d, we found the Bloch theorem, which states that the wave function can be given by

$$\psi(x) = \exp(iKx)u_K(x), \tag{8.151}$$

with K real and the $u_K(x) = u_K(x+d)$ periodic Bloch function.

The time-evolution operator is defined by

$$|\psi(x,t)\rangle = \hat{U}(x,t)|\psi(x,0)\rangle, \tag{8.152}$$

and it satisfies the equation

$$i\hbar\frac{\mathrm{d}}{\mathrm{d}t}\hat{U}(x,t) = \hat{H}(x,t)\hat{U}(x,t). \tag{8.153}$$

We define the operator

$$\hat{\Lambda}_T(x) = \hat{U}^{-1}(x,t)\hat{U}(x,t+T), \tag{8.154}$$

from which we have

$$\hat{U}(x,t+T) = \hat{U}(x,t)\hat{\Lambda}_T(x). \tag{8.155}$$

Since the operator $\hat{\Lambda}_T(x)$ is a product of unitary operators, it has to be unitary. Consequently, we can write

$$\hat{\Lambda}_T(x) = \exp(-i\hat{w}(x)T/\hbar), \tag{8.156}$$

where $\hat{w}(x)$ is Hermitian. Now, we write the time-evolution operator as

$$\hat{U}(x,t) = \Phi(x,t)\exp(-i\hat{w}(x)t/\hbar). \tag{8.157}$$

Then, with the $t \to t+T$ substitution we have

$$\hat{U}(x,t+T) = \Phi(x,t+T)\exp(-i\hat{w}(x)t/\hbar)\exp(-i\hat{w}(x)T/\hbar). \tag{8.158}$$

On the other hand, from Eqs. (8.155) and (8.156) we find

$$\hat{U}(x,t+T) = \hat{U}(x,t)\hat{\Lambda}_T(x) = \Phi(x,t)\exp(-i\hat{w}(x)t/\hbar)\exp(-i\hat{w}(x)T/\hbar), \tag{8.159}$$

i.e. $\Phi(x,t)$ is periodic with period T, $\Phi(x,t+T) = \Phi(x,t)$. So, we have found that the eigenstate of the time-periodic Hamiltonian can be written as

$$|\psi(x,t)\rangle = \hat{U}(x,t)|\psi(x,0)\rangle = \Phi(x,t)\exp(-i\hat{w}(x)t/\hbar)|\psi(x,0)\rangle. \tag{8.160}$$

If we take $|\psi(x,0)\rangle = |\nu_\alpha(x)\rangle$, with $\nu_\alpha(x)$ being an eigenstate of the Hermitian operator $\hat{w}(x)$ with quasi-energy eigenvalue ϵ_α, the corresponding time dependent wave function becomes

$$|\psi_\alpha(x,t)\rangle = \exp(-i\epsilon_\alpha t/\hbar)\Phi(x,t)|\nu_\alpha(x)\rangle = \exp(-i\epsilon_\alpha t/\hbar)|\phi_\alpha(x,t)\rangle, \quad (8.161)$$

with Floquet state $|\phi_\alpha(x,t)\rangle = |\Phi(x,t)\rangle|\nu_\alpha(x)\rangle$ periodic in time with period T. This is the *Floquet theorem.*

The Floquet states can be calculated from an eigenvalue equation. Substituting Eq. (8.161) into the Schrödinger equation (8.149) we obtain

$$\begin{aligned} i\hbar\frac{\mathrm{d}}{\mathrm{d}t}|\psi_\alpha(x,t)\rangle &= i\hbar\frac{\mathrm{d}}{\mathrm{d}t}\exp(-i\epsilon_\alpha t/\hbar)|\phi_\alpha(x,t)\rangle \\ &= \epsilon_\alpha \exp(-i\epsilon_\alpha t/\hbar)|\phi_\alpha(x,t)\rangle + i\hbar\exp(-i\epsilon_\alpha t/\hbar)\frac{\partial}{\partial t}|\phi_\alpha(x,t)\rangle \\ &= \hat{H}(x,t)\exp(-i\epsilon_\alpha t/\hbar)|\phi_\alpha(x,t)\rangle, \end{aligned} \quad (8.162)$$

or with the Floquet Hamiltonian

$$\hat{H}^{(F)}(t) = \hat{H}(t) - i\hbar\frac{\partial}{\partial t}, \quad (8.163)$$

we have an eigenvalue equation

$$\hat{H}^{(F)}(t)|\phi_\alpha(x,t)\rangle = \epsilon_\alpha|\phi_\alpha(x,t)\rangle. \quad (8.164)$$

We should notice that we have the whole family of Floquet states

$$\begin{aligned} |\psi_\alpha(x,t)\rangle &= \exp(-i\epsilon_\alpha t/\hbar)|\phi_\alpha(x,t)\rangle \\ &= \exp(-i(\epsilon_\alpha + n\hbar\omega)t/\hbar)\exp(in\hbar\omega t)|\phi_\alpha(x,t)\rangle \\ &= \exp(-i\epsilon_{\alpha n}t/\hbar)|\phi_{\alpha n}(x,t)\rangle, \end{aligned} \quad (8.165)$$

where $\epsilon_{\alpha n} = \epsilon_\alpha + n\hbar\omega$, $|\phi_{\alpha n}(x,t)\rangle = \exp(in\hbar\omega t)|\phi_\alpha(x,t)\rangle$, $\omega = 2\pi/T$ and $n = 0, \pm1, \pm2, \ldots$. They yield identical solutions with different quasi-energy.

It is convenient to introduce the combined Hilbert space of spatial and time-wise coordinates. In the spatial part, the inner product is defined by

$$\langle\phi_n|\phi_m\rangle = \int \mathrm{d}x\,\phi_n^*(x)\phi_m(x), \quad (8.166)$$

while, in the time coordinate, we have

$$(m,n) = \frac{1}{T}\int_0^T \mathrm{d}t\,\exp[i(n-m)\omega t]. \quad (8.167)$$

So, the Floquet states satisfy the orthogonality

$$\langle\langle\psi_{\alpha n}|\psi_{\beta m}\rangle\rangle = \frac{1}{T}\int_0^T \mathrm{d}t \int \mathrm{d}x \psi_{\alpha n}^*(x,t)\psi_{\beta m}(x,t) = \delta_{\alpha\beta}\delta_{nm} \tag{8.168}$$

and completeness

$$\sum_\alpha \sum_n \psi_{\alpha n}^*(x,t)\psi_{\alpha n}(x',t') = \delta(x-x')\delta(t-t') \tag{8.169}$$

relations.

We can introduce the basis states in x, $\{|\varphi_k(x)\rangle\}$, which together with Fourier vectors $\langle t|n\rangle = \exp(in\omega t)$ from a basis in the extended Hilbert space of (x,t). We can represent the solution in Eq. (8.164) on this extended basis such that

$$|\phi_\alpha(x,t)\rangle\rangle = \sum_k \sum_{n=-\infty}^{\infty} c_{\alpha,kn}|\varphi_k(x)\rangle \exp(in\omega t). \tag{8.170}$$

Then, with

$$\langle\langle\varphi_j m|\hat{H}^{(F)}|\varphi_k n\rangle\rangle = \frac{1}{T}\int_0^T \mathrm{d}t \langle\varphi_j|\hat{H}(t)|\varphi_k\rangle \exp(-i(m-n)\omega t) + n\hbar\omega\delta_{nm}\delta_{jk} \tag{8.171}$$

we obtain the eigenvalue equation

$$\sum_{k=1}^{\infty}\sum_{n=-\infty}^{\infty} (\langle\langle\varphi_j m|\hat{H}^{(F)}|\varphi_k n\rangle\rangle) c_{\alpha,kn} = \epsilon_\alpha c_{\alpha,jm}. \tag{8.172}$$

The general solution of Eq. (8.149) is a combination of Floquet states

$$|\psi(x,t)\rangle = \sum_\alpha a_\alpha \exp(-i\epsilon_\alpha t/\hbar)|\phi_\alpha(x,t)\rangle. \tag{8.173}$$

Having the Floquet eigenvalues ϵ_α and Floquet eigenstates $|\phi_\alpha(x,t)\rangle$ we can construct the physical solution of the time-dependent Hamiltonian with periodic time dependence by matching to the initial condition.

Further Reading

1. R.P. Feynman, R.B. Leighton, M. Sands, *The Feynman Lectures on Physics, Vol. III: Quantum Mechanics*, vol. 1 (Basic books, 2011)
2. L.T.A. Ho, L.F. Chibotaru, A simple derivation of the Landau-Zener formula. Phys. Chem. Chem. Phys. **16**(15), 6942–6945 (2014)
3. C. Wittig, The Landau-Zener formula. J. Phys. Chem. B **109**(17), 8428–8430 (2005)

Chapter 9
Electromagnetic Radiation

In this section, we study quantum systems, in particular atoms, interacting with the electromagnetic field. We model an atom by the Hamiltonian

$$\hat{H}_0 = \frac{\hat{\boldsymbol{p}}^2}{2m} + \hat{V}_0(\boldsymbol{r}), \tag{9.1}$$

and describe the electromagnetic field by the potential $A_\mu = (\boldsymbol{A}, \Phi)$. By adopting the minimal coupling $p_\mu \to p_\mu - q/c\, A_\mu$, with $q = -e$, we find

$$\begin{aligned} \hat{H} &= \frac{1}{2m}\left(\hat{\boldsymbol{p}} + \frac{e}{c}\boldsymbol{A}\right)^2 - e\Phi + \hat{V}_0(r) \\ &= \hat{H}_0 - e\Phi + \frac{e}{2mc}(\hat{\boldsymbol{p}}\boldsymbol{A} + \boldsymbol{A}\hat{\boldsymbol{p}}) + \frac{e^2}{2mc^2}\boldsymbol{A}^2. \end{aligned} \tag{9.2}$$

A little calculation reveals that

$$\hat{\boldsymbol{p}}\boldsymbol{A}\psi = -i\hbar\nabla(\boldsymbol{A}\psi) = (-i\hbar\nabla\boldsymbol{A})\psi + \boldsymbol{A}(-i\hbar\nabla\psi), \tag{9.3}$$

or

$$\hat{\boldsymbol{p}}\boldsymbol{A} = \boldsymbol{A}\hat{\boldsymbol{p}} - i\hbar(\nabla\boldsymbol{A}). \tag{9.4}$$

The electromagnetic field at the location of the atom is a radiation field, it does not come from charges. So, we can take $\Phi = 0$ and we can adopt the Coulomb gauge $\nabla A = 0$. Consequently, we have $\hat{\boldsymbol{p}}\boldsymbol{A} = \boldsymbol{A}\hat{\boldsymbol{p}}$. We can also neglect the $\boldsymbol{A}^2$ term, since it is proportional to $1/c^2$. This way we arrive at the Hamiltonian that describes the atom in the electromagnetic radiation field

$$\hat{H} = \hat{H}_0 + \frac{e}{mc}\boldsymbol{A}\hat{\boldsymbol{p}}. \tag{9.5}$$

Z. Papp, *Mastering Quantum Mechanics*,
https://doi.org/10.1007/978-3-032-09011-9_9

9.1 Quantization of Electromagnetic Field

9.1.1 Energy of the Electromagnetic Field

The electric and magnetic field in the vacuum can be derived from the vector potential $\boldsymbol{A}$

$$\boldsymbol{E} = -\frac{1}{c}\frac{\partial \boldsymbol{A}}{\partial t} \quad \text{and} \quad \boldsymbol{B} = \nabla \times \boldsymbol{A}. \tag{9.6}$$

The potential $\boldsymbol{A}$ satisfies the wave equation and the Coulomb gauge condition

$$\Delta \boldsymbol{A} - \frac{1}{c^2}\frac{\partial^2 \boldsymbol{A}}{\partial t^2} = 0 \quad \text{and} \quad \nabla \boldsymbol{A} = 0. \tag{9.7}$$

We assume that the radiation field is in a finite volume, which we take as a box with size $V = L^3$. Then, we expand $\boldsymbol{A}$ in terms of plane waves

$$\boldsymbol{A}(\boldsymbol{r}, t) = \frac{1}{\sqrt{V}} \sum_k \sum_{s=1,2} (q_{k,s}(t)\boldsymbol{u}_{k,s}(\boldsymbol{r}) + q^*_{k,s}(t)\boldsymbol{u}^*_{k,s}(\boldsymbol{r})), \tag{9.8}$$

with

$$\boldsymbol{u}_{k,s}(\boldsymbol{r}) = \boldsymbol{\varepsilon}_{k,s} \exp(i\boldsymbol{k}\boldsymbol{r}), \tag{9.9}$$

where $\boldsymbol{\varepsilon}_{k,s}$ denotes the polarization vector. The waves are transverse, so $(\boldsymbol{\varepsilon}_{k,1}, \boldsymbol{\varepsilon}_{k,2}, \boldsymbol{k})$ form a right-handed system. The $\boldsymbol{u}_{k,s}(\boldsymbol{r})$ vectors also satisfy the orthogonality conditions

$$\frac{1}{V}\int \mathrm{d}V\, \boldsymbol{u}_{k,s}(\boldsymbol{r}) \cdot \boldsymbol{u}^*_{k',s'}(\boldsymbol{r}) = \delta_{k,k'}\delta_{ss'}, \quad \text{or} \quad \frac{1}{V}\int \mathrm{d}V\, \boldsymbol{u}_{k,s}(\boldsymbol{r}) \cdot \boldsymbol{u}_{k',s'}(\boldsymbol{r}) = \delta_{k,-k'}\delta_{ss'}. \tag{9.10}$$

We impose the condition that the field is periodic and vanishes on the boundaries. Consequently, we have

$$k_x, k_y, k_z = \frac{2\pi}{L} n_k, \quad \text{with} \quad n_k = \pm 1, \pm 2, \dots . \tag{9.11}$$

Substituting Eqs. (9.8) into (9.7), we find

$$\ddot{q}_k + \omega_k^2 q_k = 0 \qquad \text{and} \qquad \ddot{q}^*_k + \omega_k^2 q^*_k = 0, \tag{9.12}$$

where $\omega_k = kc$, with $k = \sqrt{k_x^2 + k_y^2 + k_z^2}$. Solving these equations we obtain the time dependence

$$q_{k,s}(t) = q_{k,s}(0)\exp(-i\omega_k t) \quad \text{and} \quad q^*_{k,s}(t) = q^*_{k,s}(0)\exp(i\omega_k t). \tag{9.13}$$

To calculate the energy of the electromagnetic field, we need to have the fields

$$
\begin{aligned}
\boldsymbol{E} &= -\frac{1}{c}\frac{\partial \boldsymbol{A}}{\partial t} = -\frac{1}{\sqrt{V}c}\sum_{k}\sum_{s=1,2}\left(\dot{q}_{k,s}(t)\boldsymbol{u}_{k,s}(\boldsymbol{r}) + \dot{q}^*_{k,s}(t)\boldsymbol{u}^*_{k,s}(\boldsymbol{r})\right) \\
&= -\frac{i}{\sqrt{V}c}\sum_{k}\sum_{s=1,2}\omega_k\left(q_{k,s}(t)\boldsymbol{u}_{k,s}(\boldsymbol{r}) - q^*_{k,s}(t)\boldsymbol{u}^*_{k,s}(\boldsymbol{r})\right), \\
\boldsymbol{B} &= \nabla \times \boldsymbol{A} = \frac{1}{\sqrt{V}}\sum_{k}\sum_{s=1,2}\left(q_{k,s}(t)(\nabla \times \boldsymbol{u}_{k,s}(\boldsymbol{r})) + q^*_{k,s}(t)(\nabla \times \boldsymbol{u}^*_{k,s}(\boldsymbol{r}))\right).
\end{aligned}
\tag{9.14}
$$

The Hamiltonian of the radiation field is given by

$$
H_r = \frac{1}{8\pi}\int_V \mathrm{d}V\,(|\boldsymbol{B}|^2 + |\boldsymbol{E}|^2). \tag{9.15}
$$

In calculating the contributions from $|\boldsymbol{B}|^2$, we have terms like this

$$
\int \mathrm{d}V\,(\nabla \times \boldsymbol{u}_{k,s}) \cdot (\nabla \times \boldsymbol{u}^*_{k',s'}). \tag{9.16}
$$

So, using the formula $\nabla \cdot (\boldsymbol{a} \times \boldsymbol{b}) = \boldsymbol{b} \cdot (\nabla \times \boldsymbol{a}) - \boldsymbol{a} \cdot (\nabla \times \boldsymbol{b})$ with $\boldsymbol{b} = (\nabla \times \boldsymbol{u}_{k,s})$ and $\boldsymbol{a} = \boldsymbol{u}^*_{k',s'}$, we find

$$
\begin{aligned}
\int \mathrm{d}V\,(\nabla \times \boldsymbol{u}_{k,s}) \cdot (\nabla \times \boldsymbol{u}^*_{k',s'}) &= \int \mathrm{d}V\,\nabla[\boldsymbol{u}^*_{k,s} \times (\nabla \times \boldsymbol{u}_{k',s'})] \\
&\quad + \int \mathrm{d}V\,\boldsymbol{u}^*_{k,s}[\nabla \times (\nabla \times \boldsymbol{u}_{k',s'})].
\end{aligned}
\tag{9.17}
$$

We can turn the first volume integral into a surface integration using the divergence theorem, and then we can see that it goes away due to the boundary conditions that the field vanishes on the surface of the box. In the second integral, we can use the formula $\nabla \times (\nabla \times \boldsymbol{u}) = \nabla(\nabla \boldsymbol{u}) - \nabla^2 \boldsymbol{u}$. Then we realize that the first term vanishes again, since from $\nabla \boldsymbol{A} = 0$ follows that $\nabla \boldsymbol{u} = 0$ as well. Finally, we get

$$
\int_V \mathrm{d}V\,(\nabla \times \boldsymbol{u}_{k,s}) \cdot (\nabla \times \boldsymbol{u}^*_{k',s'}) = -\int_V \mathrm{d}V\,\boldsymbol{u}_{k,s} \cdot \nabla^2 \boldsymbol{u}^*_{k',s'} = V\left(\frac{\omega_k}{c}\right)^2 \delta_{k,k'}\delta_{s,s'}. \tag{9.18}
$$

Terms like these provide us with the magnetic contribution to the radiation energy

$$
\begin{aligned}
\int_V \mathrm{d}V\,|\boldsymbol{B}|^2 &= \left(\frac{\omega_k}{c}\right)^2 \delta_{k,k'}\delta_{s,s'} \\
&\quad \left[q_{k,s}(t)q_{-k',s'}(t) + q_{k,s}(t)q^*_{k',s'}(t) + q^*_{k,s}(t)q_{k',s'}(t) + q^*_{k,s}(t)q^*_{-k',s'}(t)\right].
\end{aligned}
\tag{9.19}
$$

In a similar fashion, for the electric contribution to the radiation energy, we have

$$\int_V \mathrm{d}V \left(\frac{1}{c}\frac{\partial}{\partial t} q_{k,s}(t)\boldsymbol{u}_{ks}(\boldsymbol{r})\right)\cdot\left(\frac{1}{c}\frac{\partial}{\partial t} q^*_{k',s'}(t)\boldsymbol{u}^*_{k's'}(\boldsymbol{r})\right) = V\left(\frac{\omega_k}{c}\right)^2 \delta_{k,k'}\delta_{s,s'} q_{k,s}(t) q^*_{k',s'}(t), \tag{9.20}$$

which results in

$$\int_V \mathrm{d}V\, |\boldsymbol{E}|^2 = \left(\frac{\omega_k}{c}\right)^2 \delta_{k,k'}\delta_{s,s'}$$
$$\left[-q_{k,s}(t)q_{-k',s'}(t) + q_{k,s}(t)q^*_{k',s'}(t) + q^*_{k,s}(t)q_{k',s'}(t) - q^*_{k,s}(t)q^*_{-k',s'}(t)\right]. \tag{9.21}$$

Putting everything together, some terms cancel out and finally we have

$$H_r = \frac{1}{8\pi}\int_V \mathrm{d}V\, (|\boldsymbol{B}|^2 + |\boldsymbol{E}|^2) = \sum_k \sum_{s=1,2} \frac{1}{2\pi}\left(\frac{\omega_k}{c}\right)^2 q_{k,s}q^*_{k,s}. \tag{9.22}$$

We can write the energy of the radiation field in a more familiar form. Introducing

$$Q_{ks} = \frac{1}{c\sqrt{4\pi}}(q_{ks} + q^*_{ks}) \quad \text{and} \quad P_{ks} = \dot{Q}_{ks} = \frac{-i\omega_k}{c\sqrt{4\pi}}(q_{ks} - q^*_{ks}), \tag{9.23}$$

we find, from Eq. (9.12), that Q_{ks} satisfy an oscillator equation

$$\ddot{Q}_{ks} + \omega_k^2 Q_{ks} = 0. \tag{9.24}$$

This equation can also be derived from the Hamiltonian

$$H_{ks} = \frac{1}{2}\left(P_{ks}^2 + \omega_k^2 Q_{ks}^2\right), \tag{9.25}$$

with the help of the canonical equations

$$\dot{Q}_{ks} = \frac{\partial H_{ks}}{\partial P_{ks}} \quad \text{and} \quad \dot{P}_{ks} = -\frac{\partial H_{ks}}{\partial Q_{ks}}. \tag{9.26}$$

From Eqs. (9.23), we can see that

$$\frac{\omega_k^2}{c^2 2\pi} q_{k,s}q^*_{k,s} = \frac{1}{2}\left(P_{ks}^2 + \omega_k^2 Q_{ks}^2\right). \tag{9.27}$$

Then, we realize that the Hamiltonian of the radiation field is a sum of uncoupled harmonic oscillator Hamiltonians

$$H_r = \sum_k \sum_{s=1,2} H_{ks} = \sum_k \sum_{s=1,2} \frac{1}{2}\left(P_{ks}^2 + \omega_k^2 Q_{ks}^2\right). \tag{9.28}$$

In a similar way, we can also determine the electromagnetic momentum vector, the volume integral of the Poynting vector $\boldsymbol{S} = \boldsymbol{E} \times \boldsymbol{B}$,

$$\boldsymbol{G} = \frac{1}{4\pi c}\int_V \mathrm{d}V\, \boldsymbol{E} \times \boldsymbol{B} = \sum_k \sum_{s=1,2} \frac{\boldsymbol{k}}{kc}\frac{1}{2}\left(P_{ks}^2 + \omega_k^2 Q_{ks}\right) = \sum_k \sum_{s=1,2} \frac{\boldsymbol{k}}{k}\frac{H_{ks}}{c}. \tag{9.29}$$

9.1.2 Harmonic Oscillator Quantization

In the transition to quantum mechanics, we impose, in the spirit of the correspondence principle, commutation relations on the harmonic oscillator coordinates

$$[\hat{Q}_{ks}, \hat{Q}_{k's'}] = [\hat{P}_{ks}, \hat{P}_{k's'}] = 0, \quad \text{and} \quad [\hat{Q}_{ks}, \hat{P}_{k's'}] = i\hbar\delta_{kk'}\delta_{ss'}. \tag{9.30}$$

Then, the Hamilton operator of the radiation field becomes

$$\hat{H}_r = \sum_k \sum_{s=1,2} \hat{H}_{ks} = \sum_k \sum_{s=1,2} \frac{1}{2}\left(\hat{P}_{ks}^2 + \omega_k^2 \hat{Q}_{ks}^2\right). \tag{9.31}$$

Here the $\hat{H}_{ks}$ are one-dimensional harmonic oscillators, with the corresponding energy eigenvalues

$$E_{ks} = \hbar\omega_k(n_{ks} + 1/2), \quad n_{ks} = 0, 1, \ldots. \tag{9.32}$$

Thus, the total energy of the electromagnetic field becomes

$$E_r = \sum_k \sum_{s=1,2} \hbar\omega_k \left(n_{ks} + \frac{1}{2}\right). \tag{9.33}$$

For the momentum of the electromagnetic field, we get

$$\hat{\boldsymbol{G}} = \sum_k \sum_{s=1,2} \hbar\boldsymbol{k} n_{ks}. \tag{9.34}$$

The zero-point energy contribution is symmetric in all directions, so its contribution to the total momentum drops out.

We can see that one quantum of light has $p = \hbar\boldsymbol{k}$ momentum and its energy changes with $\hbar\omega$ energy quanta. In this regard, it looks just like a particle as it has both energy and momentum. It is called the *photon*. We can combine the energy and the momentum of the photon into a momentum four-vector

$$p_1 = \hbar k_x, \quad p_2 = \hbar k_y, \quad p_3 = \hbar k_z, \quad p_4 = \frac{i}{c}\hbar\omega_k. \tag{9.35}$$

In general, the length of the momentum four-vector is related to the mass

$$p_i p_i = -m^2c^2 = \hbar^2(k^2 - \omega_k^2/c^2) = 0, \tag{9.36}$$

i.e. the mass of the photon is zero and

$$\boldsymbol{p} = \hbar \boldsymbol{k} = \frac{\hbar\omega_k}{c^2}\boldsymbol{c}, \quad \text{where } \boldsymbol{c} = c\hat{e}_k. \tag{9.37}$$

We can elaborate the quantized radiation field formalism a little further. Analogously to the traditional harmonic oscillator case, we can introduce the operators

$$\hat{a}_{ks} = \frac{1}{\sqrt{2\hbar\omega_k}}\left(\omega_k \hat{Q}_{ks} + i\hat{P}_{ks}\right) \quad \text{and} \quad \hat{a}^\dagger_{ks} = \frac{1}{\sqrt{2\hbar\omega_k}}\left(\omega_k \hat{Q}_{ks} - i\hat{P}_{ks}\right), \tag{9.38}$$

and find immediately that

$$\left[\hat{a}_{ks}, \hat{a}_{k's'}\right] = \left[\hat{a}^\dagger_{ks}, \hat{a}^\dagger_{k's'}\right] = 0 \quad \text{and} \quad \left[\hat{a}_{ks}, \hat{a}^\dagger_{k's'}\right] = \delta_{kk'}\delta_{ss'}. \tag{9.39}$$

From Eq. (9.31) the radiation Hamiltonian becomes

$$\hat{H}_r = \sum_k \sum_{s=1,2} \hbar\omega_k(\hat{N}_{ks} + 1/2), \quad \text{with} \quad \hat{N}_{ks} = \hat{a}^\dagger_{ks}\hat{a}_{ks}, \tag{9.40}$$

where $\hat{N}$ is the photon number operator. It has the same properties we found before with the one-dimensional harmonic oscillators

$$\left[\hat{a}_{ks}, \hat{N}_{k's'}\right] = \delta_{kk'}\delta_{ss'}\hat{a}_{ks} \quad \text{and} \quad \left[\hat{a}^\dagger_{ks}, \hat{N}_{k's'}\right] = -\delta_{kk'}\delta_{ss'}\hat{a}^\dagger_{ks}. \tag{9.41}$$

We can consider quantum states in occupation number representation. In the vacuum state $|0\rangle$ there are no photons, $n_{ks} = 0$ for all possible k and s. The n-photon state is created by a repeated application of $\hat{a}^\dagger_{ks}$

$$|n_{ks}\rangle = \frac{1}{\sqrt{n_{ks}!}}(\hat{a}^\dagger_{ks})^{n_{ks}}|0\rangle. \tag{9.42}$$

Similarly as before, we find

$$\hat{a}_{ks}|n_{ks}\rangle = \sqrt{n_{ks}}|n_{ks} - 1\rangle \quad \text{and} \quad \hat{a}^\dagger_{ks}|n_{ks}\rangle = \sqrt{n_{ks}+1}|n_{ks}+1\rangle. \tag{9.43}$$

The time dependence of $\hat{a}_{ks}$ and $\hat{a}^{\dagger}_{ks}$ are determined by the Heisenberg equations

$$\frac{d\hat{a}_{ks}}{dt} = \frac{1}{i\hbar}\left[\hat{a}_{ks}, \hat{H}_r\right] = -i\omega_k \hat{a}_{ks} \quad \text{and} \quad \frac{d\hat{a}^{\dagger}_{ks}}{dt} = \frac{1}{i\hbar}\left[\hat{a}^{\dagger}_{ks}, \hat{H}_r\right] = i\omega_k \hat{a}^{\dagger}_{ks}, \tag{9.44}$$

which are solved by

$$\hat{a}_{ks}(t) = \hat{a}_{ks}(0)\exp(-i\omega_k t) \quad \text{and} \quad \hat{a}^{\dagger}_{ks}(t) = \hat{a}^{\dagger}_{ks}(0)\exp(i\omega_k t). \tag{9.45}$$

As a result, we can express the vector potential $\hat{\boldsymbol{A}}(\boldsymbol{r}, t)$, the electric field $\hat{\boldsymbol{E}}(\boldsymbol{r}, t)$, and the magnetic field $\hat{\boldsymbol{B}}(\boldsymbol{r}, t)$ in terms of the $\hat{a}$ and $\hat{a}^{\dagger}$ operators

$$\begin{aligned}
\hat{\boldsymbol{A}}(\boldsymbol{r}, t) &= \sqrt{\frac{2\pi c^2\hbar}{V}}\sum_{k}\sum_{s=1,2}\frac{1}{\sqrt{\omega_k}}\left(\varepsilon_{ks}\hat{a}_{ks}(t)\exp(i\boldsymbol{kr}) + \varepsilon^{*}_{ks}\hat{a}^{\dagger}_{ks}(t)\exp(-i\boldsymbol{kr})\right), \\
\hat{\boldsymbol{E}}(\boldsymbol{r}, t) &= i\sqrt{\frac{2\pi\hbar}{V}}\sum_{k}\sum_{s=1,2}\sqrt{\omega_k}\left(\varepsilon_{ks}\hat{a}_{ks}(t)\exp(i\boldsymbol{kr}) - \varepsilon^{*}_{ks}\hat{a}^{\dagger}_{ks}(t)\exp(-i\boldsymbol{kr})\right), \\
\hat{\boldsymbol{B}}(\boldsymbol{r}, t) &= i\sqrt{\frac{2\pi c^2\hbar}{V}}\sum_{k}\sum_{s=1,2}\frac{1}{\sqrt{\omega_k}}\left((\boldsymbol{k}\times\varepsilon_{ks})\hat{a}_{ks}(t)\exp(i\boldsymbol{kr}) - (\boldsymbol{k}\times\varepsilon^{*}_{ks})\hat{a}^{\dagger}_{ks}(t)\exp(-i\boldsymbol{kr})\right).
\end{aligned} \tag{9.46}$$

Calculating the commutator between the electric field an the number operator, we find

$$\begin{aligned}
\left[\hat{\boldsymbol{E}}_{ks}, \hat{N}_{k's'}\right] &= i\sqrt{\frac{2\pi\hbar\omega_k}{V}}\Big[\varepsilon_{ks}(\hat{a}_{ks}\hat{a}^{\dagger}_{k's'}\hat{a}_{k's'} - \hat{a}^{\dagger}_{k's'}\hat{a}_{k's'}\hat{a}_{ks})\exp(i\boldsymbol{kr}) \\
&\quad -\varepsilon^{*}_{ks}(\hat{a}^{\dagger}_{ks}\hat{a}^{\dagger}_{k's'}\hat{a}_{k's'} - \hat{a}^{\dagger}_{k's'}\hat{a}_{k's'}\hat{a}^{\dagger}_{ks})\exp(-i\boldsymbol{kr})\Big] \\
&= i\sqrt{\frac{2\pi\hbar\omega_k}{V}}\left(\varepsilon_{ks}\hat{a}_{ks}\exp(i\boldsymbol{kr}) + \varepsilon^{*}_{ks}\hat{a}^{\dagger}_{ks}\exp(-i\boldsymbol{kr})\right) = i\frac{\omega_k}{c}\boldsymbol{A}_{ks}\delta_{kk'}\delta_{ss'},
\end{aligned} \tag{9.47}$$

where

$$\hat{\boldsymbol{A}}_{ks} = c\sqrt{\frac{2\pi\hbar\omega_k}{V}}\varepsilon_{ks}\left(\hat{a}_{ks}\exp(i\boldsymbol{kr}) + \hat{a}^{\dagger}_{ks}\exp(-i\boldsymbol{kr})\right) \tag{9.48}$$

is a partial wave component of the vector potential.

We have found that the electric field operator does not commute with the photon number operator. Consequently, they cannot be measured simultaneously. So, we have to be careful with the photon picture. A photon has energy and momentum, just like an ordinary particle. However, if we want to take the particle picture seriously, we need to localize it to a small volume, such that the electric field is singular there and gradually vanishing outside. This would require a sharp measurement of $\boldsymbol{E}$, but then the number of photons become uncertain. Conversely, if the number of photons are known, the energy and the momentum of the electromagnetic field cannot be localized to a small region of space. So, we cannot assign a classical particle picture to photon with location and path. The location and path lose sense for photons. The particle feature of the electromagnetic field is manifested in the fact that its energy and momentum are made up from discrete number of energy and momentum quanta.

We have the interesting result that the energy is a sum of oscillator quantum energies $\hbar\omega_k$. In the vacuum state, there are no oscillator quanta, i.e. all n_{ks}'s are zero. However, all oscillators with frequency ω_k possess a zero-point energy $\hbar\omega_k/2$. Since, in principle, all ω_k are possible, the vacuum energy is infinite. This, however, usually does not lead to observable effects as we measure things with respect to the vacuum. Also, we have never observed a *half photon* with energy $\hbar\omega/2$, so the vacuum energy is mostly ignored. The zero-point oscillation of the electromagnetic field shows up, however, in particular cases, like the Lamb shift and the Casimir effect.

9.2 Interaction with Atoms

The interaction term between an atom and the electromagnetic field is given by

$$
\begin{aligned}
\hat{V}(t) &= \frac{e}{mc}\hat{\boldsymbol{A}}(\boldsymbol{r},t)\hat{\boldsymbol{p}} \\
&= \frac{e}{mc}\sqrt{\frac{2\pi c^2\hbar}{V}}\times \\
&\sum_k\sum_{s=1,2}\frac{1}{\sqrt{\omega_k}}\left(\varepsilon_{ks}\hat{a}_{ks}(0)\exp[i(\boldsymbol{kr}-\omega_k t)]+\varepsilon_{ks}^*\hat{a}_{ks}^\dagger(0)\exp[-i(\boldsymbol{kr}-\omega_k t)]\right)\hat{\boldsymbol{p}} \\
&= \sum_k\sum_{s=1,2}\left(\hat{v}_{ks}\exp(-i\omega_k t)+\hat{v}_{ks}^\dagger\exp(i\omega_k t)\right),
\end{aligned}
\tag{9.49}
$$

where

$$
\hat{v}_{ks} = \frac{e}{m}\sqrt{\frac{2\pi\hbar}{V\omega_k}}\,\hat{a}_{ks}(0)\exp(i\boldsymbol{kr})\varepsilon_{ks}\hat{\boldsymbol{p}} \quad \text{and} \quad \hat{v}_{ks}^\dagger = \frac{e}{m}\sqrt{\frac{2\pi\hbar}{V\omega_k}}\,\hat{a}_{ks}^\dagger(0)\exp(-i\boldsymbol{kr})\varepsilon_{ks}^*\hat{\boldsymbol{p}}. \tag{9.50}
$$

We study the interaction between the atom and the electromagnetic field by using first order perturbation theory. As an initial state, we take $|\Phi_i\rangle = |\psi_i\rangle|n_{ks}\rangle$, where $|\psi_i\rangle$ is an atomic wave function and $|n_{ks}\rangle$ represents n_{ks} photons with wave numbers k and polarization s. The final state is given by $|\Phi_f\rangle = |\psi_f\rangle|n_{ks}\pm 1\rangle$, depending on whether we consider emission or absorption. The operator $\hat{v}$ contains $\hat{a}$ that eliminates a photon, which is absorbed by the atom. Hence, $\hat{v}$ leads to absorption and similarly $\hat{v}^\dagger$ generates emission.

For the matrix elements of a photon emission process we find

$$\begin{aligned}\langle\Phi_f|\hat{v}_{ks}^\dagger|\Phi_i\rangle &= \frac{e}{m}\sqrt{\frac{2\pi\hbar}{\omega_k V}}\langle\psi_f|\exp(-i\boldsymbol{kr})\boldsymbol{\varepsilon}_{ks}^*\cdot\hat{\boldsymbol{p}}|\psi_i\rangle\langle n_{ks}+1|\hat{a}_{ks}^\dagger|n_{ks}\rangle \\ &= \frac{e}{m}\sqrt{\frac{2\pi\hbar}{\omega_k V}}\sqrt{n_{ks}+1}\langle\psi_f|\exp(-i\boldsymbol{kr})\boldsymbol{\varepsilon}_{ks}^*\cdot\hat{\boldsymbol{p}}|\psi_i\rangle,\end{aligned} \tag{9.51}$$

and similarly, for the absorption matrix elements we have

$$\begin{aligned}\langle\Phi_f|\hat{v}_{ks}|\Phi_i\rangle &= \frac{e}{m}\sqrt{\frac{2\pi\hbar}{\omega_k V}}\langle\psi_f|\exp(i\boldsymbol{kr})\boldsymbol{\varepsilon}_{ks}\cdot\hat{\boldsymbol{p}}|\psi_i\rangle\langle n_{ks}-1|\hat{a}_{ks}|n_{ks}\rangle \\ &= \frac{e}{m}\sqrt{\frac{2\pi\hbar}{\omega_k V}}\sqrt{n_{ks}}\langle\psi_f|\exp(i\boldsymbol{kr})\boldsymbol{\varepsilon}_{ks}\cdot\hat{\boldsymbol{p}}|\psi_i\rangle.\end{aligned} \tag{9.52}$$

Then, for the emission and absorption rates in the first order perturbation approach we obtain

$$\begin{aligned}\Gamma_{i\to f}^{emi} &= \frac{4\pi^2e^2}{m^2\omega_k V}(n_{ks}+1)|\langle\psi_f|\exp(-i\boldsymbol{kr})\boldsymbol{\varepsilon}_{ks}^*\cdot\hat{\boldsymbol{p}}|\psi_i\rangle|^2\delta(E_f-E_i+\hbar\omega_k), \\ \Gamma_{i\to f}^{abs} &= \frac{4\pi^2e^2}{m^2\omega_k V}n_{ks}|\langle\psi_f|\exp(i\boldsymbol{kr})\boldsymbol{\varepsilon}_{ks}\cdot\hat{\boldsymbol{p}}|\psi_i\rangle|^2\delta(E_f-E_i-\hbar\omega_k).\end{aligned} \tag{9.53}$$

We should note that in this formalism we have emission even in the absence of the electromagnetic field, i.e. when $n_{ks}=0$. This is spontaneous emission. We should also emphasize that the emission rate is enhanced if there are a lot of photons present in the system. This phenomena is the Light Amplification by Stimulated Emission of Radiation (LASER).

9.2.1 Electromagnetic Transition Rates

In the matrix elements in Eq. (9.53), the range of the integration extends up to the size of the object, about 10^{-10} m for atoms and 10^{-15} m for nuclei. On the other hand, $k=2\pi/\lambda$, where λ is the wavelength, depends on the transition energy. So, for visible, ultraviolet or even X-rays, the wavelength is much bigger than the typical size of the atom. Similarly, for γ-rays, the wavelength is much bigger than the size of the nucleus. Therefore, $kr \ll 1$, and the approximation

$$\exp(\pm i\boldsymbol{kr})\simeq 1 \tag{9.54}$$

is justified. This is the *electric dipole approximation* and the transition is the *electric dipole*, or $E1$, transition.

To evaluate the matrix elements, we can utilize the relation

$$\left[\boldsymbol{r}, \hat{H}_0\right] = \left[\boldsymbol{r}, \frac{\hat{\boldsymbol{p}}^2}{2m} + \hat{V}(\boldsymbol{r})\right] = \left[\boldsymbol{r}, \frac{\hat{\boldsymbol{p}}^2}{2m}\right] = \left[x + y + z, \frac{1}{2m}(\hat{p}_x^2 + \hat{p}_y^2 + \hat{p}_z^2)\right] = \frac{i\hbar}{m}\hat{\boldsymbol{p}}. \tag{9.55}$$

Then, for typical matrix elements in the electric dipole approximation, we have

$$\begin{aligned}\langle\psi_f|\exp(-i\boldsymbol{k}\boldsymbol{r})\boldsymbol{\varepsilon}^*_{ks}\cdot\hat{\boldsymbol{p}}|\psi_i\rangle \simeq \langle\psi_f|\boldsymbol{\varepsilon}^*_{ks}\cdot\hat{\boldsymbol{p}}|\psi_i\rangle &= \frac{m}{i\hbar}\langle\psi_f|\boldsymbol{\varepsilon}^*_{ks}\cdot\left[\boldsymbol{r}, \hat{H}_0\right]|\psi_i\rangle \\ &= \frac{m}{i\hbar}(E_i - E_f)\langle\psi_f|\boldsymbol{\varepsilon}^*_{ks}\cdot\boldsymbol{r}|\psi_i\rangle \\ &= i m\,\omega_{fi}\langle\psi_f|\boldsymbol{\varepsilon}^*_{ks}\cdot\boldsymbol{r}|\psi_i\rangle,\end{aligned} \tag{9.56}$$

which turns the emission and absorption rates into

$$\begin{aligned}\Gamma^{emi}_{i\to f} &= \frac{4\pi^2 e^2\omega^2_{fi}}{\omega_k V}(n_{ks}+1)|\langle\psi_f|\boldsymbol{\varepsilon}^*_{ks}\cdot\boldsymbol{r}|\psi_i\rangle|^2\delta(E_f - E_i + \hbar\omega_k), \\ \Gamma^{abs}_{i\to f} &= \frac{4\pi^2 e^2\omega^2_{fi}}{\omega_k V}n_{ks}|\langle\psi_f|\boldsymbol{\varepsilon}_{ks}\cdot\boldsymbol{r}|\psi_i\rangle|^2\delta(E_f - E_i - \hbar\omega_k).\end{aligned} \tag{9.57}$$

In order to calculate the transition matrix elements, we express $\boldsymbol{r}$ in spherical coordinates

$$\boldsymbol{r} = r(\sin\theta\cos\phi, \sin\theta\sin\phi, \cos\theta). \tag{9.58}$$

Consequently,

$$\boldsymbol{\varepsilon}_{ks}\cdot\boldsymbol{r} = r(\varepsilon_x\sin\theta\cos\phi + \varepsilon_y\sin\theta\sin\phi + \varepsilon_z\cos\theta). \tag{9.59}$$

Recalling the relations with spherical harmonics, $\sin\theta\cos\phi = -\sqrt{2\pi/3}(Y_{1,1} - Y_{1,-1})$, $\sin\theta\sin\phi = i\sqrt{2\pi/3}(Y_{1,1} + Y_{1,-1})$ and $\cos\theta = \sqrt{4\pi/3}Y_{1,0}$, we find that

$$\boldsymbol{\varepsilon}_{ks}\cdot\boldsymbol{r} = \sqrt{\frac{4\pi}{3}}r\left(\frac{-\varepsilon_x + i\varepsilon_y}{\sqrt{2}}Y_{1,1} + \frac{\varepsilon_x + i\varepsilon_y}{\sqrt{2}}Y_{1,-1} + \varepsilon_z Y_{1,0}\right). \tag{9.60}$$

Sandwiching this between the initial state $\psi_i = R_{n_i l_i}(r)Y_{l_i,m_i}(\Omega)$ and the final state $\psi_f = R_{n_f l_f}(r)Y_{l_f,m_f}(\Omega)$, we obtain

$$\begin{aligned}\langle\psi_f|\boldsymbol{\varepsilon}_{ks}\cdot\boldsymbol{r}|\psi_i\rangle &= \sqrt{\frac{4\pi}{3}}\int_0^\infty \mathrm{d}r\, r^3 R^*_{n_f l_f}(r)R_{n_i l_i}(r) \\ &\times \int \mathrm{d}\Omega\, Y^*_{l_f,m_f}(\theta,\phi)\left(\frac{-\varepsilon_x + i\varepsilon_y}{\sqrt{2}}Y_{1,1}(\theta,\phi) + \frac{\varepsilon_x + i\varepsilon_y}{\sqrt{2}}Y_{1,-1}(\theta,\phi) + \varepsilon_z Y_{1,0}(\theta,\phi)\right)Y_{l_i,m_i}(\theta,\phi).\end{aligned} \tag{9.61}$$

Here, we have the integral of three $Y_{l,m}$ functions, which has been calculated before in Eq. (4.231)

$$\int \mathrm{d}\Omega \, Y^*_{l_f,m_f} Y_{1,m'} Y_{l_i,m_i} = \langle l_f m_f | Y_{1,m'} | l_i m_i \rangle$$
$$= \sqrt{\frac{3(2l_i+1)}{4\pi(2l_f+1)}} \langle l_i, 1; 0, 0 | l_f, 0 \rangle \langle l_i, 1; m_i, m' | l_f, m_f \rangle. \tag{9.62}$$

These Clebsch-Gordan coefficients determine the possible transitions in the electric dipole approximation:

- The transition is forbidden unless $m_i + m' = m_f$, where $m' = -1, 0, 1$. Therefore

$$m_f - m_i = -1, 0, 1. \tag{9.63}$$

- The vector addition rule applies, so $|l_i - 1| \le l_f \le |l_i + 1|$, or

$$l_f - l_i = -1, 0, 1. \tag{9.64}$$

- The coefficient $\langle l_i, 1; 0, 0 | l_f, 0 \rangle$ vanishes unless $(-1)^{l_i+1-l_f} = 1$, or $(-1)^{l_i-l_f} = -1$, or

$$l_i - l_f = \text{ odd integer }. \tag{9.65}$$

  Consequently, the initial and final states must have different parities, $\pi_f \pi_i = -1$, and there is no transition if $l_i = l_f = 0$.

If dipole transition is forbidden, we may consider the next term in Eq. (9.54). Then the transition rates will be proportional to the matrix element

$$\langle \psi_f | (\boldsymbol{k} \cdot \boldsymbol{r})(\boldsymbol{\varepsilon} \cdot \hat{\boldsymbol{p}}) | \psi_i \rangle. \tag{9.66}$$

Here we can utilize the relation $(\boldsymbol{a} \times \boldsymbol{b}) \cdot (\boldsymbol{c} \times \boldsymbol{d}) = (\boldsymbol{a} \cdot \boldsymbol{c})(\boldsymbol{b} \cdot \boldsymbol{d}) - (\boldsymbol{b} \cdot \boldsymbol{c})(\boldsymbol{a} \cdot \boldsymbol{d})$, i.e.

$$(\boldsymbol{k} \cdot \boldsymbol{r})(\boldsymbol{\varepsilon} \cdot \hat{\boldsymbol{p}}) = (\boldsymbol{\varepsilon} \cdot \boldsymbol{r})(\boldsymbol{k} \cdot \hat{\boldsymbol{p}}) + (\boldsymbol{k} \times \boldsymbol{\varepsilon})(\boldsymbol{r} \times \hat{\boldsymbol{p}}). \tag{9.67}$$

If we take the coordinate system such that $\boldsymbol{k}$ points along the x-axis and $\boldsymbol{\varepsilon}$ along the y-axis, the first matrix element becomes

$$\langle \psi_f | (\boldsymbol{\varepsilon} \cdot \boldsymbol{r})(\boldsymbol{k} \cdot \hat{\boldsymbol{p}}) | \psi_i \rangle = k \langle \psi_f | \hat{y} \hat{p}_x | \psi_i \rangle. \tag{9.68}$$

Then, with the help of Eq. (9.55), we find

$$\langle \psi_f | (\boldsymbol{\varepsilon} \cdot \boldsymbol{r})(\boldsymbol{k} \cdot \hat{\boldsymbol{p}}) | \psi_i \rangle = \frac{km}{i\hbar} \langle \psi_f | \hat{y} \left[\hat{x}, \hat{H}_0 \right] | \psi_i \rangle \tag{9.69}$$

which results in

$$\langle\psi_f|(\boldsymbol{\varepsilon}\cdot\boldsymbol{r})(\boldsymbol{k}\cdot\hat{\boldsymbol{p}})|\psi_i\rangle = \frac{km}{i\hbar}(E_f - E_i)\langle\psi_f|\hat{x}\hat{y}|\psi_i\rangle. \tag{9.70}$$

The term $\hat{x}\hat{y}$ looks like an element of the quadrupole operator, therefore the generated transition is called as the electric quadrupole $E2$ transition. This operator can be expressed as linear combinations of $Y_{2,m}$ functions. Consequently, Eq. (9.62) results in the $E2$ selection rules:

- The transition is forbidden unless $m_i + m' = m_f$, where $m' = -2, -1, 0, 1, 2$. Therefore
$$m_f - m_i = -2, -1, 0, 1, 2. \tag{9.71}$$
- The vector addition rule applies, so $|l_i - 2| \leq l_f \leq |l_i + 2|$, or
$$l_f - l_i = -2, -1, 0, 1, 2. \tag{9.72}$$
- The coefficient $\langle l_i, 2; 0, 0|l_f, 0\rangle$ vanishes unless $(-1)^{l_i+2-l_f} = 1$, or $(-1)^{l_i-l_f} = 1$, or
$$l_i - l_f = \text{ even integer }. \tag{9.73}$$

  Therefore, the initial and final states must have the same parities and there is no transition if $l_i = l_f = 0$.

In the other part of Eq. (9.67) we can discover the orbital angular momentum operator $\hat{\boldsymbol{L}}$. We should also keep in mind that, besides the coupling between the orbital angular momentum and the magnetic field, there is also a coupling between the electron spin and the magnetic field, Eq. (5.101),

$$\hat{H}' = \frac{e}{2mc}\hat{\boldsymbol{\mu}}\cdot\boldsymbol{B}, \tag{9.74}$$

where $\hat{\boldsymbol{\mu}} = g_L\hat{\boldsymbol{L}} + g_S\hat{\boldsymbol{S}}$ with $g_L = 1$ and $g_S = 2$. So, the total matrix element would be a combination of terms

$$\langle\psi_f|\hat{\boldsymbol{L}}|\psi_i\rangle \quad \text{and} \quad \langle\psi_f|\hat{\boldsymbol{S}}|\psi_i\rangle. \tag{9.75}$$

The transition generated by these matrix elements is the magnetic dipole $M1$ transition. Here we have the matrix elements of a rank-1 tensor operators, just like by $E1$ transitions. So, the same angular momentum selection rules apply. The only difference is that the initial and the final states should have the same parity, $\pi_f\pi_i = 1$. By $E1$ transitions $\hat{\boldsymbol{r}}$ is odd, but by $M1$ transitions $\hat{\boldsymbol{L}}$ and $\hat{\boldsymbol{S}}$ are even parity operators.

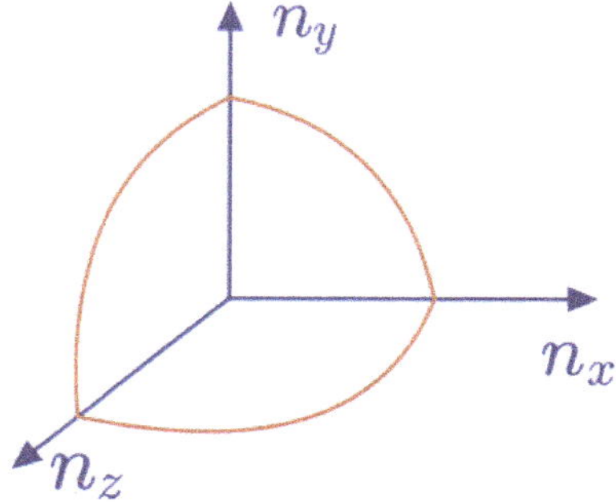

Fig. 9.1 The acceptable n's are on a quadrant of a half sphere

9.2.2 Spontaneous Emission

If $n_{ks} = 0$, no absorption can happen, but emission is possible. This is spontaneous emission and the emission rate reads

$$\Gamma_{i \to f}^{emi} = \frac{4\pi^2 \omega_{fi}^2}{\omega_k V} |\boldsymbol{\varepsilon}_{ks}^* \cdot \boldsymbol{d}|^2 \delta(E_f - E_i + \hbar\omega_k), \tag{9.76}$$

where $\boldsymbol{d}_{fi} = \langle \psi_f | \boldsymbol{d} | \psi_i \rangle = -e \langle \psi_f | \boldsymbol{r} | \psi_i \rangle$ is the electric dipole vector.

To calculate the decay rate of the atomic state, we need to find the density of final states. Using the relations $p = \hbar k = \hbar\omega_k / c = \hbar 2\pi/(cT) = \hbar 2\pi/\lambda$ and $L = n\lambda/2$, we have $n = 2L/\lambda = Lp/(\hbar\pi)$ and $\mathrm{d}n = L/(\hbar\pi)\mathrm{d}p$. The physically acceptable n's are such that $0 \le n_x, n_y, n_z$ and $n = \sqrt{n_x^2 + n_y^2 + n_z^2}$, therefore they are on the surface of the quadrant of a half sphere (Fig. 9.1). The volume is given by $Z(n) = 1/8 \cdot 4n^3\pi/3$ and the surface is

$$\mathrm{d}Z(n) = \frac{\pi}{2} n^2 \mathrm{d}n = \frac{\pi}{2} \frac{L^2 p^2}{\hbar^2 \pi^2} \frac{L}{\hbar\pi} \mathrm{d}p = \frac{V p^2}{(2\pi\hbar)^3} \mathrm{d}p \, 4\pi, \tag{9.77}$$

or, if we restrict to a $[p, p + dp]$ interval and $\mathrm{d}\Omega$ solid angle, we have

$$\mathrm{d}Z(n) = \frac{V p^2}{(2\pi\hbar)^3} \mathrm{d}p \mathrm{d}\Omega = \frac{V \hbar^3 \omega_k^2}{(2\pi\hbar)^3 c^3} \mathrm{d}\omega_k \mathrm{d}\Omega = \frac{V \omega_k^2}{(2\pi c)^3} \mathrm{d}\omega_k \mathrm{d}\Omega. \tag{9.78}$$

Then, for the emission rate into $\mathrm{d}\Omega$ solid angle, we obtain

$$\begin{aligned}
\mathrm{d}W_{i\to f}^{emi} &= \frac{V}{(2\pi)^3c^3}\mathrm{d}\Omega\int \mathrm{d}\omega_k\,\omega_k^2\,\Gamma_{i\to f}^{emi} \\
&= \frac{\omega_{fi}^2}{2\pi c^3}|\boldsymbol{\varepsilon}_s^*\cdot\boldsymbol{d}_{fi}|^2\,\mathrm{d}\Omega\int \mathrm{d}\omega_k\,\omega_k\,\delta(E_f - E_i + \hbar\omega_k) \\
&= \frac{\omega_{fi}^3}{2\pi\hbar c^3}|\boldsymbol{\varepsilon}_s^*\cdot\boldsymbol{d}_{fi}|^2\,\mathrm{d}\Omega.
\end{aligned} \tag{9.79}$$

We need to sum over polarization directions

$$\sum_{s=1,2}|\boldsymbol{\varepsilon}_s^*\cdot\boldsymbol{d}_{fi}|^2 = |\varepsilon_1^*(d_{fi})_1|^2 + |\varepsilon_2^*(d_{fi})_2|^2 = |\boldsymbol{d}_{fi}|^2 - |\varepsilon_3^*(d_{fi})_3|^2 = \frac{2}{3}|\boldsymbol{d}_{fi}|^2, \tag{9.80}$$

where we assumed that there is no preferred direction

$$\langle|\varepsilon_1^*(d_{fi})_1|2\rangle = \langle|\varepsilon_2^*(d_{fi})_2|^2\rangle = \langle|\varepsilon_3^*(d_{fi})_3|^2\rangle = \frac{1}{3}\langle|\boldsymbol{d}_{fi}|^2\rangle. \tag{9.81}$$

Altogether, we have

$$dW_{i\to f}^{emi} = \frac{\omega_{fi}^3}{3\pi\hbar c^3}\langle|\boldsymbol{d}_{fi}|^2\rangle\,\mathrm{d}\Omega, \tag{9.82}$$

or, integrating over 4π, we find

$$W_{i\to f}^{emi} = \frac{4\omega_{fi}^3 e^2}{3\pi\hbar c^3}|\langle\psi_f|\boldsymbol{r}|\psi_i|\rangle^2. \tag{9.83}$$

The radiated power is given by

$$I = \hbar\omega_{fi}W_{i\to f}^{emi}. \tag{9.84}$$

For the lifetime of an excited state, we need to consider transitions to all possible final states

$$\tau = \frac{1}{W} = \frac{1}{\sum_f W_{i\to f}^{emi}}. \tag{9.85}$$

9.2.3 Vacuum Expectation of the Electromagnetic Field

It is obvious that the electric and magnetic field operators do not commute

$$\left[\hat{\boldsymbol{E}}(\boldsymbol{r},t),\,\hat{\boldsymbol{B}}(\boldsymbol{r},t)\right] \neq 0. \tag{9.86}$$

Therefore, they cannot be measured simultaneously and we cannot expect that, in the vacuum state, both $\hat{\boldsymbol{E}}$ and $\hat{\boldsymbol{B}}$ vanish simultaneously. Clearly, the vacuum expectation values of the fields vanish

$$\langle 0|\hat{\boldsymbol{E}}(\boldsymbol{r},t)|0\rangle = \langle 0|\hat{\boldsymbol{B}}(\boldsymbol{r},t)|0\rangle = 0. \tag{9.87}$$

It is also straightforward to see that

$$\langle 0|\hat{\boldsymbol{E}}_{ks}^2|0\rangle = \langle 0|\hat{\boldsymbol{B}}_{ks}^2|0\rangle = \frac{2\pi\hbar\omega_k}{V}, \tag{9.88}$$

since only the $\hat{a}\hat{a}^\dagger$ terms contribute.

We have found that, in the vacuum state, the expectation value of the field vectors is zero, but the expectation value of the squared vectors is finite. Then, the corresponding energy density is given by

$$U = \frac{1}{8\pi}\int_V \mathrm{d}V \sum_k \sum_{s=1,2} \langle 0|(\hat{\boldsymbol{E}}_{ks}^2 + \hat{\boldsymbol{B}}_{ks}^2)|0\rangle = \sum_k \sum_{s=1,2} \frac{1}{2}\hbar\omega_k. \tag{9.89}$$

This means that the zero-point energy of the vacuum comes from the uncertainty of the electromagnetic field. Usually, the zero-point energy of the vacuum does not lead to observable effects, but the fluctuation of the electromagnetic field generates observable shifts in atomic energy levels. This is called the *Lamb shift*. It is a genuine quantum electrodynamical effect, but Bethe gave a rather elementary and intuitive physical explanation for it. The electric field in the vacuum is given by

$$\boldsymbol{E} = \sum_{k,s} \boldsymbol{E}_{ks}^{(0)} \cos(\omega_k t + \phi_{ks}), \quad \text{with} \quad |\boldsymbol{E}_{ks}^{(0)}| = \sqrt{\frac{4\pi\hbar\omega_k}{V}}. \tag{9.90}$$

This amplitude gives the expectation value of Eq. (9.88). The electron feels the force by the electric field and follows the equation

$$m\,\delta\ddot{\boldsymbol{r}} = -e\boldsymbol{E}, \tag{9.91}$$

which results in a harmonic motion

$$\delta\boldsymbol{r}(t) = \sum_{k,s} \frac{e}{m\omega_k^2} \boldsymbol{E}_{ks}^0 \cos(\omega_k t + \phi_{ks}). \tag{9.92}$$

The vacuum fluctuations of the electric field shakes the electron around its equilibrium point $\boldsymbol{r}_0$. As a result, the electron charge is smeared and this will change the time averaged potential energy

$$\overline{\Delta V} = -e[\Phi(\boldsymbol{r}_0 + \delta \boldsymbol{r}) - \Phi(\boldsymbol{r}_0)] \approx -e\left[\frac{\partial \Phi}{\partial \boldsymbol{r}}\delta \bar{\boldsymbol{r}} + \frac{1}{2}\sum_{ij}\frac{\partial^2 \Phi}{\partial x_i \partial x_j}\overline{\delta x_i \delta x_j}\right], \tag{9.93}$$

where Φ is the electric potential due to the nuclear charge. Since the fluctuations are isotropic, we have

$$\overline{\delta x} = \overline{\delta y} = \overline{\delta z} = 0 \quad \text{and} \quad \overline{\delta x^2} = \overline{\delta y^2} = \overline{\delta z^2} = \frac{1}{3}\overline{\delta r^2}, \tag{9.94}$$

which gives

$$\overline{\Delta V} = -\frac{1}{6} e\, \overline{\delta r^2}\, \nabla^2 \Phi\ . \tag{9.95}$$

From the Poisson equation we have $\nabla^2 \Phi = -4\pi\rho$, where ρ is the charge distribution of the nucleus. We are considering atomic levels, so we can assume a point-like nucleus

$$\overline{\Delta V} = \frac{2\pi}{3} e^2 \boldsymbol{\delta}(0)\, \overline{\delta r^2}. \tag{9.96}$$

Then, from Eqs. (9.90) and (9.92) we find

$$\overline{\delta r^2} = \frac{2\pi \hbar e^2}{m^2 V}\sum_{k,s}\frac{1}{\omega_k^3}. \tag{9.97}$$

We can replace the summation of the frequencies by integration. According to Eq. (9.77), the number of frequencies in the $[\omega, \omega + \mathrm{d}\omega]$ interval, considering the two polarization directions, is given by

$$\mathrm{d}Z(\omega) = \frac{V\omega^2 \mathrm{d}\omega}{\pi^2 c^3}. \tag{9.98}$$

Then, Eq. (9.97) becomes

$$\overline{\delta r^2} = \frac{2\hbar e^2}{\pi m^2 c^3}\int_{\omega_1}^{\omega_2}\frac{\mathrm{d}\omega}{\omega} = \frac{2\hbar e^2}{\pi m^2 c^3}\log\frac{\omega_2}{\omega_1}. \tag{9.99}$$

This result is clearly diverging if we take $\omega_1 = 0$ or $\omega_2 = \infty$. However, we can easily understand that there are other physical processes that suppress the extreme low and extreme high frequencies. If the ω is much lower than the frequency corresponding to the binding energy of the electron, and therefore the wavelength is much longer than the atomic radius, the effect of vibration is negligible and atomic binding dominates. On the other hand, if the vibrational energy is comparable to the rest energy of the electron, the increased momentum results in a suppression of the vibrational amplitude. So, it is reasonable to take

$$\frac{\hbar\omega_1}{2} \simeq \frac{me^4}{2\hbar^2 n^2} \quad \text{and} \quad \frac{\hbar\omega_2}{2} \simeq mc^2, \tag{9.100}$$

where n is the principal quantum number of the hydrogen atom.

As a result, the perturbative correction, due to the vacuum oscillations of the electromagnetic field, becomes

$$\Delta E = \langle\psi_n|\overline{\Delta V}|\psi_n\rangle = \frac{2\pi e^2}{3}\overline{\delta r^2}\,|\psi_n(0)|^2. \tag{9.101}$$

The $\psi_n(0)$ differs from zero only for s states, and in this case

$$|\psi_n(0)|^2 = \frac{1}{\pi n^3 a_0^3}, \tag{9.102}$$

where $a_0 = \hbar^2/me^2$ is the Bohr radius. Putting all the pieces together, we find the energy shift

$$\Delta E \simeq \frac{4\hbar e^4}{3\pi m^2 c^3 a_0^3 n^3} \log \frac{2\hbar^2 c^2 n^2}{e^4}. \tag{9.103}$$

In the Dirac theory, the energy levels of the hydrogen atom depend only on the principal quantum number n and the angular momentum j. Consequently the $2s_{1/2}$ and $2p_{1/2}$ states are degenerate. The vacuum oscillations shift the energy up for the $2s_{1/2}$ state, thus removing the degeneracy (Fig. 9.2). This has been observed in the Lamb-Retherford experiment. In this experiment, hydrogen was bombarded by electrons and excited the atom into the $2s_{1/2}$ level. This is a metastable state, since transition to the $1s$ state is forbidden. Microwaves of the frequency corresponding to the $2s_{1/2} - 2p_{1/2}$ energy difference were radiated on the $2s_{1/2}$ state hydrogen inducing the transition $2s_{1/2} - 1s_{1/2}$, a second order transition through the $2p_{1/2}$ state. As the microwave radiation was applied, the corresponding $2s_{1/2} - 1s_{1/2}$ energy line in the spectrum was observed. They found the energy difference of the $2s_{1/2} - 2p_{1/2}$ states to be around $4.3 \cdot 10^{-6}$ eV, which is in reasonable agreement with the above formula.

9.2.4 Natural Broadening of Spectral Lines

Now, we investigate the time-dependence of transition processes. Assume that an atom is in an energy level $E_b^{(a)}$, with no photons in the space. The atom decays down to the energy level $E_a^{(a)}$, with $E_a^{(a)} < E_b^{(a)}$, by emitting a photon with energy $\hbar\omega_0 = E_b^{(a)} - E_a^{(a)}$. We describe the process by

$$|\Psi(t)\rangle = \sum_n c_n(t)\exp(-iE_n t/\hbar)|\psi_n\rangle, \tag{9.104}$$

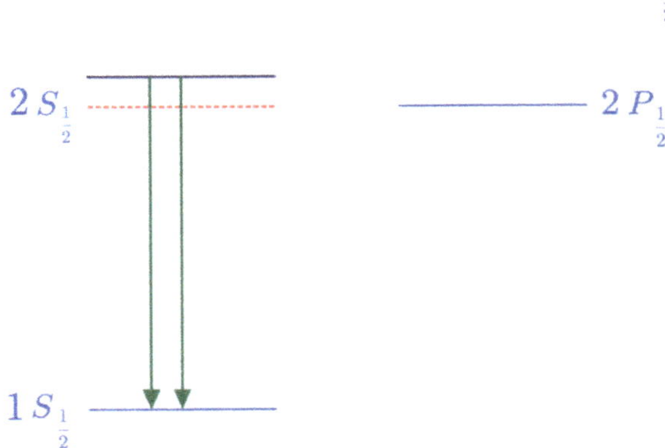

Fig. 9.2 Energy levels of the hydrogen atom. The $2s_{1/2}$ state is shifted up slightly. The microwave radiation corresponding to the energy gap $2s_{1/2} - 2p_{1/2}$ induces a second order transition to the $1s_{1/2}$ ground state

where $|\psi_n\rangle$ is the product of atomic and radiation states and E_n is the sum of atomic and radiation energies. From the time-dependent Schrödinger equation, we have

$$\dot{c}_n(t) = -\frac{i}{\hbar}\sum_l \exp[i(E_n - E_l)t/\hbar]\langle\psi_n|\hat{H}'|\psi_l\rangle c_l(t), \tag{9.105}$$

where $\hat{H}'$ generates the transition.

The coefficients of the initial state and the final states are denoted by $c_{b,0}$ and $c_{a,k}$, respectively, where the first subscript indicates the atomic level and the second one is the photon index. An atom with energy $E_b^{(a)}$ can decay into energy level $E_a^{(a)}$ with several k photon states. On the other hand, an $|a, k\rangle$ final state can be populated only from the $|b, 0\rangle$ state. So, the equations for the time-dependence of these two states are given by

$$\begin{aligned}
\dot{c}_{b,0} &= -\frac{i}{\hbar}\sum_k H'_{b0,ak}c_{a,k}\exp\left[i(E_b^{(a)} - E_a^{(a)} - \hbar\omega_k)t/\hbar\right] \\
&= -\frac{i}{\hbar}\sum_k H'_{b0,ak}c_{a,k}\exp\left[i(\omega_0 - \omega_k)t\right], \\
\dot{c}_{a,k} &= -\frac{i}{\hbar}H'_{ak,b0}c_{b,0}\exp\left[i(E_a^{(a)} - E_b^{(a)} + \hbar\omega_k)t/\hbar\right] \\
&= -\frac{i}{\hbar}H'_{ak,b0}c_{b,0}\exp\left[i(\omega_0 + \omega_k)t\right],
\end{aligned} \tag{9.106}$$

with initial conditions

$$c_{b,0}(0) = 1 \quad \text{and} \quad c_{a,k}(0) = 0. \tag{9.107}$$

It is a reasonable assumption, due to *Wigner* and *Weisskopf*, that $c_{b,0}$ decays exponentially

$$c_{b,0}(t) = \exp(-\gamma t/2). \tag{9.108}$$

Then, we can solve for $c_{a,k}$

$$c_{a,k}(t) = \frac{1}{\hbar} H'_{ak,b0} \frac{\exp[-\gamma t/2 - i(\omega_0 - \omega_k)t] - 1}{\omega_0 - \omega_k - i\gamma/2}. \tag{9.109}$$

From the equation for $\dot{c}_{b,0}(t)$, we have

$$\begin{aligned} -\frac{\gamma}{2} \exp(-\gamma t/2) = &-\frac{i}{\hbar} \sum_k H'_{b0,ak} \frac{1}{\hbar} H'_{ak,b0} \\ &\times \frac{\exp[-\gamma t/2 - i(\omega_0 - \omega_k)t] - 1}{\omega_0 - \omega_k - i\gamma/2} \exp[i(\omega_0 - \omega_k)t], \end{aligned} \tag{9.110}$$

and from that we obtain

$$\gamma = \frac{2i}{\hbar^2} \sum_k |H'_{ak,b0}|^2 \frac{1 - \exp[\gamma t/2 + i(\omega_0 - \omega_k)t]}{\omega_0 - \omega_k - i\gamma/2}, \tag{9.111}$$

or

$$\gamma = \frac{2i}{\hbar^2} \int \mathrm{d}\omega\, \rho(\hbar\omega)\, |H'_{a\omega,b0}|^2 \frac{1 - \exp[\gamma t/2 + i(\omega_0 - \omega)t]}{\omega_0 - \omega - i\gamma/2}, \tag{9.112}$$

where $\rho(\hbar\omega)$ is the density of photon final states.

We can safely assume that $\gamma \ll \omega_0$, and thus we can neglect γ in the integrand

$$\gamma \simeq \frac{2i}{\hbar^2} \int \mathrm{d}\omega\, \rho(\hbar\omega)\, |H'_{a\omega,b0}|^2 \frac{1 - \exp[i(\omega_0 - \omega)t]}{\omega_0 - \omega}. \tag{9.113}$$

We can break the numerator in the integrand into real and imaginary parts

$$1 - \exp[i(\omega_0 - \omega)t] = 1 - \cos(\omega_0 - \omega)t - i\sin(\omega_0 - \omega)t. \tag{9.114}$$

The real part contributes to the imaginary part of γ, which as we can see in Eq. (9.109), shifts the value of ω_0 slightly, and since $\gamma \ll \omega_0$, it can be neglected. So, by calculating the integral with the imaginary part, we obtain the real part of γ. The sine function oscillates heavily if ω is not in the vicinity of ω_0. Therefore the only considerable contribution comes from the $\omega \simeq \omega_0$ region. With the notation $(\omega_0 - \omega)t = x$, we have

$$\int_0^\infty \frac{\sin(\omega_0 - \omega)t}{\omega_0 - \omega} \mathrm{d}\omega = \int_{-\infty}^{\omega_0 t} \frac{\sin x}{x} \mathrm{d}x. \tag{9.115}$$

Assuming a large t limit, and considering that $\sin x/x$ differs considerably from zero only around $x = 0$, we can push the upper limit of the integration to infinity to find that the integral equals π. So, as a result, we have

$$\gamma \simeq \gamma_0 = \frac{2\pi}{\hbar}|H'_{a\omega_0,b0}|^2 \rho(\hbar\omega_0). \tag{9.116}$$

The frequency distribution of the emitted light is given by the large t limit of $|c_{a,k}|^2$. Then, from Eq. (9.109), we obtain

$$|c_{a,\omega}(+\infty)|^2 \simeq \frac{|H'_{a\omega_0,b0}|^2}{\hbar^2} \frac{1}{(\omega-\omega_0)^2 + \gamma^2/4}. \tag{9.117}$$

The frequency distribution peaks at ω_0, and spreads around ω_0 such that at $\omega = \omega_0 \pm \gamma/2$ the probability is halved. So, γ measures the natural broadening of the spectral lines. This broadening comes from quantum mechanics. Other physical phenomena also contribute to the broadening, like the thermal motion of atoms through the Doppler effect and atomic collisions.

9.3 Black-Body Radiation

Quantum Mechanics was born out of the study of black-body radiation. Now, it is interesting to consider this problem again from a somewhat different perspective. Let us consider a system, where N_a atoms are in E_a and N_b atoms are in E_b energy levels with $E_a < E_b$. We assume that there are n_ω photons in the system with frequency $\omega = (E_b - E_a)/\hbar$. We have found before that

$$\Gamma^{emi}_{b\to a} \sim n_\omega + 1 \quad \text{and} \quad \Gamma^{abs}_{a\to b} \sim n_\omega. \tag{9.118}$$

We also assume that the system is in thermal equilibrium, so the average number of *up* transitions equals the average number of *down* transitions

$$N_a n_\omega = N_b(n_\omega + 1). \tag{9.119}$$

In thermal equilibrium, the energy of the atoms follows the Boltzmann distribution

$$N \sim \exp\left(-\frac{E}{kT}\right). \tag{9.120}$$

Therefore,

$$\frac{N_a}{N_b} = \frac{n_\omega + 1}{n_\omega} = \exp\left(\frac{E_b - E_a}{kT}\right) = \exp\left(\frac{\hbar\omega}{kT}\right). \tag{9.121}$$

Reorganizing, we obtain the average number of photons with frequency ω

$$n_\omega = \frac{1}{\exp(\hbar\omega/kT) - 1}. \tag{9.122}$$

This is the *Planck* formula for the photon number distribution.

In Eq. (9.98), we determined the number of photons in the $[\omega, \omega + d\omega]$ interval and $\mathrm{d}\Omega$ solid angle. Considering the two possible polarizations and the 4π total angle, we find the number of states in the $[\omega, \omega + d\omega]$ interval,

$$\mathrm{d}Z(\omega) = V\omega^2 d\omega/(\pi^2 c^3). \tag{9.123}$$

The energy density of black body radiation is the energy over the volume, i.e. the product of the number of photons, the energy of each photon, and the density of states

$$u(\omega) = \frac{V\omega^2/(c^3\pi^2)}{V} n_\omega \hbar\omega = \frac{\hbar\omega^3}{\pi^2 c^3} \frac{1}{\exp(\hbar\omega/kT) - 1}. \tag{9.124}$$

This is the *Planck* formula for the energy density of the black-body radiation.

The maximum of $u(\omega)$ is given by the condition $\mathrm{d}u(\omega)/\mathrm{d}\omega = 0$, which leads to the implicit relation

$$x \frac{1}{1 - \exp(-x)} - 3 = 0, \tag{9.125}$$

where $x = \hbar\omega/(kT)$. The numerical solution of Eq. (9.125) is $x = 2.821439372$. We have found that the frequency, where the radiated energy takes the maximum intesity, is proportional to the temperature

$$\omega_{max} = x \frac{k}{\hbar} T. \tag{9.126}$$

This is the Wien displacement law. Wien derived this relation by using thermodynamic arguments. He concluded that ω/T is adiabatic invariant and the energy distribution should be proportional to $\omega^3 F(\omega/T)$, where F is a universal function.

The total radiation energy is the integral over the whole frequency range

$$u = \int_0^\infty u(\omega)\mathrm{d}\omega = \frac{\hbar}{\pi^2 c^3} \int_0^\infty \mathrm{d}\omega \frac{\omega^3}{\exp(\hbar\omega/kT) - 1} = \sigma T^4, \tag{9.127}$$

where

$$\sigma = \frac{k^4}{\pi^2 c^3 \hbar^3} \int_0^\infty \mathrm{d}x \frac{x^3}{e^x - 1} = \frac{k^4}{\pi^2 c^3 \hbar^3} \frac{\pi^4}{15}. \tag{9.128}$$

This is the *Stefan-Boltzmann* law for the temperature dependence of the total radiated energy.

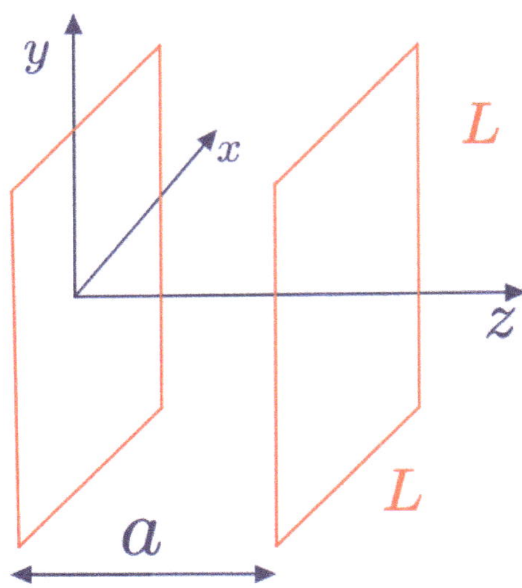

Fig. 9.3 Parallel plates with a small gap

9.4 Casimir Effect

Let us consider two parallel rectangular plates of size L, close to each other, at distance $a << L$ (Fig. 9.3). The energy of the electromagnetic field in the vacuum between the plates is given by

$$E(a) = \sum_{k,s} \frac{1}{2}\hbar\omega_k. \tag{9.129}$$

Here, k can take values that meet the boundary conditions. As we decrease a, fewer k values meet the conditions resulting in a decrease of the energy. This leads to an attractive force between the plates

$$F = -\frac{\partial}{\partial a}E(a). \tag{9.130}$$

Let us calculate the energy of the electromagnetic field between the plates. The plates are perpendicular to the z direction. The frequency of the nodes are

$$\omega_k = c\sqrt{k_x^2 + k_y^2 + k_z^2}, \tag{9.131}$$

with the condition that the field vanishes on the boundaries

$$k_x = \frac{\pi n_x}{L}, \quad k_y = \frac{\pi n_y}{L}, \quad k_z = \frac{\pi n_z}{a}, \quad \text{where} \quad n_x, n_y, n_z = 0, 1, 2, \ldots. \tag{9.132}$$

So, the ground state energy is

$$E = \frac{\hbar}{2} \sum_{n_x,n_y,n_z}^{\infty} \epsilon(\boldsymbol{n})\omega_n, \tag{9.133}$$

where $\epsilon(\boldsymbol{n}) = 2$, unless one of the n_i components vanish, then $\epsilon(\boldsymbol{n}) = 1$. This is a consequence of the Coulomb gauge, $\boldsymbol{\nabla} \cdot \boldsymbol{A} = 0$, where we have

$$k_x A_x + k_y A_y + k_z A_z = 0. \tag{9.134}$$

If one of the components of k is zero, the two others cannot be independent, so, we have only one independent polarization direction. Since L is large, in the $L \to \infty$ limit, we can consider n_x and n_y as continuous variables and the sum can be replaced by integrals. Separating out the $n_z = 0$ node, we have

$$\begin{aligned}
E &\approx \hbar c \int_{n_x=0}^{\infty} \mathrm{d}n_x \int_{n_y=0}^{\infty} \mathrm{d}n_y \left[\frac{1}{2}\sqrt{k_x^2 + k_y^2} + \sum_{n_z=1}^{\infty} \sqrt{k_x^2 + k_y^2 + \frac{\pi^2 n_z^2}{a^2}} \right] \\
&\approx \hbar c \frac{L^2}{\pi^2} \int_0^{\infty} \mathrm{d}k_x \int_0^{\infty} \mathrm{d}k_y \left[\frac{1}{2}\sqrt{k_x^2 + k_y^2} + \sum_{n_z=1}^{\infty} \sqrt{k_x^2 + k_y^2 + \frac{\pi^2 n_z^2}{a^2}} \right] \\
&\approx \hbar c \frac{L^2}{4\pi^2} \int_{-\infty}^{\infty} \mathrm{d}k_x \int_{-\infty}^{\infty} \mathrm{d}k_y \left[\frac{1}{2}\sqrt{k_x^2 + k_y^2} + \sum_{n_z=1}^{\infty} \sqrt{k_x^2 + k_y^2 + \frac{\pi^2 n_z^2}{a^2}} \right] \\
&\approx \hbar c \frac{L^2}{2\pi} \int_0^{\infty} \mathrm{d}k_{||}\, k_{||} \left[\frac{1}{2} k_{||} + \sum_{n_z=1}^{\infty} \sqrt{k_{||}^2 + \frac{\pi^2 n_z^2}{a^2}} \right].
\end{aligned} \tag{9.135}$$

In this derivation, we utilized the fact that E is an even function of k_x and k_y and we changed the integration variable to

$$k_{||} = \sqrt{k_x^2 + k_y^2}. \tag{9.136}$$

Next, we calculate the energy for the L^3 volume and extract the energy of an $L^2 a$ volume. In this case z direction is treated the same way as the x or y directions

$$\begin{aligned}
E_{L^3} \approx \int_{n_x=0}^{\infty} \int_{n_y=0}^{\infty} \int_{n_z=0}^{\infty} \mathrm{d}^3 n\, \hbar\omega_n &= \frac{\hbar c L^3}{(2\pi)^3} \int_{-\infty}^{\infty} \int_{-\infty}^{\infty} \int_{-\infty}^{\infty} \mathrm{d}^3 k\, k \\
&= \frac{\hbar c L^3}{2\pi^2} \int_0^{\infty} \mathrm{d}k\, k^3.
\end{aligned} \tag{9.137}$$

Then, for the vacuum energy of the $L^2 a$ volume, we have

$$E_0 \approx \frac{\hbar c L^2 a}{2\pi^2} \int_0^\infty \mathrm{d}k\, k^3. \tag{9.138}$$

Both energy expressions in Eqs. (9.135) and (9.138) are highly divergent. We may try to regularize them to find out in which order the energies tend to infinity. In Eq. (9.135), in the first part of the integral, we change the integration variable by substitution of $k_{||} \to \omega$, while in the second part of the integral, we apply the substitution $\sqrt{k_{||}^2 + \pi^2 n_z^2/a^2} \to \omega$. We also multiply the integrands with the function $\exp(-\epsilon\omega)$, with $\epsilon > 0$, and at the end of the procedure we take the $\epsilon \to 0$ limit.

So, the energy density of the box with the plates becomes

$$\begin{aligned} \frac{E(\epsilon)}{L^2} &= \frac{\hbar c}{2\pi} \left[\frac{1}{2} \int_0^\infty \mathrm{d}\omega\, \omega^2 \exp(-\epsilon\omega) + \sum_{n_z=1}^\infty \int_{n_z\pi/a}^\infty \mathrm{d}\omega\, \omega^2 \exp(-\epsilon\omega) \right] \\ &= \frac{\hbar c}{2\pi} \frac{\mathrm{d}^2}{\mathrm{d}\epsilon^2} \left[\frac{1}{2} \int_0^\infty \mathrm{d}\omega\, \exp(-\epsilon\omega) + \sum_{n_z=1}^\infty \int_{n_z\pi/a}^\infty \mathrm{d}\omega\, \exp(-\epsilon\omega) \right] \\ &= \frac{\hbar c}{2\pi} \frac{\mathrm{d}^2}{\mathrm{d}\epsilon^2} \left[\frac{1}{2\epsilon} + \frac{1}{\epsilon} \sum_{n_z=1}^\infty \exp(-n_z\pi\epsilon/a) \right]. \end{aligned} \tag{9.139}$$

For the sum in the second term, we have

$$\begin{aligned} \sum_{n_z=1}^\infty \exp(-n_z\pi\epsilon/a) &= \sum_{n_z=0}^\infty (\exp(-\pi\epsilon/a))^{n_z} - 1 \\ &= \frac{1}{1-\exp(-\pi\epsilon/a)} - 1 = \frac{1}{\exp(\pi\epsilon/a) - 1}. \end{aligned} \tag{9.140}$$

Finally, we obtain

$$\frac{E(\epsilon)}{L^2} = \frac{\hbar c}{2\pi} \frac{\mathrm{d}^2}{\mathrm{d}\epsilon^2} \left[\frac{1}{2\epsilon} + \frac{1}{\epsilon} \frac{1}{\exp(\pi\epsilon/a) - 1} \right]. \tag{9.141}$$

In order to evaluate the second term in Eq. (9.141) we recall the series expansion

$$\frac{x}{e^x - 1} = \sum_{k=0}^\infty B_k \frac{x^k}{k!}, \tag{9.142}$$

where B_k's are the Bernoulli numbers. To determine these coefficients, we take the Taylor series of the exponential to find

$$
\begin{aligned}
& x/(x + x^2/2! + x^3/3! + x^4/4! + x^5/5! + \cdots) \\
& = B_0 + B_1 x + B_2 x^2/2! + B_3 x^3/3! + B_4 x^4/4! + \cdots .
\end{aligned}
\tag{9.143}
$$

Then, we can establish the set of equations for the first few Bernoulli numbers

$$
\begin{aligned}
1 &= B_0 \\
0 &= B_0 \frac{1}{2!} + B_1 \\
0 &= B_0 \frac{1}{3!} + B_1 \frac{1}{2!} + B_2 \frac{1}{2!} \\
0 &= B_0 \frac{1}{4!} + B_1 \frac{1}{3!} + B_2 \frac{1}{2!}\frac{1}{2!} + B_3 \frac{1}{3!} \\
0 &= B_0 \frac{1}{5!} + B_1 \frac{1}{4!} + B_2 \frac{1}{2!}\frac{1}{3!} + B_3 \frac{1}{3!}\frac{1}{2!} + B_4 \frac{1}{4!},
\end{aligned}
\tag{9.144}
$$

which can be solved by

$$
B_0 = 1, \quad B_1 = -\frac{1}{2}, \quad B_2 = \frac{1}{6}, \quad B_3 = 0, \quad B_4 = -\frac{1}{30}. \tag{9.145}
$$

This means that

$$
\frac{1}{e^x - 1} = \sum_{k=0}^{\infty} \frac{B_k}{k!} x^{k-1}. \tag{9.146}
$$

Consequently, Eq. (9.141) becomes

$$
\begin{aligned}
\frac{E(\epsilon)}{L^2} &= \frac{\hbar c}{2\pi} \frac{\mathrm{d}^2}{\mathrm{d}\epsilon^2} \left[\frac{1}{2\epsilon} + \frac{1}{\epsilon} \sum_{k=0}^{\infty} \frac{B_k}{k!} \left(\frac{\pi\epsilon}{a} \right)^{k-1} \right] \\
&= \frac{\hbar c}{2\pi} \frac{\mathrm{d}^2}{\mathrm{d}\epsilon^2} \left[B_0 \frac{a}{\pi\epsilon^2} + \frac{1}{\epsilon}\left(B_1 + \frac{1}{2} \right) + \frac{1}{2} B_2 \frac{\pi}{a} + \frac{1}{24} B_4 \frac{\pi^3 \epsilon^2}{a^3} + \cdots \right] \\
&= \frac{\hbar c}{2\pi} \left[\frac{6a}{\pi\epsilon^4} B_0 + \frac{1}{12} B_4 \frac{\pi^3}{a^3} + \cdots \right] = \hbar c \frac{3a}{\pi^2\epsilon^4} - \hbar c \frac{1}{720} \frac{\pi^2}{a^3} + \cdots .
\end{aligned}
\tag{9.147}
$$

We should also introduce the same substitution and cutoff procedure for the energy of the no-plate field in Eq. (9.138)

$$
\begin{aligned}
\frac{E_0(\epsilon)}{L^2} &= \frac{\hbar c a}{2\pi^2} \int_0^{\infty} \mathrm{d}\omega\, \omega^3 \exp(-\epsilon\omega) \\
&= -\frac{\hbar c a}{2\pi^2} \frac{\mathrm{d}^3}{\mathrm{d}\epsilon^3} \int_0^{\infty} \mathrm{d}\omega\, \exp(-\epsilon\omega) \\
&= -\frac{\hbar c a}{2\pi^2} \frac{\mathrm{d}^3}{\mathrm{d}\epsilon^3} \frac{1}{\epsilon} = \hbar c \frac{3a}{\pi^2\epsilon^4}.
\end{aligned}
\tag{9.148}
$$

We can see that both energy terms in Eqs. (9.147) and (9.148) are infinite in the $\epsilon \to 0$ limit, but by taking difference, the divergent terms drop out and the difference is finite

$$\lim_{\epsilon\to 0} \Delta E(\epsilon) = \lim_{\epsilon\to 0}\left(\frac{E(\epsilon)}{L^2} - \frac{E_0(\epsilon)}{L^2}\right) = -\hbar c \frac{1}{720}\frac{\pi^2}{a^3}. \tag{9.149}$$

Then the force per unit area between the parallel conducting plates is

$$F = -\frac{\partial}{\partial a}\Delta E = -\frac{\partial}{\partial a}\left(-\hbar c \frac{1}{720}\frac{\pi^2}{a^3}\right) = -\frac{\hbar c \pi^2}{240 a^4}. \tag{9.150}$$

This is the *Casimir* force, a force from "nothing". The field with no photons is "empty", but between the parallel plates the field is "even more empty". Then the energy difference leads to an observable attractive force.

Chapter 10
Scattering Theory

10.1 Asymptotic States and the Scattering Operator

10.1.1 Møller Operators

Quantum scattering theory is concerned with situations where the interaction between the collision partners is described by a potential. The Hamiltonian governing the collision process is given by

$$\hat{H} = \hat{H}_0 + \hat{V}, \tag{10.1}$$

where $\hat{H}_0$ describes the free motion of the collision partners and $\hat{V}(r)$ is a short-range potential, approaching zero faster than r^{-3} as $r \to \infty$. This condition rules out the Coulomb potential, which needs a special treatment, but applies to a number of other problems in atomic and nuclear physics.

The time evolution of a quantum state is described by the time dependent Schrödinger equation

$$i\hbar \frac{\mathrm{d}}{\mathrm{d}t} |\psi(t)\rangle = \hat{H} |\psi(t)\rangle. \tag{10.2}$$

The formal solution of Eq. (10.2) is given by

$$|\psi(t)\rangle = \hat{U}(t) |\psi(0)\rangle, \tag{10.3}$$

where

$$\hat{U}(t) = \exp(-i\hat{H}t/\hbar) \tag{10.4}$$

is the unitary time evolution operator in the Schödinger picture.

In the remote past, as $t \to -\infty$, and in the distant future, as $t \to \infty$, the state $|\psi(t)\rangle$ asymptotically approaches the solutions of the unperturbed Hamiltonian $\hat{H}_0$. We suppose that there exist two states $|\psi_{in}\rangle \in \mathcal{H}$ and $|\psi_{out}\rangle \in \mathcal{H}$, called the *in*-asymptote and the *out*-asymptote, such that

Z. Papp, *Mastering Quantum Mechanics*,
https://doi.org/10.1007/978-3-032-09011-9_10

$$\hat{U}(t)|\psi\rangle \to \hat{U}_0(t)|\psi_{in}\rangle, \qquad \text{for} \quad t \to -\infty,$$
$$\hat{U}(t)|\psi\rangle \to \hat{U}_0(t)|\psi_{out}\rangle, \qquad \text{for} \quad t \to \infty, \tag{10.5}$$

where

$$\hat{U}_0(t) = \exp(-i\hat{H}_0 t/\hbar) \tag{10.6}$$

is the time evolution operator corresponding to the free Hamiltonian. We can also write these limits as

$$|\psi\rangle \underset{t\to-\infty}{\longrightarrow} \exp(i\hat{H}t/\hbar)\exp(-i\hat{H}_0 t/\hbar)|\psi_{in}\rangle,$$
$$|\psi\rangle \underset{t\to+\infty}{\longrightarrow} \exp(i\hat{H}t/\hbar)\exp(-i\hat{H}_0 t/\hbar)|\psi_{out}\rangle. \tag{10.7}$$

These limits should be understood in the sense of the *strong limit* or convergence in norm. The state $|\psi(t)\rangle \in \mathcal{H}$ converges strongly to a state $|\phi\rangle \in \mathcal{H}$ as $t \to \pm\infty$ if

$$\|\psi(t) - \phi\| \underset{t\to\pm\infty}{\longrightarrow} 0. \tag{10.8}$$

On the other hand, the *weak limit* means convergence in terms of matrix elements. The state $|\psi(t)\rangle \in \mathcal{H}$ converges weakly to $|\phi\rangle \in \mathcal{H}$ as $t \to \pm\infty$ if $\langle\chi|\psi(t)\rangle \to \langle\chi|\phi\rangle$ for every $|\chi\rangle \in \mathcal{H}$. Strong convergence implies weak convergence, but the converse is not generally true.

The $\hat{\Omega}^{(\pm)}$ operators, called the *Møller operators* or *wave operators*, are a pair of operators defined by the limits

$$\hat{\Omega}^{(\pm)} = \lim_{t\to\mp\infty} \hat{U}^\dagger(t)\hat{U}_0(t) = \lim_{t\to\mp\infty} \exp(i\hat{H}t/\hbar)\exp(-i\hat{H}_0 t/\hbar). \tag{10.9}$$

The operators $\hat{\Omega}^{(\pm)}$, being limits of unitary operators, are isometric, or norm-preserving operators, $\|\hat{\Omega}^{(\pm)}\phi\| = \|\phi\|$, for all $|\phi\rangle \in \mathcal{H}$. They map the Hilbert space of asymptotic states, which is spanned by the eigenstates of $\hat{H}_0$, to $\mathcal{H}_c$, the states spanned by the continuous eigenstates of $\hat{H}$. The Hamiltonian $\hat{H}$ may have bound eigenstates with discrete energies, which vanish at asymptotic distances, and thus they cannot develop from an *in* state and cannot develop into an *out* state. Therefore, $\hat{\Omega}^{(\pm)}$ maps the Hilbert space spanned by the eigenstates of $\hat{H}_0$ only onto a part of the Hilbert space $\mathcal{H}_c \subset \mathcal{H}$. So, as we can see in Fig. 10.1, we can write $\hat{\Omega}^{(\pm)\dagger}\hat{\Omega}^{(\pm)} = 1$, which is an identity map $\mathcal{H} \to \mathcal{H}$. The converse, however, is in general not true, $\hat{\Omega}^{(\pm)}\hat{\Omega}^{(\pm)\dagger} \neq 1$. They preserve the length, but map subspace into subspace, i.e. $\mathcal{H}_c \to \mathcal{H}_c$. So, $\hat{\Omega}^{(\pm)}$ is not unitary, only isometric. In the absence of bound states, $\hat{\Omega}^{(\pm)}$ is unitary. With the help of Møller operators, we can write the limits as

$$\lim_{t\to\mp\infty} \|\psi - \exp(i\hat{H}t/\hbar)\exp(-i\hat{H}_0 t/\hbar)\psi_{in/out}\| = 0, \tag{10.10}$$

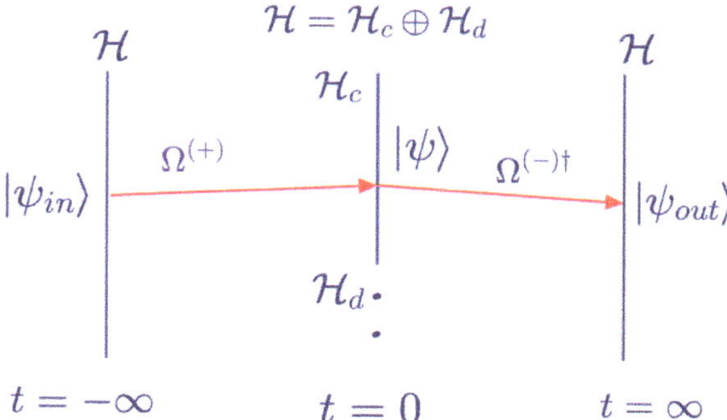

Fig. 10.1 The Møller operator $\Omega^{(+)}$ maps the *in* states of $\mathcal{H}$ into $\mathcal{H}_c$ and $\Omega^{(-)\dagger}$ maps $\mathcal{H}_c$ into the *out* states of $\mathcal{H}$

or

$$|\psi\rangle = \hat{\Omega}^{(\pm)}|\psi_{in/out}\rangle. \tag{10.11}$$

For any $|\psi\rangle \in \mathcal{H}_c$, there exist asymptotic states $|\psi_{in}\rangle \in \mathcal{H}$ and $|\psi_{out}\rangle \in \mathcal{H}$ such that

$$|\psi_{in/out}\rangle = \hat{\Omega}^{(\pm)\dagger}|\psi\rangle, \tag{10.12}$$

i.e. the adjoint maps $\hat{\Omega}^{(\pm)\dagger}$ bring the state $|\psi\rangle$ to the asymptotic states $|\psi_{in/out}\rangle$.

Suppose, the Hamiltonian $\hat{H}$ supports bound states $|\psi_n\rangle$, which are square integrable and vanish at asymptotic distances. Therefore, the adjoint mapping brings them to zero

$$\hat{\Omega}^{(\pm)\dagger}|\psi_n\rangle = 0. \tag{10.13}$$

We can also see that for a scattering state $|\psi\rangle$,

$$\langle\psi_n|\psi\rangle = \langle\psi_n|\hat{\Omega}^{(\pm)}\psi_{in/out}\rangle = \langle\hat{\Omega}^{(\pm)\dagger}\psi_n|\psi_{in/out}\rangle = 0, \tag{10.14}$$

i.e. the bound and scattering states are orthogonal. The corresponding subspaces of bound states with discrete spectrum and scattering states with continuous spectrum are called $\mathcal{H}_d$ and $\mathcal{H}_c$, respectively, and $\mathcal{H} = \mathcal{H}_d \oplus \mathcal{H}_c$.

10.1.2 Scattering States

The Møller operators link states of the remote past and distant future to the present. Therefore,

$$|\psi_{out}\rangle = \hat{\Omega}^{(-)\dagger}|\psi\rangle = \hat{\Omega}^{(-)\dagger}\hat{\Omega}^{(+)}|\psi_{in}\rangle = \hat{S}|\psi_{in}\rangle, \tag{10.15}$$

where

$$\hat{S} = \hat{\Omega}^{(-)\dagger}\hat{\Omega}^{(+)} \tag{10.16}$$

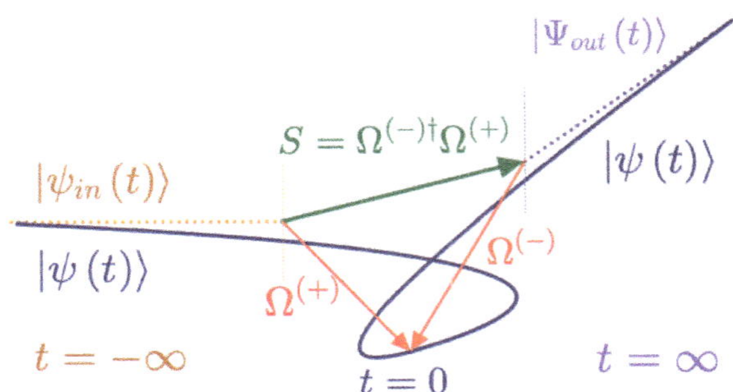

Fig. 10.2 The Møller operators connect the asymptotic states to the scattering state

is the *scattering operator*. It is a norm-preserving operator that maps the *in* Hilbert space to the *out* Hilbert space (Fig. 10.2), i.e. the $\hat{S}$ operator is unitary. If the quantum system enters into the interaction with asymptote $|\psi_{in}\rangle$, then it leaves with asymptote $|\psi_{out}\rangle = \hat{S}|\psi_{in}\rangle$. Since only the asymptotic free motion can be observed in a scattering experiment, the single operator $\hat{S}$ contains all the information of experimental interest and experimentally accessible. If we know how to calculate $\hat{S}$, the scattering problem is solved.

The Møller operators are invariant with respect to a finite time shift

$$\begin{aligned}\hat{\Omega}^{(\pm)} &= \lim_{t\to\mp\infty} \exp(i\hat{H}t/\hbar)\exp(-i\hat{H}_0 t/\hbar) \\ &= \lim_{t\to\mp\infty} \exp(i\hat{H}(t+\tau)/\hbar)\exp(-i\hat{H}_0(t+\tau)/\hbar) \\ &= \exp(i\hat{H}\tau/\hbar)\hat{\Omega}^{(\pm)}\exp(-i\hat{H}_0\tau/\hbar),\end{aligned} \tag{10.17}$$

hence,

$$\exp(-i\hat{H}\tau/\hbar)\hat{\Omega}^{(\pm)} = \hat{\Omega}^{(\pm)}\exp(-i\hat{H}_0\tau/\hbar). \tag{10.18}$$

Differentiating this with respect to τ, and then setting $\tau = 0$, we obtain the *intertwining relations*

$$\hat{H}\hat{\Omega}^{(\pm)} = \hat{\Omega}^{(\pm)}\hat{H}_0, \quad \text{or} \quad \hat{\Omega}^{(\pm)\dagger}\hat{H} = \hat{H}_0\hat{\Omega}^{(\pm)\dagger}. \tag{10.19}$$

With the help of these relations, we obtain

$$\hat{S}\hat{H}_0 = \hat{\Omega}^{(-)\dagger}\hat{\Omega}^{(+)}\hat{H}_0 = \hat{\Omega}^{(-)\dagger}\hat{H}\hat{\Omega}^{(+)} = \hat{H}_0\hat{\Omega}^{(-)\dagger}\hat{\Omega}^{(+)} = \hat{H}_0\hat{S}, \tag{10.20}$$

i.e. the operator $\hat{S}$ commutes with $\hat{H}_0$. Since $|\psi_{out}\rangle = \hat{S}|\psi_{in}\rangle$,

$$E_{out} = \langle\psi_{out}|\hat{H}_0|\psi_{out}\rangle = \langle\psi_{in}|\hat{S}^\dagger\hat{H}_0\hat{S}|\psi_{in}\rangle = \langle\psi_{in}|\hat{H}_0|\psi_{in}\rangle = E_{in}, \tag{10.21}$$

meaning that the energy is conserved during the scattering process.

Now, we assume that the free particle Hamiltonian is $\hat{H}_0 = \hat{\boldsymbol{p}}^2/2m$ with energy eigenvalue $E = \hbar^2 k^2/2m$. Its eigenstates can be chosen to be the plane wave continuum basis

$$\psi_{\boldsymbol{k}}(\boldsymbol{r}) = \langle \boldsymbol{r}|\boldsymbol{k}\rangle = \frac{1}{(2\pi)^{3/2}} \exp(i\boldsymbol{kr}), \tag{10.22}$$

and with that, we have the resolution of the identity

$$I = \int \mathrm{d}\boldsymbol{k}\, |\boldsymbol{k}\rangle\langle \boldsymbol{k}|. \tag{10.23}$$

We can expand the *in* and *out* wave packets

$$|\psi_{in/out}\rangle = \int \mathrm{d}\boldsymbol{k}\, |\boldsymbol{k}\rangle\langle \boldsymbol{k}|\psi_{in/out}\rangle = \int \mathrm{d}\boldsymbol{k}\, \psi_{in/out}(\boldsymbol{k})|\boldsymbol{k}\rangle, \tag{10.24}$$

and find that

$$|\psi_{out}\rangle = \hat{S}|\psi_{in}\rangle = \hat{S}\int \mathrm{d}\boldsymbol{k}\, |\boldsymbol{k}\rangle\langle \boldsymbol{k}|\psi_{in}\rangle = \int \mathrm{d}\boldsymbol{k}\, \hat{S}|\boldsymbol{k}\rangle\psi_{in}(\boldsymbol{k}), \tag{10.25}$$

or

$$\langle \boldsymbol{k}'|\psi_{out}\rangle = \psi_{out}(\boldsymbol{k}') = \int \mathrm{d}\boldsymbol{k}\, \langle \boldsymbol{k}'|\hat{S}|\boldsymbol{k}\rangle\psi_{in}(\boldsymbol{k}). \tag{10.26}$$

We do not really know how the asymptotic states are prepared, but as we can see in Eq. (10.26), the way $|\psi_{in}\rangle$ has been prepared, is carried over to $|\psi_{out}\rangle$, which we again cannot really measure. Therefore, we should formulate the scattering processes in terms of $|\boldsymbol{k}\rangle$ and $\langle \boldsymbol{k}'|\hat{S}|\boldsymbol{k}\rangle$. This means that we study how the basis states $|\boldsymbol{k}\rangle$ are mapped during the scattering process and try to infer the physically relevant information from that mapping. Then we can write

$$\langle \boldsymbol{k}'|\hat{S}|\boldsymbol{k}\rangle = \langle \boldsymbol{k}'|\hat{\Omega}^{(-)\dagger}\hat{\Omega}^{(+)}|\boldsymbol{k}\rangle = \langle \boldsymbol{k}'-|\boldsymbol{k}+\rangle, \tag{10.27}$$

where

$$|\boldsymbol{k}\pm\rangle = \hat{\Omega}^{(\pm)}|\boldsymbol{k}\rangle \tag{10.28}$$

are called the *scattering states*. The $|\boldsymbol{k}\pm\rangle$ states are eigenstates of the Hamiltonian $\hat{H}$

$$\hat{H}|\boldsymbol{k}\pm\rangle = \hat{H}\hat{\Omega}^{(\pm)}|\boldsymbol{k}\rangle = \hat{\Omega}^{(\pm)}\hat{H}_0|\boldsymbol{k}\rangle = \hat{\Omega}^{(\pm)}E|\boldsymbol{k}\rangle = E\hat{\Omega}^{(\pm)}|\boldsymbol{k}\rangle = E|\boldsymbol{k}\pm\rangle. \tag{10.29}$$

The scattering states are normalized the same way as the free states

$$\langle \boldsymbol{k}'\pm|\boldsymbol{k}\pm\rangle = \langle \boldsymbol{k}|\hat{\Omega}^{(\pm)\dagger}\hat{\Omega}^{(\pm)}|\boldsymbol{k}\rangle = \langle \boldsymbol{k}'|\boldsymbol{k}\rangle = \delta^{(3)}(\boldsymbol{k}'-\boldsymbol{k}). \tag{10.30}$$

10.1.3 Resolvent Operators

To deal with the Møller operators we are going to apply the Euler- or ε-limit

$$\begin{aligned}
\lim_{\varepsilon\to 0}\int_{-\infty}^{0} \mathrm{d}t\, \varepsilon \exp(\varepsilon t) g(t) &= \lim_{\varepsilon\to 0}\int_{-\infty}^{0} \mathrm{d}t\, \frac{\mathrm{d}}{\mathrm{d}t}[\exp(\varepsilon t)]\, g(t) \\
&= \lim_{\varepsilon\to 0}\left[[\exp(\varepsilon t) g(t)]_{-\infty}^{0} - \int_{-\infty}^{0} \mathrm{d}t\, \exp(\varepsilon t)\frac{\mathrm{d}g}{\mathrm{d}t}\right] \\
&= \lim_{\varepsilon\to 0}\left[[\exp(\varepsilon t) g(t)]_{-\infty}^{0}\right] - \int_{-\infty}^{0} \mathrm{d}t\, \frac{\mathrm{d}g}{\mathrm{d}t} \\
&= g(0) - 0 - g(0) + g(-\infty) = \lim_{t\to-\infty} g(t)\,. \qquad (10.31)
\end{aligned}$$

In the ε-limit, the Møller operators become

$$\begin{aligned}
\hat{\Omega}^{(\pm)} &= \lim_{t\to\mp\infty} \exp(i\hat{H}t/\hbar)\exp(-i\hat{H}_0 t/\hbar) \\
&= \lim_{\varepsilon\to 0} \pm\varepsilon/\hbar \int_{\mp\infty}^{0} \mathrm{d}t\;\; \exp(\pm\varepsilon t/\hbar)\exp(i\hat{H}t/\hbar)\exp(-i\hat{H}_0 t/\hbar) \\
&= \lim_{\varepsilon\to 0} \pm\varepsilon/\hbar \int_{\mp\infty}^{0} \mathrm{d}t\;\; \exp(i\hat{H}t/\hbar)\exp(-i(\hat{H}_0 \pm i\varepsilon)t/\hbar), \qquad (10.32)
\end{aligned}$$

and applying them on a state $|\boldsymbol{k}\rangle$, we obtain

$$\begin{aligned}
|\boldsymbol{k}\pm\rangle = \hat{\Omega}^{(\pm)}|\boldsymbol{k}\rangle &= \lim_{\varepsilon\to 0} \pm\varepsilon/\hbar \int_{\mp\infty}^{0} \mathrm{d}t\;\; \exp(i\hat{H}t/\hbar)\exp(-i(E \pm i\varepsilon)t/\hbar)|\boldsymbol{k}\rangle \\
&= \lim_{\varepsilon\to 0} \pm\varepsilon/\hbar \int_{\mp\infty}^{0} \mathrm{d}t\;\; \exp(-i(E \pm i\varepsilon - \hat{H})t/\hbar)|\boldsymbol{k}\rangle \\
&= \lim_{\varepsilon\to 0} \pm i\varepsilon(E \pm i\varepsilon - \hat{H})^{-1}|\boldsymbol{k}\rangle. \qquad (10.33)
\end{aligned}$$

Here, we have the Green's, or resolvent, operator $G(z)$, defined as the inverse operator

$$\hat{G}(z) = (z - \hat{H})^{-1}, \qquad (10.34)$$

for any z, real or complex, for which the inverse exists. Then, Eq. (10.33) becomes

$$|\boldsymbol{k}\pm\rangle = \hat{\Omega}^{(\pm)}|\boldsymbol{k}\rangle = \lim_{\varepsilon\to 0} \pm i\varepsilon \hat{G}(E_k \pm i\varepsilon)|\boldsymbol{k}\rangle. \quad (10.35)$$

In a similar way we can define the adjoint of the Møller operators

$$\begin{aligned}
\hat{\Omega}^{(\pm)\dagger} &= \lim_{t\to\mp\infty} \exp(i\hat{H}_0 t/\hbar)\exp(-i\hat{H}t/\hbar) \\
&= \lim_{\varepsilon\to 0} \pm\varepsilon/\hbar \int_{\mp\infty}^{0} \mathrm{d}t\ \exp(\pm\varepsilon t/\hbar)\exp(i\hat{H}_0 t/\hbar)\exp(-i\hat{H}t/\hbar) \\
&= \lim_{\varepsilon\to 0} \pm\varepsilon/\hbar \int_{\mp\infty}^{0} \mathrm{d}t\ \exp(i\hat{H}_0 t/\hbar)\exp(-i(\hat{H}\pm i\varepsilon)t/\hbar),
\end{aligned} \quad (10.36)$$

and applying to $|\boldsymbol{k}\pm\rangle$, we get

$$\begin{aligned}
|\boldsymbol{k}\rangle = \hat{\Omega}^{(\pm)\dagger}|\boldsymbol{k}\pm\rangle &= \lim_{\varepsilon\to 0} \pm\varepsilon/\hbar \int_{\mp\infty}^{0} \mathrm{d}t\ \exp(i\hat{H}_0 t/\hbar)\exp(-i(E\pm i\varepsilon)t/\hbar)|\boldsymbol{k}\pm\rangle \\
&= \lim_{\varepsilon\to 0} \pm\varepsilon/\hbar \int_{\mp\infty}^{0} \mathrm{d}t\ \exp(-i(E\pm i\varepsilon-\hat{H}_0)t/\hbar)|\boldsymbol{k}\pm\rangle \\
&= \lim_{\varepsilon\to 0} \pm i\varepsilon(E\pm i\varepsilon-\hat{H}_0)^{-1}|\boldsymbol{k}\pm\rangle \\
&= \lim_{\varepsilon\to 0} \pm i\varepsilon\hat{G}_0(E\pm i\varepsilon)|\boldsymbol{k}\pm\rangle,
\end{aligned} \quad (10.37)$$

where $\hat{G}_0(z) = (z-\hat{H}_0)^{-1}$ is the free Green's operator.

We can easily derive some useful relations for Green's operators. To derive the first resolvent relation, or Hilbert identity, we consider

$$z - z' = (z-\hat{H}) - (z'-\hat{H}) = \hat{G}^{-1}(z) - \hat{G}^{-1}(z'). \quad (10.38)$$

Then multiplying by $\hat{G}(z')\cdot\hat{G}(z)$ we obtain

$$\hat{G}(z') - \hat{G}(z) = (z'-z)\,\hat{G}(z')\cdot\hat{G}(z). \quad (10.39)$$

Rearranging, we find

$$\frac{\hat{G}(z')-\hat{G}(z)}{z'-z} = \hat{G}(z')\cdot\hat{G}(z), \quad (10.40)$$

and taking the $z'\to z$ limit we arrive at the relation

$$\frac{\mathrm{d}}{\mathrm{d}z}\hat{G}(z) = \hat{G}^2(z). \quad (10.41)$$

This means that $\hat{G}(z)$ is differentiable, i.e. an analytic function of z, whenever it exists.

Analogously to the identity for numbers,

$$\frac{1}{A} - \frac{1}{B} = \frac{B - A}{AB}, \tag{10.42}$$

we have identities for operators

$$\begin{aligned} \hat{A}^{-1} &= \hat{B}^{-1} + \hat{B}^{-1}(\hat{B} - \hat{A})\hat{A}^{-1} \\ &= \hat{B}^{-1} + \hat{A}^{-1}(\hat{B} - \hat{A})\hat{B}^{-1}. \end{aligned} \tag{10.43}$$

If we take

$$\hat{A} = z - \hat{H} \quad \text{and} \quad \hat{B} = z - \hat{H}_0, \tag{10.44}$$

we obtain the second resolvent relation

$$\hat{G}(z) = \hat{G}_0(z) + \hat{G}_0(z)\hat{V}\hat{G}(z), \tag{10.45}$$

or

$$\hat{G}(z) = \hat{G}_0(z) + \hat{G}(z)\hat{V}\hat{G}_0(z). \tag{10.46}$$

Green's operators possess several interesting properties. They satisfy the operator equation

$$(z - \hat{H})\hat{G}(z) = \hat{G}(z)(z - \hat{H}) = I. \tag{10.47}$$

Assume that the spectrum of $\hat{H}$ is discrete, with eigenvalues E_n and normalized eigenstates $|E_n\rangle$, such that $\sum_n |E_n\rangle\langle E_n| = I$. Then, we obtain the representation

$$\hat{G}(z) = (z - \hat{H})^{-1} \sum_n |E_n\rangle\langle E_n| = \sum_n \frac{|E_n\rangle\langle E_n|}{z - E_n}, \tag{10.48}$$

which is well defined if $z \neq E_n$. We can see that $\hat{G}(z)$ has poles at the eigenvalues of $\hat{H}$ and the residuum is the projection operator $\hat{P}_n = |E_n\rangle\langle E_n|$.

If $\hat{H}$ has a continuous spectrum, the picture is slightly different. With $\hat{H}|\lambda\rangle = \lambda|\lambda\rangle$, we find

$$\hat{G}(z) = (z - \hat{H})^{-1} \int \mathrm{d}\lambda\, |\lambda\rangle\langle\lambda| = \int \mathrm{d}\lambda\, \frac{|\lambda\rangle\langle\lambda|}{z - \lambda}. \tag{10.49}$$

Here, $\hat{G}(z)$ has a continuous singularity, i.e. a branch cut, for the range of integration at $z = \lambda$. We can calculate the discontinuity as

$$\begin{aligned}\operatorname{disc}(\hat{G}(E)) &= \lim_{\varepsilon\to+0}[\hat{G}(E+i\varepsilon)-\hat{G}(E-i\varepsilon)] \\ &= \lim_{\varepsilon\to+0}\int \mathrm{d}\lambda \left(\frac{1}{E-\lambda+i\varepsilon}-\frac{1}{E-\lambda-i\varepsilon}\right)|\lambda\rangle\langle\lambda|.\end{aligned} \tag{10.50}$$

We should recall that

$$\lim_{\varepsilon\to+0}\int \mathrm{d}x\,\frac{f(x)}{x-x_0\pm i\varepsilon} = \mathcal{P}\int \mathrm{d}x\,\frac{f(x)}{x-x_0} \mp i\pi f(x_0), \tag{10.51}$$

where $\mathcal{P}$ is the Cauchy principal value integral, which, for a singularity at x_0, is defined by

$$\lim_{\varepsilon\to 0^+}\left[\int^{x_0-\varepsilon} f(x)\,\mathrm{d}x + \int_{x_0+\varepsilon} f(x)\,\mathrm{d}x\right], \tag{10.52}$$

and abbreviated sometimes as

$$\lim_{\varepsilon\to+0}\frac{1}{x-x_0\pm i\epsilon} = \mathcal{P}\frac{1}{x-x_0} \mp i\pi\delta(x-x_0). \tag{10.53}$$

Then, for Eq. (10.50), we find that

$$\begin{aligned}\operatorname{disc}(\hat{G}(E)) &= \int \mathrm{d}\lambda \left\{\left[\mathcal{P}(E-\hat{H})^{-1} - i\pi\delta(E-\hat{H})\right] - \left[\mathcal{P}(E-\hat{H})^{-1} + i\pi\delta(E-\hat{H})\right]\right\}|\lambda\rangle\langle\lambda| \\ &= -2i\pi\int \mathrm{d}\lambda\;\delta(E-\hat{H})|\lambda\rangle\langle\lambda| = -2\pi i|E\rangle\langle E|,\end{aligned} \tag{10.54}$$

i.e. the discontinuity of $\hat{G}(E)$ is proportional to the projection $|E\rangle\langle E|$.

10.1.4 Integral Equations for Scattering States

To derive equations for scattering states, we apply one of the resolvent identities, Eq. (10.46),

$$\begin{aligned}|\boldsymbol{k}\pm\rangle &= \lim_{\varepsilon\to 0}\pm i\varepsilon\hat{G}(E\pm i\varepsilon)|\boldsymbol{k}\rangle \\ &= \lim_{\varepsilon\to 0}\pm i\varepsilon[\hat{G}_0(E\pm i\varepsilon)+\hat{G}(E\pm i\varepsilon)\hat{V}\hat{G}_0(E\pm i\varepsilon)]|\boldsymbol{k}\rangle \\ &= \lim_{\varepsilon\to 0}\frac{\pm i\varepsilon}{E\pm i\varepsilon - E}[1+\hat{G}(E\pm i\varepsilon)\hat{V}]|\boldsymbol{k}\rangle \\ &= |\boldsymbol{k}\rangle + \hat{G}(E\pm i0)\hat{V}|\boldsymbol{k}\rangle,\end{aligned} \tag{10.55}$$

and by using the other form, Eq. (10.45), we obtain

$$
\begin{aligned}
|\boldsymbol{k}\pm\rangle &= \lim_{\varepsilon\to 0} \pm i\varepsilon[\hat{G}_0(E\pm i\varepsilon) + \hat{G}_0(E\pm i\varepsilon)\hat{V}\hat{G}(E\pm i\varepsilon)]|\boldsymbol{k}\rangle \\
&= \lim_{\varepsilon\to 0}\left[\frac{\pm i\varepsilon}{E\pm i\varepsilon - E}|\boldsymbol{k}\rangle + (\pm i\varepsilon)\hat{G}_0(E\pm i\varepsilon)\hat{V}\hat{G}(E\pm i\varepsilon)|\boldsymbol{k}\rangle\right] \\
&= |\boldsymbol{k}\rangle + \hat{G}_0(E\pm i0)\hat{V}|\boldsymbol{k}\pm\rangle. \qquad (10.56)
\end{aligned}
$$

Equation (10.56) is the *Lippmann-Schwinger equation* for scattering states. We can see in Eq. (10.55) that the full scattering state is a sum of an incident plane wave $|\boldsymbol{k}\rangle$ and a perturbation $\hat{G}\hat{V}|\boldsymbol{k}\rangle$, caused by the potential $\hat{V}$. We refer to these equations as integral equations because $\hat{H}_0$ is usually represented by a differential operator and therefore its inverse operator, the Green's operator, involves an integration.

From Eqs. (10.56) and (10.37) we can find that

$$
|\boldsymbol{k}\rangle = \hat{\Omega}^{(\pm)\dagger}|\boldsymbol{k}\pm\rangle = (1 - \hat{G}_0(E\pm i0)\hat{V})|\boldsymbol{k}\pm\rangle. \qquad (10.57)
$$

As we have found before, the inverse mapping $\hat{\Omega}^{(\pm)\dagger}$ defines the bound states

$$
0 = \hat{\Omega}^{(\pm)\dagger}|\psi_n\rangle = (1 - \hat{G}_0(E)\hat{V})|\psi_n\rangle, \qquad (10.58)
$$

which represent a homogeneous Lippmann-Schwinger equation for $|\psi_n\rangle$

$$
|\psi_n\rangle = \hat{G}_0(E)\hat{V}|\psi_n\rangle. \qquad (10.59)
$$

To summarize, we obtained the inhomogeneous and homogeneous Lippmann-Schwinger equations for scattering and bound states, respectively. From Eq. (10.58), we can see that the homogeneous equation is solvable if the Fredholm determinant vanishes

$$
D(E) = \det(1 - \hat{G}_0(E)\hat{V}) = 0\,. \qquad (10.60)
$$

A formal solution of the inhomogeneous equation is given by

$$
|\boldsymbol{k}\pm\rangle = (1 - \hat{G}_0(E\pm i0)\hat{V})^{-1}|\boldsymbol{k}\rangle, \qquad (10.61)
$$

and the operator $(1 - \hat{G}_0(E)\hat{V})$ is invertible if the Fredholm determinant $D(E) \neq 0$. The Fredholm determinant is either zero or non-zero. Therefore, if the homogeneous equation has a non-trivial solution, i.e. not identically zero, then the corresponding inhomogeneous equation is either not solvable or its solution is not unique. Conversely, if the inhomogeneous equation has a unique solution for any inhomogeneity, the corresponding homogenous equation has only a trivial solution. This is the *Fredholm alternative theorem.* As a consequence, the homogeneous equation has no solution at scattering energies, hence there are no bound states embedded in the continuum.

10.1.5 Structure of the Scattering Operator

The primary aim of scattering theory is to determine the matrix elements of the $\hat{S}$ operator between free states

$$\begin{aligned} S_{k'k} &= \langle \boldsymbol{k}'|\hat{S}|\boldsymbol{k}\rangle = \langle \boldsymbol{k}'|\hat{\Omega}^{(-)\dagger}\hat{\Omega}^{(+)}|\boldsymbol{k}\rangle = \langle \boldsymbol{k}'|\hat{\Omega}^{(-)\dagger}|\boldsymbol{k}+\rangle \\ &= \lim_{t\to\infty} \langle \boldsymbol{k}'|\exp(iH_0t/\hbar)\exp(-iHt/\hbar)|\boldsymbol{k}+\rangle = \lim_{t\to\infty} \exp[i(E'-E)t/\hbar]\langle \boldsymbol{k}'|\boldsymbol{k}+\rangle \\ &= \lim_{t\to\infty}\lim_{\varepsilon\to 0} i\varepsilon\ \exp[i(E'-E)t/\hbar]\langle \boldsymbol{k}'|\hat{G}(E+i\varepsilon)|\boldsymbol{k}\rangle. \end{aligned} \tag{10.62}$$

The second resolvent relation for the Green's operator, Eq. (10.45), gives

$$\begin{aligned} S_{k'k} &= \lim_{t\to\infty}\lim_{\varepsilon\to 0} i\varepsilon \exp[i(E'-E)t]\langle \boldsymbol{k}'|[\hat{G}_0(E+i\varepsilon)+\hat{G}_0(E+i\varepsilon)\hat{V}\hat{G}(E+i\varepsilon)]|\boldsymbol{k}\rangle \\ &= \lim_{t\to\infty}\lim_{\varepsilon\to 0} i\varepsilon \exp[i(E'-E)t]\langle \boldsymbol{k}'|(E+i\varepsilon-\hat{H}_0)^{-1}[1+\hat{V}\hat{G}(E+i\varepsilon)]|\boldsymbol{k}\rangle \\ &= \lim_{t\to\infty}\lim_{\varepsilon\to 0} i\varepsilon \exp[i(E'-E)t]\left\{\frac{\delta(\boldsymbol{k}'-\boldsymbol{k})}{i\varepsilon}+\frac{\langle \boldsymbol{k}'|\hat{V}|\boldsymbol{k}+\rangle}{(E+i\varepsilon-E')i\varepsilon}\right\} \\ &= \delta(\boldsymbol{k}'-\boldsymbol{k}) - \lim_{t\to\infty}\lim_{\varepsilon\to 0}\frac{\exp[i(E'-E)t]}{E'-E-i\varepsilon}\langle \boldsymbol{k}'|\hat{V}|\boldsymbol{k}+\rangle. \end{aligned} \tag{10.63}$$

We can easily prove that

$$\lim_{t\to\infty}\lim_{\varepsilon\to 0}\frac{\exp(i\omega t)}{\omega-i\varepsilon} = 2\pi i\delta(\omega). \tag{10.64}$$

To see this, we just need to apply the formula on an analytic rapidly decaying arbitrary function $\phi(\omega)$, and perform the integral

$$\lim_{t\to\infty}\lim_{\varepsilon\to 0}\oint_C \frac{i\omega t}{\omega-i\epsilon}\phi(\omega)\mathrm{d}\omega, \tag{10.65}$$

where the contour C runs along the real axis and closes on the upper half infinity semicircle. According to the Riemann-Lebesgue lemma, if $\int \mathrm{d}x\,|f(x)| < \infty$, we have

$$\lim_{|\alpha|\to\infty}\int \mathrm{d}x f(x)\exp(i\alpha x) = 0. \tag{10.66}$$

Consequently, in Eq. (10.65), the integral on the semicircle vanishes. Then, due to the Cauchy's residue theorem we have

$$\lim_{t\to\infty}\lim_{\varepsilon\to 0}\int_{-\infty}^{\infty}\frac{i\omega t}{\omega-i\epsilon}\phi(\omega)\mathrm{d}\omega = 2\pi i\phi(0), \tag{10.67}$$

which can be abbreviated by Eq. (10.64).

This yields, for the $\hat{S}$ operator matrix element,

$$S_{k'k} = \delta(\boldsymbol{k}' - \boldsymbol{k}) - 2\pi i \delta(E' - E)\langle \boldsymbol{k}'|\hat{V}|\boldsymbol{k}+\rangle. \tag{10.68}$$

The first part gives no scattering. The second term describes scattering with an energy conservation constraint.

It is useful to define the $\hat{T}$ operator

$$\langle \boldsymbol{k}'|\hat{V}|\boldsymbol{k}+\rangle = \langle \boldsymbol{k}'|\hat{T}|\boldsymbol{k}\rangle, \tag{10.69}$$

and write

$$S_{k'k} = \delta(\boldsymbol{k}' - \boldsymbol{k}) - 2\pi i \delta(E_{k'} - E_k)\langle \boldsymbol{k}'|\hat{T}|\boldsymbol{k}\rangle. \tag{10.70}$$

The matrix element $\langle \boldsymbol{k}'|\hat{T}|\boldsymbol{k}\rangle$ is defined here between equal energy states, called the *on-shell T-matrix*.

By using Eq. (10.55), we may write

$$\langle \boldsymbol{k}'|\hat{T}|\boldsymbol{k}\rangle = \langle \boldsymbol{k}'|\hat{V}|\boldsymbol{k}+\rangle = \langle \boldsymbol{k}'|\hat{V}|\boldsymbol{k}\rangle + \langle \boldsymbol{k}'|\hat{V}\hat{G}\hat{V}|\boldsymbol{k}\rangle. \tag{10.71}$$

Hence, the $\hat{T}$ operator for an arbitrary complex variable z becomes

$$\hat{T}(z) = \hat{V} + \hat{V}\hat{G}(z)\hat{V}. \tag{10.72}$$

From the hermiticity of $\hat{H}$, we get the analytic properties of $\hat{T}$

$$\hat{H} = \hat{H}^\dagger, \quad \hat{G}(z^*) = [\hat{G}(z)]^\dagger, \quad \text{and thus} \quad \hat{T}(z^*) = [\hat{T}(z)]^\dagger. \tag{10.73}$$

Multiplying Eq. (10.72) by $\hat{G}_0$, we find

$$\hat{G}_0\hat{T} = \hat{G}_0\hat{V} + \hat{G}_0\hat{V}\hat{G}\hat{V} = (\hat{G}_0 + \hat{G}_0\hat{V}\hat{G})\hat{V} = \hat{G}\hat{V}, \tag{10.74}$$

and then obtain

$$\hat{T}(z) = \hat{V} + \hat{V}\hat{G}(z)\hat{V} = \hat{V} + \hat{V}\hat{G}_0(z)\hat{T}, \tag{10.75}$$

which is the *Lippmann-Schwinger equation* for the $\hat{T}$ operator.

This relation can serve as a starting point for many approximation methods in scattering theory. When $\hat{V}$ is *sufficiently weak*, it is reasonable to expect a solution by iteration

$$\hat{T} = \hat{V} + \hat{V}\hat{G}_0\hat{V} + \hat{V}\hat{G}_0\hat{V}\hat{G}_0\hat{V} + \cdots. \tag{10.76}$$

This is the *Born series*, and keeping only the first term amounts to the *first Born approximation*.

10.1.6 Asymptotic Behavior of the Scattering Wave Function

Now, we consider the scattering function in coordinate representation

$$\langle \boldsymbol{r}|\boldsymbol{k}\pm\rangle = \langle \boldsymbol{r}|\boldsymbol{k}\rangle + \lim_{\varepsilon\to 0}\int \mathrm{d}\boldsymbol{r}'\,\langle \boldsymbol{r}|\hat{G}_0(E\pm i\varepsilon)|\boldsymbol{r}'\rangle \hat{V}(\boldsymbol{r}')\langle \boldsymbol{r}'|\boldsymbol{k}\pm\rangle, \tag{10.77}$$

and try to link its asymptotic behavior to scattering observables. The coordinate representation of the free Green's operator reads

$$\begin{aligned}
\langle \boldsymbol{r}|\hat{G}_0(z)|\boldsymbol{r}'\rangle &= \int \mathrm{d}\boldsymbol{k}\,\langle \boldsymbol{r}|\hat{G}_0(z)|\boldsymbol{k}\rangle\langle \boldsymbol{k}|\boldsymbol{r}'\rangle = \int \mathrm{d}\boldsymbol{k}\,\langle \boldsymbol{r}|(z-\hat{H}_0)^{-1}|\boldsymbol{k}\rangle\langle \boldsymbol{k}|\boldsymbol{r}'\rangle \\
&= \frac{1}{(2\pi)^3}\int \mathrm{d}\boldsymbol{k}\,\frac{\exp[i\boldsymbol{k}(\boldsymbol{r}-\boldsymbol{r}')]}{z-\hbar^2k^2/(2m)} \\
&= \frac{2\pi}{(2\pi)^3}\int_0^\infty \mathrm{d}k\,\frac{k^2}{z-\hbar^2k^2/(2m)}\int_0^\pi \mathrm{d}\theta\,\sin\theta\,\exp(ik|\boldsymbol{r}-\boldsymbol{r}'|\cos\theta) \\
&= -\frac{1}{2\pi^2}\int_0^\infty \mathrm{d}k\,\frac{mk^2}{\hbar^2k^2-2mz}\int_{-1}^{1}\mathrm{d}u\,\exp(ik|\boldsymbol{r}-\boldsymbol{r}'|u) \\
&= \frac{im}{2\pi^2|\boldsymbol{r}-\boldsymbol{r}'|}\int_0^\infty \mathrm{d}k\,k\frac{\exp(ik|\boldsymbol{r}-\boldsymbol{r}'|)-\exp(-ik|\boldsymbol{r}-\boldsymbol{r}'|)}{\hbar^2k^2-2mz} \\
&= \frac{im}{2\pi^2\hbar^2|\boldsymbol{r}-\boldsymbol{r}'|}\int_{-\infty}^{\infty}\mathrm{d}k\,\frac{k\exp(ik|\boldsymbol{r}-\boldsymbol{r}'|)}{k^2-2mz/\hbar^2} \\
&= \frac{im}{2\pi^2\hbar^2|\boldsymbol{r}-\boldsymbol{r}'|}\int_{-\infty}^{\infty}\mathrm{d}k\,\frac{k\exp(ik|\boldsymbol{r}-\boldsymbol{r}'|)}{(k-\sqrt{2mz/\hbar^2})(k+\sqrt{2mz/\hbar^2})}.
\end{aligned} \tag{10.78}$$

Due to the Riemann-Lebesgue lemma, we can close the Cupper half k-plane as its contribution vanishes (Fig. 10.3). Then, if Im $(z) > 0$, the integrand has a single pole at $k = \sqrt{2mz/\hbar^2}$ and we can evaluate the integral by using the Cauchy's residue theorem. Note that

$$z = E \pm i\varepsilon = \frac{\hbar^2k^2}{2m} \pm i\varepsilon \quad \text{and} \quad \sqrt{2mz/\hbar^2} \to \pm k, \tag{10.79}$$

then we find

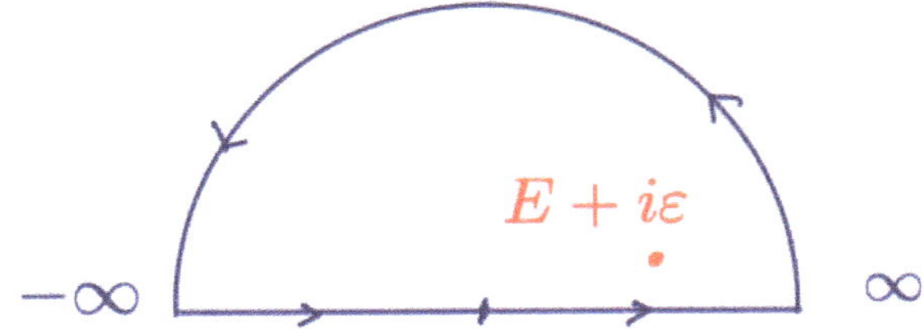

Fig. 10.3 Contour integration with pole at $E + i\varepsilon$

$$\begin{aligned}\langle \boldsymbol{r}|\hat{G}_0(E \pm i\varepsilon)|\boldsymbol{r}'\rangle &= \frac{im}{2\pi^2\hbar^2|\boldsymbol{r}-\boldsymbol{r}'|}2\pi i \frac{k\exp(ik|\boldsymbol{r}-\boldsymbol{r}'|)}{2k} \\ &= \frac{-m}{2\pi\hbar^2}\frac{\exp(\pm ik\,|\boldsymbol{r}-\boldsymbol{r}'|)}{|\boldsymbol{r}-\boldsymbol{r}'|}.\end{aligned} \tag{10.80}$$

With this representation of the free Green's operator, the Lippmann-Schwinger equation for a scattering state reads

$$\langle \boldsymbol{r}|\boldsymbol{k}\pm\rangle = \frac{1}{(2\pi)^{3/2}}\exp(i\boldsymbol{k}\,\boldsymbol{r}) - \frac{m}{2\pi\hbar^2}\int \mathrm{d}\boldsymbol{r}'\,\frac{\exp(\pm ik\,|\boldsymbol{r}-\boldsymbol{r}'|)}{|\boldsymbol{r}-\boldsymbol{r}'|}\hat{V}(\boldsymbol{r}')\langle \boldsymbol{r}'|\boldsymbol{k}\pm\rangle. \tag{10.81}$$

The potential is of short-range type, and thus $\boldsymbol{r}'$ is confined to a finite region of the configuration space (Fig. 10.4). So, for asymptotic values of $\boldsymbol{r}$, we have $|\boldsymbol{r}-\boldsymbol{r}'| \gg 0$, and we can make the approximation

$$|\boldsymbol{r}-\boldsymbol{r}'| \approx r\left(1 - \frac{\boldsymbol{r}\cdot\boldsymbol{r}'}{r^2} + \left(\frac{r'}{r}\right)^2\right) \approx r - \boldsymbol{e}_r r' + \left(\frac{r'}{r}\right)^2. \tag{10.82}$$

Then, the asymptotic behavior of the scattering-state wave function becomes

$$\begin{aligned}\langle \boldsymbol{r}|\boldsymbol{k}\pm\rangle &\underset{r\to\infty}{\longrightarrow} \frac{1}{(2\pi)^{3/2}}\exp(i\boldsymbol{k}\,\boldsymbol{r}) - \frac{m}{2\pi\hbar^2}\frac{\exp(\pm ik\,r)}{r}\int \mathrm{d}\boldsymbol{r}'\exp(\mp ik\boldsymbol{e}_r\,\boldsymbol{r}')\hat{V}(\boldsymbol{r}')\langle \boldsymbol{r}'|\boldsymbol{k}\pm\rangle \\ &\underset{r\to\infty}{\longrightarrow} \frac{1}{(2\pi)^{3/2}}\exp(i\boldsymbol{k}\,\boldsymbol{r}) - \frac{\exp(\pm ik\,r)}{r}\frac{m(2\pi)^{3/2}}{2\pi\hbar^2}\int \mathrm{d}\boldsymbol{r}'\langle \pm k\boldsymbol{e}_r|\boldsymbol{r}'\rangle\hat{V}(\boldsymbol{r}')\langle \boldsymbol{r}'|\boldsymbol{k}\pm\rangle \\ &\underset{r\to\infty}{\longrightarrow} \frac{1}{(2\pi)^{3/2}}\left[\exp(i\boldsymbol{k}\,\boldsymbol{r}) - (2\pi)^2 m/\hbar^2\langle \pm k\boldsymbol{e}_r|\hat{V}|\boldsymbol{k}\pm\rangle\frac{\exp(\pm ik\,r)}{r}\right].\end{aligned} \tag{10.83}$$

We can recognize the matrix elements of the T operator in the asymptotic form of the scattering wave function. In particular, for $\langle \boldsymbol{r}|\boldsymbol{k}+\rangle$ with $\boldsymbol{k}' = k\hat{\boldsymbol{r}}$, we find

$$\langle \boldsymbol{r}|\boldsymbol{k}+\rangle \sim \exp(i\boldsymbol{k}\,\boldsymbol{r}) + f(\boldsymbol{k}' \leftarrow \boldsymbol{k})\frac{\exp(ik\,r)}{r}, \quad \text{as} \quad r \to \infty, \tag{10.84}$$

where

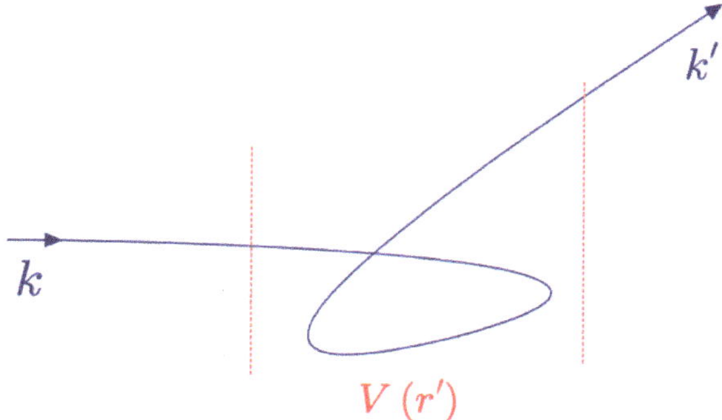

Fig. 10.4 The short range potential $\hat{V}$ confines the range of $\boldsymbol{r}'$

$$f(\boldsymbol{k}' \leftarrow \boldsymbol{k}) = -\frac{(2\pi)^2 m}{\hbar^2}\langle \boldsymbol{k}'|\hat{V}|\boldsymbol{k}+\rangle = -\frac{(2\pi)^2 m}{\hbar^2}\langle \boldsymbol{k}'|\hat{T}|\boldsymbol{k}\rangle \tag{10.85}$$

is the *scattering amplitude.*

10.1.7 Scattering Cross Section

We can see that the asymptotic form of $\langle \boldsymbol{r}|\boldsymbol{k}+\rangle$ consists of an incident plane wave that is unaffected by the potential and an outgoing scattered spherical wave, whose coefficient is the scattering amplitude. Let us assume that $\boldsymbol{k}$ is in the $\boldsymbol{e}_z$ direction. Then the incident flux is given by

$$j_{in} = \frac{i\hbar}{2m}\left(\exp(ikz)\frac{\partial}{\partial z}\exp(-ikz) - \exp(-ikz)\frac{\partial}{\partial z}\exp(ikz)\right) = \frac{\hbar k}{m}, \tag{10.86}$$

while the scattered flux is

$$j_s = \frac{i\hbar}{2m}|f|^2\left[\frac{\exp(ikr)}{r}\frac{\partial}{\partial r}\left(\frac{\exp(-ikr)}{r}\right) - \frac{\exp(-ikr)}{r}\frac{\partial}{\partial r}\left(\frac{\exp(ikr)}{r}\right)\right] = \frac{|f|^2}{r^2}\frac{\hbar k}{m}. \tag{10.87}$$

Thus,

$$\frac{j_s}{j_{in}} = \frac{|f|^2}{r^2}, \tag{10.88}$$

and the total probability current scattered into a solid angle $\mathrm{d}\Omega$ per unit incident flux is

$$\frac{|f|^2}{r^2}r^2\,\mathrm{d}\Omega = |f(\boldsymbol{k}' \leftarrow \boldsymbol{k})|^2\,\mathrm{d}\Omega = \frac{\mathrm{d}\sigma(\theta,\phi)}{\mathrm{d}\Omega}\,\mathrm{d}\Omega. \tag{10.89}$$

Here, $\mathrm{d}\sigma(\theta,\phi)/\mathrm{d}\Omega$ is the cross section of the scattering into the direction (θ,ϕ), the angle between $\boldsymbol{k}$ and $\boldsymbol{k}'$. Then, for the total cross section we have

$$\sigma(\boldsymbol{k}) = \int \frac{\mathrm{d}\sigma(\theta,\phi)}{\mathrm{d}\Omega}\,\mathrm{d}\Omega = \int |f(\boldsymbol{k}' \leftarrow \boldsymbol{k})|^2\,\mathrm{d}\Omega. \tag{10.90}$$

The total cross section is related to the forward scattering amplitude. From the unitarity of the $\hat{S}$ operator,

$$\hat{S}^\dagger \hat{S} = 1, \tag{10.91}$$

and using $\hat{S} = 1 + \hat{R}$, we readily have

$$\hat{R} + \hat{R}^\dagger = -\hat{R}^\dagger \hat{R}. \tag{10.92}$$

Taking this relation between free states, we find

$$\langle \boldsymbol{k}'|\hat{R}|\boldsymbol{k}\rangle + \langle \boldsymbol{k}'|\hat{R}^\dagger|\boldsymbol{k}\rangle = -\langle \boldsymbol{k}'|\hat{R}^\dagger \hat{R}|\boldsymbol{k}\rangle, \tag{10.93}$$

or

$$\langle \boldsymbol{k}'|\hat{R}|\boldsymbol{k}\rangle + \langle \boldsymbol{k}|\hat{R}^\dagger|\boldsymbol{k}'\rangle^* = -\int \mathrm{d}\boldsymbol{k}''\langle \boldsymbol{k}'|\hat{R}^\dagger|\boldsymbol{k}''\rangle\langle \boldsymbol{k}''|\hat{R}|\boldsymbol{k}\rangle = -\int \mathrm{d}\boldsymbol{k}''\langle \boldsymbol{k}''|\hat{R}|\boldsymbol{k}'\rangle^*\langle \boldsymbol{k}''|\hat{R}|\boldsymbol{k}\rangle. \tag{10.94}$$

From Eqs. (10.70) and (10.85) we can see that

$$\langle \boldsymbol{k}'|\hat{R}|\boldsymbol{k}\rangle = -2\pi i\delta(E_{k'} - E_k)\langle \boldsymbol{k}'|\hat{T}|\boldsymbol{k}\rangle = \frac{i\hbar^2}{2\pi m}\delta(E_{k'} - E_k) f(\boldsymbol{k}' \leftarrow \boldsymbol{k}), \tag{10.95}$$

and then from Eq. (10.94) follows

$$\begin{aligned}&\frac{i\hbar^2}{2\pi m}\delta(E_{k'} - E_k)[f(\boldsymbol{k}' \leftarrow \boldsymbol{k}) - f(\boldsymbol{k} \leftarrow \boldsymbol{k}')^*]\\ &\quad = -\frac{\hbar^4}{(2\pi m)^2}\int \mathrm{d}\boldsymbol{k}''\,\delta(E_{k''} - E_{k'}) f(\boldsymbol{k}'' \leftarrow \boldsymbol{k}')^*\delta(E_{k''} - E_k) f(\boldsymbol{k}'' \leftarrow \boldsymbol{k}).\end{aligned} \tag{10.96}$$

The function $\delta(E_{k''} - E_{k'})$ sets the other δ function to $\delta(E_{k'} - E_k)$, and it cancels out. So, finally we obtain

$$f(\boldsymbol{k}' \leftarrow \boldsymbol{k}) - f(\boldsymbol{k} \leftarrow \boldsymbol{k}')^* = \frac{i\hbar^2}{2\pi m}\int \mathrm{d}\boldsymbol{k}''\,\delta(E_{k''} - E_{k'}) f(\boldsymbol{k}'' \leftarrow \boldsymbol{k}')^* f(\boldsymbol{k}'' \leftarrow \boldsymbol{k}). \tag{10.97}$$

If we take $\boldsymbol{k}' = \boldsymbol{k}$, i.e. we consider the forward direction, then on the left side the real parts cancel out, and division by $2i$ results in

$$\mathrm{Im}\, f(\boldsymbol{k} \leftarrow \boldsymbol{k}) = \frac{\hbar^2}{4\pi m}\int \mathrm{d}\Omega_{k''}\int \mathrm{d}k''\,k''^2\delta(E_{k''} - E_k) f(\boldsymbol{k}'' \leftarrow \boldsymbol{k})^* f(\boldsymbol{k}'' \leftarrow \boldsymbol{k}). \tag{10.98}$$

Using the relations $E_{k''} = \hbar^2 k''^2/(2m)$ and $\mathrm{d}E_{k''} = \hbar^2 k''/m\,\mathrm{d}k''$, and integrating with respect to $E_{k''}$, we eliminate the δ term and find

$$\begin{aligned}\operatorname{Im} f(\boldsymbol{k} \leftarrow \boldsymbol{k}) &= \frac{k}{4\pi} \int \mathrm{d}\Omega_{k''} f(\boldsymbol{k}'' \leftarrow \boldsymbol{k})^* f(\boldsymbol{k}'' \leftarrow \boldsymbol{k}) \\ &= \frac{k}{4\pi} \int \mathrm{d}\Omega_{k''} |f(\boldsymbol{k}'' \leftarrow \boldsymbol{k})|^2, \qquad \text{with } |k''| = |k|.\end{aligned} \tag{10.99}$$

Consequently,

$$\operatorname{Im} f(\boldsymbol{k} \leftarrow \boldsymbol{k}) = \frac{k}{4\pi} \sigma(\boldsymbol{k}). \tag{10.100}$$

This is the *optical theorem*, it relates the imaginary part of the forward scattering amplitude to the total cross section. This relation is the consequence of the unitarity of the scattering operator.

10.2 Spherical Wave Expansion

10.2.1 Free States

For further investigating scattering processes, we consider the spherical basis, the common eigenstates of $\{H_0, L^2, L_z\}$. The basis states $\{|E, l, m\rangle\}$ satisfy the normalization condition

$$\langle E', l', m' | E, l, m\rangle = \delta(E' - E)\delta_{ll'}\delta_{mm'} \tag{10.101}$$

and the completeness relation

$$I = \sum_{lm} \int \mathrm{d}E \, |E, l, m\rangle\langle E, l, m|. \tag{10.102}$$

In coordinate space, they are given by

$$\psi_{Elm}(\boldsymbol{r}) = \langle \boldsymbol{r} | E, l, m\rangle = i^l \left(\frac{2m}{\pi \hbar^2 k}\right)^{1/2} \frac{1}{r} \hat{j}_l(kr) Y_{l,m}(\boldsymbol{e}_r), \tag{10.103}$$

where $k = \sqrt{2mE/\hbar^2}$, the normalization constant i^l is chosen for convenience, and $Y_{l,m}(\boldsymbol{e}_r)$ are the spherical harmonics.

The Riccati-Bessel functions are given by

$$\hat{j}_l(z) \equiv z j_l(z) \equiv \sqrt{\frac{\pi z}{2}} J_{l+1/2}(z) = z^{l+1} \sum_{n=0}^{\infty} \frac{(-z^2/2)^n}{n!(2l + 2n + 1)!!}, \tag{10.104}$$

where $j_l(z)$ is the ordinary spherical Bessel function and $J_{l+1/2}(z)$ is the ordinary Bessel function. The double factorial is defined by $n!! = \prod_{k=0}^{[n/2]-1} (n - 2k)$ and $(-2n - 1)!! = (-1)^n/(2n - 1)!!$, with $(-1)!! = 1$. The first three Riccati-Bessel functions are

$$\hat{j}_0(z) = \sin z, \quad \hat{j}_1(z) = \frac{1}{z}\sin z - \cos z, \quad \hat{j}_2(z) = \left(\frac{3}{z^2} - 1\right)\sin z - \frac{3}{z}\cos z\,. \tag{10.105}$$

The Riccati-Bessel functions are normalized as

$$\int_0^\infty \mathrm{d}r\, \hat{j}_l(k'r)\hat{j}_l(kr) = \frac{\pi}{2}\delta(k'-k). \tag{10.106}$$

From Eq. (10.104), we can see that

$$\hat{j}_l(-z) = (-)^{l+1}\hat{j}_l(z). \tag{10.107}$$

They are real for real arguments, and behave like

$$\hat{j}_l(z) \sim z^{l+1}/(2l+1)!!, \quad \text{as} \quad z \to 0, \tag{10.108}$$

and

$$\hat{j}_l(z) \sim \sin(z - l\pi/2), \quad \text{as} \quad z \to \infty. \tag{10.109}$$

The Riccati-Bessel functions $\hat{j}_l(z)$ are the regular solutions of the differential equation

$$z^2\frac{\mathrm{d}^2 y_l(z)}{\mathrm{d}z^2} + \left(z^2 - l(l+1)\right) y_l(z) = 0\,. \tag{10.110}$$

Another solution, linearly independent of $\hat{j}_l(z)$, is the *Riccati-Neumann function*

$$\hat{n}_l(z) \equiv z n_l(z) = (-1)^l\sqrt{\frac{\pi z}{2}}J_{-l-1/2}(z) = z^{-l}\sum_{n=0}^{\infty}\frac{(-z^2/2)^n(2l-2n-1)!!}{n!}, \tag{10.111}$$

where $n_l(z)$ is the ordinary spherical Neumann function. The first few Riccati-Neumann functions are

$$\hat{n}_0(z) = \cos z, \quad \hat{n}_1(z) = \frac{1}{z}\cos z + \sin z, \quad \hat{n}_2(z) = \left(\frac{3}{z^2} - 1\right)\cos z + \frac{3}{z}\sin z. \tag{10.112}$$

From the series expansion Eq. (10.111), we can see that

$$\hat{n}_l(-z) = (-)^l\hat{n}_l(z). \tag{10.113}$$

They are real for real arguments and behave like

$$\hat{n}_l(z) \sim z^{-l}(2l-1)!!, \quad \text{as} \quad z \to 0, \tag{10.114}$$

and

$$\hat{n}_l(z) \sim \cos(z - l\pi/2), \quad \text{as} \quad z \to \infty. \tag{10.115}$$

Other notable solutions are the *Riccati-Hankel functions*; a linear combination of Riccati-Bessel and Riccati-Neumann functions,

$$\hat{h}_l^{\pm}(z) = \hat{n}_l(z) \pm i\,\hat{j}_l(z). \tag{10.116}$$

They have the asymptotic form

$$\hat{h}_l^{\pm}(z) \sim \exp[\pm i(z - l\pi/2)], \quad \text{as} \quad z \to \infty. \tag{10.117}$$

The momentum representation of $|E, l, m\rangle$ can be inferred from the relation

$$\begin{aligned}
\psi_{Elm}(k) = \langle \boldsymbol{k}|E, l, m\rangle &= \int \mathrm{d}\boldsymbol{r}\,\langle \boldsymbol{k}|\boldsymbol{r}\rangle\langle \boldsymbol{r}|E, l, m\rangle \\
&= \int \mathrm{d}\boldsymbol{r}\frac{1}{(2\pi)^{3/2}}\exp(-i\boldsymbol{k}\boldsymbol{r})\langle \boldsymbol{r}|E, l, m\rangle \\
&= \frac{\hbar}{\sqrt{mk}}\delta(E - E_k)Y_{l,m}(\boldsymbol{e}_k).
\end{aligned} \tag{10.118}$$

Now we have

$$\begin{aligned}
|\boldsymbol{k}\rangle &= \sum_{lm}\int \mathrm{d}E\,|E, l, m\rangle\langle E, l, m|\boldsymbol{k}\rangle \\
&= \frac{\hbar}{\sqrt{mk}}\sum_{lm}\int \mathrm{d}E\,|E, l, m\rangle\delta(E - E_k)Y_{l,m}^{*}(\boldsymbol{e}_k) \\
&= \frac{\hbar}{\sqrt{mk}}\sum_{lm} Y_{l,m}^{*}(\boldsymbol{e}_k)|E_k, l, m\rangle,
\end{aligned} \tag{10.119}$$

which, due to Eq. (4.160), results in

$$\begin{aligned}
\langle \boldsymbol{r}|\boldsymbol{k}\rangle = \frac{1}{(2\pi)^{3/2}}\exp(i\boldsymbol{k}\boldsymbol{r}) &= \frac{\hbar}{\sqrt{mk}}\sum_{lm} Y_{l,m}^{*}(\boldsymbol{e}_k)\langle \boldsymbol{r}|E_k, l, m\rangle \\
&= \sqrt{\frac{2}{\pi}}\frac{1}{kr}\sum_{lm} i^l\,\hat{j}_l(kr)Y_{l,m}(\boldsymbol{e}_r)Y_{l,m}^{*}(\boldsymbol{e}_k) \\
&= \frac{1}{(2\pi)^{3/2}}\frac{1}{kr}\sum_{l}(2l + 1)i^l\,\hat{j}_l(kr)P_l(\boldsymbol{e}_k \cdot \boldsymbol{e}_r).
\end{aligned} \tag{10.120}$$

10.2.2 Phase Shift

The scattering operator $\hat{S}$, as a function of $\hat{H}$, commutes with $\hat{H}$. If the potential is spherical, then $\hat{H}$, and consequently $\hat{S}$, are scalar operators and commute with the angular momentum operator $\hat{\boldsymbol{L}}$. Then, according to the Wigner-Eckart theorem, $\langle l,m|\hat{S}|l,m\rangle$ is independent of m. Therefore, $|E,l,m\rangle$ is an eigenstate of $\hat{S}$ with eigenvalue $s_l(E)$

$$\hat{S}|E,l,m\rangle = s_l(E)|E,l,m\rangle. \tag{10.121}$$

Since $\hat{S}$ is unitary, $s_l(E)$ has a modulus of one and we can write

$$s_l(E) = \exp(2i\delta_l(E)), \tag{10.122}$$

with δ_l real, called the *phase shift*.

We have found before, in Eq. (10.70), that

$$\begin{aligned}\langle \boldsymbol{k}'|S|\boldsymbol{k}\rangle &= \delta^{(3)}(\boldsymbol{k}'-\boldsymbol{k}) - 2\pi i\delta(E_{k'}-E_k)\langle \boldsymbol{k}'|T|\boldsymbol{k}\rangle \\ &= \delta^{(3)}(\boldsymbol{k}'-\boldsymbol{k}) + \frac{i\hbar^2}{2\pi m}\delta(E_{k'}-E_k)f(\boldsymbol{k}'\leftarrow\boldsymbol{k}).\end{aligned} \tag{10.123}$$

Therefore,

$$\begin{aligned}\frac{i\hbar^2}{2\pi m}\delta(E_{k'}-E_k)f(\boldsymbol{k}'\leftarrow\boldsymbol{k}) &= \langle \boldsymbol{k}'|(S-1)|\boldsymbol{k}\rangle = \sum_{lm}\int \mathrm{d}E\langle \boldsymbol{k}'|(S-1)|E,l,m\rangle\langle E,l,m|\boldsymbol{k}\rangle \\ &= \sum_{lm}\int \mathrm{d}E(s_l(E)-1)\langle \boldsymbol{k}'|E,l,m\rangle\langle E,l,m|\boldsymbol{k}\rangle \\ &= \sum_{lm}\int \mathrm{d}E\frac{\hbar^2(s_l(E)-1)}{m\sqrt{k'k}}\delta(E-E_{k'})\delta(E-E_k)Y_{l,m}(\boldsymbol{e}_{k'})Y^*_{l,m}(\boldsymbol{e}_k) \\ &= \frac{\hbar^2}{mk}\delta(E_{k'}-E_k)\sum_{lm}Y_{l,m}(\boldsymbol{e}_{k'})Y^*_{l,m}(\boldsymbol{e}_k)[s_l(E)-1],\end{aligned} \tag{10.124}$$

and consequently

$$f(\boldsymbol{k}'\leftarrow\boldsymbol{k}) = \frac{2\pi}{ik}\sum_{lm}Y_{l,m}(\boldsymbol{e}_{k'})Y^*_{l,m}(\boldsymbol{e}_k)(s_l(E_k)-1). \tag{10.125}$$

By utilizing the addition properties of spherical harmonics, we arrive at the final result

$$f(\boldsymbol{k}'\leftarrow\boldsymbol{k}) = f(E_k,\theta) = \sum_{l=0}^{\infty}(2l+1)f_l(E_k)P_l(\cos\vartheta), \tag{10.126}$$

where θ is the angle between $\boldsymbol{k}$ and $\boldsymbol{k}'$ and

$$f_l(E_k) \equiv \frac{s_l(E_k) - 1}{2ik} = \frac{\exp(2i\delta_l) - 1}{2ik} = \frac{\exp(i\delta_l)}{k} \sin \delta_l. \tag{10.127}$$

So, we have found a relation between the partial wave scattering amplitude and the phase shift.

We can also relate the phase shift to the asymptotic form of the scattering wave function. The partial wave scattering states are defined by the action of the Møller operator on the $|E, l, m\rangle$ states

$$|E, l, m+\rangle = \hat{\Omega}^{(+)}|E, l, m\rangle. \tag{10.128}$$

Since $\hat{H}\hat{\Omega}^{(+)} = \hat{\Omega}^{(+)}\hat{H}_0$ and $[\hat{\Omega}^{(+)}, \hat{\boldsymbol{L}}] = 0$, the states $|E, l, m+\rangle$ are the common eigenstates of $\hat{H}$ and $\hat{L}$

$$\hat{H}|E, l, m+\rangle = E|E, l, m+\rangle \quad \text{and} \quad \hat{L}^2|E, l, m+\rangle = l(l+1)\hbar^2|E, l, m+\rangle, \tag{10.129}$$

with the normalization condition

$$\begin{aligned} \langle E', l', m'+|E, l, m+\rangle &= \langle E', l', m'|\Omega^{(+)\dagger}\Omega^{(+)}|E, l, m\rangle \\ &= \langle E', l', m'|E, l, m\rangle = \delta(E' - E)\delta_{ll'}\delta_{mm'}. \end{aligned} \tag{10.130}$$

Then, similarly to Eq. (10.103), we can write

$$\langle \boldsymbol{r}|E, l, m+\rangle = i^l \sqrt{\frac{2m}{\pi\hbar^2 k}} \frac{1}{r} \psi_{l,k}^{(+)}(r) Y_{l,m}(\boldsymbol{e}_r), \tag{10.131}$$

where $\psi_{l,k}(r)$ is a physical solution for the radial wave equation

$$\frac{\mathrm{d}^2\psi_{l,k}^{(+)}(r)}{\mathrm{d}r^2} + \left[k^2 - U_l(r)\right]\psi_{l,k}^{(+)}(r) = 0, \tag{10.132}$$

with

$$U_l(r) \equiv \frac{l(l+1)}{r^2} + V'(r), \quad \text{and} \quad V'(r) \equiv 2m/\hbar^2\, V(r). \tag{10.133}$$

The radial wave functions are normalized the same way as the Riccati-Bessel functions $\hat{j}_l(kr)$

$$\int_0^\infty \mathrm{d}r\, \psi_{l,k'}^{(+)*}(r)\psi_{l,k}^{(+)}(r) = \frac{\pi}{2}\delta(k' - k). \tag{10.134}$$

From

$$|\boldsymbol{k}\rangle = \sum_{lm} \int \mathrm{d}E \, |Elm\rangle\langle Elm|\boldsymbol{k}\rangle, \tag{10.135}$$

we have

$$|\boldsymbol{k}+\rangle = \Omega^{(+)}|\boldsymbol{k}\rangle = \sum_{lm} \int \mathrm{d}E \, |Elm+\rangle\langle Elm|\boldsymbol{k}\rangle, \tag{10.136}$$

or in coordinate representation, we find

$$\langle \boldsymbol{r}|\boldsymbol{k}+\rangle = \sum_{lm} \int \mathrm{d}E \, \langle \boldsymbol{r}|Elm+\rangle\langle Elm|\boldsymbol{k}\rangle. \tag{10.137}$$

Then, by using Eqs. (10.118), (10.131) and (4.160) we obtain

$$\begin{aligned} \langle \boldsymbol{r}|\boldsymbol{k}+\rangle &= \frac{\hbar}{\sqrt{mk}} \left(\frac{2m}{\pi \hbar^2 k}\right)^{1/2} \sum_{l,m} i^l \, \frac{1}{r} \, \psi_{l,k}^{(+)}(r) Y_{l,m}^*(\boldsymbol{e}_k) Y_{l,m}(\boldsymbol{e}_r) \\ &= \frac{1}{(2\pi)^{3/2}} \frac{1}{kr} \sum_l (2l+1) i^l \psi_{l,k}^{(+)}(r) P_l(\boldsymbol{e}_k \cdot \boldsymbol{e}_r). \end{aligned} \tag{10.138}$$

For large r, the scattering functions behave like

$$\langle \boldsymbol{r}|\boldsymbol{k}+\rangle \to \frac{1}{(2\pi)^{3/2}} \left[\exp(i\boldsymbol{k}\boldsymbol{r}) + f(k\boldsymbol{e}_r \leftarrow \boldsymbol{k}) \frac{\exp(ikr)}{r}\right]. \tag{10.139}$$

Then, by utilizing Eqs. (10.126) and (10.120), we find that

$$\langle \boldsymbol{r}|\boldsymbol{k}+\rangle \to \frac{1}{(2\pi)^{3/2}} \frac{1}{kr} \sum_l (2l+1)[i^l \hat{j}_l(kr) + kf_l(k)\exp(ikr)] P_l(\boldsymbol{e}_k \cdot \boldsymbol{e}_r). \tag{10.140}$$

By comparing Eqs. (10.138) and (10.140), we can conclude that

$$\psi_{l,k}^{(+)}(r) \to \hat{j}_l(kr) + kf_l(k)\exp[i(kr - l\pi/2)], \tag{10.141}$$

and by matching to Eqs. (10.109) and (10.127), we can infer that

$$\begin{aligned} &\psi_{l,k}^{(+)}(r) \to \sin(kr - l\pi/2) + (s_l - 1)/(2i) \, \exp[i(kr - l\pi/2)] \\ &\to i/2 \, \{\exp[-i(kr - l\pi/2)] - \exp(2i\delta_l) \, \exp[i(kr - l\pi/2)]\} \\ &\to i/2 \exp(i\delta_l) \, \{\exp(-i\delta_l)\exp[-i(kr - l\pi/2)] - \exp(i\delta_l) \, \exp[i(kr - l\pi/2)]\} \\ &\to i/2 \exp(i\delta_l) \, \{\exp[-i(kr - l\pi/2 + \delta_l)] - \exp[i(kr - l\pi/2 + \delta_l)]\} \\ &\to \exp(i\delta_l) \sin(kr - l\pi/2 + \delta_l) \, . \end{aligned} \tag{10.142}$$

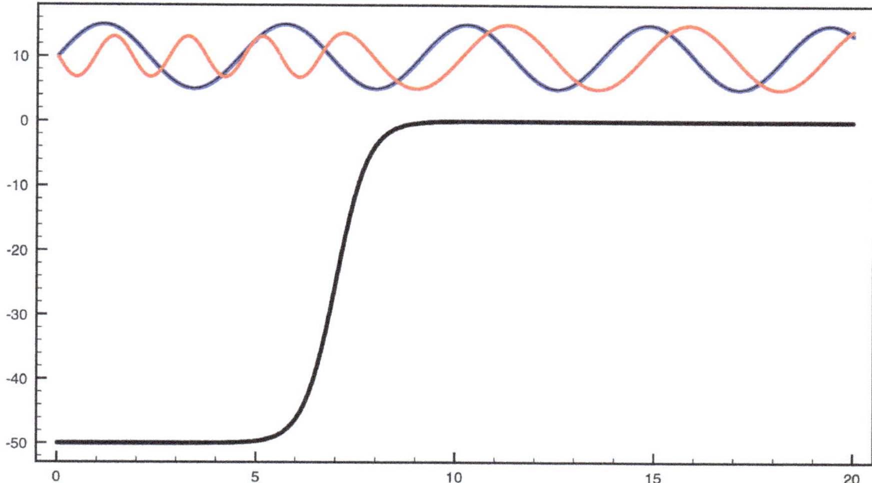

Fig. 10.5 Due to the short range potential, in the asymptotic region, the phase of the wave function is shifted with respect to the free wave

As we can see, the scattering phase shift shows up in the asymptotic behavior of the scattering wave function. It tells us that due to the potential, how much the phase of the scattering function is shifted with respect to the free wave (see Fig. 10.5).

We have learned before that in the case of bound states the physically relevant quantity, the energy, shows up in the asymptotic behavior of the wave function. In this respect the scattering states are similar. Here, the physically relevant quantity is the phase shift, which also shows up in the asymptotic behavior of the wave function.

10.2.3 The Jost Function

For the radial wave functions $\psi_{l,k}(r)$, we have the boundary conditions at $r \to 0$ and at $r \to \infty$. We can learn a lot of physics by studying a similar solution to the radial Schrödinger equation, called the *regular solution* $\phi_l(k, r)$, specified by a single point boundary condition

$$\lim_{r \to 0} \phi_l(k, r)/\hat{j}_l(kr) = 1, \quad \text{or} \quad \phi_l(k, r) \sim \hat{j}_l(kr) \sim (kr)^{l+1}/(2l+1)!! \quad \text{as } r \to 0. \tag{10.143}$$

From the definition it follows that $\phi_l(k, r)$ and $\hat{j}_l(kr)$ have analogous analytic properties, i.e. $\phi_l(k, r)$ as a function of k, is analytic on the whole complex plane, and

$$\phi_l(-k, r) = (-)^{l+1} \phi_l(k, r). \tag{10.144}$$

Furthermore, for real k^2 and for real potentials, $\phi_l(k, r)$ is real.

In the $r \to \infty$ limit, $\phi(k, r)$ should approach to some linear combination of the solutions of the free radial equation, i.e. a linear combination of $\hat{h}_l^-$ and $\hat{h}_l^+$. Considering that $\phi_l(k, r)$ is real, we can write

$$\phi_l(k, r) \underset{r\to\infty}{\longrightarrow} \frac{i}{2}[\tilde{f}_l(k)\exp(-i(kr - l\pi/2)) - \tilde{f}_l^*(k)\exp(i(kr - l\pi/2))], \tag{10.145}$$

where $\tilde{f}_l(k)$ is the *Jost function*. Comparing this relation to the asymptotic behavior of $\psi_{l,k}^{(+)}(r)$ of Eq. (10.142), we can see that

$$\phi_l(k, r) = \tilde{f}_l(k)\psi_{l,k}^{(+)}(r), \tag{10.146}$$

and

$$s_l(k) = \frac{\tilde{f}_l^*(k)}{\tilde{f}_l(k)}. \tag{10.147}$$

Furthermore, since

$$s_l(k) = \exp(2i\delta(k)) = \frac{\exp(i\delta(k))}{\exp(-i\delta(k))}, \tag{10.148}$$

we have

$$\tilde{f}_l(k) = |\tilde{f}_l(k)|\exp(-i\delta(k)). \tag{10.149}$$

The solutions of the radial Schrödinger equation are defined by boundary conditions with real k values. However, they can be extended for those complex k regions where the boundary conditions are well defined. Even more, they can be extended to other regions of the complex plane by the method of analytic continuation. In this respect, we can use the Schwarz reflection principle (Fig. 10.6). If $F(z)$ is analytic on a region R of the complex z-plane that includes a segment of the real axis, and if $F(z)$ is real on the real axis, then $F(z)$ can be analytically continued into the region R^*, and in $R \cup R^*$ the function is defined by

$$F(z^*) = [F(z)]^*. \tag{10.150}$$

Similarly, if $F(z)$ is real on the imaginary axis, the analytic continuation reads

$$F(-z^*) = [F(z)]^*. \tag{10.151}$$

The regular solution $\phi_l(k, r)$ is real for real k values. Therefore, the analytic continuation on the complex k-plane is given by

$$\phi_l(k^*, r) = \phi_l^*(k, r). \tag{10.152}$$

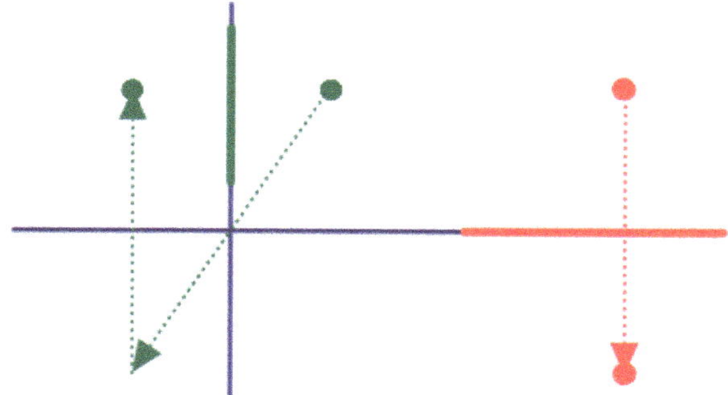

Fig. 10.6 The Schwarz reflection principle allows analytic continuation by mirroring on the real and imaginary axis

From this relation and from the analytic properties of $\hat{h}_l^{\pm}$ in Eq. (10.145), we can infer the relation

$$\tilde{f}_l(-k^*) = \tilde{f}_l^*(k). \tag{10.153}$$

Consequently, for real k we have

$$s_l(k) = \frac{\tilde{f}_l(-k)}{\tilde{f}_l(k)}, \tag{10.154}$$

and

$$\tilde{f}_l(\pm k) = |\tilde{f}_l(\pm k)| \exp(-i\delta_l(\pm k)), \quad \text{with} \quad \delta_l(k) = -\delta_l(-k). \tag{10.155}$$

Bound States

We have found before from one-dimensional scattering that the bound states are related to the poles of the scattering quantities, like the reflection and transmission coefficients. In three dimensional scattering, the corresponding quantity is the s_l, whose poles are related to the zeros of $\tilde{f}_l(k)$. We are going to see that the zeros of the Jost function $\tilde{f}_l(k)$ in the upper half k-plane are pure imaginary simple zeros that correspond to bound states.

Let $\bar{k} = k_r + i\kappa$, with k_r and κ real and positive, and assume that $\tilde{f}_l(\bar{k}) = 0$ and $\tilde{f}_l(-\bar{k})$ is analytic at $\bar{k}$. Then $\tilde{f}_l(-\bar{k}) \neq 0$, it cannot vanish, otherwise $\phi_l(k, r)$ would be zero everywhere. Now, we have

$$\phi_l(\bar{k}, r) \underset{r\to\infty}{\longrightarrow} \tilde{f}_l(-\bar{k}) \exp(i\bar{k}r) \underset{r\to\infty}{\longrightarrow} \tilde{f}_l(-\bar{k}) \exp(-\kappa r + ik_r r). \tag{10.156}$$

So, $\phi_l(\bar{k}, r)$ vanishes at the both ends of the $[0, \infty)$ interval,

$$\phi_l(\bar{k}, r) \underset{r\to 0}{\longrightarrow} 0 \quad \text{and} \quad \phi_l(\bar{k}, r) \underset{r\to\infty}{\longrightarrow} 0, \tag{10.157}$$

it is continuous in between, and therefore it is square integrable

$$0 < \int_0^\infty \mathrm{d}r |\phi_l(\bar{k}, r)|^2 < \infty. \tag{10.158}$$

Consequently, $\phi_l(\bar{k}, r)$ is a normalizable solution of the radial wave equation with angular momentum l, and represents a bound state.

The function $\phi_l(\bar{k}, r)$ satisfies the radial Schrödinger equation

$$\frac{\mathrm{d}^2\phi_l(\bar{k}, r)}{\mathrm{d}r^2} + \left[\bar{k}^2 - \frac{l(l+1)}{r^2} - V^{'}(r)\right]\phi_l(\bar{k}, r) = 0, \tag{10.159}$$

and the complex conjugate reads

$$\frac{\mathrm{d}^2\phi_l^*(\bar{k}, r)}{\mathrm{d}r^2} + \left[(\bar{k}^2)^* - \frac{l(l+1)}{r^2} - V^{'}(r)\right]\phi_l^*(\bar{k}, r) = 0. \tag{10.160}$$

If we multiply Eq. (10.159) by $\phi_l^*(\bar{k}, r)$, Eq. (10.160) by $\phi_l(\bar{k}, r)$, and subtract them, we find

$$\frac{\mathrm{d}}{\mathrm{d}r}\left(\phi_l^*(\bar{k}, r)\frac{\mathrm{d}\phi_l(\bar{k}, r)}{\mathrm{d}r} - \frac{\mathrm{d}\phi_l^*(\bar{k}, r)}{\mathrm{d}r}\phi_l(\bar{k}, r)\right) + 2i\,\mathrm{Im}\,(\bar{k}^2)|\phi_l(\bar{k}, r)|^2 = 0. \tag{10.161}$$

Integrating over r from 0 to ∞ and recalling the boundary conditions in Eq. (10.157), we obtain

$$2i\,\mathrm{Im}\,(\bar{k}^2)\int_0^\infty \mathrm{d}r |\phi_l(\bar{k}, r)|^2 = 0. \tag{10.162}$$

The integral is the norm of a bound state and cannot vanish. Therefore Im $(\bar{k}^2)$ has to be zero, and consequently, $\bar{k}$ has to be pure imaginary, i.e. $\bar{k} = i\kappa$ with $\kappa > 0$ real. The corresponding energy is $E = -\hbar^2\kappa^2/(2m)$ and the state behaves like $\phi_l(\bar{k}, r) \sim \exp(-\kappa r)$ as $r \to \infty$.

The Jost function $\tilde{f}_l(k)$ cannot have any zeros on the real k axis. If $\tilde{f}_l(k)$ were zero for real k, then $\phi_{l,k}$ would vanish. Since, in the $|k| \to \infty$ limit all solutions become undistorted plane waves,

$$\lim_{|k|\to\infty} \tilde{f}_l(k) = 1. \tag{10.163}$$

Levinson Theorem

The *Levinson theorem* highlights an intimate connection between the phase shift and the bound states. For a spherically symmetric potential $V(r)$, we have

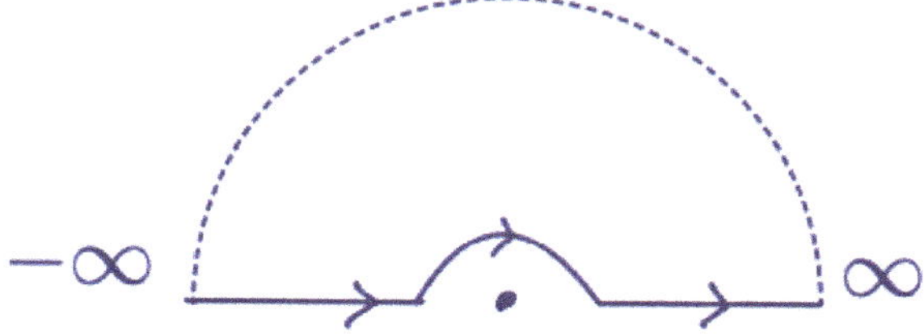

Fig. 10.7 Contour C for evaluating the integral in the proof of the Levinson theorem

$$\delta_l(0) - \delta_l(\infty) = N_l\pi, \tag{10.164}$$

where N_l is the number of bound states for angular momentum l. If $l = 0$, and the system has a zero energy bound state, we have an extra term

$$\delta_0(0) - \delta_0(\infty) = (N_0 + 1/2)\pi. \tag{10.165}$$

To prove this relation, we evaluate the contour integral

$$I = \frac{1}{2\pi i}\oint_C \mathrm{d}k \frac{1}{\tilde{f}_l(k)}\frac{\mathrm{d}\tilde{f}_l(k)}{\mathrm{d}k}, \tag{10.166}$$

where the contour C is taken along the real k axis and it is closed over the imaginary positive infinity (Fig. 10.7). On the upper half k-plane, the integrand is analytic, except for possible zeros of $\tilde{f}_l$ on the imaginary k-axis that correspond to bound states.

According to the *Argument Theorem*, this integral is equal to $N - P$, where N and P are the number of zeros and poles of $\tilde{f}_l(k)$, respectively. Obviously, $\tilde{f}_l(k)$ cannot have poles because in this case the regular solution would be infinite. So, the integral just gives the number of bound states.

The integral can be written as

$$I = \frac{1}{2\pi i}\oint_C \mathrm{d}k \frac{\mathrm{d}}{\mathrm{d}k}(\log \tilde{f}_l(k)) = \frac{1}{2\pi i}\int_{-\infty}^{\infty} \mathrm{d}k \frac{\mathrm{d}}{\mathrm{d}k}(\log \tilde{f}_l(k)), \tag{10.167}$$

since, from Eq. (10.163), it follows that the contribution from the large semicircle is zero. On the positive real axis,

$$\tilde{f}_l(k) = |\tilde{f}_l(k)| \exp(-i\delta_l(k)), \tag{10.168}$$

and thus

$$\log \tilde{f}_l(k) = \log |\tilde{f}_l(k)| - i\delta_l(k). \tag{10.169}$$

On the negative real axis, from the relation $\tilde{f}_l(-k) = \tilde{f}_l^*(k)$, we find that

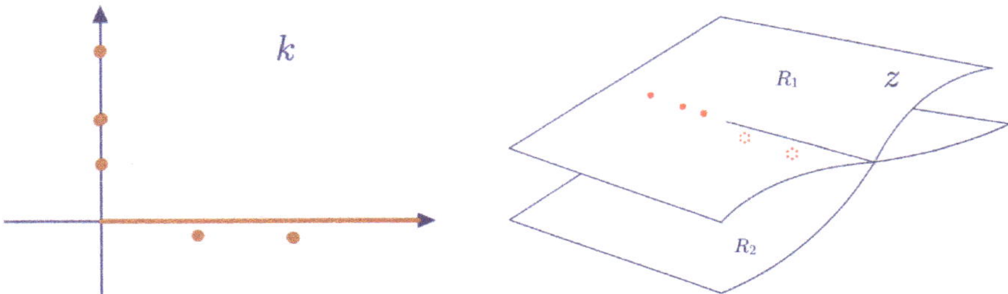

Fig. 10.8 The zeros of $\tilde{f}_l$ on the imaginary k-axis correspond to bound states, while the zeros under the positive k-axis correspond to resonant states. On the double layer Riemann z-sheet the bound state zeros are located on the upper, or physical, R_1 Riemann level, while the zeros corresponding to resonances are on the second, or unphysical, R_2 Riemann sheet

$$\log \tilde{f}_l(-k) = \log|\tilde{f}_l(k)| + i\delta_l(k). \tag{10.170}$$

Evaluating the integral, we see that some terms cancel out, and finally we have

$$N_l = I = \frac{1}{2\pi i}\left\{ [\log|\tilde{f}_l(k)| + i\delta_l(k)]\Big|_{-\infty}^{0} + [\log|\tilde{f}_l(k)| - i\delta_l(k)]_0^\infty \right\} = \frac{2i}{2\pi i}(\delta(0) - \delta(\infty)), \tag{10.171}$$

which proves our assertion in Eq. (10.164). In the $l = 0$ case, if $f_0(0) = 0$, a small semicircle around zero gives another $\pi/2$ term.

Resonances

Now, we consider the possible zeros of $\tilde{f}_l(k)$ in the lower half imaginary k-plane, close to the real axis (Fig. 10.8).

These zeros can be written in the form $\tilde{f}_l(k_0) = 0$, with $k_0 = k_r - i\kappa, k_r > 0$, and $\kappa > 0$. For these k values, $\phi_l(k, r)$ are not normalizable and cannot represent bound states. Due to the analyticity of the Jost functions, if $\tilde{f}_l(k_0) = 0$, then $\tilde{f}_l(-k_0^*) = \tilde{f}_l(-k_r - i\kappa) = 0$ as well. If k_0 is a simple zero of $\tilde{f}_l(k)$, then for real k in the vicinity of k_0, we can approximate

$$\tilde{f}_l(k) \approx (k - k_r + i\kappa)\, C, \quad \text{where} \quad C = \rho \exp(-i\theta), \tag{10.172}$$

and then

$$\tilde{f}_l(-k) \approx \tilde{f}_l^*(k^*) \approx (k - k_r - i\kappa)\, C^*. \tag{10.173}$$

As a result,

$$s_l(k) = \frac{\tilde{f}_l(-k)}{\tilde{f}_l(k)} \approx \frac{C^*}{C}\frac{k - k_r - i\kappa}{k - k_r + i\kappa} = \exp(2i\theta)\frac{k - k_r - i\kappa}{k - k_r + i\kappa}. \tag{10.174}$$

For simplicity, we assume $\theta = 0$, and then, from Eq. (10.122), we have

$$\exp(2i\delta_l(k)) \approx \frac{k - k_r - i\kappa}{k - k_r + i\kappa}. \tag{10.175}$$

Since k is real, we can easily infer that

$$\tan \delta_l(p) \approx \frac{-\kappa}{k - k_r}, \tag{10.176}$$

and then,

$$\sigma_l(k) \sim |f_l|^2 \sim \sin^2 \delta_l(k) \approx \frac{\kappa^2}{(k - k_r)^2 + \kappa^2}. \tag{10.177}$$

We have found that the partial wave scattering cross section $\sigma_l(k)$ peaks at k_r. The imaginary part κ describes the width of the peak. So, the zeros of the Jost function on the negative imaginary k-plane close to the real axis correspond to *resonances*. Equation (10.176) shows that if $\kappa \ll k_r$, as k ranges across the real axis from $k \ll k_r$ to $k_r \ll k$, the scattering phase shift $\delta_l(k)$ changes sharply, roughly from 0 to π, going through $\pi/2$ at exactly $k = k_r$. This is a key feature of resonances.

In terms of energy variables, $E_0 = \hbar^2 k_0^2/(2m) = E_r - i\Gamma/2$, then Eq. (10.177) becomes

$$\sin^2 \delta_l(E) = \frac{\Gamma^2/4}{(E - E_r)^2 + \Gamma^2/4}. \tag{10.178}$$

This is the *Breit-Wigner formula* for a resonance, where E_r is the resonance energy and Γ is the width.

Since $E \propto k^2$, when E is a complex variable in $\tilde{f}_l(k(E))$ or $s_l(k(E))$, the Riemann surface of these functions is a two-sheeted surface cut along the positive real axis. The first sheet corresponding to Im $(k) > 0$ is called the "physical sheet". The bound states are associated with zeros of $\tilde{f}_l(k(E))$ on the negative real axis of the physical sheet, while the resonances are related to the zeros on the second "unphysical" sheet, right below the positive real axis. The asymptotic form of the resonant wave function reads

$$u_l \approx \exp(ik_0 r)\exp(-iE_0 t) \approx \exp[i\,(k_r r - E_r t)] \cdot \exp(\kappa r)\exp(-\Gamma t/2). \tag{10.179}$$

This describes a plane wave with exponentially growing amplitude, which, however, exponentially decays in time. So, resonant states are states with finite lifetime, i.e. "metastable" states.

10.2.4 Time Delay

The asymptotic form of the radial scattering wave function with stationary time dependence is given by

$$\psi_{l,k}(r) \underset{r\to\infty}{\longrightarrow} [\exp(-i(kr - l\pi/2)) - \exp(2i\delta_l)\exp(i(kr - l\pi/2))]\exp(-iEt), \tag{10.180}$$

with $E = \hbar^2 k^2/(2m)$. The total wave function of the wave packet is given as an integral over a momentum distribution $\phi(k)$, i.e.

$$\psi(r,t) = \int \mathrm{d}k\ \psi_{l,k}(r,t)\phi(k), \tag{10.181}$$

where we assume that $\phi(k)$ is strongly peaked around $\bar{k}$. Only a small region of k around $\bar{k}$ contributes significantly to the integral. We approximate $E(k)$ around $\bar{k}$,

$$E(k) \approx \bar{E} + \bar{v}(k - \bar{k}), \quad \text{with} \quad \bar{E} = E(\bar{k}) \quad \text{and} \quad \bar{v} = \left.\frac{\mathrm{d}E}{\mathrm{d}k}\right|_{\bar{k}} = \frac{\hbar^2 \bar{k}}{m}. \tag{10.182}$$

The incoming wave at large r distances behaves like

$$\psi_{in}(r,t) \sim \int \mathrm{d}k\ \exp(-i(kr + Et - l\pi/2))\phi(k) \approx \exp(-i(\bar{k}r + \bar{E}t))i^l \int \mathrm{d}q\ \exp(-iq(r + \bar{v}t))\tilde{\phi}(q), \tag{10.183}$$

where $q = k - \bar{k}$. In deriving this formula, we changed the integration from k to q and we replaced $\phi(k)$ by $\tilde{\phi}(q)$, which peaks at $q = 0$. The integral is the Fourier transform of the momentum-space function $\tilde{\phi}(q)$, so we can write

$$\psi_{in}(r,t) \sim \exp(-i(\bar{k}r + \bar{E}t))\Phi(r + \bar{v}t). \tag{10.184}$$

If $\tilde{\phi}(q)$ is a narrowly peaked Gaussian-like wave packet

$$\tilde{\phi}(q) \sim \exp(-b^2 q^2/2), \tag{10.185}$$

then in coordinate representation it looks like

$$\Phi(x) \sim \exp(-x^2/(2b^2)), \tag{10.186}$$

and $\psi_{in}(r,t)$ looks like a wave packet traveling with velocity $\bar{v}$.

We can have a similar treatment for the outgoing wave. We approximate the phase shift as

$$\delta_l(k) \approx \delta_l(\bar{k}) + (k - \bar{k}) \left.\frac{\mathrm{d}\delta_l}{\mathrm{d}k}\right|_{\bar{k}} . \tag{10.187}$$

Then, for large r, the outgoing wave takes the form

$$\begin{aligned}\psi_{out}(r,t) &\sim -\int \mathrm{d}k \exp(i(kr + Et - l\pi/2)) \exp(2i\delta_l)\phi(k) \\ &\sim -\exp(i\bar{k}r - i\bar{E}t)\exp(2i\delta_l(\bar{k}))(-i)^l \int \mathrm{d}q \exp(-iq[-(r - \bar{v}t + \Delta r)])\tilde{\phi}(q),\end{aligned} \tag{10.188}$$

where

$$\Delta r = 2 \left.\frac{\mathrm{d}\delta_l}{\mathrm{d}k}\right|_{\bar{k}} = 2\bar{v} \left.\frac{\mathrm{d}\delta_l}{\mathrm{d}E}\right|_{\bar{E}} . \tag{10.189}$$

Similarly as before, we find

$$\psi_{out}(r,t) \sim \exp(i\bar{k}r - i\bar{E}t)\exp(2i\delta_l(\bar{k}))\, \Phi[-(r - \bar{v}t + \Delta r)]. \tag{10.190}$$

The outgoing wave packet is localized around $r = \bar{v}t - \Delta r$, and travels away from the origin with group velocity $\bar{v}$. It follows the incoming wave packet with distance Δr, or by a time delay

$$\tau = \frac{\Delta r}{\bar{v}} = 2 \left.\frac{\mathrm{d}\delta_l}{\mathrm{d}E}\right|_{\bar{E}} . \tag{10.191}$$

Around resonant energies, the phase shift changes rapidly and $\mathrm{d}\delta_l/\mathrm{d}E$ is large, indicating a large time delay. In the case of resonant scattering, the wave packet spends a considerably long time in the potential before it goes into the outgoing channel.

10.3 Scattering in Coulomb-Like Potential

Coulomb scattering is extremely important since most of the particles in atomic and nuclear physics are charged particles. Thus far, we have assumed that the particles at asymptotic distances move freely. This is, however, not the case for the Coulomb potential. Therefore Coulomb scattering needs special treatment.

10.3.1 Classical Coulomb Scattering

Before highlighting the peculiarities of Coulomb scattering in quantum mechanics, we may have a look at the corresponding classical mechanical problem. We consider the Hamilton function

$$H = \frac{p^2}{2m} + V_{eff}(r) = E = \frac{1}{2}mv_0^2, \tag{10.192}$$

where

$$V_{eff}(r) = V(r) + \frac{l^2}{2mr^2}, \tag{10.193}$$

with angular momentum l and asymptotic velocity v_0. Reorganizing, we find

$$p = \sqrt{2m[E - V_{eff}(r)]} = m\frac{\mathrm{d}r}{\mathrm{d}t}. \tag{10.194}$$

For large r values, when $V_{eff}(r) \ll E$, we have

$$\frac{\mathrm{d}r}{\mathrm{d}t} = \sqrt{\frac{2}{m}(E - V_{eff}(r))} = v_0\sqrt{1 - \frac{V_{eff}(r)}{E}} \approx v_0\left(1 - \frac{1}{2}\frac{V_{eff}(r)}{E}\right). \tag{10.195}$$

Now, we assume that

$$V_{eff}(r) \underset{r\to\infty}{\longrightarrow} \frac{f}{r^\alpha}, \quad \text{with} \quad \alpha \geq 1. \tag{10.196}$$

We solve the equation by using the method of *dominant balance*. In the first step, we take only the dominant term on the right hand side of Eq. (10.195)

$$\frac{\mathrm{d}r}{\mathrm{d}t} \simeq v_0, \tag{10.197}$$

which gives

$$r \simeq r_0 + v_0 t, \tag{10.198}$$

with initial position r_0.

In the next step, we assume

$$r \simeq r_0 + v_0 t + r_1, \tag{10.199}$$

where r_1 is an unknown function. Substituting into Eq. (10.195) we find

$$v_0 + \frac{\mathrm{d}r_1}{\mathrm{d}t} \simeq v_0 - \frac{v_0}{2E}\frac{f}{(v_0 t)^\alpha}. \tag{10.200}$$

Now, if $\alpha > 1$, we obtain

$$r_1 \sim t^{-\alpha+1}. \tag{10.201}$$

For large t, and consequently for large r, this function vanishes

$$\lim_{t\to\infty} r_1 \to 0. \tag{10.202}$$

So, the asymptotic motion for these types of potentials is a free motion described by Eq. (10.198).

The $\alpha = 1$ case is genuinely different. Here

$$\frac{\mathrm{d}r_1}{\mathrm{d}t} \simeq -\frac{f}{2E}\frac{1}{t}, \tag{10.203}$$

which results in

$$r \simeq r_0 + v_0 t - \frac{f}{2E}\log t. \tag{10.204}$$

The $\log t$, just like t, tends to infinity as t goes to infinity, and it is not negligible compared to t. Therefore, the motion in a Coulomb-like potential is never a free motion, not even at asymptotic distances.

10.3.2 Quantum Coulomb Scattering

We consider a particle moving in a Coulomb-like potential

$$\hat{V} = \hat{V}^{(C)} + \hat{V}^{(s)}, \tag{10.205}$$

where $\hat{V}^{(C)} = Z/r$ is the pure Coulomb potential and $\hat{V}^{(s)}$ is a short-range potential that falls off exponentially at large distances. Then, the Hamiltonian is given by

$$\hat{H} = \hat{H}_0 + \hat{V}^{(C)} + \hat{V}^{(s)} = \hat{H}^{(C)} + \hat{V}^{(s)}, \tag{10.206}$$

where $\hat{H}^{(C)}$ is the pure Coulomb Hamiltonian. We define the corresponding Green's operators

$$\hat{G}(z) = (z - \hat{H})^{-1} \quad \text{and} \quad \hat{G}^{(C)}(z) = (z - \hat{H}^{(C)})^{-1}, \tag{10.207}$$

which satisfy the resolvent relations

$$\hat{G}(z) = \hat{G}^{(C)}(z) + \hat{G}^{(C)}(z)\hat{V}^{(s)}\hat{G}(z) \quad \text{and} \quad \hat{G}(z) = \hat{G}^{(C)}(z) + \hat{G}(z)\hat{V}^{(s)}\hat{G}^{(C)}(z). \tag{10.208}$$

The same way as in the non-Coulomb case, we can link the S matrix elements to the Green's operator

$$S_{k'k} = \lim_{t\to\infty}\lim_{\varepsilon\to 0} i\varepsilon e^{i(E'-E)t/\hbar}\langle \boldsymbol{k}'|\hat{G}(E+i\varepsilon)|\boldsymbol{k}\rangle. \tag{10.209}$$

Now, we should use Eq. (10.208) to find

$$\begin{aligned} S_{k'k} &= \lim_{t\to\infty}\lim_{\varepsilon\to 0} i\varepsilon e^{i(E'-E)t/\hbar}\langle \boldsymbol{k}'|\hat{G}^{(C)}(E+i\varepsilon)|\boldsymbol{k}\rangle \\ &\quad + \lim_{t\to\infty}\lim_{\varepsilon\to 0} i\varepsilon e^{i(E'-E)t/\hbar}\langle \boldsymbol{k}'|\hat{G}^{(C)}(E+i\varepsilon)\hat{V}^{(s)}\hat{G}(E+i\varepsilon)|\boldsymbol{k}\rangle \\ &= S^{(C)}_{k'k} + S^{(Cs)}_{k'k}, \end{aligned} \tag{10.210}$$

where the first term is the pure Coulomb S matrix and the second term is the Coulomb modified short range one.

To evaluate the first term, we insert $\hat{G}_0(\hat{G}_0)^{-1}$ and by using Eq. (10.64), we obtain

$$\begin{aligned} S^{(C)}_{k'k} &= \lim_{t\to\infty}\lim_{\varepsilon\to 0} i\varepsilon e^{i(E'-E)t/\hbar}\langle \boldsymbol{k}'|\hat{G}_0(E+i\varepsilon)(\hat{G}_0(E+i\varepsilon))^{-1}\hat{G}^{(C)}(E+i\varepsilon)|\boldsymbol{k}\rangle \\ &= \lim_{t\to\infty}\lim_{\varepsilon\to 0} \frac{i\varepsilon}{E-i\varepsilon-E'} e^{i(E'-E)t/\hbar}\langle \boldsymbol{k}'|(\hat{G}_0(E+i\varepsilon))^{-1}\hat{G}^{(C)}(E+i\varepsilon)|\boldsymbol{k}\rangle \\ &= -2\pi i\delta(E'-E)\langle \boldsymbol{k}'^{(C)} - |\boldsymbol{k}\rangle = -2\pi i\delta(E'-E)\langle \boldsymbol{k}'|\boldsymbol{k}^{(C)}+\rangle, \end{aligned} \tag{10.211}$$

where

$$|\boldsymbol{k}^{(C)}\pm\rangle = \lim_{\varepsilon\to 0}\hat{G}^{(C)}(E\pm i\varepsilon)(\hat{G}_0(E\pm i\varepsilon))^{-1}|\boldsymbol{k}\rangle = \lim_{\varepsilon\to 0} \pm i\varepsilon\hat{G}^{(C)}(E\pm i\varepsilon)|\boldsymbol{k}\rangle \tag{10.212}$$

is the Coulomb scattering function. It should be noticed that in $S^{(C)}_{k'k}$, due to the long-range nature of the Coulomb potential, the $\delta(\boldsymbol{k}'-\boldsymbol{k})$ term is absent.

The evaluation of the second term is analogous,

$$\begin{aligned} S^{(Cs)}_{k'k} &= \lim_{t\to\infty}\lim_{\varepsilon\to 0} i\varepsilon e^{i(E'-E)t/\hbar}\langle \boldsymbol{k}'|\hat{G}_0(E+i\varepsilon)(\hat{G}_0(E+i\varepsilon))^{-1}\hat{G}^{(C)}(E+i\varepsilon)V^{(s)}\hat{G}(E+i\varepsilon)|\boldsymbol{k}\rangle \\ &= \lim_{t\to\infty}\lim_{\varepsilon\to 0} e^{i(E'-E)t/\hbar}\frac{i\varepsilon}{E+i\varepsilon-E'}\langle \boldsymbol{k}'|(\hat{G}_0(E+i\varepsilon))^{-1}\hat{G}^{(C)}(E+i\varepsilon)V^{(s)}\hat{G}(E+i\varepsilon)|\boldsymbol{k}\rangle \\ &= -2\pi i\delta(E'-E)\langle \boldsymbol{k}'^{(C)} - |V^{(s)}|\boldsymbol{k}+\rangle, \end{aligned} \tag{10.213}$$

where

$$|\boldsymbol{k}+\rangle = \lim_{\varepsilon\to 0} i\varepsilon\hat{G}(E+i\varepsilon)|\boldsymbol{k}\rangle \tag{10.214}$$

is the scattering function. By using Eq. (10.208) we can easily derive the Lippmann-Schwinger equation

$$|\boldsymbol{k}+\rangle = |\boldsymbol{k}^{(C)}+\rangle + \hat{G}^{(C)}(E+i0)V^{(s)}|\boldsymbol{k}+\rangle. \tag{10.215}$$

As we learned before, the scattering quantities are hidden in the asymptotic behavior of the scattering functions. The scattering problem with the pure Coulomb potential is analytically solvable in parabolic coordinates (see p. 424 in [2]). The parabolic coordinates (ξ, ζ, φ), in terms of spherical coordinates (r, θ, φ), are defined by

$$\xi = r(1-\cos\theta) = r - z, \qquad \zeta = r(1+\cos\theta) = r + z, \qquad \varphi = \varphi. \tag{10.216}$$

The Schrödinger equation with Coulomb potential reads

$$\left[-\frac{4}{\xi+\zeta}\left(\frac{\partial}{\partial\xi}\xi\frac{\partial}{\partial\xi}+\frac{\partial}{\partial\zeta}\zeta\frac{\partial}{\partial\zeta}\right)-\frac{1}{\xi\zeta}\frac{\partial^2}{\partial\varphi^2}+\frac{\gamma k}{\xi+\zeta}\right]\langle \boldsymbol{r}|\boldsymbol{k}^{(C)}+\rangle = k^2\langle \boldsymbol{r}|\boldsymbol{k}^{(C)}+\rangle. \tag{10.217}$$

If $\boldsymbol{e}_z$ is the direction of the incoming wave we can drop the $\partial^2/\partial\varphi^2$ term because of the axial symmetry of the problem. We can seek the solution in the form

$$\langle \boldsymbol{r}|\boldsymbol{k}^{(C)}+\rangle = \exp(ikz)f(\xi) \tag{10.218}$$

and obtain

$$\xi f'' + (1-ik\xi)f' - \gamma k f = 0, \tag{10.219}$$

where $\gamma = Zm/(\hbar^2 k)$. Eq. (10.219) is a differential equation for the ${}_1F_1$ confluent hypergeometric function, whose regular solution is

$$f(\xi) \simeq {}_1F_1(-i\gamma;1;ik\xi). \tag{10.220}$$

From the large ξ asymptotic behavior of the ${}_1F_1$ confluent hypergeometric function we can infer that the asymptotic form of the Coulomb scattering wave function is a Coulomb distorted plane wave and a Coulomb distorted spherical wave

$$\langle \boldsymbol{r}|\boldsymbol{k}^{(C)}+\rangle \simeq \exp[i(kz+\gamma\log(r-z))] + \frac{f^{(C)}(\theta)}{r}\exp[i(kr-\gamma\log 2kr)], \tag{10.221}$$

where

$$f^{(C)}(\theta) = \frac{\gamma}{2p\sin^2(\theta/2)}\exp[-i\gamma\log\sin^2(\theta/2)+2i\eta_0], \tag{10.222}$$

with $\eta_0 = \arg\Gamma(1+i\gamma)$. This results in the cross section for the pure Coulomb field, the Rutherford formula

$$\frac{d\sigma(\theta)}{d\Omega} = |f^{(C)}(\theta)|^2 = \frac{\gamma^2}{4k^2\sin^4(\theta/2)} = \left(\frac{Z}{4E\sin^2(\theta/2)}\right)^2. \tag{10.223}$$

If the short range potential $\hat{V}^{(s)}$ is spherically symmetric, we can expand the Coulomb modified scattering amplitude in partial waves

$$f^{(Cs)}(\theta) = \frac{1}{k}\sum_l (2l+1) f_l\, P_l(\cos\theta). \tag{10.224}$$

Here,

$$f_l = \langle\psi_l^{(C)(-)}|\hat{V}^{(s)}|\psi_l^+\rangle = \frac{1}{k}\exp[i(2\eta_l+\delta_l)]\sin\delta_l, \tag{10.225}$$

where $\eta_l = \arg\Gamma(l+1+i\gamma)$ is the Coulomb phase shift and δ_l is the Coulomb modified phase shift. The partial wave Coulomb wave function is given by

$$\psi_{l,k}^{(C)(+)}(r) = \frac{1}{2}(2kr)^{l+1}\exp(-1/2\gamma\pi)\frac{\Gamma(l+1+i\gamma)}{(2l+1)!}\exp(ikr)\,{}_1F_1(l+1+i\gamma;2l+2;-2ikr). \tag{10.226}$$

The total wave function is the solution of the Lippmann-Schwinger equation

$$|\psi_l^{(+)}\rangle = |\psi_l^{(C)(+)}\rangle + \hat{g}_l^{(C)}(E+i0)\hat{V}_l^{(s)}|\psi_l^{(+)}\rangle, \tag{10.227}$$

with $g_l^{(C)}$ being the partial wave Coulomb Green's operator. The phase shifts show up in the asymptotic form of the wave function

$$\psi_{l,k}^{(+)}(r) \underset{r\to\infty}{\longrightarrow} \sin(kr - l\pi/2 - \gamma\log 2kr + \eta_l + \delta_l). \tag{10.228}$$

10.3.3 Solution in Discrete Basis Representation

The Coulomb problem is quite heavy in mathematics and requires strong involvement of special functions. The Coulomb Hamiltonian is given by

$$\hat{h}_l^{(C)} = -\frac{\hbar^2}{2m}\left(\frac{\mathrm{d}^2}{\mathrm{d}r^2} - \frac{l(l+1)}{r^2}\right) + \frac{Z}{r}, \tag{10.229}$$

and the Schrödinger eigenvalue problem for the hydrogen atom reads

$$\left[-\frac{\hbar^2}{2m}\left(\frac{\mathrm{d}^2}{\mathrm{d}r^2} - \frac{l(l+1)}{r^2}\right) - \frac{e^2}{r}\right]\psi_l(r) = E\psi_l(r). \tag{10.230}$$

The energy spectrum consists of an infinite number of negative energy bound states that are accumulating at $E = 0$ and a continuum number of positive energy scattering states. The bound states together with the scattering states form a basis in the angular momentum l subspace.

The Coulomb-Sturmian basis is much simpler. The Coulomb-Sturmian functions are the Sturm-Liouville solutions of the Coulomb Hamiltonian. In Sturm-Liouville problems, the energy eigenvalue is fixed and the depth of the potential is varied. The corresponding differential equation reads

$$\left(-\frac{\mathrm{d}^2}{\mathrm{d}r^2} + \frac{l(l+1)}{r^2} - \frac{2b(n+l+1)}{r}\right)\langle r|nl;b\rangle = -b^2\langle r|nl;b\rangle, \tag{10.231}$$

Here, b is a parameter, n is the radial quantum number and

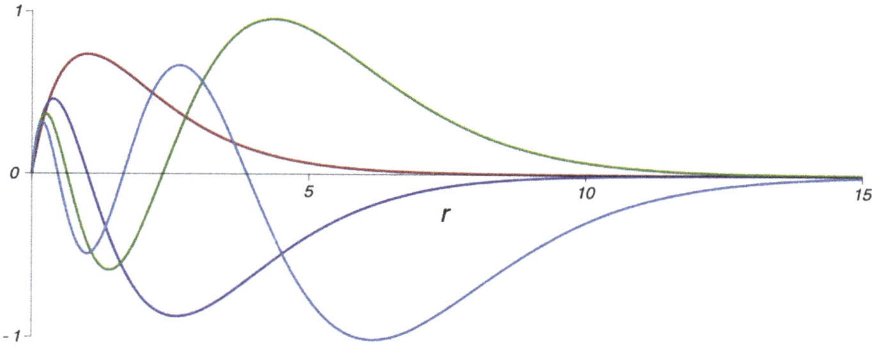

Fig. 10.9 The first few Coulomb-Sturmian functions with $l = 0$ and $b = 1$

$$\langle r|nl;b\rangle = \left(\frac{\Gamma(n+1)}{\Gamma(n+2l+2)}\right)^{1/2} \exp(-br)(2br)^{l+1} L_n^{(2l+1)}(2br), \tag{10.232}$$

where L is the associated Laguerre polynomial. We can see that the Coulomb-Sturmian functions (Fig. 10.9), unlike the hydrogen functions in Fig. 5.4, have the same exponential fall off $\exp(-br)$. With $\langle r|\widetilde{nl;b}\rangle = \langle r|nl;b\rangle/r$, the Coulomb-Sturmian functions are orthogonal

$$\langle nl;b|\widetilde{n'l;b}\rangle = \delta_{nn'} \tag{10.233}$$

and complete

$$\lim_{N\to\infty} \sum_{n=0}^{N} |\widetilde{nl;b}\rangle\langle nl;b| = I. \tag{10.234}$$

We can approximate the short range potential in Eq. (10.227)

$$V_l^{(s)} \approx \sum_{nn'}^{N} |\widetilde{nl;b}\rangle\langle nl;b|V_l^{(s)}|n'l;b\rangle\langle\widetilde{n'l;b}|, \tag{10.235}$$

and then the Lippmann-Schwinger equation becomes

$$|\psi_l^{(+)}\rangle = |\psi_l^{(C)(+)}\rangle + \sum_{nn'}^{N} g_l^{(C)}(E+i0)|\widetilde{nl;b}\rangle \underline{V}_{l,nn'}^{(s)} \langle\widetilde{n'l;b}|\psi_l^{(+)}\rangle, \tag{10.236}$$

with $\underline{V}_{l,nn'}^{(s)} = \langle nl;b|V_l^{(s)}|n'l;b\rangle$, which can always be calculated, at least numerically. We should notice that the summation in n' goes up to N only. Therefore, to solve the equation, we need to apply $\langle\widetilde{n''l,b}|$ only up to $n'' = N$. This results in the matrix equation

$$\underline{\psi}_l^{(+)} = \underline{\psi}_l^{(C)(+)} + \underline{g}_l^{(C)}(E + i0)\underline{V}_l^{(s)}\underline{\psi}_l^{(+)}, \tag{10.237}$$

where $\underline{\psi}_l^{(+)} = \langle \widetilde{nl, b}|\psi_l^{(+)}\rangle$, $\underline{\psi}_l^{(C)(+)} = \langle \widetilde{nl, b}|\psi_l^{(C)(+)}\rangle$ and $\underline{g}_l^{(C)} = \langle \widetilde{nl, b}|\hat{g}_l^{(C)}|\widetilde{n'l, b}\rangle$. The equation is solved by

$$\underline{\psi}_l^{(+)} = [(\underline{g}_l^{(C)})^{-1} - \underline{V}_l^{(s)}]^{-1}(\underline{g}_l^{(C)})^{-1}\underline{\psi}_l^{(C)(+)}. \tag{10.238}$$

The crucial point in this approach is the calculation of the matrix elements of the Coulomb Green's operator. This is possible in the Coulomb-Sturmian basis in an exact and analytic manner. In this basis, the operator $\hat{J} = z - \hat{h}_l^C$ is represented by an infinite symmetric tridiagonal matrix. By using Eq. (10.231) and some known integrals involving Laguerre polynomials, we can readily derive

$$\langle nl; b|(z - \hat{h}_l^{(C)})|n'l; b\rangle = \underline{J}_{nn'} = \begin{cases} \dfrac{k^2 - b^2}{2m/\hbar^2\, b}(n + l + 1) - Z & \text{for } n' = n \\ -\dfrac{k^2 + b^2}{4m/\hbar^2\, b}\sqrt{(n+1)(n+2l+2)} & \text{for } n' = n + 1 \\ -\dfrac{k^2 + b^2}{4m/\hbar^2\, b}\sqrt{n(n+2l+1)} & \text{for } n' = n - 1 \\ 0 & \text{otherwise,} \end{cases} \tag{10.239}$$

where z is a complex number and $k = \sqrt{2m/\hbar^2\, z}$.

The Coulomb Green's operator is defined by the relation

$$(z - \hat{h}_l^{(C)})\hat{g}_l^{(C)}(z) = I, \tag{10.240}$$

which in matrix representation looks like

$$\sum_{n'}^{N+1} \langle nl; b|(z - \hat{h}_l^{(C)})|n'l; b\rangle \langle \widetilde{n'l; b}|\hat{g}_l^{(C)}|\widetilde{ml; b}\rangle = \delta_{nm}, \tag{10.241}$$

or

$$\begin{pmatrix} J_{0,0} & J_{0,1} & 0 & 0 & \dots \\ J_{1,0} & J_{1,1} & J_{1,2} & 0 & \dots \\ 0 & J_{2,1} & J_{2,2} & J_{2,3} & \dots \\ 0 & 0 & J_{3,2} & J_{3,3} & \dots \\ \vdots & \ddots & \ddots & \ddots & \ddots \end{pmatrix} \begin{pmatrix} g_{0,0} & g_{0,1} & g_{0,2} & g_{0,3} & \cdots \\ g_{1,0} & g_{1,1} & g_{1,2} & g_{1,3} & \cdots \\ g_{2,0} & g_{2,1} & g_{2,2} & g_{2,3} & \cdots \\ g_{3,0} & g_{3,1} & g_{3,2} & g_{3,3} & \cdots \\ \vdots & \ddots & \ddots & \ddots & \ddots \end{pmatrix} = \begin{pmatrix} 1 & 0 & 0 & 0 & \dots \\ 0 & 1 & 0 & 0 & \dots \\ 0 & 0 & 1 & 0 & \dots \\ 0 & 0 & 0 & 1 & \dots \\ \vdots & \ddots & \ddots & \ddots & \ddots \end{pmatrix}. \tag{10.242}$$

We should keep in mind that in Eq. (10.241), $n, m = 0, 1, \dots, N$. Then, due to the tridiagonal structure of $\underline{J}_{nn'}$, $n' = 0, 1, \dots, N, N + 1$. For $n = N$, only one extra term beyond N comes up

$$J_{N,N-1}g_{N-1,m} + J_{N,N}g_{N,m} + J_{N,N+1}g_{N+1,m} = \delta_{N,m}. \tag{10.243}$$

We can eliminate this extra term by redefining $J_{N,N}$

$$J_{N,N-1}g_{N-1,m} + (J_{N,N} + J_{N,N+1}g_{N+1,m}/g_{N,m})g_{N,m} = \delta_{N,m}. \tag{10.244}$$

The ratio $g_{N+1,m}/g_{N,m}$ can be obtained from the next three-term relation

$$J_{N+1,N}g_{N,m} + J_{N+1,N+1}g_{N+1,m} + J_{N+1,N+2}g_{N+2,m} = 0. \tag{10.245}$$

Rearranging, we find

$$\left(-\frac{1}{J_{N+1,N}}\frac{g_{N+1,m}}{g_{N,m}}\right)^{-1} = J_{N+1,N+1} - J_{N+1,N+2}\left(-\frac{1}{J_{N+2,N+1}}\frac{g_{N+2,m}}{g_{N+1,m}}\right)J_{N+2,N+1}, \tag{10.246}$$

or with

$$C_{N+1} = -\frac{1}{J_{N+1,N}}\frac{g_{N+1,m}}{g_{N,m}}, \tag{10.247}$$

we have

$$C_{N+1}^{-1} = J_{N+1,N+1} - J_{N+1,N+2}\frac{1}{C_{N+2}^{-1}}J_{N+2,N+1}. \tag{10.248}$$

We can calculate C_{N+1}^{-1} if C_{N+2}^{-1} is at our disposal, and we can calculate C_{N+2}^{-1} if we know C_{N+3}^{-1}, and so on. Putting all this together, we get a continued fraction

$$\begin{aligned} C_{N+1}^{-1} = J_{N+1,N+1} - J_{N+1,N+2}(J_{N+2,N+2} - J_{N+2,N+3}(J_{N+3,N+3} - \\ \dots)^{-1}J_{N+3,N+2})^{-1}J_{N+2,N+1}. \end{aligned} \tag{10.249}$$

This continued fraction does not depend on m, consequently, the correction term is the same for all m's. Therefore, we can write Eq. (10.244) as

$$(\underline{J}^{(N)} - \delta_{i,N}\delta_{j,N}J_{N,N+1}C_{N+1}J_{N+1,N})\underline{g}^{(N)} = I^{(N)}. \tag{10.250}$$

We have found that the modified $\underline{J}^{(N)}$ is the inverse of $\underline{g}^{(N)}$ and the modification amounts to adding a continued fraction term to the bottom-right corner

$$\underline{g}^{(N)} = (\underline{J}^{(N)} - \delta_{i,N}\delta_{j,N}J_{N,N+1}C_{N+1}J_{N+1,N})^{-1}. \tag{10.251}$$

This result is rather general, we have utilized only the tridiagonal feature of the J matrix. In the case of the Coulomb Green's matrix, we can link this continued fraction to the ratio of ${}_2F_1$ hypergeometric functions

$$C_N = -\frac{4m/\hbar^2 b}{(b-ik)^2(N+l+i\gamma)} \frac{{}_2F_1(-l+i\gamma, N+1; N+l+2+i\gamma; (b+ik)^2/(b-ik)^2)}{{}_2F_1(-l+i\gamma, N; N+l+1+i\gamma; (b+ik)^2/(b-ik)^2)}. \tag{10.252}$$

We also need the overlap of Coulomb-Sturmian and Coulomb functions $\underline{\psi}_l^{(C)(\pm)} = \langle \widetilde{nl; b} | \psi_l^{(C)(\pm)} \rangle$. We can separate off the Coulomb phase, $\psi_l^{(C)(\pm)} = \exp(\pm i\eta_l)\phi_l^{(C)}$, with $\phi_l^{(C)}$, a real function. Then, we can calculate $\underline{\phi}_l^{(C)} = \langle \widetilde{nl; b} | \phi_l^{(C)} \rangle$ from the discontinuity of the Coulomb Green's matrix using Eq. (10.54). Considering Eq. (10.131) we obtain,

$$-\frac{\hbar^2 k}{4im}\left(\underline{g}_l^{(C)}(E+i0) - \underline{g}_l^{(C)}(E-i0)\right) = \underline{\phi}_l^{(C)}(\underline{\phi}_l^{(C)})^T, \tag{10.253}$$

and from that, the desired matrix elements $\underline{\psi}_l^{(C)(\pm)}$, can easily be inferred.

Having the Coulomb modified scattering state solution in Coulomb-Sturmian basis representation, the evaluation of the partial wave scattering amplitude is straightforward. From Eq. (10.225) we have

$$f_l = \frac{1}{k}\exp[i(2\eta_l + \delta_l)]\sin\delta_l = \langle \psi_l^{(C)(-)} | V_l^{(s)} | \psi_l^{(+)} \rangle = \underline{\psi}_l^{(C)(-)} \underline{V}_l^{(s)} \underline{\psi}_l^{(+)}. \tag{10.254}$$

For calculating bound and resonant states, we can solve the homogeneous version of Eq. (10.227)

$$|\psi_l\rangle = \hat{g}_l^{(C)}(E) V_l^{(s)} |\psi_l\rangle \tag{10.255}$$

for real and complex energies, respectively. In this case, the Fredholm determinant should vanish

$$D(E) = |1 - \underline{g}_l^{(C)}(E)\underline{V}_l^{(s)}| = |\underline{g}_l^{(C)}(E)||(\underline{g}_l^{(C)}(E))^{-1} - \underline{V}_l^{(s)}| = 0. \tag{10.256}$$

We should notice that in this approach, the wave function is not a linear combination of the basis functions. Rather, as can be seen in Eq. (10.236), it is a linear combination of $\langle r | g_l^{(C)} | \widetilde{nl; b} \rangle$, which has the correct Coulomb-like asymptotic behavior. The method is also applicable for the non-Coulomb case. One just needs to take $Z = 0$ in all formulae.

10.4 Few-Body Problems

10.4.1 Two-Body Problems

Consider particles α and β with masses m_α and m_β located at $\boldsymbol{r}_\alpha$ and $\boldsymbol{r}_\beta$, respectively. If we assume that the interaction depends on the relative distance $\boldsymbol{r}_\alpha - \boldsymbol{r}_\beta$, the two-particle Hamiltonian reads

$$\hat{H} = \frac{\hat{\boldsymbol{p}}_\alpha^2}{2m_\alpha} + \frac{\hat{\boldsymbol{p}}_\beta^2}{2m_\beta} + \hat{v}(\boldsymbol{r}_\alpha - \boldsymbol{r}_\beta). \tag{10.257}$$

With center-of-mass coordinates

$$\boldsymbol{R} = \frac{m_\alpha \boldsymbol{r}_\alpha + m_\beta \boldsymbol{r}_\beta}{m_\alpha + m_\beta} \quad \text{and} \quad \boldsymbol{r} = \boldsymbol{r}_\alpha - \boldsymbol{r}_\beta, \tag{10.258}$$

we can rewrite this Hamiltonian as

$$\hat{H} = \frac{\hat{\boldsymbol{P}}^2}{2M} + \frac{\hat{\boldsymbol{p}}^2}{2\mu} + \hat{v}(\boldsymbol{r}), \tag{10.259}$$

where $M = m_\alpha + m_\beta$, $\mu = m_\alpha m_\beta / M$ is the reduced mass, $\hat{\boldsymbol{P}} = -i\hbar\partial/\partial R$ and $\hat{\boldsymbol{p}} = -i\hbar\partial/\partial r$. We can see that the motion in coordinate $\boldsymbol{R}$ is a straightforward free motion which does not present much physics. If we put the observer in the center of mass by imposing the condition $m_\alpha \boldsymbol{r}_\alpha + m_\beta \boldsymbol{r}_\beta = 0$, the center-of-mass kinetic energy vanishes and the Hamiltonian becomes

$$\hat{H} = \frac{\hat{\boldsymbol{p}}^2}{2\mu} + \hat{v}(\boldsymbol{r}). \tag{10.260}$$

So, the quantum mechanical two-body problem is reduced to a problem of one particle in relative coordinate $\boldsymbol{r}$, with reduced mass μ .

10.4.2 Three-Body Problems

The physics gets much more interesting if we have three interacting particles. Let us denote particles α, β and γ, with masses m_α, m_β and m_γ, located at $\boldsymbol{r}_\alpha$, $\boldsymbol{r}_\beta$ and $\boldsymbol{r}_\gamma$, with individual momenta $\hat{\boldsymbol{k}}_\alpha$, $\hat{\boldsymbol{k}}_\beta$ and $\hat{\boldsymbol{k}}_\gamma$, respectively. We can introduce the reduced masses

$$M = m_\alpha + m_\beta + m_\gamma, \quad \mu_1^\alpha = \frac{m_\beta m_\gamma}{m_\beta + m_\gamma}, \quad \mu_2^\alpha = \frac{m_\alpha(m_\beta + m_\gamma)}{m_\alpha + m_\beta + m_\gamma}, \tag{10.261}$$

the Jacobi coordinates, (Fig. 10.10),

$$\boldsymbol{R} = \frac{m_\alpha \boldsymbol{r}_\alpha + m_\beta \boldsymbol{r}_\beta + m_\gamma \boldsymbol{r}_\gamma}{M}, \quad \boldsymbol{\xi}_\alpha = \boldsymbol{r}_\beta - \boldsymbol{r}_\gamma, \quad \boldsymbol{\eta}_\alpha = \boldsymbol{r}_\alpha - \frac{m_\beta \boldsymbol{r}_\beta + m_\gamma \boldsymbol{r}_\gamma}{m_\beta + m_\gamma}, \tag{10.262}$$

and the corresponding Jacobi momenta

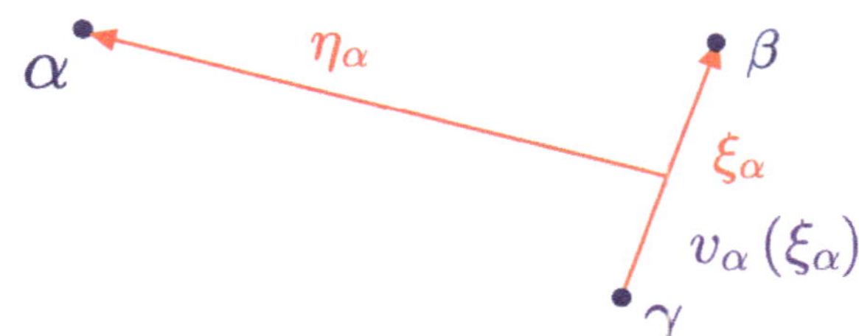

Fig. 10.10 The three-body Jacobi coordinates associated to particle α

$$\hat{\boldsymbol{P}} = \hat{\boldsymbol{k}}_\alpha + \hat{\boldsymbol{k}}_\beta + \hat{\boldsymbol{k}}_\gamma, \quad \hat{\boldsymbol{p}}_\alpha = \frac{m_\gamma \hat{\boldsymbol{k}}_\beta - m_\beta \hat{\boldsymbol{k}}_\gamma}{m_\beta + m_\gamma}, \quad \hat{\boldsymbol{q}}_\alpha = \frac{(m_\beta + m_\gamma)\hat{\boldsymbol{k}}_\alpha - m_\alpha(\hat{\boldsymbol{k}}_\beta + \hat{\boldsymbol{k}}_\gamma)}{M}. \tag{10.263}$$

We can transform one set of Jacobi coordinates into another

$$\boldsymbol{\xi}_\beta = a_1 \boldsymbol{\xi}_\alpha + a_2 \boldsymbol{\eta}_\alpha \quad \text{and} \quad \boldsymbol{\eta}_\beta = a_3 \boldsymbol{\xi}_\alpha + a_4 \boldsymbol{\eta}_\alpha, \tag{10.264}$$

where

$$a_1 = -\frac{m_\beta}{m_\beta + m_\gamma}, \quad a_2 = -(-)^\epsilon, \quad a_3 = (-)^\epsilon \frac{m_\gamma M}{(m_\alpha + m_\gamma)(m_\beta + m_\gamma)}, \quad a_4 = -\frac{m_\alpha}{m_\alpha + m_\gamma}, \tag{10.265}$$

and ϵ is even/odd if (α, β, γ) is an even/odd permutation of $(1, 2, 3)$.

We denote potentials

$$\hat{v}_\alpha(\xi_\alpha) = \hat{v}(\beta, \gamma), \quad \hat{v}_\beta(\xi_\beta) = \hat{v}(\gamma, \alpha), \quad \text{and} \quad \hat{v}_\gamma(\xi_\gamma) = \hat{v}(\alpha, \beta), \tag{10.266}$$

i.e. $v_\alpha(\xi_\alpha)$ is the potential between particles β and γ. We assume that the potentials are of short-range type.

The three-particle Hamiltonian is given by

$$\hat{H} = \frac{\hat{\boldsymbol{k}}_\alpha^2}{2m_\alpha} + \frac{\hat{\boldsymbol{k}}_\beta^2}{2m_\beta} + \frac{\hat{\boldsymbol{k}}_\gamma^2}{2m_\gamma} + v_\alpha + v_\beta + v_\gamma. \tag{10.267}$$

This Hamiltonian, in terms of Jacobi momenta, becomes

$$\hat{H} = \frac{\hat{\boldsymbol{P}}^2}{2M} + \hat{H}_0 + \hat{v}_\alpha + \hat{v}_\beta + \hat{v}_\gamma, \tag{10.268}$$

where

$$\hat{H}_0 = \frac{\hat{\boldsymbol{p}}_\alpha^2}{2\mu_1^\alpha} + \frac{\hat{\boldsymbol{q}}_\alpha^2}{2\mu_2^\alpha} = \frac{\hat{\boldsymbol{p}}_\beta^2}{2\mu_1^\beta} + \frac{\hat{\boldsymbol{q}}_\beta^2}{2\mu_2^\beta} = \frac{\hat{\boldsymbol{p}}_\gamma^2}{2\mu_1^\gamma} + \frac{\hat{\boldsymbol{q}}_\gamma^2}{2\mu_2^\gamma}. \tag{10.269}$$

Here again, we can omit the kinetic energy term associated to the center of mass coordinate and work with the Hamiltonian

$$\hat{H} = \hat{H}_0 + \hat{v}_\alpha + \hat{v}_\beta + \hat{v}_\gamma. \tag{10.270}$$

A new feature of three-body systems is that they may have three genuinely different types of asymptotic configurations, i.e. three different types of asymptotic channels. The α-channel is when particles β and γ form a bound state and the particle α is far away. The corresponding asymptotic channel Hamiltonian is given by

$$\hat{H}_\alpha = \hat{H}_0 + \hat{v}_\alpha = \frac{\hat{\boldsymbol{p}}_\alpha^2}{2\mu_1^\alpha} + \frac{\hat{\boldsymbol{q}}_\alpha^2}{2\mu_2^\alpha} + \hat{v}_\alpha = \hat{h}_{\xi_\alpha} + \hat{h}_{\eta_\alpha}, \tag{10.271}$$

where

$$\hat{h}_{\xi_\alpha} = \frac{\hat{\boldsymbol{p}}_\alpha^2}{2\mu_1^\alpha} + \hat{v}_\alpha \quad \text{and} \quad \hat{h}_{\eta_\alpha} = \frac{\hat{\boldsymbol{q}}_\alpha^2}{2\mu_2^\alpha}. \tag{10.272}$$

We have analogous formulae for the β and the γ channels. Another possibility is that all three particles are far apart, but this situation can be described, for example, as particle α is far away and particles β and γ are in a scattering state. The total Hamiltonian can be written as

$$\hat{H} = \hat{H}_\alpha + \hat{V}^\alpha = \hat{H}_\beta + \hat{V}^\beta = \hat{H}_\gamma + \hat{V}^\gamma, \tag{10.273}$$

with $\hat{V}^\alpha = \hat{v}_\beta + \hat{v}_\gamma$, $\hat{V}^\beta = \hat{v}_\alpha + \hat{v}_\gamma$ and $\hat{V}^\gamma = \hat{v}_\beta + \hat{v}_\alpha$.

The asymptotic channel states are the eigenstates of the channel Hamiltonian

$$\hat{H}_\alpha |\Phi_\alpha\rangle = E|\Phi_\alpha\rangle. \tag{10.274}$$

Here $|\Phi_\alpha\rangle = |\phi_\alpha(\xi_\alpha)\rangle|\chi_\alpha(\eta_\alpha)\rangle$ with

$$\hat{h}_{\xi_\alpha}|\phi_\alpha(\xi_\alpha)\rangle = \epsilon_\alpha|\phi_\alpha(\xi_\alpha)\rangle \quad \text{and} \quad \hat{h}_{\eta_\alpha}|\chi_\alpha(\eta_\alpha)\rangle = q_\alpha^2/2\mu_2^\alpha|\chi_\alpha(\eta_\alpha)\rangle, \tag{10.275}$$

i.e. $|\phi_\alpha(\xi_\alpha)\rangle$ is a bound state in coordinate ξ_α, $|\chi_\alpha(\eta_\alpha)\rangle$ is a scattering state in η_α, and $E = \epsilon_\alpha + q_\alpha^2/2\mu_2^\alpha$.

The channel Møller operators are defined by

$$\begin{aligned} \hat{\Omega}_\alpha^{(\pm)} &= \lim_{t\to\mp\infty} \exp(i\hat{H}t/\hbar)\exp(-i\hat{H}_\alpha t/\hbar) \\ &= \lim_{\varepsilon\to 0} \pm\varepsilon/\hbar \int_{\mp\infty}^{0} \mathrm{d}t \exp(\pm\varepsilon t/\hbar)\exp(i\hat{H}t/\hbar)\exp(-i\hat{H}_\alpha t/\hbar). \end{aligned} \tag{10.276}$$

This operator maps the asymptotic channel state to the scattering state

$$|\Psi^{(+)}\rangle = \hat{\Omega}_\alpha^{(+)}|\Phi_\alpha\rangle = \lim_{\varepsilon\to 0} i\varepsilon\hat{G}(E + i\varepsilon)|\Phi_\alpha\rangle, \tag{10.277}$$

where $\hat{G}(z) = (z - \hat{H})^{-1}$ is the total Green's operator. With the help of intertwining relation $\hat{H}\hat{\Omega}_\alpha^{(\pm)} = \hat{\Omega}_\alpha^{(\pm)}\hat{H}_\alpha$, we can easily show that $|\Psi^{(\pm)}\rangle$ is an eigenstate of $\hat{H}$

$$\hat{H}|\Psi^{(+)}\rangle = E|\Psi^{(+)}\rangle. \tag{10.278}$$

We can define three Green's operators

$$\hat{G}_\lambda(z) = (z - \hat{H}_\lambda)^{-1}, \tag{10.279}$$

for the corresponding channels $\lambda = \alpha, \beta, \gamma$. With the help of resolvent relations

$$\hat{G}(z) = \hat{G}_\lambda(z) + \hat{G}_\lambda(z)\hat{V}^\lambda\hat{G}(z) \quad \text{and} \quad \hat{G}(z) = \hat{G}_\lambda(z) + \hat{G}(z)\hat{V}^\lambda\hat{G}_\lambda(z), \tag{10.280}$$

we can derive integral equations

$$\begin{aligned}|\Psi^{(+)}\rangle &= \lim_{\varepsilon\to 0} i\varepsilon\hat{G}(E+i\varepsilon)|\Phi_\alpha\rangle \\ &= \lim_{\varepsilon\to 0} i\varepsilon\hat{G}_\lambda(E+i\varepsilon)|\Phi_\alpha\rangle + \lim_{\varepsilon\to 0} i\varepsilon\hat{G}_\lambda(E+i\varepsilon)\hat{V}^\lambda\hat{G}(E+i\varepsilon)|\Phi_\alpha\rangle.\end{aligned} \tag{10.281}$$

If $\lambda \neq \alpha$, the first term vanishes and we get

$$|\Psi^{(+)}\rangle = \delta_{\lambda\alpha}|\Phi_\alpha\rangle + \hat{G}_\lambda(E+i0)\hat{V}^\lambda|\Psi^{(+)}\rangle, \quad \text{with} \quad \lambda = \alpha, \beta, \gamma, \tag{10.282}$$

or writing out explicitly, we have

$$\begin{aligned}|\Psi^{(+)}\rangle &= |\Phi_\alpha\rangle + \hat{G}_\alpha(E+i0)(\hat{v}_\beta + \hat{v}_\gamma)|\Psi^{(+)}\rangle \\ |\Psi^{(+)}\rangle &= \hat{G}_\beta(E+i0)(\hat{v}_\alpha + \hat{v}_\gamma)|\Psi^{(+)}\rangle \\ |\Psi^{(+)}\rangle &= \hat{G}_\gamma(E+i0)(\hat{v}_\alpha + \hat{v}_\beta)|\Psi^{(+)}\rangle.\end{aligned} \tag{10.283}$$

These are the triad, or the fundamental set of Lippmann-Schwinger equations. They, together, uniquely determine the solution. The first equation in Eqs. (10.283) is an inhomogeneous integral equation that puts down the asymptotic state $|\Phi_\alpha\rangle$ and the two others ensure that there are no incoming waves in channels β and γ.

Still, it is hard to impose so many boundary conditions on a single wave function. However, we have one more resolvent relation

$$\hat{G}(z) = \hat{G}_0(z) + \hat{G}_0(z)(\hat{v}_\alpha + \hat{v}_\beta + \hat{v}_\gamma)\hat{G}(z). \tag{10.284}$$

Applying this relation in Eq. (10.277), we obtain

$$
\begin{aligned}
|\Psi^{(+)}\rangle &= \lim_{\varepsilon\to 0} i\varepsilon\hat{G}(E+i\varepsilon)|\Phi_\alpha\rangle \\
&= \lim_{\varepsilon\to 0} i\varepsilon\hat{G}_0(E+i\varepsilon)|\Phi_\alpha\rangle + \lim_{\varepsilon\to 0} i\varepsilon\hat{G}_0(E+i\varepsilon)(\hat{v}_\alpha+\hat{v}_\beta+\hat{v}_\gamma)\hat{G}(E+i\varepsilon)|\Phi_\alpha\rangle \\
&= \lim_{\varepsilon\to 0} \hat{G}_0(E+i\varepsilon)\hat{v}_\alpha|\Psi^{(+)}\rangle + \lim_{\varepsilon\to 0} \hat{G}_0(E+i\varepsilon)\hat{v}_\beta|\Psi^{(+)}\rangle + \lim_{\varepsilon\to 0} \hat{G}_0(E+i\varepsilon)\hat{v}_\gamma|\Psi^{(+)}\rangle.
\end{aligned}
\tag{10.285}
$$

We can see that the total wave function is naturally broken into three components

$$
|\Psi^{(+)}\rangle = |\psi_\alpha\rangle + |\psi_\beta\rangle + |\psi_\gamma\rangle, \tag{10.286}
$$

where

$$
|\psi_\lambda\rangle = \hat{G}_0(E+i0)\hat{v}_\lambda|\Psi^{(+)}\rangle, \quad \lambda = \alpha, \beta, \gamma. \tag{10.287}
$$

These components are the *Faddeev components*. The potential $v_\alpha(\xi_\alpha)$ acting on $|\Psi^{(+)}\rangle$, suppresses configurations when ξ_α gets large, i.e. when particles β and γ move apart. The Faddeev decomposition breaks the wave function into components such that each $|\psi_\lambda\rangle$ possesses only λ-type asymptotic behavior, while the other type of asymptotic behaviors are filtered out.

If we multiply the first equation in Eqs. (10.283) by $\hat{G}_0\hat{v}_\alpha$, the second by $\hat{G}_0\hat{v}_\beta$, and the third by $\hat{G}_0\hat{v}_\gamma$, we find

$$
\begin{aligned}
\hat{G}_0\hat{v}_\alpha|\Psi^{(+)}\rangle &= \hat{G}_0 v_\alpha|\Phi_\alpha\rangle + \hat{G}_0\hat{v}_\alpha\hat{G}_\alpha(E+i0)(\hat{v}_\beta+\hat{v}_\gamma)|\Psi^{(+)}\rangle, \\
\hat{G}_0\hat{v}_\beta|\Psi^{(+)}\rangle &= \hat{G}_0\hat{v}_\beta\hat{G}_\beta(E+i0)(\hat{v}_\alpha+\hat{v}_\gamma)|\Psi^{(+)}\rangle, \\
\hat{G}_0\hat{v}_\gamma|\Psi^{(+)}\rangle &= \hat{G}_0\hat{v}_\gamma\hat{G}_\gamma(E+i0)(\hat{v}_\alpha+\hat{v}_\beta)|\Psi^{(+)}\rangle.
\end{aligned}
\tag{10.288}
$$

From the two forms of the resolvent relations in Eq. (10.280) we have

$$
\hat{G}_0 v_\lambda \hat{G}_\lambda = \hat{G}_\lambda v_\lambda \hat{G}_0. \tag{10.289}
$$

Moreover

$$
\hat{G}_0 v_\alpha|\Phi_\alpha\rangle = |\Phi_\alpha\rangle, \tag{10.290}
$$

since this is equivalent to

$$
(E-\hat{H}_0)|\Phi_\alpha\rangle = v_\alpha|\Phi_\alpha\rangle, \quad \text{or} \quad E|\Phi_\alpha\rangle = (\hat{H}_0+\hat{v}_\alpha)|\Phi_\alpha\rangle. \tag{10.291}
$$

So, Eqs. (10.288) become

$$
\begin{aligned}
|\psi_\alpha\rangle &= |\Phi_\alpha\rangle + \hat{G}_\alpha(E+i0)\hat{v}_\alpha\left(|\psi_\beta\rangle+|\psi_\gamma\rangle\right), \\
|\psi_\beta\rangle &= \hat{G}_\beta(E+i0)\hat{v}_\beta\left(|\psi_\alpha\rangle+|\psi_\gamma\rangle\right), \\
|\psi_\gamma\rangle &= \hat{G}_\gamma(E+i0)\hat{v}_\gamma\left(|\psi_\alpha\rangle+|\psi_\beta\rangle\right).
\end{aligned}
\tag{10.292}
$$

These are the *Faddeev integral equations*.

In the case of bound or resonant states, the inhomogeneous term in the first equation is absent.

We can rewrite the integral equations into differential equation form

$$\begin{aligned}(E - \hat{H}_0 - \hat{v}_\alpha)|\psi_\alpha\rangle &= \hat{v}_\alpha \left(|\psi_\beta\rangle + |\psi_\gamma\rangle\right), \\ (E - \hat{H}_0 - \hat{v}_\beta)|\psi_\beta\rangle &= \hat{v}_\beta \left(|\psi_\alpha\rangle + |\psi_\gamma\rangle\right), \\ (E - \hat{H}_0 - \hat{v}_\gamma)|\psi_\gamma\rangle &= \hat{v}_\gamma \left(|\psi_\alpha\rangle + |\psi_\beta\rangle\right),\end{aligned} \tag{10.293}$$

with boundary conditions such that only $|\psi_\alpha\rangle$ has an incoming $|\Phi_\alpha\rangle$ wave, other components have outgoing wave only.

Identical Particles

The possible identity of particles gives rise to some constraints on the wave function. Assume, for example, that particles β and γ are identical and we denote the operator of their mutual exchange by $\hat{P}_\alpha$. Then,

$$\hat{P}_\alpha|\Psi\rangle = p_\alpha|\Psi\rangle, \tag{10.294}$$

with $p_\alpha = 1$ for bosons and $p_\alpha = -1$ for fermions. If l_α is the orbital angular momentum associated with ξ_α, and particles β and γ have spins S_β and S_γ that are coupled to s_α and isospins T_β and T_γ that are coupled to t_α, respectively, then their exchange results in a phase

$$p_\alpha = (-)^{l_\alpha + s_\alpha - S_\beta - S_\gamma + t_\alpha - T_\beta - T_\gamma}. \tag{10.295}$$

From the definition of the Faddeev components, Eq. (10.287), it follows that

$$\hat{P}_\alpha|\psi_\beta\rangle = \hat{P}_\alpha\hat{G}_0\hat{v}_\beta|\Psi\rangle = \hat{G}_0\hat{P}_\alpha\hat{v}_\beta\hat{P}_\alpha p_\alpha|\Psi\rangle. \tag{10.296}$$

Considering that $\hat{v}_\beta$ is the interaction between particles α and γ, while $\hat{P}_\alpha$ exchanges particles β and γ, we have

$$\hat{P}_\alpha\hat{v}_\beta\hat{P}_\alpha = \hat{v}_\gamma. \tag{10.297}$$

Consequently, Eq. (10.296) becomes

$$\hat{P}_\alpha|\psi_\beta\rangle = p_\alpha\hat{G}_0\hat{v}_\gamma|\Psi\rangle = p_\alpha|\psi_\gamma\rangle, \tag{10.298}$$

and in very much the same way, $\hat{P}_\alpha|\psi_\gamma\rangle = p_\alpha|\psi_\beta\rangle$. Since $\hat{P}_\alpha\hat{v}_\alpha\hat{P}_\alpha = \hat{v}_\alpha$, we have

$$\hat{P}_\alpha|\psi_\alpha\rangle = \hat{G}_0\hat{P}_\alpha\hat{v}_\alpha\hat{P}_\alpha p_\alpha|\Psi\rangle = p_\alpha|\psi_\alpha\rangle. \tag{10.299}$$

So, if particles β and γ are identical, then the angular momentum channels for $|\psi_\alpha\rangle$ have to be selected such that $p_\alpha = 1$ for bosons and $p_\alpha = -1$ for fermions. Then the equation for the Faddeev component $|\psi_\alpha\rangle$ reads

$$|\psi_\alpha\rangle = \hat{G}_\alpha \hat{v}_\alpha(|\psi_\beta\rangle + |\psi_\gamma\rangle) = (1 + p_\alpha \hat{P}_\alpha)\hat{G}_\alpha \hat{v}_\alpha |\psi_\beta\rangle. \tag{10.300}$$

We should notice that $1 + p_\alpha \hat{P}_\alpha$ is just twice the symmetrizing or anti-symmetrizing operator. So, if the angular basis for $|\psi_\alpha\rangle$ is selected such that it ensures the correct symmetry, the value of $1 + p_\alpha \hat{P}_\alpha$ is just 2.

Putting everything together, the three-component Faddeev equation with the proper selection of angular momentum channels simplifies to a set of two-component equations

$$\begin{aligned} |\psi_\alpha\rangle &= |\Phi_\alpha\rangle + 2\hat{G}_\alpha(E + i0)\hat{v}_\alpha |\psi_\beta\rangle, \\ |\psi_\beta\rangle &= |\Phi_\alpha\rangle + \hat{G}_\beta(E + i0)\hat{v}_\beta \left(|\psi_\alpha\rangle + p_\alpha \hat{P}_\alpha |\psi_\beta\rangle\right). \end{aligned} \tag{10.301}$$

If all three particles are identical, the interchange of particles α and β, analogously to the $\beta - \gamma$ exchange, transforms the Faddeev components

$$\hat{P}_\gamma |\psi_\gamma\rangle = p_\gamma |\psi_\gamma\rangle, \quad \hat{P}_\gamma |\psi_\alpha\rangle = p_\gamma |\psi_\beta\rangle, \quad \text{and} \quad \hat{P}_\gamma |\psi_\beta\rangle = p_\gamma |\psi_\alpha\rangle. \tag{10.302}$$

Two consecutive exchanges results in a cyclic permutation of particles

$$\hat{P}_{\alpha\beta\gamma} = \hat{P}_\gamma \hat{P}_\alpha, \tag{10.303}$$

which gives

$$\hat{P}_{\alpha\beta\gamma}|\psi_\alpha\rangle = \hat{P}_\gamma \hat{P}_\alpha |\psi_\alpha\rangle = p_\alpha \hat{P}_\gamma |\psi_\alpha\rangle = p_\alpha p_\gamma |\psi_\beta\rangle = |\psi_\beta\rangle, \tag{10.304}$$

since $p_\alpha = p_\gamma$ and $p_\alpha p_\gamma = 1$. This further reduces Eqs. (10.301) to a single equation

$$|\psi_\alpha\rangle = |\Phi_\alpha\rangle + 2\hat{G}_\alpha(E + i0)\hat{v}_\alpha \hat{P}_{\alpha\beta\gamma}|\psi_\alpha\rangle = |\Phi_\alpha\rangle + 2\hat{G}_\alpha(E + i0)\hat{v}_\alpha |\psi_\beta\rangle. \tag{10.305}$$

We have found that the identity of particles greatly simplifies the Faddeev equations. If we impose the correct symmetry with respect to two-body exchanges, for example by selecting the appropriate angular momentum basis, the structure of the equations imposes the proper symmetry for all the three particles. This is a great advantage of the Faddeev formalism. For example, light quarks are spin-$1/2$ and isospin-$1/2$ particles and the three-quark wave function describing baryons should be symmetric with respect to spatial, spin and isospin exchanges, since the wave function is anti-symmetric in color variables. Consequently, l_α, s_α and t_α should be selected such that the condition

$$p_\alpha = (-)^{l_\alpha + s_\alpha + t_\alpha} = 1 \tag{10.306}$$

is satisfied, since $S_\beta + S_\gamma + T_\beta + T_\gamma = 2$. On the other hand, in three nucleon systems, like the ${}^3H$ or ${}^3He$, the total wave function has to be anti-symmetric and therefore the condition

$$p_\alpha = (-)^{l_\alpha + s_\alpha + t_\alpha} = -1 \tag{10.307}$$

should be fulfilled.

Partial Wave Decomposition

We have seen before that partial wave decomposition turns equations into more manageable forms. This is also the case with three-body equations. We define the bipolar angular basis by

$$|\boldsymbol{\xi}_\alpha \boldsymbol{\eta}_\alpha\rangle = |\xi_\alpha \eta_\alpha L M l_\alpha \lambda_\alpha\rangle \mathcal{Y}^{L,M}_{l_\alpha,\lambda_\alpha}, \tag{10.308}$$

where

$$\mathcal{Y}^{L,M}_{l_\alpha,\lambda_\alpha} = \sum_{mm'} \langle l_\alpha, \lambda_\alpha; m, m' | L, M\rangle Y_{l_\alpha,m}(\boldsymbol{e}_{\xi_\alpha}) Y_{\lambda_\alpha,m'}(\boldsymbol{e}_{\eta_\alpha}). \tag{10.309}$$

We can transform from one set of bipolar bases from the α configuration to the β configuration by the identity operator

$$\begin{aligned} I = \sum_{LMl_\alpha\lambda_\alpha l_\beta\lambda_\beta} \int_0^\infty \mathrm{d}\xi_\alpha \int_0^\infty \mathrm{d}\eta_\alpha \int_{-1}^1 \mathrm{d}u_\alpha\, \xi_\alpha^2 \eta_\alpha^2 \\ |\xi_\alpha \eta_\alpha L M l_\alpha \lambda_\alpha\rangle A(u_\alpha, x_\alpha, L, l_\alpha, \lambda_\alpha, l_\beta, \lambda_\beta) \langle \xi_\beta \eta_\beta L M l_\beta \lambda_\beta|. \end{aligned} \tag{10.310}$$

Here, u_α is the cosine of the angle between $\boldsymbol{e}_{\xi_\alpha}$ and $\boldsymbol{e}_{\eta_\alpha}$, and $x_\alpha = \xi_\alpha/\eta_\alpha$. The variables ξ_β and η_β are expressed in terms of the variables in α coordinates

$$\begin{aligned} \xi_\beta &= (a_1^2 \xi_\alpha^2 + a_2^2 \eta_\alpha^2 + 2a_1 a_2 \xi_\alpha \eta_\alpha u_\alpha)^{1/2} \\ \eta_\beta &= (a_3^2 \xi_\alpha^2 + a_4^2 \eta_\alpha^2 + 2a_3 a_4 \xi_\alpha \eta_\alpha u_\alpha)^{1/2}, \end{aligned} \tag{10.311}$$

where the a_i coefficients are defined in Eq. (10.265). For the coupling function we have

$$
\begin{aligned}
A(u_\alpha, x_\alpha, L, l_\alpha, \lambda_\alpha, l_\beta, \lambda_\beta) = & \frac{8\pi^2}{2L+1} \sum_{M m_\alpha m'_\alpha m_\beta m'_\beta} \langle l_\alpha, \lambda_\alpha; m_\alpha, m'_\alpha | L, M\rangle \langle l_\beta, \lambda_\beta; m_\beta, m'_\beta | L, M\rangle \\
& \times \Gamma_{l_\alpha m_\alpha} \Gamma_{\lambda_\alpha m'_\alpha} \Gamma_{l_\beta m_\beta} \Gamma_{\lambda_\beta m'_\beta} \\
& \times \exp[-iM(\boldsymbol{e}_{\eta_\beta}, \boldsymbol{e}_{\eta_\alpha}) - im_\alpha(\boldsymbol{e}_{\eta_\alpha}, \boldsymbol{e}_{\xi_\alpha}) + im_\beta(\boldsymbol{e}_{\eta_\beta}, \boldsymbol{e}_{\xi_\beta})],
\end{aligned}
\tag{10.312}
$$

where $(\boldsymbol{e}_\eta, \boldsymbol{e}_\xi)$ denotes the angle between the respective unit vectors and the Γ coefficients are given by

$$
\Gamma_{lm} = \frac{(-)^{(l+m)/2}}{2^l((l+m)/2)!((l-m)/2)!} \left[\frac{2l+1}{4\pi} (l+m)!(l-m)! \right]^{1/2}, \tag{10.313}
$$

if $l - m$ is even and vanish otherwise.

The transformation gets simplified between the same type of Jacobi coordinates, i.e. when $\beta = \alpha$. In this case $(\boldsymbol{e}_{\eta_\beta}, \boldsymbol{e}_{\eta_\alpha}) = (\boldsymbol{e}_{\eta_\alpha}, \boldsymbol{e}_{\eta_\alpha}) = 0$ and $(\boldsymbol{e}_{\eta_\beta}, \boldsymbol{e}_{\xi_\beta}) = (\boldsymbol{e}_{\eta_\alpha}, \boldsymbol{e}_{\xi_\alpha})$. Consequently, the exponential term in Eq. (10.312) becomes $\cos[(m'_\beta - m_\alpha)(\boldsymbol{e}_{\eta_\alpha}, \boldsymbol{e}_{\xi_\alpha})]$.

10.4.3 Faddeev Approach to Long-Range Potentials

In this section, we briefly outline how the Faddeev approach can be modified if the potentials are of long-range type. We follow the method we adopted for the two-body case. As before, we split the potential into long-range and short-range parts

$$
\hat{v}_\alpha = \hat{v}_\alpha^{(l)} + \hat{v}_\alpha^{(s)}, \tag{10.314}
$$

and treat the long-range parts on equal footing with the kinetic energy. Then, the three-body Hamiltonian becomes

$$
\hat{H} = \hat{H}^{(l)} + \hat{v}_\alpha^{(s)} + \hat{v}_\beta^{(s)} + \hat{v}_\gamma^{(s)}, \tag{10.315}
$$

with long-range Hamiltonian

$$
\hat{H}^{(l)} = \hat{H}_0 + \hat{v}_\alpha^{(l)} + \hat{v}_\beta^{(l)} + \hat{v}_\gamma^{(l)}. \tag{10.316}
$$

Then, for the asymptotic Hamiltonian, we have

$$
\hat{H}_\alpha^{(l)} = \hat{H}^{(l)} + \hat{v}_\alpha^{(s)}, \tag{10.317}
$$

and the asymptotic state is given by

$$
\hat{H}_\alpha^{(l)} |\Phi_\alpha^{(l)}\rangle = E_\alpha |\Phi_\alpha^{(l)}\rangle. \tag{10.318}
$$

The corresponding Green's operators should also be defined accordingly

$$\hat{G}^{(l)}(z) = (z - \hat{H}^{(l)})^{-1} \quad \text{and} \quad \hat{G}_\alpha^{(l)}(z) = (z - \hat{H}_\alpha^{(l)})^{-1}. \tag{10.319}$$

Analogously to the short-range case, the Faddeev decomposition takes the form

$$|\psi_\alpha\rangle = \hat{G}^{(l)}\hat{v}_\alpha^{(s)}|\Psi\rangle, \tag{10.320}$$

and for the Faddeev components, we arrive at the modified Faddeev equations

$$\begin{aligned} |\psi_\alpha\rangle &= |\Phi_\alpha^{(l)}\rangle + \hat{G}_\alpha^{(l)}\hat{v}_\alpha^{(s)}(|\psi_\beta\rangle + |\psi_\gamma\rangle), \\ |\psi_\beta\rangle &= \qquad\quad \hat{G}_\beta^{(l)}\hat{v}_\beta^{(s)}(|\psi_\gamma\rangle + |\psi_\alpha\rangle), \\ |\psi_\gamma\rangle &= \qquad\quad \hat{G}_\gamma^{(l)}\hat{v}_\gamma^{(s)}(|\psi_\alpha\rangle + |\psi_\beta\rangle). \end{aligned} \tag{10.321}$$

The splitting of the potential into long-range and short-range parts is rather arbitrary. However, if in the Faddeev decomposition in Eq. (10.320) the Green's operator $\hat{G}^{(l)}$ has a pole, it may generate a non-vanishing $|\psi_\alpha\rangle$, even from vanishing $|\Psi\rangle$. This would result in spurious solutions of the Faddeev equations that are not the solutions of the original problem.

Since $v_\alpha^{(s)}$ vanishes as $\xi_\alpha \to \infty$, its action on $|\psi_\beta\rangle$ suppresses all but α-type asymptotic components. So, if $\hat{H}_\alpha^{(l)}$ does not have any non-α-type asymptotic structure, $\hat{G}_\alpha^{(l)}$ generates only α-type asymptotic components for $|\psi_\alpha\rangle$. This way, the Faddeev decomposition remains an asymptotic filtering and as a consequence, each Faddeev component will have only one type of asymptotic structure.

A Coulomb-like potential influences the character of the asymptotic motion. Additionally, if the Coulomb potential is attractive, it supports infinitely many bound states accumulating at the two-body threshold. We should separate the potential into short-range and long-range parts in such a way that the asymptotic filtering character of the Faddeev decomposition prevails.

We should notice that the potentials are defined in three-body configuration space. The two-body asymptotic region Ω_α is defined by the condition

$$(\xi_\alpha, \eta_\alpha) \in \Omega_\alpha, \quad \text{if} \quad \xi_\alpha < \xi_\alpha^{(0)}(1 + \eta_\alpha/\eta_\alpha^{(0)})^{1/\nu} \tag{10.322}$$

with $\xi_\alpha^{(0)} > 0$, $\eta_\alpha^{(0)} > 0$, and $\nu > 2$ (see p. 134 in [1]). The Coulomb potentials enter into the formalism as three-body potentials

$$v_\alpha^{(C)}(\xi_\alpha, \eta_\alpha) = v_\alpha^{(C)}(\xi_\alpha)\mathbf{1}_{\eta_\alpha} = v_\alpha^{(s)}(\xi_\alpha, \eta_\alpha) + v_\alpha^{(l)}(\xi_\alpha, \eta_\alpha), \tag{10.323}$$

where the separation is defined by a splitting function

$$v_\alpha^{(s)}(\xi_\alpha, \eta_\alpha) = v_\alpha^{(C)}(\xi_\alpha)\zeta(\xi_\alpha, \eta_\alpha) \quad \text{and} \quad v_\alpha^{(l)}(\xi_\alpha, \eta_\alpha) = v_\alpha^{(C)}(\xi_\alpha)[1 - \zeta(\xi_\alpha, \eta_\alpha)]. \tag{10.324}$$

The splitting function is rather arbitrary provided it satisfies the condition

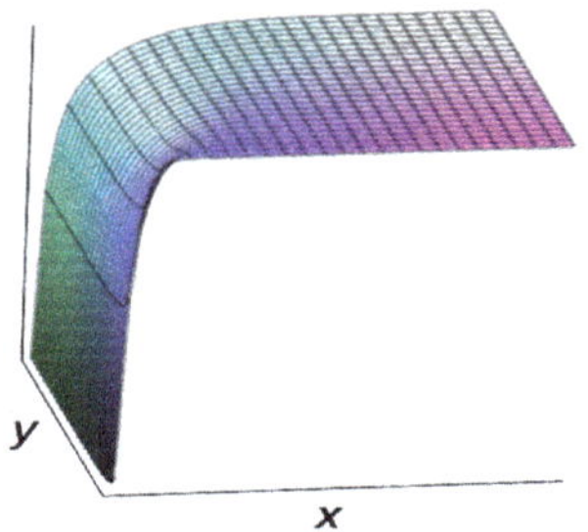

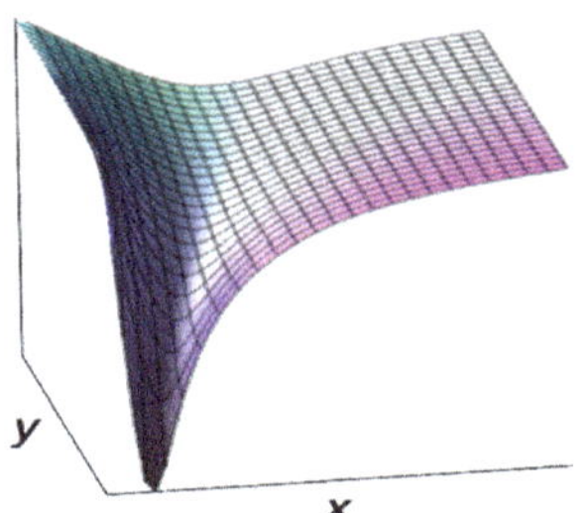

Fig. 10.11 A short- and long-range potentials $v^{(s)}$ and $v^{(l)}$ associated with an attractive Coulomb potential

$$\zeta_\alpha(\xi_\alpha, \eta_\alpha) \xrightarrow[(\xi_\alpha,\eta_\alpha)\to\infty]{} \begin{cases} 1, & \text{if } (\xi_\alpha, \eta_\alpha) \in \Omega_\alpha, \\ 0, & \text{otherwise.} \end{cases} \tag{10.325}$$

For example, the form

$$\zeta(\xi, \eta) = \frac{2}{1 + \exp((\xi/\xi^{(0)})^\nu/(1 + \eta/\eta^{(0)}))}, \tag{10.326}$$

with ξ_0 and η_0 are about the size of the reaction region and $\nu > 2$ meets the requirements. This procedure is the Merkuriev's splitting (Fig. 10.11).

Now, the channel Hamiltonian is given by

$$\hat{H}_\alpha^{(l)} = \hat{H}^{(l)} + \hat{v}_\alpha^{(s)} = \hat{H}_0 + \hat{v}_\alpha + \hat{v}_\beta^{(l)} + \hat{v}_\gamma^{(l)}, \tag{10.327}$$

and its eigenstates are denoted as $\Phi_\alpha^{(l)}$. Then, for the channel Green's operator we have

$$\hat{G}_\alpha^{(l)}(z) = (z - \hat{H}^{(l)} - \hat{v}_\alpha^{(s)})^{-1} = (z - \hat{H}^{(0)} - \hat{v}_\alpha - \hat{v}_\beta^{(l)} - \hat{v}_\gamma^{(l)})^{-1}, \tag{10.328}$$

which is linked to $\hat{G}(z)$ by the resolvent relation

$$\hat{G}(z) = \hat{G}_\alpha^{(l)}(z) + \hat{G}_\alpha^{(l)}(z)(\hat{v}_\beta^{(s)} + \hat{v}_\gamma^{(s)})\hat{G}(z). \tag{10.329}$$

These lead to the modified Faddeev equations defined in Eqs. (10.321).

Obviously, with Merkuriev splitting, the channel Hamiltonian $\hat{H}_\alpha^{(l)}$ of Eq. (10.327) supports α-type asymptotic channels. We can see, however, that it does not support β- or γ-type asymptotic channels. If $\hat{v}_\beta^{(C)}$ is attractive, the long range part $\hat{v}_\beta^{(l)}$ is a valley along the y_β coordinate. However, as $y_\beta \to \infty$, the valley is getting wider and shallower, so in the limit, it does not support bound states and therefore does not generate β-type asymptotic channels. The same analysis holds for γ-type asymptotic channels.

Although the Hamiltonian $\hat{H}_\alpha^{(l)}$ is much simpler than the total Hamiltonian, it is still a genuine three-body operator. However, it can be linked to the much simpler channel Hamiltonian

$$\hat{\tilde{H}}_\alpha(\xi_\alpha, \eta_\alpha) = \hat{H}_0(\xi_\alpha, \eta_\alpha) + \hat{v}_\alpha(\xi_\alpha) + \hat{u}_\alpha^{(l)}(\eta_\alpha), \tag{10.330}$$

where the $\hat{u}_\alpha^{(l)}(\eta_\alpha)$ auxiliary potential is such that $\hat{U}_\alpha^{(l)}(\xi_\alpha, \eta_\alpha) = \hat{v}_\beta^{(l)} + \hat{v}_\gamma^{(l)} - \hat{u}_\alpha^{(l)}$ is of short-range type as $\eta_\alpha \to \infty$. Basically, $\hat{u}_\alpha^{(l)}(\eta_\alpha)$ should be chosen such that it compensates the Coulomb tail of $\hat{v}_\beta^{(l)} + \hat{v}_\gamma^{(l)}$ in the η_α coordinate.

We can easily establish the relation

$$\hat{H}_\alpha^{(l)} = \hat{\tilde{H}}_\alpha + \hat{U}_\alpha^{(l)}, \tag{10.331}$$

which involves the resolvent equation

$$\hat{G}_\alpha^{(l)}(z) = \hat{\tilde{G}}_\alpha(z) + \hat{\tilde{G}}_\alpha(z)\hat{U}_\alpha^{(l)}\hat{G}_\alpha^{(l)}(z), \tag{10.332}$$

where $\hat{\tilde{G}}_\alpha(z) = (z - \hat{\tilde{H}}_\alpha)^{-1}$. The state $|\Phi_\alpha^{(l)}\rangle$ satisfies a similar Lippmann-Schwinger equation

$$|\Phi_\alpha^{(l)}\rangle = |\tilde{\Phi}_\alpha\rangle + \hat{\tilde{G}}_\alpha\hat{U}_\alpha^{(l)}|\Phi_\alpha^{(l)}\rangle, \tag{10.333}$$

where $|\tilde{\Phi}_\alpha\rangle$ is the eigenstate of $\tilde{H}_\alpha$.

The kinetic energy is separable in terms of variables $\boldsymbol{\xi}_\alpha$ and $\boldsymbol{\eta}_\alpha$, consequently $\tilde{H}_\alpha$ is also separable, $\hat{\tilde{H}}_\alpha = \hat{h}_{\xi_\alpha} + \hat{h}_{\eta_\alpha}$, where

$$\hat{h}_{\xi_\alpha} = \hat{h}_{\xi_\alpha}^{(0)} + \hat{v}_\alpha(\boldsymbol{\xi}_\alpha) \quad \text{and} \quad \hat{h}_{\eta_\alpha} = \hat{h}_{\eta_\alpha}^{(0)} + \hat{u}_\alpha^{(l)}(\boldsymbol{\eta}_\alpha). \tag{10.334}$$

Therefore, the solution $\hat{\tilde{H}}_\alpha|\tilde{\Phi}_\alpha\rangle = E_\alpha|\tilde{\Phi}_\alpha\rangle$ is given by

$$\langle \boldsymbol{\xi}_\alpha \boldsymbol{\eta}_\alpha|\tilde{\Phi}_\alpha\rangle = \langle \boldsymbol{\xi}_\alpha|\phi_\alpha\rangle\langle \boldsymbol{\eta}_\alpha|\tilde{\chi}_\alpha\rangle, \tag{10.335}$$

where $|\phi_\alpha\rangle$ is a bound eigenstate of $\hat{h}_{\xi_\alpha}$ and $|\tilde{\chi}_\alpha\rangle$ is a scattering eigenstate of $\hat{h}_{\eta_\alpha}$.

The asymptotic Hamiltonian is the same as in the short-range case

$$\hat{H}_\alpha = \hat{h}_{\xi_\alpha}^{(0)} + \hat{v}_\alpha(\boldsymbol{\xi}_\alpha) + \hat{h}_{\eta_\alpha}^{(0)}, \tag{10.336}$$

and its i-th eigenstate is given by

$$\langle \boldsymbol{\xi}_\alpha \boldsymbol{\eta}_\alpha|\Phi_{\alpha i}\rangle = \langle \boldsymbol{\xi}_\alpha|\phi_{\alpha i}\rangle\langle \boldsymbol{\eta}_\alpha|\chi_\alpha^{(0)}\rangle, \tag{10.337}$$

where $|\chi_\alpha^{(0)}\rangle$ is a free state in coordinate $\boldsymbol{\eta}_\alpha$.

We can view this three-body scattering process as the consecutive transitions $\hat{H}_\alpha \to \hat{\hat{H}}_\alpha \to \hat{H}_\alpha^{(l)} \to \hat{H}$. At first, the system goes from the asymptotic state governed by $\hat{H}_\alpha$ to the intermediate state governed by $\hat{\hat{H}}_\alpha$. This is a two-body scattering by the potential $\hat{u}_\alpha$. Then, the system goes from $\hat{\hat{H}}_\alpha$ to $\hat{H}_\alpha^{(l)}$ by a two-body multichannel scattering generated by $\hat{U}_\alpha^{(l)}$, and finally, goes over to a state defined by $\hat{H}$ via genuine three-body scattering.

The S-matrix element is given by

$$S_{\beta j,\alpha i} = \lim_{t\to\infty}\lim_{\epsilon\to 0} i\epsilon e^{i(E_{\beta j}-E_{\alpha i})t/\hbar}\langle\Phi_{\beta j}|\hat{G}(E_\alpha + i\epsilon)|\Phi_{\alpha i}\rangle, \tag{10.338}$$

and by utilizing Eq. (10.329), we find

$$S_{\beta j,\alpha i} = \lim_{t\to\infty}\lim_{\epsilon\to 0} i\epsilon e^{i(E_{\beta j}-E_{\alpha i})t/\hbar}\langle\Phi_\beta|\hat{G}_\alpha^{(l)}(E_\alpha + i\epsilon) + \hat{G}_\alpha^{(l)}(E_\alpha + i\epsilon)\hat{V}^\alpha\hat{G}(E_\alpha + i\epsilon)|\Phi_\alpha\rangle. \tag{10.339}$$

The first term, with the help of the resolvent relation Eq. (10.332), can further be broken into two parts

$$\begin{aligned} S^{(1)}_{\beta j,\alpha i} &= \lim_{t\to\infty}\lim_{\epsilon\to 0} i\epsilon e^{i(E_{\beta j}-E_{\alpha i})t/\hbar}\langle\Phi_\beta|\hat{\hat{G}}_\alpha(E_\alpha + i\epsilon) + \hat{\hat{G}}_\alpha(E_\alpha + i\epsilon)\hat{U}_\alpha^{(l)}\hat{G}_\alpha^{(l)}(E_\alpha + i\epsilon)|\Phi_\alpha\rangle \\ &= \lim_{t\to\infty}\lim_{\epsilon\to 0} i\epsilon e^{i(E_{\beta j}-E_{\alpha i})t/\hbar}\langle\Phi_\beta|\hat{\hat{G}}_\alpha(E_\alpha + i\epsilon)|\Phi_\alpha\rangle \\ &+ \lim_{t\to\infty}\lim_{\epsilon\to 0} i\epsilon e^{i(E_{\beta j}-E_{\alpha i})t/\hbar}\langle\Phi_\beta|\hat{\hat{G}}_\alpha(E_\alpha + i\epsilon)\hat{U}_\alpha^{(l)}\hat{G}_\alpha^{(l)}(E_\alpha + i\epsilon)|\Phi_\alpha\rangle. \end{aligned} \tag{10.340}$$

In these terms, if $\hat{\hat{G}}_\alpha$ acting on the left does not result in an eigenstate, it vanishes in the $\epsilon \to 0$ limit, resulting in a $\delta_{\beta,\alpha}$ term. So, the first term is the S matrix of a two-body scattering by the potential $u_\alpha^{(l)}$

$$S^{(1,1)}_{\beta j,\alpha i} = \delta_{\beta\alpha}\delta_{ji}S(\hat{u}_\alpha^{(l)}). \tag{10.341}$$

If $u_\alpha^{(l)}$ is absent, $S(u_\alpha^{(l)}) = 1$. The second term in Eq. (10.340) describes a two-body multichannel scattering by the potential $U_\alpha^{(l)}$

$$S^{(1,2)}_{\beta j,\alpha i} = -2\pi i\,\delta_{\beta\alpha}\delta(E_{\beta j} - E_{\alpha i})\langle\tilde{\Phi}^{(-)}_{\alpha j}|\hat{U}_\alpha^{(l)}|\Phi^{(l)(+)}_{\alpha i}\rangle. \tag{10.342}$$

The third term, which is the second term in Eq. (10.339), takes account of the complete three-body dynamics

$$S^{(3)}_{\beta j,\alpha i} = \lim_{t\to\infty}\lim_{\epsilon\to 0} i\epsilon e^{i(E_{\beta j}-E_{\alpha i})t/\hbar}\langle\Phi_{\beta j}|\hat{G}_\alpha^{(l)}(E_\alpha + i\epsilon)\hat{V}^\alpha\hat{G}(E_\alpha + i\epsilon)|\Phi_{\alpha i}\rangle. \tag{10.343}$$

The $\hat{G}(E_\alpha + i\epsilon)|\Phi_{\alpha i}\rangle$ term gives $|\Psi_\alpha^{(+)}\rangle$. Due to the triad of Lippmann-Schwinger equations, Eqs. (10.283), $\hat{G}_\alpha^{(l)}\hat{V}^\alpha|\Psi_\alpha^{(+)}\rangle$ can be replaced by $\hat{G}_\beta^{(l)}\hat{V}^\beta|\Psi_\alpha^{(+)}\rangle$. Then, $\hat{G}_\beta^{(l)}$ acting on the left gives

$$\begin{aligned} S^{(3)}_{\beta j,\alpha i} &= -2\pi i\delta(E_{\beta j} - E_{\alpha i})\langle\Phi_{\beta j}^{(l)(-)}|\hat{V}^\beta|\Psi_{\alpha i}^{(+)}\rangle \\ &= -2\pi i\delta(E_{\beta j} - E_{\alpha i})\langle\Phi_{\beta j}^{(l)(-)}|(\hat{v}_\alpha^{(s)} + \hat{v}_\gamma^{(s)})|\Psi_{\alpha i}^{(+)}\rangle. \end{aligned} \tag{10.344}$$

Now, considering that

$$|\Phi_{\beta j}^{(l)}\rangle = \hat{G}^{(l)}\hat{v}_\beta^{(s)}|\Phi_{\beta j}^{(l)}\rangle, \tag{10.345}$$

and the definition of the Faddeev components, we obtain

$$\begin{aligned} S^{(3)}_{\beta j,\alpha i} &= -2\pi i\delta(E_{\beta j} - E_{\alpha i})\langle\Phi_{\beta j}^{(l)(-)}|\hat{v}_\beta^{(s)}\hat{G}^{(l)}(\hat{v}_\alpha^{(s)} + \hat{v}_\gamma^{(s)})|\Psi_{\alpha i}^{(+)}\rangle \\ &= -2\pi i\delta(E_{\beta j} - E_{\alpha i})\langle\Phi_{\beta j}^{(l)(-)}|\hat{v}_\beta^{(s)}|\sum_{\lambda\neq\beta}\psi_\lambda^{(+)}\rangle. \end{aligned} \tag{10.346}$$

So, the final result is

$$\begin{aligned} S_{\beta j,\alpha i} &= \delta_{\beta\alpha}\delta_{ji}S(\hat{u}_\alpha^{(l)}) \\ &\quad - 2\pi i\,\delta(E_{\beta j} - E_{\alpha i})\left[\delta_{\beta\alpha}\langle\tilde{\Phi}_{\alpha j}^{(-)}|\hat{U}_\alpha^{(l)}|\Phi_{\alpha i}^{(l)(+)}\rangle + \langle\Phi_{\beta j}^{(l)(-)}|\hat{v}_\beta^{(s)}|\sum_{\lambda\neq\beta}\psi_\lambda^{(+)}\rangle\right]. \end{aligned} \tag{10.347}$$

Faddeev Equations in Discrete Basis Representation

The three-body Hilbert space is a direct product of two-body Hilbert spaces in coordinates ξ_α and η_α. An appropriate basis can be defined as an angular momentum coupled direct product of two-body Coulomb-Sturmian bases,

$$|n\nu l\lambda\rangle_\alpha = |nl; b\rangle_\alpha \otimes |\nu\lambda; b\rangle_\alpha, \quad (n, \nu = 0, 1, 2, \ldots), \tag{10.348}$$

where $|nl; b\rangle_\alpha$ and $|\nu\lambda; b\rangle_\alpha$ are associated with coordinates ξ_α and η_α, respectively. If we have a closer look at the Faddeev equations in Eq. (10.321), we can notice that, for example, $\hat{v}_\alpha^{(s)}$ facing the left meets $\hat{G}_\alpha^{(l)}$, that can naturally be represented in α-type coordinates. On the other hand, facing the right, it meets $|\psi_\beta\rangle$ or $|\psi_\gamma\rangle$, and for them, the natural coordinates are the β or γ Jacobi coordinates. Therefore, it is reasonable to represent the potential operators in such a way that it fits for the neighborhood,

$$v_\alpha^{(s)} \approx \sum |\widetilde{n\nu l\lambda}\rangle_\alpha \underline{v}^{(s)}_{\alpha\beta}{}_\beta\langle\widetilde{n'\nu' l'\lambda'}| \quad \text{and} \quad v_\alpha^{(s)} \approx \sum |\widetilde{n\nu l\lambda}\rangle_\alpha \underline{v}^{(s)}_{\alpha\gamma}{}_\gamma\langle\widetilde{n'\nu' l'\lambda'}|, \tag{10.349}$$

depending on the Faddeev components on the right hand side. The three-body matrix elements $\underline{v}^{(s)}_{\alpha\beta} = {}_{\alpha}\langle n\nu l\lambda|v^{(s)}_{\alpha}|n'\nu'l'\lambda'\rangle_{\beta}$ can be calculated numerically with the help of the transformation of the Jacobi coordinates. The same way as before in the two-body case, this approximation turns the Faddeev equation into a matrix equation with the Green's matrix $\underline{G}^{(l)}_{\alpha} = {}_{\alpha}\langle\widetilde{n\nu l\lambda}|\hat{G}^{(l)}_{\alpha}|\widetilde{n'\nu'l'\lambda'}\rangle_{\alpha}$.

This Green's operator is still rather complicated, but by using Eq. (10.332) and approximating the potential

$$U^{(l)}_{\alpha} \approx \sum |\widetilde{n\nu l\lambda}\rangle_{\alpha}\underline{U}^{(l)}_{\alpha}\,{}_{\alpha}\langle\widetilde{n'\nu'l'\lambda'}|, \tag{10.350}$$

with $\underline{U}^{(l)}_{\alpha} = {}_{\alpha}\langle n\nu l\lambda|U^{(l)}_{\alpha}|n'\nu'l'\lambda'\rangle_{\alpha}$, the Lippmann-Schwinger equation becomes a matrix equation with $\underline{\tilde{G}}_{\alpha} = {}_{\alpha}\langle\widetilde{n\nu l\lambda}|\hat{\tilde{G}}_{\alpha}|\widetilde{n'\nu'l'\lambda'}\rangle_{\alpha}$.

We have learned before that a function f of a Hermitian operator can be calculated by a Cauchy-type complex contour integral

$$f(\hat{h}) = \frac{1}{2\pi i}\oint_C \mathrm{d}z' f(z')(z' - \hat{h})^{-1} = \frac{1}{2\pi i}\oint_C \mathrm{d}z' f(z')\hat{g}(z'), \tag{10.351}$$

where the contour C should be taken in a counterclockwise direction in such a way that it encircles the spectrum of $\hat{h}$, and f should be analytic on that area including the boundary. In this relation, the argument of f on the left is the operator $\hat{h}$, while on the right, it is a complex number z'.

The Green's operator $\hat{\tilde{G}}_{\alpha}$ is a resolvent of the sum of two commuting two body Hamiltonians

$$\hat{\tilde{G}}_{\alpha}(z) = (z - \hat{h}_{\xi_{\alpha}} - \hat{h}_{\eta_{\alpha}})^{-1}, \tag{10.352}$$

i.e. it is a function of the Hermitian operator $\hat{h}_{\xi_{\alpha}}$. Therefore, we can write

$$\begin{aligned}\hat{\tilde{G}}_{\alpha}(z) = (z - \hat{h}_{\xi_{\alpha}} - \hat{h}_{\eta_{\alpha}})^{-1} &= \frac{1}{2\pi i}\oint_C \mathrm{d}z'(z - z' - \hat{h}_{\eta_{\alpha}})^{-1}(z' - \hat{h}_{\xi_{\alpha}})^{-1}\\ &= \frac{1}{2\pi i}\oint_C \mathrm{d}z'\hat{g}_{\eta_{\alpha}}(z - z')\hat{g}_{\xi_{\alpha}}(z'),\end{aligned} \tag{10.353}$$

where $\hat{g}_{\xi_{\alpha}}$ and $\hat{g}_{\eta_{\alpha}}$ are two-body Green's operators. So, we can calculate the three-body Green's operator $\hat{\tilde{G}}_{\alpha}(z)$ as a complex contour integral of two-body Green's operators. Consequently, its matrix representation becomes a contour integral with two-body Green's matrices

$$\underline{\tilde{G}}_{\alpha}(E + i\varepsilon) = \frac{1}{2\pi i}\oint_C \mathrm{d}z'\underline{g}_{\eta_{\alpha}}(E + i\varepsilon - z')\underline{g}_{\xi_{\alpha}}(z'). \tag{10.354}$$

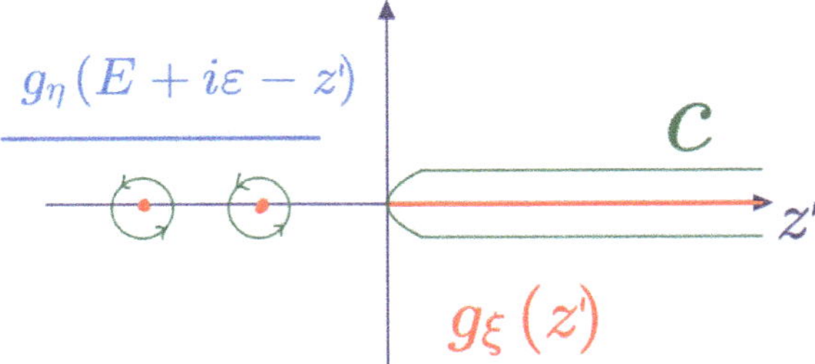

Fig. 10.12 The analytic structure of $\hat{g}_\eta(E + i\varepsilon - z')\hat{g}_\xi(z')$ as a function of z' with $\varepsilon > 0$. The operator $\hat{g}_\xi(z')$ has bound state poles at negative energies and a branch-cut singularity on the $[0, \infty)$ interval. The contour C should avoid the branch-cut singularity of $\hat{g}_\eta(E + i\varepsilon - z')$ even in the $\varepsilon \to 0$ limit

A typical contour is shown in Fig. 10.12. Here, $\hat{h}_{\xi_\alpha}$ has two bound states that are encircled separately, resulting in projections onto those bound states, and the contour around the branch-cut goes to infinity where the integrand vanishes and the contour can be closed.

Further Reading

1. L.D. Faddeev, S.P. Merkuriev, *Quantum Scattering Theory for Several Particle Systems*, vol. 11. (Springer Science & Business Media, 2013)
2. R.G. Newton, *Scattering Theory of Waves and Particles* (Springer Science & Business Media, 2013)

Chapter 11
Relativistic Quantum Mechanics

Modern physics is based on two pillars, relativity and quantum mechanics. In transition from non-relativistic to relativistic quantum mechanics we have to reconsider some ideas. From the Heisenberg uncertainty relation and from the finiteness of c, we have for the uncertainty of the location of a particle with mass m

$$\Delta x \simeq \frac{\hbar}{\Delta p} \simeq \frac{\hbar}{mc} \simeq \lambda_c, \tag{11.1}$$

where λ_c is the *Compton wave length*. This means that a relativistic particle cannot be localized more than λ_c. From this, it also follows that

$$\Delta t \simeq \frac{\Delta x}{c} \simeq \frac{\hbar}{mc^2}, \tag{11.2}$$

i.e. the time is not well defined either. This is in stark contradiction with the non-relativistic theory, where the time is a parameter and can be given with infinite accuracy. On the other hand, in a relativistic theory the space and the time coordinates should be treated on equal footing. So the space coordinates should lose their status as variables to become parameters, similar to time in the non-relativistic theory.

11.1 Klein-Gordon Equation

We can start with the length-invariance of the four-momentum vector

$$p_1^2 + p_2^2 + p_3^2 + p_4^2 = -m^2c^2, \tag{11.3}$$

Z. Papp, *Mastering Quantum Mechanics*,
https://doi.org/10.1007/978-3-032-09011-9_11

and quantize it with the correspondence

$$\hat{p}_i \to -i\hbar \frac{\partial}{\partial x_i}, \tag{11.4}$$

where $x_1 = x$, $x_2 = y$, $x_3 = z$, and $x_4 = ict$. This results in the equation

$$-\hbar^2 \left(\frac{\partial^2 \psi}{\partial x_1^2} + \frac{\partial^2 \psi}{\partial x_2^2} + \frac{\partial^2 \psi}{\partial x_3^2} + \frac{\partial^2 \psi}{\partial x_4^2} \right) = -m^2 c^2 \psi, \tag{11.5}$$

or in a more concise form, we have

$$\nabla^2 \psi - \frac{1}{c^2} \frac{\partial^2 \psi}{\partial t^2} = \frac{m^2 c^2}{\hbar^2} \psi. \tag{11.6}$$

This is the *Klein-Gordon equation*. By using the correspondence $i\hbar\partial/\partial t \to E$, we obtain the time-independent Klein-Gordon equation

$$\nabla^2 \psi + \frac{1}{c^2 \hbar^2} \left(E^2 - m^2 c^4 \right) \psi = 0. \tag{11.7}$$

We can see that Eq. (11.5) is basically the length of a four-vector, which is Lorentz invariant. Therefore, ψ has to be a scalar, or pseudo-scalar. The wave function ψ has no preferred direction and therefore can only describe particles with zero spin, such as pions or kaons.

The continuity equation can be derived analogously to the non-relativistic case. Multiplying Eq. (11.6) by ψ^*, and the complex conjugate equation by ψ, then subtracting them, we obtain

$$\frac{\partial \rho}{\partial t} + \nabla \cdot \boldsymbol{j} = 0, \tag{11.8}$$

with

$$\rho = \frac{i\hbar}{2mc^2} \left(\psi^* \frac{\partial \psi}{\partial t} - \psi \frac{\partial \psi^*}{\partial t} \right) \quad \text{and} \quad \boldsymbol{j} = \frac{i\hbar}{2m} \left(\psi \nabla \psi^* - \psi^* \nabla \psi \right). \tag{11.9}$$

The $\boldsymbol{j}$ is just like in the non-relativistic case, but the ρ is different. Using $E \to i\hbar\partial/\partial t$ we can rewrite ρ as

$$\rho = \frac{1}{2mc^2} \left(\psi^* E \psi + \psi E^* \psi^* \right) = \Re \left(\psi^* \frac{E}{mc^2} \psi \right), \tag{11.10}$$

which, in the non-relativistic limit as $E \simeq mc^2$, becomes $\rho \to \psi^* \psi$. This suggests that we can interpret ρ as a probability density and $\boldsymbol{j}$ as a probability current. But ψ and $\partial \psi / \partial t$ can be given independently, therefore ρ can be negative as well. So, a probabilistic interpretation of ρ is problematic.

However, if we multiply Eq. (11.9) by the charge q, we get

$$\rho = \frac{iq\hbar}{2mc^2}\left(\psi^*\frac{\partial\psi}{\partial t} - \psi\frac{\partial\psi^*}{\partial t}\right) \quad \text{and} \quad \boldsymbol{j} = \frac{iq\hbar}{2m}\left(\psi\nabla\psi^* - \psi^*\nabla\psi\right), \tag{11.11}$$

which can be interpreted as a *charge density* and a *current density*, respectively. Now ρ can either be positive or negative, only the total charge is conserved.

The plane wave

$$\psi(\boldsymbol{r}, t) = A\exp[i(\boldsymbol{pr} - Et)/\hbar] \tag{11.12}$$

is a solution of Eq. (11.6) if

$$E = \pm E_p, \quad \text{with} \quad E_p = \sqrt{c^2\boldsymbol{p}^2 + m^2c^4}. \tag{11.13}$$

With the same momentum, we have two possible solutions, namely

$$\psi_+ = A_+\exp[i(\boldsymbol{pr} - E_pt)/\hbar] \quad \text{and} \quad \psi_- = A_-\exp[i(\boldsymbol{pr} + E_pt)/\hbar], \tag{11.14}$$

and these solutions provide us with the charge density

$$\rho_\pm = \pm E_p\frac{q}{mc^2}|A_\pm|^2. \tag{11.15}$$

We have found that the solution of the Klein-Gordon equation may have positive and negative energies and the sign of the energy is associated with the charge density. The relativistic treatment of spin-zero particles reveals an additional feature, the charge, as an internal degree of freedom.

11.1.1 Hamiltonian Form

In order to further exhibit the charge degree of freedom, we write the Klein-Gordon equation in a Hamiltonian form. If we add a potential by using minimal coupling, $p_i \to p_i - q/c\,A_i$, with $\boldsymbol{A} = 0$ and $q/c\,\Phi = \hat{V}$, we have

$$\left(i\hbar\frac{\partial}{\partial t} - \hat{V}\right)^2\psi = \hat{\boldsymbol{p}}^2c^2\psi + m^2c^4\psi, \tag{11.16}$$

or for stationary states we find

$$\left(E - \hat{V}\right)^2\psi = \hat{\boldsymbol{p}}^2c^2\psi + m^2c^4\psi. \tag{11.17}$$

If we look seriously at Eq. (11.16), we realize that this equation is not a true quantum mechanical equation. The 5th postulate of quantum mechanics tells us that the time evolution is governed by the Hamiltonian through the Schrödinger equation, which is first order in the time derivative. The above equation contains second order time derivatives. So, to determine the quantum state, a knowledge of the wave function is not enough, we need to know the time derivative as well.

However, it is possible to rewrite the Klein-Gordon equation into a Hamiltonian form [1]. If we split the Klein-Gordon wave function ψ into components such that

$$\psi = \phi + \chi \quad \text{and} \quad \left(i\hbar\frac{\partial}{\partial t} - \hat{V}\right)\psi = mc^2(\phi - \chi), \tag{11.18}$$

we have

$$\begin{aligned} \phi &= \frac{1}{2}\left[\left(1 - \frac{\hat{V}}{mc^2}\right)\psi + \frac{i\hbar}{mc^2}\frac{\partial}{\partial t}\psi\right], \\ \chi &= \frac{1}{2}\left[\left(1 + \frac{\hat{V}}{mc^2}\right)\psi - \frac{i\hbar}{mc^2}\frac{\partial}{\partial t}\psi\right]. \end{aligned} \tag{11.19}$$

Then, we can rewrite Eq. (11.16) as

$$\begin{aligned} \left(i\hbar\frac{\partial}{\partial t} - \hat{V}\right)(\phi + \chi) &= mc^2(\phi - \chi), \\ \left(i\hbar\frac{\partial}{\partial t} - \hat{V}\right)(\phi - \chi) &= \left[\frac{\hat{\boldsymbol{p}}^2}{m} + mc^2\right](\phi + \chi). \end{aligned} \tag{11.20}$$

By adding and subtracting these two equations, we obtain a system of coupled equations for the components, the *Feshbach-Villars* (FV0) equations, that are first order in the time derivative

$$\begin{aligned} i\hbar\frac{\partial\phi}{\partial t} &= \frac{\hat{\boldsymbol{p}}^2}{2m}(\phi + \chi) + (mc^2 + \hat{V})\phi, \\ i\hbar\frac{\partial\chi}{\partial t} &= -\frac{\hat{\boldsymbol{p}}^2}{2m}(\phi + \chi) - (mc^2 - \hat{V})\chi. \end{aligned} \tag{11.21}$$

The corresponding stationary equations follow again from the $i\hbar\partial/\partial t \to E$ correspondence

$$\begin{aligned} E\phi &= \frac{\hat{\boldsymbol{p}}^2}{2m}(\phi + \chi) + (mc^2 + \hat{V})\phi, \\ E\chi &= -\frac{\hat{\boldsymbol{p}}^2}{2m}(\phi + \chi) - (mc^2 - \hat{V})\chi, \end{aligned} \tag{11.22}$$

and for the wave function components, we have

$$\begin{aligned}\phi &= \frac{1}{2}\left[\left(1-\frac{\hat{V}}{mc^2}\right)\psi + \frac{E}{mc^2}\psi\right],\\ \chi &= \frac{1}{2}\left[\left(1+\frac{\hat{V}}{mc^2}\right)\psi - \frac{E}{mc^2}\psi\right].\end{aligned} \tag{11.23}$$

In terms of these components, the probability density and probability current become

$$\rho = \frac{i\hbar}{2mc^2}\left(\psi^*\frac{\partial\psi}{\partial t} - \psi\frac{\partial\psi^*}{\partial t}\right) - \frac{\hat{V}}{mc^2}\psi^*\psi = \phi^*\phi - \chi^*\chi, \tag{11.24}$$

$$\begin{aligned}j &= \frac{i\hbar}{2m}\left(\psi\nabla\psi^* - \psi^*\nabla\psi\right)\\ &= \frac{i\hbar}{2m}\left[(\phi\nabla\phi^* - \phi^*\nabla\phi) + (\chi\nabla\chi^* - \chi^*\nabla\chi) + (\phi\nabla\chi^* - \chi^*\nabla\phi) + (\chi\nabla\phi^* - \phi^*\nabla\chi)\right].\end{aligned} \tag{11.25}$$

By defining

$$\Psi = \begin{pmatrix}\phi\\ \chi\end{pmatrix} \tag{11.26}$$

and

$$\hat{H}_{FV0} = (\tau_3 + i\tau_2)\frac{\hat{\boldsymbol{p}}^2}{2m} + \tau_3 mc^2 + \hat{V}, \tag{11.27}$$

where τ_i denote the Pauli matrices

$$\tau_1 = \begin{pmatrix}0 & 1\\ 1 & 0\end{pmatrix}, \quad \tau_2 = \begin{pmatrix}0 & -i\\ i & 0\end{pmatrix}, \quad \tau_3 = \begin{pmatrix}1 & 0\\ 0 & -1\end{pmatrix}, \tag{11.28}$$

we can cast Eqs. (11.21) and (11.22) into Schrödinger-like forms

$$i\hbar\frac{\partial\Psi}{\partial t} = \hat{H}_{FV0}\Psi, \qquad \text{or} \qquad E\Psi = \hat{H}_{FV0}\Psi. \tag{11.29}$$

We can also write the Hamiltonian as

$$\hat{H}_{FV0} = \begin{pmatrix}1 & 1\\ -1 & -1\end{pmatrix}\frac{\hat{\boldsymbol{p}}^2}{2m} + \begin{pmatrix}1 & 0\\ 0 & -1\end{pmatrix}mc^2 + \begin{pmatrix}1 & 0\\ 0 & 1\end{pmatrix}\hat{V}. \tag{11.30}$$

This Hamiltonian is not symmetric, and therefore it is not Hermitian in the usual sense. Instead, the expectation value of an operator $\hat{\Omega}$ in the Feshbach-Villars picture is given by

$$\langle\hat{\Omega}\rangle = \langle\Psi|\tau_3\hat{\Omega}|\Psi\rangle \tag{11.31}$$

and the adjoint operator is defined as

$$\hat{\tilde{\Omega}} = \tau_3\hat{\Omega}^\dagger\tau_3, \tag{11.32}$$

where $\hat{\Omega}^\dagger$ denotes the usual adjoint. The operator is Feshbach-Villars Hermitian if

$$\hat{\tilde{\Omega}} = \hat{\Omega}. \tag{11.33}$$

Operators that are Hermitian in this sense possess real eigenvalues. We can easily check that the Hamiltonian in Eq. (11.30) is Hermitian in this sense.

Interpretation of the Components

The plane wave solution of Eq. (11.14) in the two-component form is given by

$$|\Psi\rangle = \begin{pmatrix}\phi\\ \chi\end{pmatrix} = \begin{pmatrix}u_0\\ v_0\end{pmatrix}\exp\left[\frac{i}{\hbar}(\boldsymbol{pr} - Et)\right], \tag{11.34}$$

and the eigenvalue equation Eq. (11.29) becomes

$$E\begin{pmatrix}u_0\\ v_0\end{pmatrix} = \frac{p^2}{2m}\begin{pmatrix}1 & 1\\ -1 & -1\end{pmatrix}\begin{pmatrix}u_0\\ v_0\end{pmatrix} + mc^2\begin{pmatrix}1 & 0\\ 0 & -1\end{pmatrix}\begin{pmatrix}u_0\\ v_0\end{pmatrix}. \tag{11.35}$$

The solution is the energy eigenvalue $E = \pm E_p$. For the eigenstates, with the normalization condition $u_{0\pm}^* u_{0\pm} - v_{0\pm}^* v_{0\pm} = \pm 1$, we obtain

$$u_{0\pm} = \frac{mc^2 \pm E_p}{2\sqrt{mc^2 E_p}} \quad \text{and} \quad v_{0\pm} = \frac{mc^2 \mp E_p}{2\sqrt{mc^2 E_p}}. \tag{11.36}$$

We can see that if $|\Psi\rangle$ is a solution with $E = E_p$, then the ϕ component dominates over χ, but if $E = -E_p$, the χ component dominates over ϕ. We can also see that if $E = E_p$, the transformation

$$C|\Psi\rangle = C\begin{pmatrix}\phi\\ \chi\end{pmatrix} = \begin{pmatrix}\chi^*\\ \phi^*\end{pmatrix} = \tau_1|\Psi^*\rangle = \begin{pmatrix}0 & 1\\ 1 & 0\end{pmatrix}\begin{pmatrix}\phi^*\\ \chi^*\end{pmatrix} = |\Psi\rangle_c \tag{11.37}$$

results in a solution with $E = -E_p$. This solution is associated with an opposite charge, so we can assiciate $|\Psi\rangle_c$ as the charge conjugate partner of $|\Psi\rangle$ and C as the charge conjugation operator. The transformation $C|\Psi\rangle = |\Psi\rangle_c$ implies

$$u_{0+} \to v_{0-}, \quad v_{0+} \to u_{0-}, \quad \hat{\boldsymbol{p}} \to -\hat{\boldsymbol{p}}, \quad E \to -E. \tag{11.38}$$

We have found before that the parity transformation $\mathcal{P}$ transforms $\boldsymbol{p} \to -\boldsymbol{p}$ and the time reversal operator $\mathcal{T} : t \to -t$ is equivalent to $E \to -E$ and a complex conjugation. So,

$$\mathcal{CPT}|\Psi\rangle = \mathcal{CPT}\begin{pmatrix}\phi\\ \chi\end{pmatrix} = \begin{pmatrix}\chi\\ \phi\end{pmatrix} = |\bar{\Psi}\rangle \tag{11.39}$$

describes a particle with energy E and momentum $\boldsymbol{p}$, but with opposite charge. The C cancels the complex conjugation coming from $\mathcal{T}$, flips the sign of energy from negative to positive, and restores the direction of the momentum. Therefore we can call $|\bar{\Psi}\rangle$ the antiparticle of $|\Psi\rangle$, and the $|\bar{\Psi}\rangle = \mathcal{CPT}|\Psi\rangle$ the particle-antiparticle transformation.

We can further illustrate the two-component formalism by taking $\hat{V} \equiv 0$. Then,

$$\phi = \frac{1}{2}\left[\psi + \frac{E}{mc^2}\psi\right] \quad \text{and} \quad \chi = \frac{1}{2}\left[\psi - \frac{E}{mc^2}\psi\right]. \tag{11.40}$$

In the non-relativistic limit, $E \simeq mc^2$, we have $\phi \to \psi$ and $\chi \to 0$. In general, $E = mc^2/\sqrt{1-(v/c)^2}$, and as a consequence

$$\frac{|\chi|}{|\phi|} = \frac{1-\sqrt{1-(v/c)^2}}{1+\sqrt{1-(v/c)^2}}. \tag{11.41}$$

If $v \to 0$, we recover the non-relativistic limit, but as $v \to c$ we get $|\chi|/|\phi| \to 1$. We can interpret χ as a *shadow antiparticle* component which is negligible in the non-relativistic limit and becomes increasingly important at higher energies. A particle with non-zero kinetic energy always has some antiparticle component, which becomes more and more prominent as the kinetic energy increases.

11.1.2 Klein Paradox

Here we consider scattering in one dimension on a step potential

$$V(x) = \begin{cases} 0, & \text{if } x < 0, \\ V_0 > 0, & \text{if } x > 0. \end{cases} \tag{11.42}$$

The Klein-Gordon equation reads

$$(E - V(x))^2\psi = -c^2\hbar^2\frac{\partial^2}{\partial x^2}\psi + m^2c^4\psi, \tag{11.43}$$

or

$$\left\{\frac{\partial^2}{\partial x^2} + \frac{(E-V(x))^2 - m^2c^4}{c^2\hbar^2}\right\}\psi(x) = 0. \tag{11.44}$$

We assume an incident wave traveling to the right

$$\psi_i = A\exp\left[i(px - Et)/\hbar\right], \qquad \text{as} \quad x < 0, \tag{11.45}$$

with $p > 0$, and $E = \sqrt{p^2c^2 + m^2c^4} > 0$. Besides this, we have a reflected and a transmitted wave

$$\begin{aligned} \psi_r &= B\exp\left[i(-px - Et)/\hbar\right], && \text{if} \ x < 0, \\ \psi_t &= C\exp\left[i(p'x - Et)/\hbar\right], && \text{if} \ x > 0, \end{aligned} \tag{11.46}$$

respectively, where $p'^2 = [(E - V_0)^2 - m^2c^4]/c^2 = [(E - V_0 - mc^2)(E - V_0 + mc^2)]/c^2$. The continuity of ψ and ψ' at $x = 0$ gives

$$\frac{B}{A} = \frac{1 - p'/p}{1 + p'/p} \quad \text{and} \quad \frac{C}{A} = \frac{2}{1 + p'/p}. \tag{11.47}$$

We should consider the following cases:

1: $E < V_0 < E + mc^2$, i.e. the energy of the particle is below the barrier, and the barrier is not too high. Now $V_0 - E < mc^2$, therefore $p' = \sqrt{(V_0 - E)^2/c^2 - m^2c^2} = iq$, with q real and p' pure imaginary. Then the transmitted wave is exponentially decaying

$$\psi_t = C\exp[(-qx - iEt)/\hbar], \quad \text{if} \ x > 0, \tag{11.48}$$

and for the reflection coefficient we find

$$R = |\frac{B}{A}|^2 = \frac{|1 - iq/p|^2}{|1 + iq/p|^2} = 1. \tag{11.49}$$

Just like in the non-relativistic case, we have a complete reflection.

2: $V_0 < E - mc^2$, i.e. the kinetic energy of the particle is above the height of the barrier. In this case, $mc^2 < E - V_0$ and $p' = \sqrt{(V_0 - E)^2/c^2 - m^2c^2}$ remains real with $p' < p$, then the transmitted wave is an oscillating plane wave. The reflection and transmission coefficients become

$$R = |\frac{B}{A}|^2 = \frac{(p - p')^2}{(p + p')^2}, \quad T = \frac{p'}{p}|\frac{C}{A}|^2 = 4\frac{p'p}{(p + p')^2}, \quad \text{and} \quad R + T = 1. \tag{11.50}$$

Again, just like in the non-relativistic case, we have partial transmission and reflection.

3: $E + mc^2 < V_0$, i.e. the potential is very strong, even stronger than twice of the rest energy of the incoming particle. In this case, $mc^2 < V_0 - E$ and $p' = \pm\sqrt{(V_0 - E)^2/c^2 - m^2c^2}$ remains real, so there is a transmitted oscillating plane wave below the barrier and the reflection is not complete

$$R = \frac{|p - p'|^2}{|p + p'|^2} \neq 1. \tag{11.51}$$

This is in sharp contradiction to the non-relativistic case when the particle bounces back if the potential step is higher than the kinetic energy of the particle. This is the essence of the *Klein paradox*.

Let us analyze the situation in the framework of the Feshbach-Villars formalism. For $x < 0$, $V(x) = 0$, and thus,

$$\phi_i = \frac{1}{2}\left(1 + \frac{E}{mc^2}\right)\psi_i \quad \text{and} \quad \chi_i = \frac{1}{2}\left(1 - \frac{E}{mc^2}\right)\psi_i. \tag{11.52}$$

Then,

$$\rho_i = |\phi_i|^2 - |\chi_i|^2 = \frac{E}{mc^2}|A|^2 > 0 \quad \text{and} \quad j_i = \frac{p}{m}|A|^2 > 0, \tag{11.53}$$

represents a particle with positive charge traveling to the right. Similarly, for the reflected wave, we have

$$\rho_r = \frac{E}{mc^2}|B|^2 > 0 \quad \text{and} \quad j_r = -\frac{p}{m}|B|^2 < 0, \tag{11.54}$$

representing a left-traveling positive charge.

The transmitted wave is influenced by the potential

$$\phi_t = \frac{1}{2}\left(1 + \frac{E - V_0}{mc^2}\right)\psi_t \quad \text{and} \quad \chi_t = \frac{1}{2}\left(1 - \frac{E - V_0}{mc^2}\right)\psi_t, \tag{11.55}$$

and therefore

$$\rho_t = |\phi_t|^2 - |\chi_t|^2 = \frac{E - V_0}{mc^2}|C|^2 \quad \text{and} \quad j_t = \frac{p'}{m}|C|^2. \tag{11.56}$$

If $E - mc^2 > V_0$, then $p' > 0$ and we observe a wave representing positive charges traveling to the right.

However, if $V_0 > E + mc^2$, χ_t becomes dominant over ϕ_t and ρ_t becomes negative, i.e. as the particle hits the potential step, its charge for $x > 0$ becomes negative. On the other hand, from the continuity of the current we must have

$$j_i + j_r = j_t. \tag{11.57}$$

As far as the sign of p' is concerned, we have two alternatives. If $p' > 0$, then $j_t > 0$ and $|j_r| < j_i$. This would be a negative charge traveling to the right, which is a negative current, not a positive one as we assumed. With the $p' < 0$ choice, we have

$j_t < 0$ and $\rho_t < 0$. With this choice, Eq. (11.57) is satisfied with $|j_r| > |j_i|$, and thus with $R > 1$ and $T < 0$.

For the observer located at $x > 0$, the energy of the transmitted wave should be measured with respect to the local potential V_0. Here $E' = E - V_0 < 0$, so the particle behaves like a negative energy state. The wave function should be redefined as

$$\psi_t \to \tilde{\psi}_t = C' \exp\left[i(p'x - E't)/\hbar\right], \quad \text{for } x > 0. \tag{11.58}$$

Now, the Feshbach-Villars components become

$$\tilde{\phi}_t = \frac{1}{2}\left(1 + \frac{E'}{mc^2}\right)\tilde{\psi}_t \quad \text{and} \quad \tilde{\chi}_t = \frac{1}{2}\left(1 - \frac{E'}{mc^2}\right)\tilde{\psi}_t, \tag{11.59}$$

which results in

$$\rho_t = |\tilde{\phi}_t|^2 - |\tilde{\chi}_t|^2 = \frac{E'}{mc^2}|C'|^2 < 0 \quad \text{and} \quad j_t = \frac{p'}{m}|C'| < 0. \tag{11.60}$$

We can see that $\tilde{\chi}_t$ is the dominant component, i.e. in this region the state looks more like an *antiparticle*. If $x < 0$, the wave has particle-antiparticle features, with a dominant particle component. At the barrier, part of the wave bounces back, but the rest continues to travel to the right with a dominantly antiparticle component, which looks like a negative charge traveling to the left.

11.1.3 Zitterbewegung of Spin-Zero Particles

For simplicity [2], let us consider the free Hamiltonian in one dimension

$$\hat{H}_0 = (\tau_3 + i\tau_2)\frac{\hat{p}_x^2}{2m} + mc^2\tau_3. \tag{11.61}$$

An operator in the Heisenberg picture is defined by

$$\hat{\Omega}(t) = \exp(i\hat{H}_0 t/\hbar)\hat{\Omega}\exp(-i\hat{H}_0 t/\hbar), \tag{11.62}$$

where $\hat{H}_0$ is time-independent and $\hat{\Omega}$ is the Schrödinger picture operator. The time-derivative is given by the Heisenberg equation

$$\frac{\mathrm{d}\hat{\Omega}(t)}{\mathrm{d}t} = \frac{i}{\hbar}\left[\hat{H}_0, \hat{\Omega}(t)\right]. \tag{11.63}$$

The Hamiltonian $\hat{H}_0$ commutes with itself,

$$\frac{\mathrm{d}\hat{H}_0(t)}{\mathrm{d}t} = \frac{i}{\hbar}\left[\hat{H}_0, \hat{H}_0(t)\right] = 0, \tag{11.64}$$

and with the momentum operator,

$$\frac{\mathrm{d}\hat{p}_x(t)}{\mathrm{d}t} = \frac{i}{\hbar}\left[\hat{H}_0, \hat{p}_x(t)\right] = 0, \tag{11.65}$$

so the energy and the momentun are constants of motion and form a complete set of commuting observables. Therefore, in subsequent calculations, we can work in the space spanned by their common eigenstates.

For the position operator, we find

$$\hat{v}_x(t) = \frac{\mathrm{d}\hat{x}(t)}{\mathrm{d}t} = \frac{i}{\hbar}\left[\hat{H}_0, \hat{x}(t)\right] = [\tau_3(t) + i\tau_2(t)]\frac{\hat{p}_x}{m}, \tag{11.66}$$

so we can call $\hat{v}(t)$ the velocity operator. From Eq. (11.28), we can see that

$$(\tau_3 + i\tau_2)^2 = \begin{pmatrix} 1 & 1 \\ -1 & -1 \end{pmatrix}^2 = 0, \tag{11.67}$$

and consequently $\hat{v}_x(t)\hat{v}_x(t) = 0$. This indicates that in relativistic theory, contrary to the non-relativistic one, the velocity operator is not an observable.

We can also calculate the time evolution of $\hat{v}(t)$

$$\frac{\mathrm{d}\hat{v}_x(t)}{\mathrm{d}t} = \frac{i}{\hbar}\left[\hat{H}_0, \hat{v}_x(t)\right] = \frac{i}{\hbar}[\hat{H}_0\hat{v}_x(t) + \hat{v}_x(t)\hat{H}_0] - \frac{2i}{\hbar}\hat{v}_x(t)\hat{H}_0. \tag{11.68}$$

Using Eqs. (11.28) and (11.67), keeping in mind that $\tau_i\tau_j + \tau_j\tau_i = 2\delta_{i,j}$, we can readily derive

$$\hat{H}_0\hat{v}_x(t) + \hat{v}_x(t)\hat{H}_0 = 2\hat{p}_x c^2, \tag{11.69}$$

and therefore,

$$\frac{\mathrm{d}\hat{v}_x(t)}{\mathrm{d}t} = \frac{2i}{\hbar}\hat{p}_x c^2 - \frac{2i}{\hbar}\hat{v}_x(t)\hat{H}_0. \tag{11.70}$$

Since the energy and momentum are constants of motion, their operators in Eq. (11.70) can be replaced by their corresponding eigenvalues E_0 and p_x, respectively, i.e.

$$\frac{\mathrm{d}\hat{v}_x(t)}{\mathrm{d}t} = \frac{2i}{\hbar}p_x c^2 - \frac{2i}{\hbar}E_0\hat{v}_x(t). \tag{11.71}$$

Then, by integrating out, we obtain

$$\hat{v}_x(t) = p_x c^2/E_0 + \big(v_x(0) - p_x c^2/E_0\big)\exp(-2iE_0 t/\hbar). \tag{11.72}$$

We can see that although the momentum is a constant of motion, the velocity is not.

By integrating further, we find

$$\hat{x}(t) = x(0) + \frac{p_x c^2}{E_0}t - \left(v_x(0) - \frac{p_x c^2}{E_0}\right)\frac{\exp(-2iE_0 t/\hbar) - 1}{2iE_0/\hbar}. \tag{11.73}$$

The first term is the initial position. The second term describes a straight-line motion of a relativistic particle. The third term is an oscillatory motion with frequency $\simeq 2mc^2/\hbar$ and with amplitude $\simeq \hbar/(2mc)$. For a particle like the π meson, these values are in the range of $10^{21} - 10^{22}s^{-1}$ and $10^{-12} - 10^{-13}m$, respectively. The amplitude is in the range of the Compton wave length and the frequency is associated with the Compton time scale. So, it is meaningless to specify the position of a relativistic particle within that range. This extreme high frequency oscillation is like a trembling motion, called *Zitterbewegung* in German, and also in the physics literature.

11.1.4 Klein-Gordon Equation as Effective Schrödinger Equation

For practical purposes it is useful to write the Klein-Gordon equation as an effective Schrödinger equation. If we divide Eq. (11.17) by $2mc^2$, we obtain

$$\left[\frac{1}{2m}\hat{\boldsymbol{p}}^2 + \frac{mc^2}{2} - \frac{1}{2mc^2}(E^2 - 2E\hat{V} + \hat{V}^2)\right]\psi = 0. \tag{11.74}$$

By introducing the notations

$$\tilde{E} = \frac{1}{2mc^2}(E^2 - m^2c^4) \quad \text{and} \quad \hat{\tilde{V}} = \hat{V}\left(\frac{E}{mc^2} - \frac{\hat{V}}{2mc^2}\right), \tag{11.75}$$

we can write Eq. (11.74) in a more familiar form

$$\hat{\tilde{H}}\psi = \tilde{E}\psi, \quad \text{where} \quad \hat{\tilde{H}} = \frac{1}{2m}\hat{\boldsymbol{p}}^2 + \hat{\tilde{V}}. \tag{11.76}$$

With $E = mc^2 + E'$, we find

$$\tilde{E} = E'\left(1 + \frac{E'}{2mc^2}\right) \quad \text{and} \quad \hat{\tilde{V}} = \hat{V}\left(1 + \frac{E'}{mc^2} - \frac{\hat{V}}{2mc^2}\right). \tag{11.77}$$

We can clearly see that in the $c \to \infty$ limit, this equation falls back to the Schrödinger equation.

For a spherical potential $V(r)$, the Hamiltonian $\hat{\tilde{H}}$ commutes with the angular momentum operators. So, $\{\hat{\tilde{H}}, \hat{L}^2, \hat{L}_z\}$ form a complete set of commuting observables. Assuming a Coulomb plus a short range potential

$$V(r) = \frac{Ze^2}{r} + W(r), \tag{11.78}$$

the radial equation reads

$$\left[-\frac{\hbar^2}{2m}\frac{\partial^2}{\partial r^2} + \frac{\hbar^2 l(l+1)}{2mr^2} + \left(\frac{Ze^2}{r} + W\right)\left(1 + \frac{E'}{mc^2} - \frac{1}{2mc^2}\left(\frac{Ze^2}{r} + W\right)\right)\right] u(r) \\ = \tilde{E} u(r), \tag{11.79}$$

or reorganizing, and with the $\alpha = e^2/(\hbar c)$ fine-structure constant, we find

$$\left[-\frac{\hbar^2}{2m}\frac{\partial^2}{\partial r^2} + \frac{\hbar^2}{2m}\frac{l(l+1) - Z^2\alpha^2}{r^2} + \frac{Z'e^2}{r} + W'\right] u(r) = \tilde{E} u(r), \tag{11.80}$$

where

$$Z' = Z\left(1 + \frac{E'}{mc^2}\right) \quad \text{and} \quad W' = W\left(1 + \frac{E'}{mc^2} - \frac{1}{mc^2}\frac{Ze^2}{r} - \frac{1}{2mc^2}W\right). \tag{11.81}$$

If we equate $l(l+1) - Z^2\alpha^2 = \lambda(\lambda+1)$, we find that

$$\lambda = -1/2 + \sqrt{(l+1/2)^2 - Z^2\alpha^2}, \tag{11.82}$$

then Eq. (11.80) becomes

$$\left[-\frac{\hbar^2}{2m}\frac{\partial^2}{\partial r^2} + \frac{\hbar^2 \lambda(\lambda+1)}{2mr^2} + \frac{Z'e^2}{r} + W'\right] u(r) = \tilde{E} u(r). \tag{11.83}$$

Now, this equation really looks just like an ordinary radial Schrödinger equation. The behavior of the wave function u at $r = 0$ and at $r \to \infty$ is the same as in the non-relativistic case. Therefore, the numerical techniques developed previously are readily applicable.

We can easily solve the Klein-Gordon hydrogen problem. We should take $W' = 0$, and apply the correspondences $l \to \lambda$, $E \to \tilde{E}$ and $Z \to Z'$ in the non-relativistic hydrogen formula to obtain

$$E'\left(1 + \frac{E'}{2mc^2}\right) = -\frac{mZ^2\alpha^2(1 + E'/(mc^2))^2}{2\,(n_r + \lambda + 1)^2}, \tag{11.84}$$

where $n_r = 0, 1, \ldots$ is the radial quantum number. After some manipulations, we find

$$E' = mc^2 \left[\left(1 + \frac{Z^2\alpha^2}{(n_r + \lambda + 1)^2} \right)^{-1/2} - 1 \right]$$
$$= mc^2 \left[\left(1 + \frac{Z^2\alpha^2}{\left(n_r + 1/2 + \sqrt{(l+1/2)^2 - Z^2\alpha^2}\right)^2} \right)^{-1/2} - 1 \right] . \quad (11.85)$$

If $Z\alpha \ll 1$, or $Z \ll 1/\alpha \simeq 137$, the energy can be expanded in powers of Z,

$$E' = -\frac{mc^2 Z^2\alpha^2}{2n^2} - \frac{mc^2 Z^4\alpha^4}{2n^4} \left(\frac{n}{l + 1/2} - \frac{3}{4} \right) + \cdots , \quad (11.86)$$

where $n = n_r + l + 1$. We can see that the first term is the non-relativistic hydrogen binding energy. The second term is the relativistic correction. It depends on the angular momentum quantum number l, but experiments show that the correction should depend on the total quantum number j.

The Klein-Gordon equation was the first relativistic quantum mechanical equation. It is named after Oscar Klein and Walter Gordon, who proposed it in 1926. About the same time, it had been proposed by several others, like Vladimir Fock, Johann Kudar, Théophile de Donder and Frans-H. van den Dungen, and Louis de Broglie. In fact, it was Erwin Schrödinger who wrote this equation down in his notebook in 1925, but abandoned it in the favor of his non-relativistic equation because it gave an incorrect result for the hydrogen atom. By now, we know that the Klein-Gordon equation describes spin-0 particles, like pions and kaons. The Klein-Gordon equation has been abandoned in favor of the Dirac equation which provided good results for the hydrogen spectrum, and described spin-1/2 particles like electrons. Also, several interesting relativistic phenomena were discovered originally with the Dirac equation, which are also present in the Klein-Gordon equation, like the Klein paradox and the Zitterbewegung.

11.2 Dirac Equation

We have seen before that the second order time derivative poses some issues with the Klein-Gordon equation. On the other hand, relativistic invariance requires an equal treatment of spatial and time coordinates. To accomplish this, Dirac tried to take the square root of the relativistic energy-momentum relation

$$p_1^2 + p_2^2 + p_3^2 + p_4^2 = (\gamma_1 p_1 + \gamma_2 p_2 + \gamma_3 p_3 + \gamma_4 p_4) \times (\gamma_1 p_1 + \gamma_2 p_2 + \gamma_3 p_3 + \gamma_4 p_4) . \quad (11.87)$$

It is obvious that the γ's cannot be numbers and should possess the following properties

$$\gamma_i^2 = 1, \quad \text{and} \quad \gamma_i\gamma_k + \gamma_k\gamma_i = 0 \quad \text{for} \quad i \neq k. \tag{11.88}$$

The defining properties for the γ's are the anti-commutation relations

$$[\gamma_i, \gamma_k]_+ = \gamma_i\gamma_k + \gamma_k\gamma_i = 2\delta_{ik}, \tag{11.89}$$

where $[,]_+$ denotes the anti-commutator. A possible representation is given by the Hermitian matrices

$$\gamma_1 = \begin{pmatrix} 0 & 0 & 0 & -i \\ 0 & 0 & -i & 0 \\ 0 & i & 0 & 0 \\ i & 0 & 0 & 0 \end{pmatrix}, \quad \gamma_2 = \begin{pmatrix} 0 & 0 & 0 & -1 \\ 0 & 0 & 1 & 0 \\ 0 & 1 & 0 & 0 \\ -1 & 0 & 0 & 0 \end{pmatrix},$$
$$\gamma_3 = \begin{pmatrix} 0 & 0 & -i & 0 \\ 0 & 0 & 0 & i \\ i & 0 & 0 & 0 \\ 0 & -i & 0 & 0 \end{pmatrix}, \quad \gamma_4 = \begin{pmatrix} 1 & 0 & 0 & 0 \\ 0 & 1 & 0 & 0 \\ 0 & 0 & -1 & 0 \\ 0 & 0 & 0 & -1 \end{pmatrix}. \tag{11.90}$$

This representation is not unique, since

$$\gamma_i' = U^\dagger \gamma_i U, \tag{11.91}$$

with U unitary, also satisfies the requirements.

With the 4×4 identity matrix I, the relativistic four-momentum relation is given by

$$\gamma_1 p_1 + \gamma_2 p_2 + \gamma_3 p_3 + \gamma_4 p_4 - imc\, I = 0. \tag{11.92}$$

We can quantize this relation by using the correspondence $p_i \to \hat{p}_i = -i\hbar\partial/\partial x_i$. Then, for the four-component wave function

$$\psi = \begin{pmatrix} \psi_1 \\ \psi_2 \\ \psi_3 \\ \psi_4 \end{pmatrix}, \tag{11.93}$$

we get

$$\left(\gamma_1 \hat{p}_1 + \gamma_2 \hat{p}_2 + \gamma_3 \hat{p}_3 + \gamma_4 \hat{p}_4 - imc\, I\right) \psi = 0, \tag{11.94}$$

or with implicit summation notation we have

$$\left(\gamma_i \hat{p}_i - imc\, I\right) \psi = 0. \tag{11.95}$$

In the following, we are going to omit the 4×4 identity matrix I next to the mass term, so that

$$\left(\gamma_i \frac{\partial}{\partial x_i} + \frac{mc}{\hbar}\right)\psi = 0, \tag{11.96}$$

and with $\chi = mc/\hbar$, we get the usual form of the *Dirac equation*

$$\left(\gamma_i \frac{\partial}{\partial x_i} + \chi\right)\psi = 0. \tag{11.97}$$

In order to derive the continuity relation, we take the adjoint of the wave function

$$\psi^\dagger = \left(\psi_1^*, \psi_2^*, \psi_3^*, \psi_4^*\right). \tag{11.98}$$

Notice that the γ matrices are self-adjoint matrices and $\partial/\partial x_4$ is imaginary, so the adjoint wave function satisfies the equation

$$\frac{\partial \psi^\dagger}{\partial \psi^{x_1}}\gamma_1 + \frac{\partial \psi^\dagger}{\partial \psi^{x_2}}\gamma_2 + \frac{\partial \psi^\dagger}{\partial \psi^{x_3}}\gamma_3 - \frac{\partial \psi^\dagger}{\partial \psi^{x_4}}\gamma_4 + \chi\psi^\dagger = 0. \tag{11.99}$$

We can define the Dirac conjugate as

$$\bar{\psi} = \psi^\dagger \gamma_4, \tag{11.100}$$

and if we multiply Eq. (11.99) by γ_4 from the right, we get the conjugate Dirac equation

$$\frac{\partial \bar{\psi}}{\partial x_i}\gamma_i - \chi\bar{\psi} = 0. \tag{11.101}$$

By multiplying Eq. (11.101) by ψ from the right and Eq. (11.97) by $\bar{\psi}$ from the left, and then adding them, we obtain

$$\frac{\partial \bar{\psi}}{\partial x_i}\gamma_i\psi + \bar{\psi}\gamma_i\frac{\partial \psi}{\partial x_i} = 0. \tag{11.102}$$

This is a four-divergence

$$\frac{\partial}{\partial x_i}(\bar{\psi}\gamma_i\psi) = 0, \tag{11.103}$$

which can be expressed as

$$\frac{\partial}{\partial x_1}(\bar{\psi}\gamma_1\psi) + \frac{\partial}{\partial x_2}(\bar{\psi}\gamma_2\psi) + \frac{\partial}{\partial x_3}(\bar{\psi}\gamma_3\psi) + \frac{1}{ic}\frac{\partial}{\partial t}(\bar{\psi}\gamma_4\psi) = 0, \tag{11.104}$$

or

$$\frac{\partial}{\partial t}(\psi^\dagger\psi) + \sum_{k=1}^{3}\frac{\partial}{\partial x_k} ic(\bar{\psi}\gamma_k\psi) = 0. \tag{11.105}$$

By defining

$$\rho = \psi^\dagger\psi, \quad \text{and} \quad j_k = ic(\bar{\psi}\gamma_k\psi) \text{ where } k = 1, 2, 3, \tag{11.106}$$

the continuity equation reads

$$\frac{\partial\rho}{\partial t} + \nabla\cdot\boldsymbol{j} = 0. \tag{11.107}$$

Here $\rho = \psi^\dagger\psi$ is positive, so we can assign a statistical interpretation for ψ. The ρ is the probability of finding a particle at some location and j_k is the probability current.

11.2.1 Dirac Particle in Electromagnetic Field

So far, we have studied the free Dirac equation. We can introduce interactions by using the method of minimal coupling, where we replace the mechanical momentum operator by the canonical momentum operator

$$\hat{p}_i \longrightarrow \hat{p}_i - \frac{q}{c}A_i, \tag{11.108}$$

where $A_i = (\boldsymbol{A}, i\Phi)$ is the electromagnetic four-potential. This substitution leads to the equation

$$\left[\gamma_i\left(\frac{\partial}{\partial x_i} - \frac{iq}{\hbar c}A_i\right) + \chi\right]\psi = 0. \tag{11.109}$$

We can cast this equation into a Hamiltonian form. For this purpose, we multiply Eq. (11.109) from the left by $\hbar c\gamma_4$ and arrive at a Schrödinger-like equation with the Hamiltonian

$$\begin{aligned}\hat{H} &= \hbar c\sum_{k=1}^{3}\gamma_4\gamma_k\frac{\partial}{\partial x_k} - iq\sum_{k=1}^{3}\gamma_4\gamma_k A_k + q\Phi + mc^2\gamma_4\\ &= ic\sum_{k=1}^{3}\gamma_4\gamma_k\hat{p}_k - iq\sum_{k=1}^{3}\gamma_4\gamma_k A_k + q\Phi + mc^2\gamma_4.\end{aligned} \tag{11.110}$$

Introducing the notation

$$\alpha_k = i\gamma_4\gamma_k, \tag{11.111}$$

we find

$$\hat{H} = c\sum_{k=1}^{3}\alpha_k\hat{p}_k - q\sum_{k=1}^{3}\alpha_k A_k + q\Phi + mc^2\gamma_4. \tag{11.112}$$

Obviously, the α_k matrices satisfy the anti-commutation relations

$$[\alpha_i, \gamma_4]_+ = 0 \quad \text{and} \quad [\alpha_i, \alpha_k]_+ = 2\delta_{ik}, \tag{11.113}$$

and their matrix form, in our adopted representation of Eq. (11.90), is given by

$$\alpha_1 = i\gamma_4\gamma_1 = \begin{pmatrix} 0 & 0 & 0 & 1 \\ 0 & 0 & 1 & 0 \\ 0 & 1 & 0 & 0 \\ 1 & 0 & 0 & 0 \end{pmatrix} = \begin{pmatrix} 0 & \sigma_1 \\ \sigma_1 & 0 \end{pmatrix}, \tag{11.114}$$

$$\alpha_2 = i\gamma_4\gamma_2 = \begin{pmatrix} 0 & 0 & 0 & -i \\ 0 & 0 & i & 0 \\ 0 & -i & 0 & 0 \\ i & 0 & 0 & 0 \end{pmatrix} = \begin{pmatrix} 0 & \sigma_2 \\ \sigma_2 & 0 \end{pmatrix}, \tag{11.115}$$

$$\alpha_3 = i\gamma_4\gamma_3 = \begin{pmatrix} 0 & 0 & 1 & 0 \\ 0 & 0 & 0 & -1 \\ 1 & 0 & 0 & 0 \\ 0 & -1 & 0 & 0 \end{pmatrix} = \begin{pmatrix} 0 & \sigma_3 \\ \sigma_3 & 0 \end{pmatrix}, \tag{11.116}$$

where σ_i are the usual 2×2 Pauli matrices.

11.2.2 Symmetry Transformations

Lorentz Transformation

The Dirac equation is Lorentz invariant if it has the same form for observers in different inertial frames x and x'. In fact, it should be Lorentz invariant since it has been derived from an invariant quantity, the length of the four-momentum vector. Nevertheless, it is interesting to see how the Dirac wave function transforms with respect to the Lorentz transformation.

The Lorentz transformation of the coordinates and momenta is given by

$$x_i' = a_{ik}x_k \quad \text{and} \quad p_i' = a_{ik}p_k, \tag{11.117}$$

and if the transformation is infinitesimal we have

$$a_{ik} = \delta_{ik} + \epsilon_{ik}, \tag{11.118}$$

with $\epsilon_{ik} = -\epsilon_{ki}$, $|\epsilon_{ik}| << 1$, ϵ_{4k} is imaginary, and ϵ_{rs} is real for $r, s = 1, 2, 3$.

The wave function is transformed by the Lorentz transformation

$$\psi'(x_i') = \hat{U}\psi(x_i), \tag{11.119}$$

and the transformed Dirac equation should have the form

$$(\gamma_i p_i' - imc)\psi'(x_i') = 0. \tag{11.120}$$

Consequently, we should have

$$(\gamma_i a_{ik} p_k - imc)\hat{U}\psi(x_i) = 0. \tag{11.121}$$

Multiplying from the left by $\hat{U}^{-1}$ we find

$$(\hat{U}^{-1}\gamma_i \hat{U} a_{ik} p_k - imc)\psi(x_i) = 0. \tag{11.122}$$

We can see that Eq. (11.122) falls back to Eq. (11.95) if

$$\hat{U}^{-1}\gamma_i \hat{U} a_{ik} = \gamma_k. \tag{11.123}$$

By using the orthogonality relation of the Lorentz transformation matrices $a_{ki}a_{kj} = a_{ik}a_{jk} = \delta_{ij}$, we obtain

$$\hat{U}^{-1}\gamma_i \hat{U} = a_{ij}\gamma_j. \tag{11.124}$$

We consider an infinitesimal Lorentz transformation $\hat{U} = 1 + \varepsilon$ and $\hat{U}^{-1} = 1 - \varepsilon$, where ε should be infinitesimal, i.e. $\varepsilon << 1$. Furthermore, ε should be a linear function of ϵ_{jk}, i.e. $\varepsilon = M_{jk}\epsilon_{jk}$. As a result, Eq. (11.124) reads

$$(1 - \varepsilon)\gamma_i(1 + \varepsilon) = (1 - M_{jk}\epsilon_{jk})\gamma_i(1 + M_{jk}\epsilon_{jk}) = \gamma_i + \epsilon_{ik}\gamma_k, \tag{11.125}$$

or

$$\gamma_i\varepsilon - \varepsilon\gamma_i = (\gamma_i M_{jk} - M_{jk}\gamma_i)\epsilon_{jk} = \epsilon_{ik}\gamma_k. \tag{11.126}$$

Considering that ϵ_{jk} is antisymmetric, we can write the right hand side as

$$\epsilon_{ik}\gamma_k = \epsilon_{jk}\delta_{ji}\gamma_k = \frac{1}{2}\epsilon_{jk}(\delta_{ji}\gamma_k - \delta_{ki}\gamma_j). \tag{11.127}$$

Then, Eq. (11.125) becomes

$$[\gamma_i M_{jk} - M_{jk}\gamma_i - \frac{1}{2}\delta_{ji}\gamma_k + \frac{1}{2}\delta_{ki}\gamma_j]\epsilon_{jk} = 0. \tag{11.128}$$

By utilizing the properties of γ_i, Eq. (11.89), we find

$$[\gamma_i M_{jk} - M_{jk}\gamma_i - \frac{1}{4}(\gamma_j\gamma_i + \gamma_i\gamma_j)\gamma_k + \frac{1}{4}(\gamma_k\gamma_i + \gamma_i\gamma_k)\gamma_j]\epsilon_{jk} = 0, \tag{11.129}$$

or, reorganizing, we obtain

$$(\gamma_i M_{jk} - \frac{1}{4}\gamma_i\gamma_j\gamma_k - M_{jk}\gamma_i + \frac{1}{4}\gamma_j\gamma_k\gamma_i)\epsilon_{jk} = 0. \tag{11.130}$$

We can see that $M_{jk} = 1/4\,\gamma_j\gamma_k$ solves this equation and then

$$\varepsilon = \frac{1}{4}\epsilon_{ik}\gamma_i\gamma_k. \tag{11.131}$$

So, we have found that the wave function transforms with respect to the infinitesimal Lorentz transformation as

$$\psi'(x_i') \approx \psi(x_i) + \frac{1}{4}\epsilon_{ik}\gamma_i\gamma_k\psi(x_i). \tag{11.132}$$

This transformation rule is valid also for the Dirac equation with electromagnetic potentials, since the electromagnetic potentials also form a four-vector, and they transform the same way with respect to the Lorentz transformation as the four-momentum vector.

Now, we can prove that the bilinear form in Eq. (11.103), $j_i = \bar{\psi}\gamma_i\psi$, is a Lorentz four-vector. For this purpose, we need to carry out an infinitesimal Lorentz transformation

$$\begin{aligned} j_i' &= \bar{\psi}'\gamma_i\psi' = \psi'^{\dagger}\gamma_4\gamma_i\psi' = \psi^{\dagger}\gamma_4\gamma_4(1+\varepsilon^{\dagger})\gamma_4\gamma_i(1+\varepsilon)\psi \\ &= \bar{\psi}(1+\gamma_4\varepsilon^{\dagger}\gamma_4)\gamma_i(1+\varepsilon)\psi = \bar{\psi}(1-\varepsilon)\gamma_i(1+\varepsilon)\psi \\ &= \bar{\psi}\gamma_i\psi + \epsilon_{ik}\bar{\psi}\gamma_k\psi = j_i + \epsilon_{ik}j_k. \end{aligned} \tag{11.133}$$

Indeed, j_i is a four-vector, i.e. it transforms like a four-vector with respect to the Lorentz transformation.

In deriving the above relation, we assumed that $\gamma_4\varepsilon^{\dagger}\gamma_4 = -\varepsilon$. To show this, we write ε as

$$\varepsilon = \frac{1}{4}\sum_{r,s=1}^{3}\epsilon_{rs}\gamma_r\gamma_s + \frac{1}{2}\epsilon_{k4}\gamma_k\gamma_4. \tag{11.134}$$

By using Eq. (11.118) we find

$$\varepsilon^{\dagger} = \frac{1}{4}\sum_{r,s=1}^{3}\epsilon_{rs}\gamma_s\gamma_r - \frac{1}{2}\epsilon_{k4}\gamma_4\gamma_k = -\frac{1}{4}\sum_{r,s=1}^{3}\epsilon_{rs}\gamma_r\gamma_s + \frac{1}{2}\epsilon_{k4}\gamma_k\gamma_4. \tag{11.135}$$

Then multiplying by γ_4 from left and right, and after interchanging the γ matrices, we get the desired relation

$$\gamma_4 \varepsilon^{\dagger} \gamma_4 = -\varepsilon. \tag{11.136}$$

Furthermore, we can easily show that $\bar{\psi}\psi$ is a scalar and $\bar{\psi}\gamma_i\gamma_k\psi$ is a two-index tensor.

Let us introduce γ_5 as

$$\gamma_5 = \gamma_1\gamma_2\gamma_3\gamma_4 = \begin{pmatrix} 0 & 0 & -1 & 0 \\ 0 & 0 & 0 & -1 \\ -1 & 0 & 0 & 0 \\ 0 & -1 & 0 & 0 \end{pmatrix}. \tag{11.137}$$

It is easy to see that $[\gamma_i, \gamma_5]_+ = 0$ for $i = 1, 2, 3, 4$. For example,

$$[\gamma_2, \gamma_5]_+ = \gamma_2\gamma_1\gamma_2\gamma_3\gamma_4 + \gamma_1\gamma_2\gamma_3\gamma_4\gamma_2 = -\gamma_1\gamma_2^2\gamma_3\gamma_4 + \gamma_1\gamma_2^2\gamma_3\gamma_4 = 0. \tag{11.138}$$

We can also show that $\bar{\psi}\gamma_5\psi$ transforms like a scalar and $\bar{\psi}\gamma_i\gamma_5\psi$ transforms like a vector with respect to the Lorentz transformation. Indeed,

$$\begin{aligned} \bar{\psi}'\gamma_5\psi' &= \psi'^{\dagger}\gamma_4\gamma_5\psi' = \psi^{\dagger}\gamma_4\gamma_4(1+\varepsilon^{\dagger})\gamma_4\gamma_5(1+\varepsilon)\psi = \bar{\psi}(1-\varepsilon)\gamma_5(1+\varepsilon)\psi \\ &= \bar{\psi}\gamma_5\psi + \bar{\psi}(\gamma_5\varepsilon - \varepsilon\gamma_5)\psi = \bar{\psi}\gamma_5\psi, \end{aligned} \tag{11.139}$$

since $\gamma_5\epsilon_{rs}\gamma_r\gamma_s - \epsilon_{rs}\gamma_r\gamma_s\gamma_5 = \epsilon_{rs}(\gamma_5\gamma_r\gamma_s - \gamma_r\gamma_s\gamma_5) = 0$. Also,

$$\begin{aligned} \bar{\psi}'\gamma_i\gamma_5\psi' &= \psi'^{\dagger}\gamma_4\gamma_i\gamma_5\psi' = \psi^{\dagger}\gamma_4\gamma_4(1+\varepsilon^{\dagger})\gamma_4\gamma_i\gamma_5(1+\varepsilon)\psi = \bar{\psi}(1-\varepsilon)\gamma_i\gamma_5(1+\varepsilon)\psi \\ &= \bar{\psi}\gamma_i\gamma_5\psi + \bar{\psi}(\gamma_i\varepsilon - \varepsilon\gamma_i)\gamma_5\psi = \bar{\psi}\gamma_i\gamma_5\psi + \epsilon_{ik}\bar{\psi}\gamma_k\gamma_5\psi, \end{aligned} \tag{11.140}$$

i.e. $\bar{\psi}\gamma_i\gamma_5\psi$ transforms like a four-vector with respect to the Lorentz transformation.

Parity

Another symmetry the Dirac equation should follow is the invariance with respect to the mirroring of the axes

$$x_k' = -x_k \quad \text{for} \quad k = 1, 2, 3, \quad \text{and} \quad x_4' = x_4. \tag{11.141}$$

The transformed equation should have the form

$$\left(\gamma_i \frac{\partial}{\partial x_i'} + \chi\right)\psi_P = 0, \tag{11.142}$$

where

$$\frac{\partial}{\partial x_k'} = -\frac{\partial}{\partial x_k} \quad \text{for} \quad k = 1, 2, 3, \quad \text{and} \quad \frac{\partial}{\partial x_4'} = \frac{\partial}{\partial x_4}. \tag{11.143}$$

We can write Eq. (11.142) as

$$-\sum_{k=1}^{3} \gamma_k \frac{\partial \psi_P}{\partial x_k} + \gamma_4 \frac{\partial \psi_P}{\partial x_4} + \chi \psi_P = 0, \tag{11.144}$$

and we can see that with

$$\psi_P(x) = \gamma_4 \psi(x) \tag{11.145}$$

this equation goes back to the original Dirac equation. So the Dirac equation is invariant with respect to the parity transformation, and the wave function transforms according to Eq. (11.145). This invariance remains valid with the A_i electromagnetic potential as well, since they transform the same way as x_i, i.e. $A_k^P = -A_k$ for $k = 1, 2, 3$, and $A_4^P = A_4$.

We can also see that

$$\bar{\psi}_P \psi_P = (\gamma_4 \psi)^\dagger \gamma_4 (\gamma_4 \psi) = \psi^\dagger \gamma_4 \psi = \bar{\psi} \psi, \tag{11.146}$$

i.e. $\bar{\psi}\psi$ is invariant with respect to the mirroring of the axes. On the other hand,

$$\bar{\psi}_P \gamma_5 \psi_P = (\gamma_4 \psi)^\dagger \gamma_4 \gamma_5 (\gamma_4 \psi) = \psi^\dagger \gamma_5 \gamma_4 \psi = -\psi^\dagger \gamma_4 \gamma_5 \psi = -\bar{\psi} \gamma_5 \psi. \tag{11.147}$$

So, this quantity behaves like a scalar with respect to the Lorentz transformation, but changes sign with respect to the parity transformation. These type of quantities are pseudo-scalars.

Analogously, we find that

$$\begin{aligned} \bar{\psi}_P \gamma_k \psi_P &= (\gamma_4 \psi)^\dagger \gamma_4 \gamma_k (\gamma_4 \psi) = \psi^\dagger \gamma_k \gamma_4 \psi \\ &= -\psi^\dagger \gamma_4 \gamma_k \psi = -\bar{\psi} \gamma_k \psi \quad \text{for} \quad k = 1, 2, 3, \end{aligned} \tag{11.148}$$

and

$$\bar{\psi}_P \gamma_4 \psi_P = (\gamma_4 \psi)^\dagger \gamma_4 \gamma_4 (\gamma_4 \psi) = \psi^\dagger \psi = \bar{\psi} \gamma_4 \psi, \tag{11.149}$$

i.e. $\bar{\psi} \gamma_i \psi$ behaves like a four-vector with respect to the Lorentz transformation and changes sign with respect to the parity transformation. These quantities are vectors, or polar vectors.

On the other hand, for $\bar{\psi} \gamma_i \gamma_5 \psi$ with $k = 1, 2, 3$ we have

$$\begin{aligned} \bar{\psi}_P \gamma_k \gamma_5 \psi_P &= (\gamma_4 \psi)^\dagger \gamma_4 \gamma_k \gamma_5 (\gamma_4 \psi) = \psi^\dagger \gamma_4 \gamma_4 \gamma_k \gamma_5 \gamma_4 \psi \\ &= -\psi^\dagger \gamma_4 \gamma_k \gamma_4 \gamma_5 \gamma_4 \psi = \psi^\dagger \gamma_4 \gamma_k \gamma_5 \psi = \bar{\psi} \gamma_k \gamma_5 \psi, \end{aligned} \tag{11.150}$$

and with $k = 4$, we find

$$\bar{\psi}_P \gamma_4 \gamma_5 \psi_P = (\gamma_4 \psi)^\dagger \gamma_4 \gamma_4 \gamma_5 \gamma_4 \psi = \psi^\dagger \gamma_4 \gamma_5 \gamma_4 \psi = -\psi^\dagger \gamma_4 \gamma_4 \gamma_5 \psi. = -\bar{\psi} \gamma_4 \gamma_5 \psi. \tag{11.151}$$

This means that $\bar{\psi}\gamma_i\gamma_5\psi$ behaves like a vector with respect to the Lorentz transformation, but with respect to the parity transformation, the spatial coordinates do not change, only the time component is mirrored. These type of quantities are called axial, or pseudo-vectors.

Time Reversal

The time reversal transformation leaves the spatial coordinates unchanged while reversing the time coordinate

$$\begin{aligned} x_k' &= x_k \quad \text{for} \quad k = 1, 2, 3, \quad \text{and} \quad x_4' = -x_4, \\ \frac{\partial}{\partial x_k'} &= \frac{\partial}{\partial x_k} \quad \text{for} \quad k = 1, 2, 3, \quad \text{and} \quad \frac{\partial}{\partial x_4'} = -\frac{\partial}{\partial x_4}. \end{aligned} \tag{11.152}$$

The electromagnetic potential comes from moving charges, so the time-reversed potential should be

$$A_k^T(x) = -A_k(x') \quad \text{for} \quad k = 1, 2, 3, \quad \text{and} \quad A_4^T(x) = A_4(x'). \tag{11.153}$$

Keeping in mind that x_4 is complex, the time reversed and complex conjugated Dirac equation becomes

$$\left[\sum_{k=1}^{3} \gamma_k^* \left(\frac{\partial}{\partial x_k} - \frac{iq}{\hbar c} A_k^T(x) \right) + \gamma_4^* \left(\frac{\partial}{\partial x_4} - \frac{iq}{\hbar c} A_4^T(x) \right) + \chi \right] \psi^*(x') = 0. \tag{11.154}$$

Here we used the fact that $A_k^{T*} = A_k^T$ and $A_4^{T*} = -A_4^T$. Assuming that the time-reversed wave function is a combination of a unitary transformation and the complex conjugation, i.e.

$$\psi_T(x) = \hat{U}_T \psi^*(x'), \tag{11.155}$$

then plugging $\psi^*(x') = \hat{U}_T^{-1}\psi_T(x)$ into Eq. (11.154), and multiplying by $\hat{U}_T$, we obtain

$$\left[\sum_{k=1}^{3} \hat{U}_T \gamma_k^* \hat{U}_T^{-1} \left(\frac{\partial}{\partial x_k} - \frac{iq}{\hbar c} A_k^T(x) \right) + \hat{U}_T \gamma_4^* \hat{U}_T^{-1} \left(\frac{\partial}{\partial x_4} - \frac{iq}{\hbar c} A_4^T(x) \right) + \chi \right] \psi_T(x) = 0. \tag{11.156}$$

We can see that this equation is equivalent to the Dirac equation Eq. (11.109) provided

$$\hat{U}_T \gamma_i^* \hat{U}_T^{-1} = \gamma_i \quad \text{for} \quad i = 1, 2, 3, 4. \tag{11.157}$$

Then, considering the actual form of the γ matrices in Eq. (11.90), i.e. that γ_1 and γ_3 are pure imaginary while γ_2 and γ_4 are real, we find

$$\hat{U}_T\gamma_1 = -\gamma_1\hat{U}_T, \quad \hat{U}_T\gamma_2 = \gamma_2\hat{U}_T, \quad \hat{U}_T\gamma_3 = -\gamma_3\hat{U}_T, \quad \hat{U}_T\gamma_4 = \gamma_4\hat{U}_T. \tag{11.158}$$

From that we can conclude that $\hat{U}_T$ should be a product of γ_1 and γ_3. A little calculation reveals that $\hat{U}_T = \gamma_1\gamma_3$. With this choice, the time-reversed wave function

$$\psi_T(x) = \gamma_1\gamma_3\psi^*(x') \tag{11.159}$$

also satisfies the Dirac equation.

Charge Conjugation

The charge conjugation amounts of the transformation $q \to -q$. From the Dirac equation Eq. (11.109) we can see that the charge conjugation should be related to complex conjugation

$$\hat{C}\psi \equiv \psi_C(x) = \hat{U}_C\psi^*(x), \tag{11.160}$$

where $\hat{U}_C$ is some unitary operator. As the transformed electromagnetic four-potential should come from charges and currents with opposite sign, it should be negated

$$A_i^C(x) = -A_i(x). \tag{11.161}$$

So, the charge-conjugated Dirac equation should have the form

$$\left[\gamma_i\left(\frac{\partial}{\partial x_i} - \frac{iq}{\hbar c}A_i^C\right) + \chi\right]\psi_C = 0. \tag{11.162}$$

Taking the complex conjugate of Eq. (11.109) we find

$$\left[\sum_{k=1}^{3}\gamma_k^*\left(\frac{\partial}{\partial x_k} - \frac{iq}{\hbar c}A_k^C(x)\right) + (-\gamma_4^*)\left(\frac{\partial}{\partial x_4} - \frac{iq}{\hbar c}A_4^C(x)\right) + \chi\right]\psi^*(x) = 0, \tag{11.163}$$

where we have used the fact that $A_k^* = A_k$ for $k = 1, 2, 3$, $A_4^* = -A_4$, $x_k^* = x_k$ for $k = 1, 2, 3$, and $x_4^* = -x_4$. Substituting $\psi^* = \hat{U}_C^{-1}\psi_C$ into the above equation and multiplying by $\hat{U}_C$ we obtain

$$\left[\sum_{k=1}^{3}\hat{U}_C\gamma_k^*\hat{U}_C^{-1}\left(\frac{\partial}{\partial x_k} - \frac{iq}{\hbar c}A_k^C(x)\right) + \hat{U}_C(-\gamma_4^*)\hat{U}_C^{-1}\left(\frac{\partial}{\partial x_4} - \frac{iq}{\hbar c}A_4^C(x)\right) + \chi\right]\psi_C(x) = 0. \tag{11.164}$$

We can see that this equation can be brought into the form of Eq. (11.162) if

$$\hat{U}_C \gamma_k^* \hat{U}_C^{-1} = \gamma_k \quad \text{for} \quad k = 1, 2, 3, \quad \text{and} \quad \hat{U}_C(-\gamma_4^*)\hat{U}_C^{-1} = \gamma_4. \tag{11.165}$$

Considering that γ_1 and γ_3 are pure imaginary while γ_2 and γ_4 are real, the above relations become

$$\hat{U}_C \gamma_1 = -\gamma_1 \hat{U}_C, \quad \hat{U}_C \gamma_2 = \gamma_2 \hat{U}_C, \quad \hat{U}_C \gamma_3 = -\gamma_3 \hat{U}_C, \quad \hat{U}_C \gamma_4 = -\gamma_4 \hat{U}_C. \tag{11.166}$$

From the anti-commutation relations of the γ matrices it follows that $\hat{U}_C = \gamma_2$, and consequently the charge-conjugated solution is given by

$$\psi_C(x) = \gamma_2 \psi^*(x). \tag{11.167}$$

If we calculate the expectation value of various physical quantities like the energy and the momentum with ψ_C, we find that $E_C = -E$ and $\langle \hat{\boldsymbol{p}} \rangle_C = -\langle \hat{\boldsymbol{p}} \rangle$. In order that we describe a particle with the same momentum and energy, but with opposite charge, we should apply the combined $\mathcal{CPT}$ transformation

$$\psi_{CPT} = \mathcal{CPT}\psi = \mathcal{CP}\gamma_1\gamma_3\psi^* = \mathcal{C}\gamma_4\gamma_1\gamma_3\psi^* = \gamma_2\gamma_4\gamma_1\gamma_3\psi = -\gamma_5\psi. \tag{11.168}$$

The state $\psi_{CPT}(x)$ also satisfies the Dirac equation and describes a particle with energy E and momentum $\boldsymbol{p}$, but with opposite charge, i.e. it describes an antiparticle.

11.2.3 Angular Momentum

We have seen before in non-relativistic quantum mechanics that in a spherical potential the orbital angular momentum is conserved

$$\frac{\mathrm{d}\hat{\boldsymbol{L}}}{\mathrm{d}t} = \frac{i}{\hbar}\left[\hat{H}, \hat{\boldsymbol{L}}\right] = 0, \tag{11.169}$$

where $\hat{\boldsymbol{L}} = \hat{\boldsymbol{r}} \times \hat{\boldsymbol{p}}$ and $\hat{H}$ is the non-relativistic Hamiltonian with a spherical potential $V(r)$.

We consider the Dirac Hamiltonian in Eq. (11.110) with $\boldsymbol{A} = 0$ and spherical potential $q\Phi = V(r)$

$$\hat{H} = ic\sum_{k=1}^{3} \gamma_4\gamma_k \hat{p}_k + V(r) + mc^2\gamma_4 = c\sum_{k=1}^{3} \alpha_k \hat{p}_k + V(r) + mc^2\gamma_4. \tag{11.170}$$

Then,

$$\begin{aligned}\frac{d\hat{L}_z}{dt} &= \frac{i}{\hbar}\left[H, L_z\right] = \frac{i}{\hbar}\left[\hat{H}(\hat{x}_1\hat{p}_2 - \hat{x}_2\hat{p}_1) - (\hat{x}_1\hat{p}_2 - \hat{x}_2\hat{p}_1)\hat{H}\right] \\ &= -\frac{c}{\hbar}\sum_{k=1}^{3}\left[\gamma_4\gamma_k\hat{p}_k(\hat{x}_1\hat{p}_2 - \hat{x}_2\hat{p}_1) - (\hat{x}_1\hat{p}_2 - \hat{x}_2\hat{p}_1)\gamma_4\gamma_k\hat{p}_k\right],\end{aligned} \tag{11.171}$$

and utilizing the coordinate-momentum commutation relations, we find that

$$\frac{d\hat{L}_z}{dt} = ic\gamma_4(\gamma_1\hat{p}_2 - \gamma_2\hat{p}_1) = c(\alpha_1\hat{p}_2 - \alpha_2\hat{p}_1) = c(\boldsymbol{\alpha} \times \hat{\boldsymbol{p}})_z \neq 0. \tag{11.172}$$

So, we have found a strange situation with the Dirac equation where the orbital angular momentum is not conserved in a spherical potential.

We may define the operator

$$\hat{\mathbf{S}} = \frac{\hbar}{2}\hat{\boldsymbol{\Sigma}}, \tag{11.173}$$

with

$$\hat{S}_x = \frac{\hbar}{2}\Sigma_1, \quad \hat{S}_y = \frac{\hbar}{2}\Sigma_2, \quad \hat{S}_z = \frac{\hbar}{2}\Sigma_3, \tag{11.174}$$

where

$$\Sigma_1 = -i\gamma_2\gamma_3 \quad = \begin{pmatrix} 0 & 1 & 0 & 0 \\ 1 & 0 & 0 & 0 \\ 0 & 0 & 0 & 1 \\ 0 & 0 & 1 & 0 \end{pmatrix} = \begin{pmatrix} \sigma_1 & 0 \\ 0 & \sigma_1 \end{pmatrix}, \tag{11.175}$$

$$\Sigma_2 = -i\gamma_3\gamma_1 \quad = \begin{pmatrix} 0 & -i & 0 & 0 \\ i & 0 & 0 & 0 \\ 0 & 0 & 0 & -i \\ 0 & 0 & i & 0 \end{pmatrix} = \begin{pmatrix} \sigma_2 & 0 \\ 0 & \sigma_2 \end{pmatrix}, \tag{11.176}$$

$$\Sigma_3 = -i\gamma_1\gamma_2 \quad = \begin{pmatrix} 1 & 0 & 0 & 0 \\ 0 & -1 & 0 & 0 \\ 0 & 0 & 1 & 0 \\ 0 & 0 & 0 & -1 \end{pmatrix} = \begin{pmatrix} \sigma_3 & 0 \\ 0 & \sigma_3 \end{pmatrix}. \tag{11.177}$$

Let us calculate the time evolution of $\hat{\mathbf{S}}$

$$\frac{d\hat{S}_z}{dt} = \frac{i}{\hbar}\left[\hat{H}, \hat{S}_z\right] = \frac{i}{2}\left[\hat{H}, \Sigma_3\right]. \tag{11.178}$$

Evaluating the commutators, we find

$$\begin{aligned}
[\gamma_4, \Sigma_3] &= -i\gamma_4\gamma_1\gamma_2 + i\gamma_1\gamma_2\gamma_4 = i(\gamma_1\gamma_2\gamma_4 - \gamma_1\gamma_2\gamma_4) = 0, \\
[\gamma_4\gamma_3, \Sigma_3] &= -i\gamma_4\gamma_3\gamma_1\gamma_2 + i\gamma_1\gamma_2\gamma_4\gamma_3 = i(\gamma_1\gamma_2\gamma_4\gamma_3 - \gamma_1\gamma_2\gamma_4\gamma_3) = 0, \\
[\gamma_4\gamma_1, \Sigma_3] &= -i\gamma_4\gamma_1\gamma_1\gamma_2 + i\gamma_1\gamma_2\gamma_4\gamma_1 = -i\gamma_4\gamma_2 + i\gamma_2\gamma_4 = -2i\gamma_4\gamma_2, \\
[\gamma_4\gamma_2, \Sigma_3] &= -i\gamma_4\gamma_2\gamma_1\gamma_2 + i\gamma_1\gamma_2\gamma_4\gamma_2 = i\gamma_4\gamma_1 - i\gamma_1\gamma_4 = 2i\gamma_4\gamma_1,
\end{aligned} \tag{11.179}$$

which lead to

$$\frac{\mathrm{d}\hat{S}_z}{\mathrm{d}t} = -ic\gamma_4(\gamma_1 p_2 - \gamma_2 p_1) = -c(\boldsymbol{\alpha} \times \hat{\boldsymbol{p}})_z. \tag{11.180}$$

This is just the negative of the time derivative of L_z.

So, if we define $\hat{\boldsymbol{J}} = \hat{\boldsymbol{L}} + \hat{\boldsymbol{S}}$, then we obtain

$$\frac{\mathrm{d}\hat{J}_z}{\mathrm{d}t} = \frac{\mathrm{d}\hat{L}_z}{\mathrm{d}t} + \frac{\mathrm{d}\hat{S}_z}{\mathrm{d}t} = 0, \tag{11.181}$$

meaning that neither $\hat{L}_z$ nor $\hat{S}_z$ are conserved quantities, only

$$\hat{J}_z = \hat{L}_z + \hat{S}_z \tag{11.182}$$

is conserved. We can see that the eigenvalues of σ_3 are ± 1, and therefore the eigenvalues of $\hat{S}_z$ are $\pm\hbar/2$. The commutator of $\hat{S}_x$ and $\hat{S}_y$ is given by

$$\begin{aligned}
[\hat{S}_x, \hat{S}_y] &= \frac{\hbar^2}{4}(\Sigma_x\Sigma_y - \Sigma_y\Sigma_x) = \frac{\hbar^2}{4}((-i\gamma_2\gamma_3)(-i\gamma_3\gamma_1) - (-i\gamma_3\gamma_1)(-i\gamma_2\gamma_3)) \\
&= \frac{\hbar^2}{4}(-\gamma_2\gamma_3\gamma_3\gamma_1 + \gamma_3\gamma_1\gamma_2\gamma_3) = \frac{\hbar^2}{4}(-\gamma_2\gamma_1 + \gamma_1\gamma_2) = \frac{\hbar^2}{2}\gamma_1\gamma_2 \\
&= \frac{\hbar^2}{2} i(-i\gamma_1\gamma_2) = i\hbar\left(\frac{\hbar}{2}\Sigma_3\right) = i\hbar\hat{S}_z,
\end{aligned} \tag{11.183}$$

and analogously,

$$[\hat{S}_y, \hat{S}_z] = i\hbar\hat{S}_x \quad \text{and} \quad [\hat{S}_z, \hat{S}_x] = i\hbar\hat{S}_y. \tag{11.184}$$

So, $\hat{\boldsymbol{S}}$ really behaves like an angular momentum operator. We have established the commutation relations in 4×4 representation, but since the Σ_i matrices are just block-diagonal matrices with the usual Pauli σ_i matrices in the diagonal, the commutation relations are also valid in the 2×2 representation of the operators. We established the conservation of $\hat{J}_z$, but the direction of z is arbitrary, therefore this is valid for other components as well

$$[\hat{\boldsymbol{J}}, \hat{H}] = 0. \tag{11.185}$$

Rotation as Lorentz Transformation

Let us investigate the transformation of the Dirac equation with respect to an infinitesimal rotation around the $x_3 = z$ axis, i.e.

$$x'_1 = x_1 + \delta\vartheta\, x_2, \quad x'_2 = x_2 - \delta\vartheta\, x_1, \quad x'_3 = x_3, \quad \text{and} \quad x'_4 = x_4, \tag{11.186}$$

and consequently

$$\epsilon_{ik} = \begin{pmatrix} 0 & \delta\vartheta & 0 & 0 \\ -\delta\vartheta & 0 & 0 & 0 \\ 0 & 0 & 0 & 0 \\ 0 & 0 & 0 & 0 \end{pmatrix}. \tag{11.187}$$

The wave function transforms as

$$\begin{aligned} \psi'(x') &= \psi(x) + \frac{1}{4}\epsilon_{ik}\gamma_i\gamma_k\psi(x) = \psi(x) + \frac{1}{4}(\epsilon_{12}\gamma_1\gamma_2 + \epsilon_{21}\gamma_2\gamma_1)\psi(x) \\ &= \psi(x) + \frac{1}{2}\epsilon_{12}\gamma_1\gamma_2\psi(x) = \psi(x) + \frac{1}{2}\delta\vartheta\,\gamma_1\gamma_2\psi(x) \\ &= \psi(x) + \frac{i}{2}\Sigma_3\,\delta\vartheta\,\psi(x). \end{aligned} \tag{11.188}$$

On the other hand,

$$\begin{aligned} \psi(x') &= \psi(x_1 + \delta\vartheta x_2, x_2 - \delta\vartheta x_1, x_3, x_4) = \psi(x) + \frac{\partial\psi(x)}{\partial x_1}x_2\delta\vartheta - \frac{\partial\psi(x)}{\partial x_2}x_1\delta\vartheta \\ &= \psi(x) - \frac{i}{\hbar}(x_1\hat{p}_2 - x_2\hat{p}_1)\psi(x)\delta\vartheta = \psi(x) - \frac{i}{\hbar}\hat{L}_z\delta\vartheta\,\psi(x). \end{aligned} \tag{11.189}$$

Then, the change in the functional form of the wave function is given by

$$\delta\psi = \psi'(x') - \psi(x') = \frac{i}{\hbar}\left(\hat{L}_z + \frac{\hbar}{2}\Sigma_3\right)\delta\vartheta\,\psi(x) = \frac{i}{\hbar}(\hat{L}_z + \hat{S}_z)\delta\vartheta\,\psi(x). \tag{11.190}$$

So, we have found that the generator of the infinitesimal rotation about the z-axis is $\hat{J}_z = \hat{L}_z + \hat{S}_z$. The $\hat{L}_z$ is responsible for the change of the spatial coordinates and $\hat{S}_z$ generates an additional change of the components. Again, since z is arbitrary, the generator of an arbitrary rotation is $\hat{\boldsymbol{J}} = \hat{\boldsymbol{L}} + \hat{\boldsymbol{S}}$.

We should notice that the non-conservation of the orbital angular momentum is due to the $\boldsymbol{\alpha} \cdot \hat{\boldsymbol{p}}$ term in the Hamiltonian. This is the reason why $\hat{\boldsymbol{L}}$ does not commute even with the free Dirac Hamiltonian, i.e. only $\hat{\boldsymbol{J}}$ does, only $\hat{\boldsymbol{J}}$ is a conserved quantity. The Dirac equation, which is a result of an approach to reconcile relativity and quantum mechanics, demands and naturally explains the existence of the spin.

11.2.4 Free Dirac Particle

The free Dirac equation

$$\left(\gamma_i \hat{p}_i - imc\right)\psi = 0 \tag{11.191}$$

can be solved by a plane wave ansatz

$$\psi = C \exp[i(p_1x_1 + p_2x_2 + p_3x_3 + p_4x_4)/\hbar]. \tag{11.192}$$

Substituting this ansatz into Eq. (11.191), we obtain a set of linear equations

$$\begin{pmatrix} mc + ip_4 & 0 & p_3 & p_1 - ip_2 \\ 0 & mc + ip_4 & p_1 + ip_2 & -p_3 \\ -p_3 & -(p_1 - ip_2) & mc - ip_4 & 0 \\ -(p_1 + ip_2) & p_3 & 0 & mc - ip_4 \end{pmatrix} \begin{pmatrix} C_1 \\ C_2 \\ C_3 \\ C_4 \end{pmatrix} = 0, \tag{11.193}$$

which is solvable if the determinant is zero. Expanding the determinant, we obtain the relation

$$\left[p_1^2 + p_2^2 + p_3^2 + p_4^2 + m^2c^2\right]^2 = 0, \tag{11.194}$$

which is the same as we have seen before for the length of the four-momentum vector. We can satisfy Eq. (11.194) by taking

$$p_k = \frac{mv_k}{\sqrt{1 - v^2/c^2}} \quad \text{for} \quad k = 1, 2, 3, \quad \text{and} \quad p_4 = \frac{imc}{\sqrt{1 - v^2/c^2}} = \frac{i}{c}E, \tag{11.195}$$

with energy

$$E = \pm\sqrt{m^2c^4 + p^2c^2}, \tag{11.196}$$

which has positive and negative values as well.

First, we study the positive-energy states and we use polar variables

$$p_3 = p\cos\vartheta \quad \text{and} \quad p_1 \pm ip_2 = p\sin\vartheta e^{\pm i\varphi}. \tag{11.197}$$

Rewriting Eq. (11.193), we get

$$\begin{pmatrix} mc - E/c & 0 & p\cos\vartheta & p\sin\vartheta e^{-i\varphi} \\ 0 & mc - E/c & p\sin\vartheta e^{i\varphi} & -p\cos\vartheta \\ -p\cos\vartheta & -p\sin\vartheta e^{-i\varphi} & mc + E/c & 0 \\ -p\sin\vartheta e^{i\varphi} & p\cos\vartheta & 0 & mc + E/c \end{pmatrix} \begin{pmatrix} C_1 \\ C_2 \\ C_3 \\ C_4 \end{pmatrix} = 0. \tag{11.198}$$

The C_i coefficients can be determined with the help of the helicity operator. Helicity is the projection of the spin on the direction of the momentum vector

$$\hat{h} = \frac{1}{p}\sum_{k=1}^{3}\Sigma_k \hat{p}_k = -\frac{i}{p}(\gamma_2\gamma_3\hat{p}_1 + \gamma_3\gamma_1\hat{p}_2 + \gamma_1\gamma_2\hat{p}_3), \tag{11.199}$$

and some easy calculations reveal that it commutes with the free Dirac Hamiltonian

$$\hat{H} = ic\sum_{k=1}^{3}\gamma_4\gamma_k\hat{p}_k + mc^2\gamma_4. \tag{11.200}$$

In polar variables, the helicity operator is given by

$$\hat{h} = \frac{1}{p}\begin{pmatrix} p_3 & p_1 - ip_2 & 0 & 0 \\ p_1 + ip_2 & -p_3 & 0 & 0 \\ 0 & 0 & p_3 & p_1 - ip_2 \\ 0 & 0 & p_1 + ip_2 & -p_3 \end{pmatrix} = \begin{pmatrix} \cos\vartheta & \sin\vartheta e^{-i\varphi} & 0 & 0 \\ \sin\vartheta e^{i\varphi} & -\cos\vartheta & 0 & 0 \\ 0 & 0 & \cos\vartheta & \sin\vartheta e^{-i\varphi} \\ 0 & 0 & \sin\vartheta e^{i\varphi} & -\cos\vartheta \end{pmatrix}, \tag{11.201}$$

which leads to the eigenvalue equation

$$\begin{pmatrix} \cos\vartheta & \sin\vartheta e^{-i\varphi} & 0 & 0 \\ \sin\vartheta e^{i\varphi} & -\cos\vartheta & 0 & 0 \\ 0 & 0 & \cos\vartheta & \sin\vartheta e^{-i\varphi} \\ 0 & 0 & \sin\vartheta e^{i\varphi} & -\cos\vartheta \end{pmatrix}\begin{pmatrix} C_1 \\ C_2 \\ C_3 \\ C_4 \end{pmatrix} = h\begin{pmatrix} C_1 \\ C_2 \\ C_3 \\ C_4 \end{pmatrix}. \tag{11.202}$$

We can see that the first two equations determine C_1 and C_2 and the last two determine C_3 and C_4. They are the same equations, thus we need to solve only one of them. The eigenvalue equation reads

$$\begin{vmatrix} \cos\vartheta - h & \sin\vartheta e^{-i\varphi} \\ \sin\vartheta e^{i\varphi} & -\cos\vartheta - h \end{vmatrix} = 0, \tag{11.203}$$

which is solved by $h = \pm 1$. By utilizing the relations,

$$\tan\vartheta/2 = \frac{1 - \cos\vartheta}{\sin\vartheta} = \frac{\sin\vartheta}{1 + \cos\vartheta}, \tag{11.204}$$

we find, for the $h = +1$ case,

$$C_2 = \tan\vartheta/2\, e^{i\varphi} C_1 \quad \text{and} \quad C_4 = \tan\vartheta/2\, e^{i\varphi} C_3, \tag{11.205}$$

and for the $h = -1$ case, we get

$$C_2 = -\cot\vartheta/2\, e^{i\varphi} C_1 \quad \text{and} \quad C_4 = -\cot\vartheta/2\, e^{i\varphi} C_3. \tag{11.206}$$

The helicity eigenvalue equation resulted in two relations between the four coefficients, leaving only two independent. Therefore, lets say C_1 and C_3 can be arbitrarily chosen. To find further constraints, we can use the energy eigenvalue equation. Substituting into Eq. (11.198), for $h = 1$ and $h = -1$, respectively, we find

$$\begin{pmatrix} mc - E/c & 0 & p\cos\vartheta & p\sin\vartheta e^{-i\varphi} \\ 0 & mc - E/c & p\sin\vartheta e^{i\varphi} & -p\cos\vartheta \\ -p\cos\vartheta & -p\sin\vartheta e^{-i\varphi} & mc + E/c & 0 \\ -p\sin\vartheta e^{i\varphi} & p\cos\vartheta & 0 & mc + E/c \end{pmatrix} \begin{pmatrix} C_1 & C_1 \\ \tan\vartheta/2\, e^{i\varphi} C_1 & -\cot\vartheta/2\, e^{i\varphi} C_1 \\ C_3 & C_3 \\ \tan\vartheta/2\, e^{i\varphi} C_3 & -\cot\vartheta/2\, e^{i\varphi} C_3 \end{pmatrix} = 0. \tag{11.207}$$

Then, for example from the third row, we find

$$C_3 = \eta C_1, \quad \text{where} \quad \eta = \frac{pc}{E + mc^2} = \frac{E - mc^2}{pc} \quad \text{for} \quad h = 1, \tag{11.208}$$

and

$$C_3 = -\eta C_1 \quad \text{for} \quad h = -1. \tag{11.209}$$

Now, only the C_1 coefficient is optional. We can normalize ψ over a volume V

$$1 = \int_V \mathrm{d}V\, \psi^\dagger \psi = \int_V \mathrm{d}V\, C^\dagger C = |C_1|^2\,(1 + \tan^2\vartheta/2 + \eta^2 + \eta^2 \tan^2\vartheta/2)V \quad \text{for} \quad h = 1 \tag{11.210}$$

and

$$1 = \int_V \mathrm{d}V\, \psi^\dagger \psi = \int_V \mathrm{d}V\, C^\dagger C = |C_1|^2\,(1 + \cot^2\vartheta/2 + \eta^2 + \eta^2 \cot^2\vartheta/2)V \quad \text{for} \quad h = -1. \tag{11.211}$$

Then, with

$$C_1 = \frac{\cos\vartheta/2\, e^{-i\phi/2}}{\sqrt{V(1+\eta^2)}}, \quad \text{we have} \quad C_+ = \frac{1}{\sqrt{V(1+\eta^2)}} \begin{pmatrix} \cos\vartheta/2\, e^{-i\varphi/2} \\ \sin\vartheta/2\, e^{i\varphi/2} \\ \eta\cos\vartheta/2\, e^{-i\varphi/2} \\ \eta\sin\vartheta/2\, e^{i\varphi/2} \end{pmatrix} \quad \text{for} \quad h = 1, \tag{11.212}$$

and with

$$C_1 = \frac{\sin\vartheta/2\, e^{-i\phi/2}}{\sqrt{V(1+\eta^2)}}, \quad \text{we have} \quad C_- = \frac{1}{\sqrt{V(1+\eta^2)}} \begin{pmatrix} \sin\vartheta/2\, e^{-i\varphi/2} \\ -\cos\vartheta/2\, e^{i\varphi/2} \\ -\eta\sin\vartheta/2\, e^{-i\varphi/2} \\ \eta\cos\vartheta/2\, e^{i\varphi/2} \end{pmatrix} \quad \text{for} \quad h = -1. \tag{11.213}$$

These are the wave functions of a freely moving electron with various spin orientations. The solution C_+ describes a state in which the spin of the electron is parallel to the direction of the momentum, while in the C_- state the spin is opposite to the momentum. The Dirac equation at a given energy has two linearly independent solutions, which differ from each other in the orientation of the spin.

In a non-relativistic approximation, $p \approx mv$ and

$$\eta = \frac{E - mc^2}{cp} = \frac{\sqrt{c^2p^2 + m^2c^4} - mc^2}{cp} = \frac{mc^2\sqrt{1 + c^2p^2/(m^2c^4)} - mc^2}{cp} \approx \frac{1}{2}\frac{v}{c}. \tag{11.214}$$

Hence, the magnitude of the coefficients C_3 and C_4 are about v/c times that of C_1 and C_2. So, in the non-relativistic limit, $C_3 \simeq C_4 \simeq 0$, and we recover the Pauli equation with a two-component wave function.

Since the Dirac equation is linear, a general solution is a superposition of linearly independent solutions. In this case, it is a combination of different helicity states.

Let us examine the expectation value of the spin in the x direction. For the sake of simplicity, we assume that the particle moves in the direction of the z-axis, where $\vartheta = 0$ and $\varphi = 0$. The wave function takes the form

$$\psi = C \exp[i(p_3x_3 + p_4x_4)/\hbar], \tag{11.215}$$

where

$$C = aC_+ + bC_-, \tag{11.216}$$

with

$$C_+ = \frac{1}{\sqrt{V(1+\eta^2)}}\begin{pmatrix}1\\0\\\eta\\0\end{pmatrix} \quad \text{and} \quad C_- = \frac{1}{\sqrt{V(1+\eta^2)}}\begin{pmatrix}0\\-1\\0\\\eta\end{pmatrix}. \tag{11.217}$$

The coefficients a and b must be chosen to meet the normalization criteria. Here,

$$a = \cos\alpha\, e^{i\beta} \quad \text{and} \quad b = \sin\alpha\, e^{-i\beta}, \tag{11.218}$$

with α and β real numbers satisfy the requirements.

The desired expectation value is given by

$$\bar{S}_x = \langle\psi|\hat{S}_x|\psi\rangle = \frac{\hbar}{2}\int \mathrm{d}V\, C^\dagger \Sigma_1 C, \tag{11.219}$$

where

$$\Sigma_1 = -i\gamma_2\gamma_3 = \begin{pmatrix} 0 & 1 & 0 & 0 \\ 1 & 0 & 0 & 0 \\ 0 & 0 & 0 & 1 \\ 0 & 0 & 1 & 0 \end{pmatrix}. \tag{11.220}$$

Easy calculations give

$$C_+^\dagger \Sigma_1 C_+ = 0 = C_-^\dagger \Sigma_1 C_-, \tag{11.221}$$

and

$$\int \mathrm{d}V\, C_-^\dagger \Sigma_1 C_+ = \int \mathrm{d}V\, C_+^\dagger \Sigma_1 C_- = -\frac{1-\eta^2}{1+\eta^2}. \tag{11.222}$$

Consequently, we obtain

$$\bar{S}_x = \frac{\hbar}{2} \int \mathrm{d}V\, \left(a^* b C_+^\dagger \Sigma_1 C_- + b^* a C_-^\dagger \Sigma_1 C_+ \right) = -\frac{\hbar}{2} \sin 2\alpha\, \cos 2\beta\, \frac{1-\eta^2}{1+\eta^2}. \tag{11.223}$$

The expectation value of S_x is not zero and in general, $|\bar{S}_x| \leq \hbar/2$. In the non-relativistic case, $v \ll c$ and $\eta \sim 0$. As a result, any polarization direction is possible, depending on the initial conditions given by α and β. With $\beta = 0$ and $\alpha = \pm\pi/4$ we have orthogonal polarization, $\boldsymbol{S} \perp \boldsymbol{p}$ and $\bar{S}_x = \pm\hbar/2$.

However, in the extreme relativistic case, $v \sim c$ and $\eta \sim 1$. Consequently, $\hat{S}_x \sim 0$, thus the spin is parallel or antiparallel to the momentum vector. At relativistic energies, there is a very strong correlation between the momentum and the spin.

So far, we have considered only positive energy solutions. However,

$$E = -\sqrt{m^2c^4 + p^2c^2}$$

is also a possible solution. The C_i coefficients can be determined the same way and we find analogous formulae.

The massless Dirac equation, or the *Weyl equation*,

$$\gamma_i \hat{p}_i \psi = 0, \tag{11.224}$$

describes a relativistic zero-mass spin-1/2 particle. Since the mass of neutrinos are very small, the zero-mass Dirac equation is a good approximation for them.

In this case, $\eta = 1$ and the helicity states are given by

$$C_+ = \frac{1}{\sqrt{2\,V}} \begin{pmatrix} \cos\theta/2\, e^{-i\phi/2} \\ \sin\theta/2\, e^{i\phi/2} \\ \cos\theta/2\, e^{-i\phi/2} \\ \sin\theta/2\, e^{i\phi/2} \end{pmatrix} \quad \text{and} \quad C_- = \frac{1}{\sqrt{2\,V}} \begin{pmatrix} \sin\theta/2\, e^{-i\phi/2} \\ -\cos\theta/2\, e^{i\phi/2} \\ -\sin\theta/2\, e^{-i\phi/2} \\ \cos\theta/2\, e^{i\phi/2} \end{pmatrix}. \tag{11.225}$$

Recall that $\gamma_5 = \gamma_1\gamma_2\gamma_3\gamma_4$, $\gamma_5^2 = 1$ and $\left[\gamma_i, \gamma_5\right]_+ = 0$. If we multiply the zero mass Dirac equation from the left by γ_5, we find that $\gamma_5\psi$ is also a solution. We can easily see from Eq. (11.137), that

$$(1+\gamma_5)C_+ = 0, \quad (1-\gamma_5)C_+ = 2C_+, \quad (1+\gamma_5)C_- = 2C_-, \quad \text{and} \quad (1-\gamma_5)C_- = 0. \tag{11.226}$$

The operator $(1+\gamma_5)/2$ takes the $h = -1$ state into itself, so C_- is an eigenstate of $(1+\gamma_5)/2$. On the other hand, C_+ is an eigenstate of $(1-\gamma_5)/2$.

Beta decay experiments show that the neutrino ν has negative helicity, $h_\nu = -1$, i.e. it is left-handed ν_L. On the other hand, the antineutrino has positive helicity, $h_{\bar{\nu}} = +1$, i.e. it is right-handed $\bar{\nu}_R$. So, for the neutrino, the spin is antiparallel to the momentum, while for the antineutrino, the spin is parallel. The $\hat{\mathcal{P}}\hat{\mathcal{T}}$ transformation transforms ν_L to ν_R, which does not exist in nature. The charge conjugation transforms the right-handed neutrino into an antineutrino $\bar{\nu}_R$. So, it is the combined $\hat{C}\hat{\mathcal{P}}\hat{\mathcal{T}}$ transformation that transforms the neutrino into an antineutrino $\hat{C}\hat{\mathcal{P}}\hat{\mathcal{T}}|\nu_L\rangle = \hat{C}|\nu_R\rangle = |\bar{\nu}_R\rangle$.

11.2.5 Spin-Orbit Interaction

We have seen that in the Dirac formalism of relativistic quantum mechanics, there is a strong correlation between the spin and the momentum. Now we investigate the coupling between the spin and the orbital angular momentum in the non-relativistic limit. We take the vector potential $A = 0$ and consider a spherical potential $V(r)$. If we separate off the rest energy

$$E = mc^2 + E', \tag{11.227}$$

and assume that $E' \ll mc^2$, Eq. (11.112) takes the form

$$\left(c\sum_{k=1}^{3}\alpha_k\hat{p}_k + mc^2\gamma_4 + V(r)\right)\psi = (E' + mc^2)\psi. \tag{11.228}$$

We can write the four-component wave function in the form

$$\psi = \begin{pmatrix}\psi_a\\ \psi_b\end{pmatrix}, \tag{11.229}$$

where ψ_a and ψ_b are two-component functions. Utilizing the explicit form of the $\boldsymbol{\alpha}$ matrices,

$$\alpha_k = \begin{pmatrix}0 & \sigma_k\\ \sigma_k & 0\end{pmatrix}, \tag{11.230}$$

Equation (11.228) can be written as

$$\begin{aligned} \left(E' - V\right)\psi_a &= c(\boldsymbol{\sigma}\hat{\boldsymbol{p}})\psi_b, \\ \left(E' + 2\,mc^2 - V\right)\psi_b &= c(\boldsymbol{\sigma}\hat{\boldsymbol{p}})\psi_a. \end{aligned} \tag{11.231}$$

We can express ψ_b from the second equation

$$\psi_b = (E' + 2mc^2 - V)^{-1} c(\boldsymbol{\sigma}\hat{\boldsymbol{p}})\psi_a, \tag{11.232}$$

and substitute back into the first one

$$E'\psi_a = \left(c(\boldsymbol{\sigma}\hat{\boldsymbol{p}})(E' + 2\,mc^2 - V)^{-1} c(\boldsymbol{\sigma}\hat{\boldsymbol{p}}) + V\right)\psi_a. \tag{11.233}$$

As $E' - V << 2mc^2$, the term $(E' + 2mc^2 - V)^{-1}$ can be approximated by

$$(E' + 2\,mc^2 - V)^{-1} = \frac{1}{2\,mc^2}\left(1 + \frac{E' - V}{2\,mc^2}\right)^{-1} \simeq \frac{1}{2\,mc^2}\left(1 - \frac{E' - V}{2\,mc^2}\right), \tag{11.234}$$

and thus Eq. (11.233) takes the form

$$\left(\frac{1}{2\,m}(\boldsymbol{\sigma}\hat{\boldsymbol{p}})^2 - \frac{1}{4\,m^2c^2}(\boldsymbol{\sigma}\hat{\boldsymbol{p}})(E' - V)(\boldsymbol{\sigma}\hat{\boldsymbol{p}}) + V\right)\psi_a = E'\psi_a. \tag{11.235}$$

By utilizing the relations $(\boldsymbol{\sigma}\boldsymbol{a})(\boldsymbol{\sigma}\boldsymbol{b}) = \boldsymbol{a}\boldsymbol{b} + i\boldsymbol{\sigma}(\boldsymbol{a}\times\boldsymbol{b})$ and

$$\hat{\boldsymbol{p}}V = V\hat{\boldsymbol{p}} - i\hbar\nabla V(r) = V\hat{\boldsymbol{p}} - i\hbar\nabla V(\sqrt{x^2 + y^2 + z^2}) = V\hat{\boldsymbol{p}} - i\hbar\frac{\mathrm{d}V}{\mathrm{d}r}\frac{\boldsymbol{r}}{r}, \tag{11.236}$$

we find

$$\begin{aligned} (\boldsymbol{\sigma}\nabla V)(\boldsymbol{\sigma}\hat{\boldsymbol{p}}) = (\nabla V\hat{\boldsymbol{p}}) + i\sigma(\nabla V)\times\hat{\boldsymbol{p}} &= \frac{\mathrm{d}V}{\mathrm{d}r}\frac{1}{r}\boldsymbol{r}\hat{\boldsymbol{p}} + i\boldsymbol{\sigma}\frac{1}{r}\frac{\mathrm{d}V}{\mathrm{d}r}\boldsymbol{r}\times\hat{\boldsymbol{p}} \\ &= -\,-\frac{\mathrm{d}V}{\mathrm{d}r}i\hbar\frac{\partial}{\partial r} + i\frac{1}{r}\frac{\mathrm{d}V}{\mathrm{d}r}\boldsymbol{\sigma}\hat{\boldsymbol{L}}, \end{aligned} \tag{11.237}$$

and

$$(\sigma\hat{\boldsymbol{p}})(\sigma\hat{\boldsymbol{p}}) = \hat{\boldsymbol{p}}^2. \tag{11.238}$$

As a consequence, Eq. (11.235) turns into

$$\left(\frac{\hat{\boldsymbol{p}}^2}{2m} - \frac{1}{4m^2c^2}(E' - V)\hat{\boldsymbol{p}}^2 - \frac{\hbar^2}{4m^2c^2}\frac{\mathrm{d}V}{\mathrm{d}r}\frac{\partial}{\partial r} + \frac{1}{2m^2c^2}\frac{1}{r}\frac{\mathrm{d}V}{\mathrm{d}r}\hat{\boldsymbol{S}}\hat{\boldsymbol{L}} + V(r)\right)\psi_a = E'\psi_a, \tag{11.239}$$

or considering that $E' - V \approx \hat{\boldsymbol{p}}^2/2m$, we obtain

$$\left(\frac{\hat{\boldsymbol{p}}^2}{2m} + V(r) - \frac{\hat{\boldsymbol{p}}^4}{8m^3c^2} - \frac{\hbar^2}{4m^2c^2}\frac{\mathrm{d}V}{\mathrm{d}r}\frac{\partial}{\partial r} + \frac{1}{2m^2c^2}\frac{1}{r}\frac{\mathrm{d}V}{\mathrm{d}r}\hat{\mathbf{S}}\hat{\boldsymbol{L}}\right)\psi_a = E'\psi_a. \tag{11.240}$$

The first two terms represent the non-relativistic Hamiltonian. The third one comes from the relativistic correction to the kinetic energy since

$$E' = E - mc^2 = \sqrt{p^2c^2 + m^2c^4} - mc^2 \approx \frac{p^2}{2m} - \frac{p^4}{8\,m^3c^2} + \cdots . \tag{11.241}$$

The fourth term is a relativistic correction to the potential. These terms are usually small compared to corresponding non-relativistic terms. The most important correction term is the last one, the spin-orbit coupling term. This coupling between the spin and the orbital angular momentum prevents $\boldsymbol{S}$ and $\boldsymbol{L}$ from commuting with the Hamiltonian, so only $\boldsymbol{J}$ commutes.

11.2.6 Radial Dirac Equation

The Dirac equation, as it stands, is a four component equation. However, if the potential is spherical, we can achieve a significant simplification. From Eq. (4.126), for the position and momentum operators, we find

$$(\boldsymbol{\Sigma}\hat{\boldsymbol{r}})(\boldsymbol{\Sigma}\hat{\boldsymbol{p}}) = \hat{\boldsymbol{r}}\hat{\boldsymbol{p}} + i\boldsymbol{\Sigma}(\hat{\boldsymbol{r}} \times \hat{\boldsymbol{p}}) = \hat{\boldsymbol{r}}\hat{\boldsymbol{p}} + i\boldsymbol{\Sigma}\hat{\boldsymbol{L}}, \tag{11.242}$$

and then for the position and orbital angular momentum operator we have

$$(\boldsymbol{\Sigma}\hat{\boldsymbol{r}})(\boldsymbol{\Sigma}\hat{\boldsymbol{L}}) = i[(\boldsymbol{\Sigma}\hat{\boldsymbol{r}})(\hat{\boldsymbol{r}}\hat{\boldsymbol{p}}) - r^2(\boldsymbol{\Sigma}\hat{\boldsymbol{p}})]. \tag{11.243}$$

Multiplying the above relation by $-\gamma_5$ and keeping in mind that $\boldsymbol{\alpha} = -\gamma_5\boldsymbol{\Sigma}$, we can write

$$\boldsymbol{\alpha}\hat{\boldsymbol{p}} = \frac{\boldsymbol{\alpha}\hat{\boldsymbol{r}}}{r^2}\left[\hat{\boldsymbol{r}}\hat{\boldsymbol{p}} + i\boldsymbol{\Sigma}\hat{\boldsymbol{L}}\right]. \tag{11.244}$$

With the help of the operators

$$\alpha_r = \boldsymbol{\alpha}\boldsymbol{e}_r, \quad \hat{p}_r = -i\hbar\frac{1}{r}\frac{\partial}{\partial r}r = -i\hbar\left(\frac{\partial}{\partial r} + \frac{1}{r}\right), \quad \text{and} \quad \hat{K} = \gamma_4(\boldsymbol{\Sigma}\hat{\boldsymbol{L}} + \hbar), \tag{11.245}$$

where $\boldsymbol{e}_r = \boldsymbol{r}/r$, we can write the Hamiltonian of Eq. (11.170) in the following form

$$\hat{H} = c\alpha_r\hat{p}_r + \frac{ic}{r}\alpha_r\gamma_4\hat{K} + \gamma_4 mc^2 + V(r). \tag{11.246}$$

By using the relations $\hat{r} \times \hat{L} + \hat{L} \times \hat{r} = 2i\hbar\hat{r}$ and keeping in mind that $\boldsymbol{\Sigma} = -\gamma_5\boldsymbol{\alpha}$, we can prove the following commutation relations

$$\begin{aligned}
\left[\gamma_4, \hat{K}\right] &= 0, \\
\left[\alpha_r, \hat{K}\right] &= 0, \\
\left[\hat{p}_r, \hat{K}\right] &= 0, \\
\left[\alpha_r, \hat{p}_r\right] &= 0.
\end{aligned} \tag{11.247}$$

Indeed,

$$\begin{aligned}
\left[\gamma_4, \hat{K}\right] &= \left[\gamma_4, \gamma_4(\boldsymbol{\Sigma}\hat{\boldsymbol{L}} + \hbar)\right] = \gamma_4\left[\gamma_4, \boldsymbol{\Sigma}\hat{\boldsymbol{L}}\right] = -\gamma_4\left[\gamma_4, \gamma_5\boldsymbol{\alpha}\hat{\boldsymbol{L}}\right] \\
&= -\gamma_4\left[\gamma_4, \gamma_5\right]\boldsymbol{\alpha}\hat{\boldsymbol{L}} - \gamma_4\gamma_5\left[\gamma_4, \boldsymbol{\alpha}\right]\hat{\boldsymbol{L}} \\
&= -\gamma_4\gamma_4\gamma_5\boldsymbol{\alpha}\hat{\boldsymbol{L}} + -\gamma_4\gamma_5\gamma_4\boldsymbol{\alpha}\hat{\boldsymbol{L}} - \gamma_4\gamma_5\gamma_4\boldsymbol{\alpha}\hat{\boldsymbol{L}} + \gamma_4\gamma_5\boldsymbol{\alpha}\gamma_4\hat{\boldsymbol{L}} = 0, \\
\left[\alpha_r, \hat{K}\right] &= \hbar\left[\alpha_r, \gamma_4\right] + \left[\alpha_r, \gamma_4\boldsymbol{\Sigma}\hat{\boldsymbol{L}}\right] = \hbar\left[\alpha_r, \gamma_4\right] - \left[\alpha_r, \gamma_4\gamma_5\boldsymbol{\alpha}\hat{\boldsymbol{L}}\right] \\
&= \hbar\alpha_r\gamma_4 - \hbar\gamma_4\alpha_r - \alpha_r\gamma_4\gamma_5\boldsymbol{\alpha}\hat{\boldsymbol{L}} + \gamma_4\gamma_5\boldsymbol{\alpha}\hat{\boldsymbol{L}}\alpha_r \\
&= 2\hbar\alpha_r\gamma_4 + \gamma_4\gamma_5[(\boldsymbol{\alpha}\boldsymbol{e}_r)(\boldsymbol{\alpha}\hat{\boldsymbol{L}}) + (\boldsymbol{\alpha}\hat{\boldsymbol{L}})(\boldsymbol{\alpha}\boldsymbol{e}_r)] \\
&= 2\hbar\alpha_r\gamma_4 + \gamma_4\gamma_5[\boldsymbol{e}_r\hat{\boldsymbol{L}} + \hat{\boldsymbol{L}}\boldsymbol{e}_r + i\Sigma(\boldsymbol{e}_r \times \hat{\boldsymbol{L}} + \hat{\boldsymbol{L}} \times \boldsymbol{e}_r)] \\
&= 2\hbar\alpha_r\gamma_4 + \gamma_4\gamma_5 i\Sigma 2i\hbar\boldsymbol{e}_r = 2\hbar(\alpha_r\gamma_4 + \gamma_4\alpha_r) = 0, \\
\left[\hat{p}_r, \hat{K}\right] &= \left[\hat{p}_r, \gamma_4\boldsymbol{\Sigma}\hat{\boldsymbol{L}}\right] = \gamma_4\boldsymbol{\Sigma}\left[\hat{p}_r, \hat{\boldsymbol{L}}\right] = 0, \\
\left[\alpha_r, \hat{p}_r\right] &= 0, \quad \text{since } \frac{\partial \boldsymbol{e}_r}{\partial r} = 0.
\end{aligned} \tag{11.248}$$

From these commutation relations, it follows immediately that $\hat{K}$ commutes with the Hamiltonian, and thus they have common eigenstates. A little calculation reveals that

$$\begin{aligned}
\hat{K}^2 &= \gamma_4(\boldsymbol{\Sigma}\hat{\boldsymbol{L}} + \hbar)\gamma_4(\boldsymbol{\Sigma}\hat{\boldsymbol{L}} + \hbar) = (\boldsymbol{\Sigma}\hat{\boldsymbol{L}})(\boldsymbol{\Sigma}\hat{\boldsymbol{L}}) + 2\hbar(\boldsymbol{\Sigma}\hat{\boldsymbol{L}}) + \hbar^2 \\
&= \hat{\boldsymbol{L}}^2 + i\Sigma(\hat{\boldsymbol{L}} \times \hat{\boldsymbol{L}}) + 2\hbar(\boldsymbol{\Sigma}\hat{\boldsymbol{L}}) + \hbar^2 \\
&= \hat{\boldsymbol{L}}^2 + \hbar(\boldsymbol{\Sigma}\hat{\boldsymbol{L}}) + \hbar^2 = \left(\hat{\boldsymbol{L}} + \frac{\hbar}{2}\boldsymbol{\Sigma}\right)^2 + \frac{\hbar^2}{4} = \hat{\boldsymbol{J}}^2 + \frac{\hbar^2}{4},
\end{aligned} \tag{11.249}$$

since $\hat{\boldsymbol{L}} \times \hat{\boldsymbol{L}} = i\hbar\hat{\boldsymbol{L}}$ and $\boldsymbol{\Sigma}^2 = \Sigma_x^2 + \Sigma_y^2 + \Sigma_z^2 = 3$. So, the eigenvalue equation reads

$$\hat{K}^2\psi = \hbar^2\kappa^2\psi, \tag{11.250}$$

where

$$\kappa^2 = j(j+1) + 1/4 = (j+1/2)^2. \tag{11.251}$$

Consequently,

$$\hat{K}\psi = \hbar\kappa\psi, \tag{11.252}$$

with $\kappa = \pm 1, \pm 2, \ldots$, since $j = 1/2, 3/2, \ldots$. Assuming that ψ is a common eigenstate of $\hat{K}$ and the Hamiltonian in Eq. (11.246), the eigenvalue equation becomes

$$\left(E - V(r) - mc^2\gamma_4 + ci\hbar\alpha_r\left(\frac{\partial}{\partial r} + \frac{1}{r}\right) - \frac{ic}{r}\hbar\kappa\alpha_r\gamma_4\right)\psi = 0. \tag{11.253}$$

Since $\hat{K}^2$ and $\hat{\boldsymbol{J}}^2$ commute, the eigenstates of $\hat{K}$ coincide with the common eigenstates of $\hat{\boldsymbol{J}}^2$, $\hat{\boldsymbol{L}}^2$, $\hat{\boldsymbol{S}}^2$ and $\hat{J}_z$

$$\begin{aligned}
\hat{\boldsymbol{J}}^2|\Phi_{j,m}^{(\pm)}\rangle &= j(j+1)\hbar^2|\Phi_{j,m}^{(\pm)}\rangle, \qquad j = 1/2, 3/2, \ldots, \\
\hat{J}_z|\Phi_{j,m}^{(\pm)}\rangle &= m\hbar|\Phi_{j,m}^{(\pm)}\rangle, \qquad -j \le m \le j, \\
\hat{\boldsymbol{L}}^2|\Phi_{j,m}^{(\pm)}\rangle &= l_\pm(l_\pm+1)\hbar^2|\Phi_{j,m}^{(\pm)}\rangle, \quad l_\pm = j \mp 1/2 \\
\hat{\boldsymbol{S}}^2|\Phi_{j,m}^{(\pm)}\rangle &= s(s+1)\hbar^2|\Phi_{j,m}^{(\pm)}\rangle, \qquad s = 1/2\,.
\end{aligned} \tag{11.254}$$

The eigenstates $|\Phi_{j,m}^{(\pm)}\rangle$ can be constructed by coupling the orbital angular momentum and the spin

$$\Phi_{j,m}^{(\pm)} = \sum_{m_l, m_s} \langle l_\pm, 1/2; m_l, m_s | j, m\rangle Y_{l_\pm, m_l}(\vartheta, \varphi)\chi_{1/2, m_s}, \tag{11.255}$$

or explicitly,

$$\begin{aligned}
\Phi_{j,m}^{(+)} &= \sqrt{\frac{l_+ - m + 1/2}{2l_+ + 1}} Y_{l_+, m+1/2}(\vartheta, \varphi)\begin{pmatrix}0\\1\end{pmatrix} + \sqrt{\frac{l_+ + m + 1/2}{2l_+ + 1}} Y_{l_+, m-1/2}(\vartheta, \varphi)\begin{pmatrix}1\\0\end{pmatrix}, \\
\Phi_{j,m}^{(-)} &= \sqrt{\frac{l_- + m + 1/2}{2l_- + 1}} Y_{l_-, m+1/2}(\vartheta, \varphi)\begin{pmatrix}0\\1\end{pmatrix} - \sqrt{\frac{l_- - m + 1/2}{2l_- + 1}} Y_{l_-, m-1/2}(\vartheta, \varphi)\begin{pmatrix}1\\0\end{pmatrix}.
\end{aligned} \tag{11.256}$$

As we noted before, the γ, and therefore the α matrices are defined by their commutation relations and the actual representation is unique only up to a unitary transformation. Since the α_r is a scalar product of $\boldsymbol{\alpha}$ and the unit vector $\hat{e}_r$, it should satisfy the relations

$$\alpha_r^2 = \gamma_4^2 = 1 \quad \text{and} \quad \alpha_r\gamma_4 + \gamma_4\alpha_r = 0. \tag{11.257}$$

So, we have some flexibility in the actual representation, provided the above anti-commutation relations are observed. We can see that with

$$\alpha_r = \begin{pmatrix} 0 & 0 & -i & 0 \\ 0 & 0 & 0 & -i \\ i & 0 & 0 & 0 \\ 0 & i & 0 & 0 \end{pmatrix} \quad \text{and} \quad \gamma_4 = \begin{pmatrix} 1 & 0 & 0 & 0 \\ 0 & 1 & 0 & 0 \\ 0 & 0 & -1 & 0 \\ 0 & 0 & 0 & -1 \end{pmatrix}, \tag{11.258}$$

the conditions in Eq. (11.257) are satisfied. With these matrices, Eq. (11.253) becomes

$$\begin{aligned} (E - V - mc^2)\psi_1 + \hbar c \left(\frac{\partial}{\partial r} + \frac{1+\kappa}{r} \right) \psi_3 = 0, \\ (E - V - mc^2)\psi_2 + \hbar c \left(\frac{\partial}{\partial r} + \frac{1+\kappa}{r} \right) \psi_4 = 0, \\ (E - V + mc^2)\psi_3 - \hbar c \left(\frac{\partial}{\partial r} + \frac{1-\kappa}{r} \right) \psi_1 = 0, \\ (E - V + mc^2)\psi_4 - \hbar c \left(\frac{\partial}{\partial r} + \frac{1-\kappa}{r} \right) \psi_2 = 0. \end{aligned} \tag{11.259}$$

We can see that the equations are invariant with respect to the correspondence $\psi_1 \to \psi_2$ and $\psi_3 \to \psi_4$. Therefore, we can use only two equations to determine them

$$\begin{pmatrix} \psi_1 \\ \psi_3 \end{pmatrix} = \frac{1}{r} \begin{pmatrix} f(r)\Phi^{(+)}_{j,m} \\ g(r)\Phi^{(-)}_{j,m} \end{pmatrix}, \tag{11.260}$$

and for the radial functions f and g, we have

$$\begin{aligned} \frac{1}{\hbar c}(E - V - mc^2) f + \frac{\mathrm{d}g}{\mathrm{d}r} + \kappa \frac{g}{r} = 0, \\ \frac{1}{\hbar c}(E - V + mc^2) g - \frac{\mathrm{d}f}{\mathrm{d}r} + \kappa \frac{f}{r} = 0. \end{aligned} \tag{11.261}$$

If the potential V is of short-range type, the asymptotic form of these equations is given by

$$\begin{aligned} \frac{1}{\hbar c}(E - mc^2) f + \frac{\mathrm{d}g}{\mathrm{d}r} = 0, \\ \frac{1}{\hbar c}(E + mc^2) g - \frac{\mathrm{d}f}{\mathrm{d}r} = 0. \end{aligned} \tag{11.262}$$

Assuming an asymptotic behavior similar to the non-relativistic case, $f \sim \exp(\lambda r)$, the second equation shows that $g \sim \exp(\lambda r)$ as well. Then, inserting

$$\begin{pmatrix} f \\ g \end{pmatrix} \sim \begin{pmatrix} A \\ B \end{pmatrix} \exp(\lambda r) \tag{11.263}$$

into Eq. (11.262) results in

$$\lambda = -\frac{1}{\hbar c}\sqrt{(mc^2 - E)(mc^2 + E)} \quad \text{and} \quad \frac{B}{A} = -\sqrt{(mc^2 - E)/(mc^2 + E)}. \tag{11.264}$$

Another singular point is at $r = 0$. Assuming that the potential is singular like a Coulomb potential, $V \sim Z\hbar c\alpha/r$, we have for the dominant terms

$$\begin{aligned} -\frac{Z\alpha}{r} f + \frac{\mathrm{d}g}{\mathrm{d}r} + \kappa\frac{g}{r} &= 0, \\ -\frac{Z\alpha}{r} g - \frac{\mathrm{d}f}{\mathrm{d}r} + \kappa\frac{f}{r} &= 0. \end{aligned} \tag{11.265}$$

If we assume a $g \sim r^\beta$ behavior at $r \sim 0$, the first equation shows that $f \sim r^\beta$ as well. Then, from Eq. (11.265), considering that the solution should be normalizable, and consequently regular, we obtain that $\beta = \sqrt{\kappa^2 - Z^2\alpha^2}$. Having the Dirac equation as set of ordinary first order differential equations with these boundary conditions, we can solve them numerically by adopting standard techniques.

11.2.7 Magnetic Moment

In order that we can analyze the coupling between the spin and the electromagnetic field, we should bring Eq. (11.109) into a form similar to the non-relativistic Schrödinger equation. To accomplish this, we "square" the Dirac equation

$$\left[\gamma_k\left(\frac{\partial}{\partial x_k} - \frac{iq}{\hbar c}A_k\right) - \chi\right] \cdot \left[\gamma_i\left(\frac{\partial}{\partial x_i} - \frac{iq}{\hbar c}A_i\right) + \chi\right]\psi = 0. \tag{11.266}$$

To simplify notations, we use

$$\hat{D}_i = \frac{\partial}{\partial x_i} - \frac{iq}{\hbar c}A_i, \tag{11.267}$$

and carry out the multiplication

$$\begin{aligned} &(\hat{D}_1^2 + \hat{D}_2^2 + \hat{D}_3^2 + \hat{D}_4^2 - \chi^2)\psi \\ &+ (\gamma_1\gamma_2\hat{D}_1\hat{D}_2 + \gamma_2\gamma_1\hat{D}_2\hat{D}_1 + \gamma_1\gamma_3\hat{D}_1\hat{D}_3 + \gamma_3\gamma_1\hat{D}_3\hat{D}_1 \\ &+ \gamma_2\gamma_3\hat{D}_2\hat{D}_3 + \gamma_3\gamma_2\hat{D}_3\hat{D}_2 + \gamma_1\gamma_4\hat{D}_1\hat{D}_4 + \gamma_4\gamma_1\hat{D}_4\hat{D}_1 \\ &+ \gamma_2\gamma_4\hat{D}_2\hat{D}_4 + \gamma_4\gamma_2\hat{D}_4\hat{D}_2 + \gamma_3\gamma_4\hat{D}_3\hat{D}_4 + \gamma_4\gamma_3\hat{D}_4\hat{D}_3)\psi = 0\,. \end{aligned} \tag{11.268}$$

A little calculation gives

$$
\begin{aligned}
&(\gamma_1\gamma_2\hat{D}_1\hat{D}_2 + \gamma_2\gamma_1\hat{D}_2\hat{D}_1)\psi = \gamma_1\gamma_2(\hat{D}_1\hat{D}_2 - \hat{D}_2\hat{D}_1)\psi \\
&= \gamma_1\gamma_2\left[\left(\frac{\partial}{\partial x_1} - \frac{iq}{\hbar c}A_1\right)\left(\frac{\partial}{\partial x_2} - \frac{iq}{\hbar c}A_2\right) - \left(\frac{\partial}{\partial x_2} - \frac{iq}{\hbar c}A_2\right)\left(\frac{\partial}{\partial x_1} - \frac{iq}{\hbar c}A_1\right)\right]\psi \\
&= \gamma_1\gamma_2\bigg(\frac{\partial^2\psi}{\partial x_1 x_2} - \frac{iq}{\hbar c}\frac{\partial A_2}{x_1}\psi - \frac{iq}{\hbar c}A_2\frac{\partial\psi}{\partial x_1} - \frac{iq}{\hbar c}A_1\frac{\partial\psi}{\partial x_2} - \frac{q^2}{\hbar^2 c^2}A_1A_2\psi \\
&\qquad - \frac{\partial^2\psi}{\partial x_1 x_2} + \frac{iq}{\hbar c}\frac{\partial A_1}{\partial x_2}\psi + \frac{iq}{\hbar c}A_1\frac{\partial\psi}{\partial x_2} + \frac{iq}{\hbar c}A_2\frac{\partial\psi}{\partial x_1} + \frac{q^2}{\hbar^2 c^2}A_2A_1\psi\bigg) \\
&= -\frac{iq}{\hbar c}\gamma_1\gamma_2\left(\frac{\partial A_2}{\partial x_1} - \frac{\partial A_1}{\partial x_2}\right)\psi = \frac{q}{\hbar c}\Sigma_3\left(\frac{\partial A_2}{\partial x_1} - \frac{\partial A_1}{\partial x_2}\right)\psi\,.
\end{aligned}
\tag{11.269}
$$

Then, by using the connection between the field and the potential,

$$
\boldsymbol{E} = -\frac{1}{c}\frac{\partial \boldsymbol{A}}{\partial t} - \nabla\Phi \quad \text{and} \quad \boldsymbol{B} = \nabla\times\boldsymbol{A}, \tag{11.270}
$$

we find

$$
(\gamma_1\gamma_2\hat{D}_1\hat{D}_2 + \gamma_2\gamma_1\hat{D}_2\hat{D}_1)\psi = \frac{q}{\hbar c}\Sigma_3 B_z\psi, \tag{11.271}
$$

$$
(\gamma_1\gamma_3\hat{D}_1\hat{D}_3 + \gamma_3\gamma_1\hat{D}_3\hat{D}_1)\psi = \frac{q}{\hbar c}\Sigma_2 B_y\psi, \tag{11.272}
$$

$$
(\gamma_2\gamma_3\hat{D}_2\hat{D}_3 + \gamma_3\gamma_2\hat{D}_3\hat{D}_2)\psi = \frac{q}{\hbar c}\Sigma_1 B_x\psi. \tag{11.273}
$$

A similar analysis can be carried out for $\hat{D}_k\hat{D}_4$

$$
\begin{aligned}
(\gamma_1\gamma_4\hat{D}_1\hat{D}_4 + \gamma_4\gamma_1\hat{D}_4\hat{D}_1)\psi &= \gamma_1\gamma_4(\hat{D}_1\hat{D}_4 - \hat{D}_4\hat{D}_1)\psi = -\frac{iq}{\hbar c}\gamma_1\gamma_4\left(\frac{\partial A_4}{\partial x_1} - \frac{\partial A_1}{\partial x_4}\right)\psi \\
&= \frac{q}{\hbar c}\alpha_1\left(\frac{\partial i\Phi}{\partial x_1} - \frac{\partial A_1}{\partial ict}\right)\psi = -\frac{iq}{\hbar c}\alpha_1 E_x\psi,
\end{aligned}
\tag{11.274}
$$

and analogously, we have

$$
(\gamma_2\gamma_4\hat{D}_2\hat{D}_4 + \gamma_4\gamma_2\hat{D}_4\hat{D}_2)\psi = -\frac{iq}{\hbar c}\alpha_2 E_y\psi, \tag{11.275}
$$

$$
(\gamma_3\gamma_4\hat{D}_3\hat{D}_4 + \gamma_4\gamma_3\hat{D}_4\hat{D}_3)\psi = -\frac{iq}{\hbar c}\alpha_3 E_z\psi. \tag{11.276}
$$

Finally, we arrive at the equation

$$(\hat{D}_1^2 + \hat{D}_2^2 + \hat{D}_3^2 + \hat{D}_4^2 - \chi^2)\psi + \frac{q}{\hbar c}(\boldsymbol{B\Sigma} - i\boldsymbol{E\alpha})\psi = 0. \tag{11.277}$$

If we write out the $\boldsymbol{\Sigma}$ and $\boldsymbol{\alpha}$ matrices in terms of the $\boldsymbol{\sigma}$ matrices, we get

$$\begin{aligned}(\hat{D}_1^2 + \hat{D}_2^2 + \hat{D}_3^2 + \hat{D}_4^2 - \chi^2)\psi + \frac{q}{\hbar c}(\boldsymbol{B\sigma} - i\boldsymbol{E\sigma})\psi = 0,\\ (\hat{D}_1^2 + \hat{D}_2^2 + \hat{D}_3^2 + \hat{D}_4^2 - \chi^2)\psi + \frac{q}{\hbar c}(\boldsymbol{B\sigma} - i\boldsymbol{E\sigma})\psi = 0,\end{aligned} \tag{11.278}$$

i.e. the two equations are the same. We can drop one of them and we can work with a two component formalism. This is still exact, with no loss of generality, just instead of four first order equations we have two second order equations.

Here, $\hat{D}_4$ is given by

$$\hat{D}_4 = -\frac{1}{\hbar c}(E - q\Phi), \tag{11.279}$$

where $E = mc^2 + E'$. If $E' \ll mc^2$ and $q\Phi \ll mc^2$, i.e. we consider the non-relativistic limit and a weak field, we have

$$\hat{D}_4^2 = \frac{1}{\hbar^2 c^2}(E - q\Phi)^2 = \frac{1}{\hbar^2 c^2}(E' + mc^2 - q\Phi)^2 \simeq \frac{1}{\hbar^2}[m^2c^2 + 2m(E' - q\Phi)]. \tag{11.280}$$

We can infer the Hamiltonian for slow-moving particles by rewriting Eq. (11.278) into a Schrödinger form. The Hamiltonian is the same as in the non-relativistic case, except for some extra terms

$$\begin{aligned}\hat{H} &= \sum_{k=1}^{3} \frac{-\hbar^2}{2m}\left(\frac{\partial}{\partial x_k} - \frac{iq}{\hbar c}A_k\right)^2 + q\Phi - \frac{q\hbar}{2mc}(\boldsymbol{B\sigma} - i\boldsymbol{E\sigma})\\ &= \frac{1}{2m}\left(\hat{\boldsymbol{p}} - \frac{q}{c}\boldsymbol{A}\right)^2 + q\Phi - \frac{q\hbar}{2mc}(\boldsymbol{B\sigma} - i\boldsymbol{E\sigma}).\end{aligned} \tag{11.281}$$

Assuming a time-independent field with Coulomb gauge

$$(\hat{\boldsymbol{p}}\boldsymbol{A} - \boldsymbol{A}\hat{\boldsymbol{p}}) = -i\hbar\nabla\boldsymbol{A} = 0, \tag{11.282}$$

and taking $\boldsymbol{A} = 1/2\,\boldsymbol{B} \times \boldsymbol{r}$, we find

$$\begin{aligned}\hat{H} &= \frac{1}{2m}\hat{\boldsymbol{p}}^2 - \frac{q}{2mc}(\boldsymbol{B}\times\boldsymbol{r})\hat{\boldsymbol{p}} + \frac{q^2}{8mc^2}(\boldsymbol{B}\times\boldsymbol{r})^2 + q\Phi - \frac{q\hbar}{2mc}(\boldsymbol{B}\boldsymbol{\sigma} - i\boldsymbol{E}\boldsymbol{\sigma}) \\ &= \frac{1}{2m}\hat{\boldsymbol{p}}^2 - \frac{q}{2mc}(\boldsymbol{r}\times\hat{\boldsymbol{p}})\boldsymbol{B} + \frac{q^2}{8mc^2}(\boldsymbol{B}\times\boldsymbol{r})^2 + q\Phi - \frac{q\hbar}{2mc}\boldsymbol{\sigma}\boldsymbol{B} + \frac{iq\hbar}{2mc}\boldsymbol{\sigma}\boldsymbol{E} \\ &= \frac{1}{2m}\hat{\boldsymbol{p}}^2 - \frac{q}{2mc}\hat{\boldsymbol{L}}\cdot\boldsymbol{B} + \frac{q^2}{8mc^2}(\boldsymbol{B}\times\boldsymbol{r})^2 + q\Phi - \frac{q}{mc}\hat{\boldsymbol{S}}\cdot\boldsymbol{B} + \frac{iq\hbar}{2mc}\boldsymbol{\sigma}\boldsymbol{E} \\ &= \frac{1}{2m}\hat{\boldsymbol{p}}^2 - \hat{\boldsymbol{\mu}}\cdot\boldsymbol{B} + \frac{q^2}{8mc^2}(\boldsymbol{B}\times\boldsymbol{r})^2 + q\Phi + \frac{iq\hbar}{2mc}\boldsymbol{\sigma}\boldsymbol{E},\end{aligned} \tag{11.283}$$

where

$$\hat{\boldsymbol{\mu}} = \frac{q}{2mc}(\hat{\boldsymbol{L}} + 2\hat{\boldsymbol{S}}). \tag{11.284}$$

We have found that the coupling of the spin to the magnetic field is twice as strong as the coupling of the orbital angular momentum to the magnetic field. We have introduced this feature before as an experimental fact, but it comes out nicely from the Dirac equation as a consequence of the relativistic invariance.

11.2.8 Hydrogen Atom

It was an enormous success of the Dirac equation that it gave the correct result for the hydrogen atom. To see this, we should consider the two-component second order equation

$$(\hat{D}_1^2 + \hat{D}_2^2 + \hat{D}_3^2 + \hat{D}_4^2 - \chi^2)\psi + \frac{q}{\hbar c}(\boldsymbol{B}\boldsymbol{\sigma} - i\boldsymbol{E}\boldsymbol{\sigma})\psi = 0, \tag{11.285}$$

and take $\boldsymbol{B} = 0$, $\hat{D}_k = \partial/\partial x_k$ for $k = 1, 2, 3$, $\hat{D}_4 = -1/\hbar c\,(E - V)$, $\boldsymbol{E} = -\boldsymbol{\nabla}\Phi$, and $q\Phi = V$. Then, the equation becomes

$$\left(\nabla^2 + \frac{1}{\hbar^2 c^2}(E - V)^2 - \frac{m^2c^2}{\hbar^2}\right)\psi + \frac{i}{\hbar c}\boldsymbol{\nabla}V\cdot\boldsymbol{\sigma}\psi = 0. \tag{11.286}$$

If we separate off the rest energy, $E = E' + mc^2$, we find

$$\left[-\frac{\hbar^2}{2m}\nabla^2 + V\left(1 + \frac{E'}{mc^2} - \frac{V}{2mc^2}\right) - \frac{i\hbar}{2mc}\boldsymbol{\nabla}V\cdot\boldsymbol{\sigma}\right]\psi = \tilde{E}\psi, \tag{11.287}$$

where $\tilde{E} = E'\left(1 + E'/(2mc^2)\right)$.

For spherical potentials, the total wave function is a product of radial and angular terms

$$|\psi\rangle = \sum_{\pm}\frac{1}{r}u_j^{(\pm)}(r)\Phi_{j,m}^{(\pm)}, \tag{11.288}$$

where $\Phi^{(\pm)}_{j,m}$ has been defined in Eqs. (11.254). Then, with the help of the Laplacian

$$\nabla^2 = \frac{1}{r}\frac{\partial^2}{\partial r^2}r - \frac{\hat{\boldsymbol{L}}^2}{\hbar^2 r^2}, \tag{11.289}$$

and for a Coulomb plus short-range potential

$$V(r) = \frac{Ze^2}{r} + W(r), \tag{11.290}$$

Equation (11.287) becomes

$$\sum_{\pm}\left[-\frac{\hbar^2}{2m}\frac{\partial^2}{\partial r^2} + \frac{\hbar^2}{2mr^2}\left(\frac{\hat{\boldsymbol{L}}^2}{\hbar^2} - Z^2\alpha^2 + iZ\alpha\,\boldsymbol{e}_r\boldsymbol{\sigma}\right) + \frac{\tilde{Z}e^2}{r} + \tilde{W}(r) - \frac{i\hbar}{2mc}W'(r)\,\boldsymbol{e}_r\boldsymbol{\sigma}\right] u_j^{(\pm)}\Phi^{(\pm)}_{j,m} = \tilde{E}\sum_{\pm}u_j^{(\pm)}\Phi^{(\pm)}_{j,m}, \tag{11.291}$$

where $\alpha = e^2/(\hbar c)$ is the fine-structure constant, $\tilde{Z} = Z(1 + E'/(mc^2))$, $W'(r) = \mathrm{d}W(r)/\mathrm{d}r$ and

$$\tilde{W}(r) = W(r)\left(1 + \frac{E'}{mc^2} - \frac{1}{mc^2}\frac{Ze^2}{r} - \frac{1}{2mc^2}W(r)\right). \tag{11.292}$$

We should recall that the parity operator $\mathcal{P}$, which mirrors the coordinates, in polar coordinates entails the transformation $\theta \to \pi - \theta$ and $\phi \to \phi + \pi$. The spherical harmonics transform as $\mathcal{P}\,Y_{l,m} = (-)^l Y_{l,m}$ and the electron has positive intrinsic parity. Consequently,

$$\mathcal{P}\,|\Phi^{(\pm)}_{j,m}\rangle = (-)^{l_\pm}|\Phi^{(\pm)}_{j,m}\rangle, \tag{11.293}$$

i.e. the states $|\Phi^{(+)}_{j,m}\rangle$ and $|\Phi^{(-)}_{j,m}\rangle$ have opposite parities. We can also see that

$$\boldsymbol{e}_r\boldsymbol{\sigma} = \sigma_x\sin\theta\cos\phi + \sigma_y\sin\theta\sin\phi + \sigma_z\cos\theta = \begin{pmatrix}\cos\theta & e^{-i\phi}\sin\theta\\ e^{i\phi}\sin\theta & -\cos\theta\end{pmatrix} \tag{11.294}$$

is an odd operator under parity, i.e. $\mathcal{P}(\boldsymbol{e}_r\boldsymbol{\sigma}) = -\boldsymbol{e}_r\boldsymbol{\sigma}$, and also $(\boldsymbol{e}_r\boldsymbol{\sigma})^2 = 1$. Additionally, $\boldsymbol{e}_r\boldsymbol{\sigma}$ commutes with the $\hat{\boldsymbol{J}}$ angular momentum operator

$$\left[\hat{\boldsymbol{J}}, \boldsymbol{e}_r\boldsymbol{\sigma}\right] = 0. \tag{11.295}$$

We can easily verify this by using $L_z = -i\hbar\partial/\partial\phi$, $S_z = \hbar/2\,\sigma_z$ and $[\sigma_j, \sigma_k] = 2i\epsilon_{jkl}\sigma_l$. We find that

$$\left[L_z, \sigma_x \sin\theta\cos\phi + \sigma_y \sin\theta\sin\phi + \sigma_z\cos\theta\right] = -i\hbar(-\sigma_x \sin\theta\sin\phi + \sigma_y\sin\theta\cos\phi)$$
$$\left[S_z, \sigma_x \sin\theta\cos\phi + \sigma_y \sin\theta\sin\phi + \sigma_z\cos\theta\right] = i\hbar(\sigma_y\sin\theta\cos\phi - \sigma_x\sin\theta\sin\phi). \tag{11.296}$$

So, we have found that the $|\Phi^{(\pm)}_{j,m}\rangle$ states are the degenerate eigenstates of $\hat{\boldsymbol{J}}$. Since $\hat{\boldsymbol{J}}$ and $\boldsymbol{e}_r\boldsymbol{\sigma}$ commute, $\boldsymbol{e}_r\boldsymbol{\sigma}|\Phi^{(\pm)}_{j,m}\rangle$ should remain in the subspace spanned by $|\Phi^{(\pm)}_{j,m}\rangle$. On the other hand, since $\boldsymbol{e}_r\boldsymbol{\sigma}$ is an odd operator whose square is one, $|\Phi^{(\pm)}_{j,m}\rangle$ and $\boldsymbol{e}_r\boldsymbol{\sigma}|\Phi^{(\pm)}_{j,m}\rangle$ should have opposite parity, i.e. $\boldsymbol{e}_r\boldsymbol{\sigma}$ transforms the states into each other

$$(\boldsymbol{e}_r\boldsymbol{\sigma})\,|\Phi^{(\pm)}_{j,m}\rangle = |\Phi^{(\mp)}_{j,m}\rangle \quad \text{and} \quad (\boldsymbol{e}_r\boldsymbol{\sigma})^2\,|\Phi^{(\pm)}_{j,m}\rangle = (\boldsymbol{e}_r\boldsymbol{\sigma})|\Phi^{(\mp)}_{j,m}\rangle = |\Phi^{(\pm)}_{j,m}\rangle. \tag{11.297}$$

Consequently, $\langle\Phi^{(\pm)}_{j,m}|\boldsymbol{e}_r\boldsymbol{\sigma}|\Phi^{(\pm)}_{j,m}\rangle = 0$ and $\langle\Phi^{(\mp)}_{j,m}|\boldsymbol{e}_r\boldsymbol{\sigma}|\Phi^{(\pm)}_{j,m}\rangle = 1$.

Instead of the angular basis $|\Phi^{(\pm)}\rangle$, we can take their linear combinations as basis states

$$|F^{(\pm)}_{j,m}\rangle = |\Phi^{(\pm)}_{j,m}\rangle \mp i\sqrt{(j+1/2-s)/(j+1/2+s)}|\Phi^{(\mp)}_{j,m}\rangle, \tag{11.298}$$

where $s = \sqrt{(j+1/2)^2 - Z^2\alpha^2}$. Now, we have

$$|\psi\rangle = \sum_{\pm}\frac{1}{r}u^{(\pm)}_j(r)|F^{(\pm)}_{j,m}\rangle. \tag{11.299}$$

The states $|F^{(\pm)}_{j,m}\rangle$ are not orthogonal, therefore we need to define the bi-orthogonal partner

$$\langle\tilde{F}^{(\pm)}_{j,m}| = (j+1/2+s)/(2s)\left[\langle\Phi^{(\pm)}_{j,m}| \mp i\sqrt{(j+1/2-s)/(j+1/2+s)}\langle\Phi^{(\mp)}_{j,m}|)\right], \tag{11.300}$$

such that $\langle\tilde{F}^{(\pm)}_{j,m}|F^{(\pm)}_{j,m}\rangle = I$. By using Eq. (11.297), we can easily derive

$$\langle\tilde{F}^{(\pm)}_{j,m}|\boldsymbol{e}_r\boldsymbol{\sigma}|F^{(\pm)}_{j,m}\rangle = \begin{pmatrix} -iZ\alpha/s & (j+1/2)/s \\ (j+1/2)/s & iZ\alpha/s \end{pmatrix}. \tag{11.301}$$

From Eqs. (11.254) and (11.297), we can infer that the states $|F^{(\pm)}_{j,m}\rangle$ are eigenstates

$$\left(\hat{\boldsymbol{L}}^2/\hbar^2 - Z^2\alpha^2 + iZ\alpha\,\boldsymbol{e}_r\boldsymbol{\sigma}\right)\begin{pmatrix}|F^{(+)}_{j,m}\rangle \\ |F^{(-)}_{j,m}\rangle\end{pmatrix} = \begin{pmatrix}(s-1)s & 0 \\ 0 & s(s+1)\end{pmatrix}\begin{pmatrix}|F^{(+)}_{j,m}\rangle \\ |F^{(-)}_{j,m}\rangle\end{pmatrix}. \tag{11.302}$$

Indeed, for example, by using the relation $Z\alpha = \sqrt{(j+1/2)^2 - s^2}$, we find

$$
\begin{aligned}
&\left(\hat{\boldsymbol{L}}^2/\hbar^2 - Z^2\alpha^2 + iZ\alpha\, \boldsymbol{e}_r\boldsymbol{\sigma}\right)\left[\Phi^{(+)}_{j,m} - i\sqrt{(j+1/2-s)/(j+1/2+s)}\Phi^{(-)}_{j,m}\right] \\
&= [(j-1/2)(j+1/2) - (j+1/2)^2 + s^2]\Phi^{(+)}_{j,m} \\
&+ i\sqrt{(j+1/2)^2 - s^2}\,\Phi^{(-)}_{j,m} - i\sqrt{(j+1/2-s)/(j+1/2+s)} \\
&\times \left\{[(j+1/2)(j+3/2) - (j+1/2)^2 + s^2]\Phi^{(-)}_{j,m} + i\sqrt{(j-1/2)^2 - s^2}\Phi^{(+)}_{j,m}\right\} \\
&= (s^2 - s)\Phi^{(+)}_{j,m} - i\sqrt{(j+1/2-s)/(j+1/2+s)}(s^2 - s)\Phi^{(-)}_{j,m} \\
&= (s-1)s\left[\Phi^{(+)}_{j,m} - i\sqrt{(j+1/2-s)/(j+1/2+s)}\Phi^{(-)}_{j,m}\right] = (s-1)s\, F^{(+)}_{j,m}.
\end{aligned}
\tag{11.303}
$$

We can equate the eigenvalues in Eq. (11.302) such that $\lambda_\pm(\lambda_\pm + 1) = s(s \mp 1)$, then we find that $\lambda_\pm = s - 1/2 \mp 1/2$. If we substitute the wave function of Eqs. (11.299) into (11.287) and act by $\langle \tilde{F}^{(\pm)}_{j,m}|$ from the left, we obtain a set of two-component equations for the radial wave functions

$$
\begin{aligned}
&\left[-\frac{\hbar^2}{2m}\frac{\partial^2}{\partial r^2} + \frac{\hbar^2\lambda_\pm(\lambda_\pm + 1)}{2mr^2} + \frac{\tilde{Z}e^2}{r} + \underline{\tilde{W}}(r) - \frac{i\hbar}{2\,mc}\underline{W'}(r)\right]\begin{pmatrix} u^{(+)}_j(r) \\ u^{(-)}_j(r)\end{pmatrix} \\
&= \tilde{E}\begin{pmatrix} u^{(+)}_j(r) \\ u^{(-)}_j(r)\end{pmatrix},
\end{aligned}
\tag{11.304}
$$

where

$$
\underline{\tilde{W}}(r) = \langle \tilde{F}^{(\pm)}_{j,m}|\tilde{W}(r)|F^{(\pm)}_{j,m}\rangle = \begin{pmatrix} \langle \tilde{F}^{(+)}_{j,m}|\tilde{W}(r)|F^{(+)}_{j,m}\rangle & 0 \\ 0 & \langle \tilde{F}^{(-)}_{j,m}|\tilde{W}(r)|F^{(-)}_{j,m}\rangle \end{pmatrix}
\tag{11.305}
$$

and

$$
\begin{aligned}
\underline{W'}(r) &= \langle \tilde{F}^{(\pm)}_{j,m}|W'(r)\boldsymbol{e}_r\boldsymbol{\sigma}|F^{(\pm)}_{j,m}\rangle \\
&= \begin{pmatrix} -iZ\alpha/s\langle \tilde{F}^{(+)}_{j,m}|W'(r)|F^{(+)}_{j,m}\rangle & (j+1/2)/s\langle \tilde{F}^{(+)}_{j,m}|W'(r)|F^{(-)}_{j,m}\rangle \\ (j+1/2)/s\langle \tilde{F}^{(-)}_{j,m}|W'(r)|F^{(+)}_{j,m}\rangle & iZ\alpha/s\langle \tilde{F}^{(-)}_{j,m}|W'(r)|F^{(-)}_{j,m}\rangle \end{pmatrix}.
\end{aligned}
\tag{11.306}
$$

If we consider the hydrogen problem only, the terms proportional to W are absent. Then, the similarity to the non-relativistic hydrogen problem becomes striking. If we make the correspondences

$$
l \to \lambda_+ = s - 1, \quad E \to E'(1 + E'/(2mc^2)), \quad \text{and} \quad Z \to Z(1 + E'/(mc^2)),
\tag{11.307}
$$

with $s = \sqrt{(j+1/2)^2 - Z^2\alpha^2}$, in the energy formula for non-relativistic hydrogen atom Eq. (5.65), we obtain

$$E'(1 + E'/(2mc^2)) = -\frac{mc^2(s^2 - (j + 1/2)^2)}{2} \frac{(1 + E'/(mc^2))^2}{(n' + \lambda_+ + 1)^2}, \tag{11.308}$$

where $n' = 0, 1, \ldots$ is the radial quantum number. After some algebra, we find the solution

$$E'_{n',j} = mc^2 \left[\left(1 + \frac{s^2 - (j + 1/2)^2}{(n' + s)^2} \right)^{-1/2} - 1 \right], \tag{11.309}$$

or with the principal quantum number $n = n' + j + 1/2$, we have

$$E'_{n,j} = mc^2 \left[\left(1 + \frac{s^2 - (j + 1/2)^2}{(n - (j + 1/2) + s)^2} \right)^{-1/2} - 1 \right]$$
$$\text{with} \quad n = 1, 2, \ldots \text{ and } j = 1/2, 3/2, \ldots, n - 1/2\,. \tag{11.310}$$

The numerical value $\alpha \approx 1/137.036 \ll 1$, so expanding the energy in terms of α gives

$$E'_{n,j} \approx -\frac{mc^2 Z^2 \alpha^2}{2n^2} - \frac{mc^2 Z^4 \alpha^4}{2n^4} \left(\frac{n}{j + 1/2} - \frac{3}{4} \right) + \cdots. \tag{11.311}$$

The first term is the non-relativistic binding energy. The subsequent terms are the relativistic corrections. They remove the degeneracy of levels with the same n. Indeed, we observe, in agreement with experiments, a dependence in n and j.

11.2.9 Dirac Equation in Feshbach-Villars Form

We have derived Eqs. (11.278) from the free Dirac equation by adopting the technique of minimal coupling. If we take the case with $\boldsymbol{B} = 0$ and $q\Phi = V$, one of the Eqs. (11.278) reads

$$\left(-\hbar^2 c^2 \nabla^2 - \left(i\hbar \frac{\partial}{\partial t} - V \right)^2 - m^2 c^4 + i\hbar c \nabla V \boldsymbol{\sigma} \right) \psi = 0, \tag{11.312}$$

or

$$\left(i\hbar \frac{\partial}{\partial t} - V \right)^2 \psi = \boldsymbol{p}^2 c^2 \psi + m^2 c^4 \psi - i\hbar c \nabla V \boldsymbol{\sigma} \psi. \tag{11.313}$$

The Feshbach-Villars linearization amounts to splitting the wave function into components such that

$$\psi = \phi + \chi \quad \text{and} \quad (i\hbar \frac{\partial}{\partial t} - V)\psi = mc^2(\phi - \chi). \tag{11.314}$$

This leads to the set of equations

$$\begin{aligned}\left(i\hbar\frac{\partial}{\partial t}-V\right)(\phi+\chi)&=mc^2(\phi-\chi),\\ \left(i\hbar\frac{\partial}{\partial t}-V\right)(\phi-\chi)&=\frac{\hat{\boldsymbol{p}}^2}{m}(\phi+\chi)+mc^2(\phi+\chi)-\frac{i\hbar}{mc}\nabla V\,\boldsymbol{\sigma}(\phi+\chi),\end{aligned} \tag{11.315}$$

and reorganizing, we obtain

$$\begin{aligned}i\hbar\frac{\partial}{\partial t}\phi&=\frac{\hat{\boldsymbol{p}}^2}{2m}(\phi+\chi)+mc^2\phi+V\phi-\frac{i\hbar}{2mc}\nabla V\,\boldsymbol{\sigma}(\phi+\chi),\\ i\hbar\frac{\partial}{\partial t}\chi&=-\frac{\hat{\boldsymbol{p}}^2}{2m}(\phi+\chi)-mc^2\chi+V\chi+\frac{i\hbar}{2mc}\nabla V\,\boldsymbol{\sigma}(\phi+\chi).\end{aligned} \tag{11.316}$$

This is a Schrödinger-like Feshbach-Villars equation for spin-1/2 particles

$$i\hbar\frac{\partial}{\partial t}\begin{pmatrix}\phi\\ \chi\end{pmatrix}=\hat{H}_{FV1/2}\begin{pmatrix}\phi\\ \chi\end{pmatrix}, \tag{11.317}$$

with Hamiltonian

$$\hat{H}_{FV1/2}=\begin{pmatrix}1&1\\-1&-1\end{pmatrix}\frac{\hat{\boldsymbol{p}}^2}{2m}+\begin{pmatrix}1&0\\0&-1\end{pmatrix}mc^2+\begin{pmatrix}1&0\\0&1\end{pmatrix}V-\begin{pmatrix}1&1\\-1&-1\end{pmatrix}\frac{i\hbar}{2mc}\nabla V\,\boldsymbol{\sigma}. \tag{11.318}$$

Comparing this Hamiltonian to the Hamiltonian in Eq. (11.30) we can see that $\hat{H}_{FV1/2}$ is basically two Feshbach-Villars $\hat{H}_{FV0}$ Hamiltonians coupled by a spin term. The Feshbach-Villars formalism offers a unified formalism for spin-0 and spin-1/2 particles.

Further Reading

1. H. Feshbach, F. Villars, Elementary relativistic wave mechanics of spin 0 and spin 1/2 particles. Rev. Modern Phys. **30**(1), 24 (1958)
2. M.G. Fuda, E. Furlani, Zitterbewegung and the Klein paradox for spin-zero particles. Am. J. Phys. **50**(6), 545–549 (1982)

Chapter 12
Appendix: Numerical Methods for Simple Problems

In this chapter we present some basic numerical tools and the corresponding computer codes to solve simple problems in quantum mechanics. The adopted computational language is Fortran, one of the oldest and most mature language used in scientific programing. We assume that the reader has, or can acquire, some familiarity in Fortran programming, and has access to the LAPACK linear algebraic package. We present the codes in the most simple form and delegate plotting to independent software tools.

12.1 Auxiliary Computational Tools

12.1.1 Chebyshev Approximation

We need a numerical method to find derivatives, zeros, and extrema of functions. For that purpose we use the Chebyshev polynomials. The Chebyshev polynomials are solutions of the differential equation

$$(1-x^2)\,\frac{\mathrm{d}^2 T_n(x)}{\mathrm{d}x^2} - x\frac{\mathrm{d}T_n(x)}{\mathrm{d}x} + n^2 T_n(x) = 0 \quad \text{with} \quad n = 0, 1, \ldots, \tag{12.1}$$

and they are given by

$$T_n(x) = \cos(n \arccos x). \tag{12.2}$$

The first few polynomials are $T_0(x) = 1$, $T_1(x) = x$, and $T_2(x) = 2x^2 - 1$. They satisfy the recursion relation

$$T_{n+1}(x) = 2xT_n(x) - T_{n-1}(x), \tag{12.3}$$

Z. Papp, *Mastering Quantum Mechanics*,
https://doi.org/10.1007/978-3-032-09011-9_12

and they are orthonormal on the interval $[-1, 1]$,

$$\int_{-1}^{1} \mathrm{d}x \frac{T_i(x)T_j(x)}{\sqrt{1-x^2}} = \begin{cases} 0, & \text{if } i \neq j, \\ \pi/2, & \text{if } i = j \neq 0, \\ \pi, & \text{if } i = j = 0. \end{cases} \tag{12.4}$$

From Eq. (12.2) we can see that the Chebyshev polynomial $T_n(x)$ has n zeros in the $[-1, 1]$ interval at

$$x_k = \cos\left(\frac{(k+1/2)\pi}{n}\right) \quad \text{where} \quad k = 0, 1, \ldots n-1. \tag{12.5}$$

We can also see that it has $n+1$ extrema at

$$x'_k = \cos\left(\frac{k\pi}{n}\right) \quad \text{with} \quad k = 0, 1, 2, \ldots n, \tag{12.6}$$

and $|T_n(x'_k)| = 1$.

A function $f(x)$ can be approximated on the $[-1, 1]$ interval by a combination of Chebyshev polynomials,

$$f(x) \approx f_N(x) = \sum_{n=0}^{N} a_n T_n(x). \tag{12.7}$$

This approximation is *exact* at the zeros of $T_N(x)$, and the coefficients are given by

$$\begin{aligned} a_n &= \frac{2-\delta_{0n}}{N+1} \sum_{k=0}^{N} f(x_k) T_n(x_k) \\ &= \frac{2-\delta_{0n}}{N+1} \sum_{k=0}^{N} f\left[\cos\left(\frac{(k+1/2)\pi}{N+1}\right)\right] \cos\left(\frac{n\pi(k+1/2)}{N+1}\right). \end{aligned} \tag{12.8}$$

Another nice property of the Chebyshev expansion is that its derivative can easily be calculated. The Chebyshev coefficients of $f'(x)$ are given by the recursion relation

$$a'_{n-1} = (2n\, a_n + a'_{n+1})/(1+\delta_{0n}) \quad \text{with} \quad a'(N+1) = a'(N) = 0. \tag{12.9}$$

The roots of a Chebyshev-approximated function are related to the eigenvalues of a companion matrix [1]. All the roots of $f_N(x)$, both on and off the interval $[-1, 1]$, are the eigenvalues of the $N \times N$ companion matrix $\underline{A}$, which is given by

$$\underline{A}_{jk} = \begin{cases} \delta_{2,k}, & j = 1,\ k = 1, 2, \ldots, N, \\ 1/2\ \left(\delta_{j,k+1} + \delta_{j,k-1}\right), & j = 2, \ldots, N-1,\ k = 1, 2, \ldots, N, \\ -a_{k-1}/(2a_N) + 1/2\,\delta_{k,N-1}, & j = N,\ k = 1, 2, \ldots, N. \end{cases} \tag{12.10}$$

For example, for $N = 5$, the matrix takes the form

$$\underline{A} = \begin{pmatrix} 0 & 1 & 0 & 0 & 0 \\ 1/2 & 0 & 1/2 & 0 & 0 \\ 0 & 1/2 & 0 & 1/2 & 0 \\ 0 & 0 & 1/2 & 0 & 1/2 \\ -a_0/2a_5 & -a_1/2a_5 & -a_2/2a_5 & -a_3/2a_5 + 1/2 & -a_4/2a_5 \end{pmatrix}. \tag{12.11}$$

These roots can further be polished by using an iterative zero search proposed in Ref. [2].

Listing 12.1 Fortran module for locating zeros by Chebyshev approximation

```

module chebyshev

    use numtype
    implicit none
    integer, parameter :: maxch = 50
    real(dp), dimension(0:maxch) :: cheb, chder, chder2
    real(dp), dimension(maxch) :: z0
    integer :: iz0

    contains

        subroutine chebyex(func,n,a,ya,yb)
        !    func([ya,yb]) = sum_{i=0}^n  a_i T_i(x)

            real(dp), external :: func
            integer :: n
            real(dp), dimension(0:maxch) :: f, a
            real(dp) :: ya, yb, aa, bb, x, ss
            integer :: i, j

            if ( n > maxch ) stop '  n > maxch '
            aa = (yb-ya)/2; bb = (yb+ya)/2
            do i = 0, n
                x = cos(pi/(n+1)*(i+0.5_dp))
                f(i) = func(aa*x+bb)
            end do
            do j = 0, n
                ss = 0._dp
                do i = 0, n
                    ss = ss + &
                        f(i)*cos((pi/(n+1))*j*(i+0.5_dp))
                end do
                a(j) = 2._dp*ss/(n+1)
            end do
            a(0) = 0.5_dp*a(0)

        end subroutine chebyex

```

```
    subroutine chebyderiv(n,a,der,ya,yb)
    !  func'([ya,yb]) sum_{i=0}^(n-1)  der_i T_i(x)

        integer :: n
        real(dp) :: ya, yb, a(0:maxch), der(0:maxch)
        integer :: j

        der(n) = 0._dp; der(n-1) = 2*n*a(n)
        do j = n-1, 1, -1
            der(j-1) = der(j+1)+2*j*a(j)
        end do
        der(0) = der(0)/2
        der(0:n-1) = der(0:n-1)*2/(yb-ya)

    end subroutine chebyderiv

    function cheby(y,n,a,ya,yb) result(t)
    ! func(y) =  sum_{i=0}^n  a_i T_i (y)

        implicit none
        integer :: n
        real(dp) :: y, ya, yb
        real(dp) :: a(0:maxch)
        real(dp) :: aa, bb, x, t, y0, y1
        integer :: k

        aa = (yb-ya)/2; bb = (yb+ya)/2
        x = (y-bb)/aa
        y1 = 0._dp; y0 = a(n)
        do k = n-1, 0, -1
            t = y1; y1 = y0
            y0 = a(k)+2*x*y1-t
        end do
        t = y0-x*y1

    end function cheby

    subroutine chebyzero(n,a,ya,yb,z0,iz0)
    !   find zero by using Boyd's method

        integer :: n, iz0
        real(dp), dimension(0:maxch) :: a
        integer :: j
        real(dp), dimension(maxch) :: wr0, wi0, z0, wwr0
        real(dp) :: ya, yb

        call boyd(n,a,wr0,wi0)
        wwr0(1:n) = wr0(1:n)*(yb-ya)/2+(yb+ya)/2

        iz0 = 0
        do j = 1, n
            if( wi0(j) == 0._dp .and. &
              -1 <= wr0(j) .and. wr0(j) <= 1 ) then
                    iz0 = iz0+1;  z0(iz0) = wwr0(j)
            end if
        end do

        contains

            subroutine boyd(n,a,wr,wi)
            !  J. P. Boyd, J. Eng. Math. (2006) 56, 203-219

                integer :: n, j, ie
```

```
                    real(dp) :: a(0:maxch)
                    real(dp) :: wr(maxch), wi(maxch)
                    integer, parameter :: lwork=4*maxch
                    real(dp) :: aamat(maxch,maxch),  &
                        work(lwork), rwork(lwork), &
                        vl(1), vr(1)

                    if (abs(a(n)) == 0._dp) stop 'a(n)=0'
                    aamat(1:n,1:n) = 0._dp
                    aamat(1,2) = 1._dp
                    do j = 2, n-1
                        aamat(j,j-1) = 0.5_dp
                        aamat(j,j+1) = 0.5_dp
                    end do
                    aamat(n,1:n) = -a(0:n-1)/(2*a(n))
                    aamat(n,n-1) = aamat(n,n-1) + 0.5_dp

                    ie = 0
                    call dgeev('n','n',n,aamat,maxch,wr,&
                        wi,vl,1,vr,1,work,lwork,rwork,ie)
                    if( ie /= 0 ) stop ' boyd: ie /= 0 '

                end subroutine boyd

        end subroutine chebyzero

        subroutine root_polish(func,zz,dz,eps,maxf,minf,i)
        ! iterative three-point search for pole
        ! J. Zinn-Justin,  Physics Reports, 1, No 3 (1971) 55-102
        !
            complex(dp), external :: func
            real(dp) :: eps
            complex(dp) :: zz, dz, z1, z2, z3, &
                f1, f2, f3, a12, a23, a31
            integer :: i, minf, maxf

            z1 = zz+dz;    f1 = func(z1)
            z2 = zz-dz;    f2 = func(z2)
            z3 = zz;       f3 = func(z3)

            do i = 1, maxf
                a23 = (z2-z3)*f2*f3
                a31 = (z3-z1)*f1*f3
                a12 = (z1-z2)*f1*f2
                zz = (z1*a23+z2*a31+z3*a12)/(a23+a31+a12)
                if ( abs(zz-z3) < eps .and. minf < i ) exit
                z1 = z2;  f1 = f2
                z2 = z3;  f2 = f3
                z3 = zz;  f3 = func(z3)
            end do

        end subroutine root_polish

end module chebyshev
```

Listing 12.2 Auxiliary Fortran module for calculating the inverse and the determinant of symmetric matrices by using LAPACK routines

```

module matrix

    use NumType, only : dp
    integer, parameter :: nb = 16

    contains

        subroutine zsys(n,matr,lda,bx)
        ! linear equation solver with complex symmetric matrix
            implicit none
            integer :: n, lda
            complex(dp) :: matr(lda,lda), bx(lda)
            complex(dp), dimension(nb*lda) :: work
            integer :: info, ipvt(lda)

            info = 0
            call zsytrf('u',n,matr,lda,ipvt,work,nb*lda,info)
                if(info /= 0) stop 'zsytrf-zsys'
            call zsytrs('u',n,1,matr,lda,ipvt,bx,lda,info)
                if(info /= 0) stop 'zsytrs-zsys'

        end subroutine zsys

        subroutine zsyi(n,matr,lda)
        ! inverse of a complex symmetric matrix
            integer, intent(in) :: n, lda
            complex(dp) :: matr(lda,lda)
            complex(dp), dimension(nb*lda) :: work
            integer :: i, j, info, ipvt(lda)

            info=0
            call zsytrf('u',n,matr,lda,ipvt,work,nb*lda,info)
                if(info /= 0) stop 'zsytrf-zsyi'
            call zsytri('u',n,matr,lda,ipvt,work,info)
                if(info /= 0) stop 'zsytri-zsyi'
            forall ( i=1:n, j=1:n, i < j ) matr(j,i) = matr(i,j)

        end subroutine zsyi

        subroutine zsydet(n,matr,lda,det)
        ! determinant of a complex symmetric matrix
            integer :: n, lda
            complex(dp) :: matr(lda,lda), det(2)
            complex(dp), dimension(nb*lda) :: work
            integer :: info, ipvt(lda), k
            complex(dp) :: t, d

            info = 0
            call zsytrf('u',n,matr,lda,ipvt,work,nb*lda,info)
                if(info /= 0) stop 'zsytrf-zsydet'
            det(1) = 1.0_dp;    det(2) = 0.0_dp
            t = 0.0_dp
            do k = 1, n
                d = matr(k,k)
                if ( ipvt(k) <= 0 ) then
                    if ( t == 0.0_dp ) then
                        t = matr(k,k+1); d = (d/t)*matr(k+1,k+1) - t
                    else
                        d = t;  t = 0.0_dp
```

```
                end if
            end if
            det(1) = d*det(1)
            if ( det(1) /= 0.0_dp ) then
                do while ( abs(det(1)) <= 1 )
                    det(1) = 10*det(1); det(2) = det(2) - 1
                end do
                do while ( abs(det(1)) >= 10 )
                    det(1) = det(1)/10; det(2) = det(2) + 1
                end do
            end if
        end do

    end subroutine zsydet

    subroutine dsyi(n,matr,lda)
    ! inverse of a symmetric real matrix

        integer, intent(in) :: n, lda
        real(dp) :: matr(lda,lda)
        real(dp), dimension(nb*lda) :: work
        integer :: i, j, info, ipvt(lda)

        info=0
        call dsytrf('u',n,matr,lda,ipvt,work,nb*lda,info)
            if(info /= 0) stop 'dsytrf-dsyi'
        call dsytri('u',n,matr,lda,ipvt,work,info)
            if(info /= 0) stop 'dsytri-dsyi'
        forall ( i=1:n, j=1:n, i < j ) matr(j,i) = matr(i,j)

    end subroutine dsyi

end module matrix
```

12.1.2 Runge-Kutta Methods

The backbone of the numerical solution of differential equations are the Runge-Kutta methods. We solve a set of first order differential equations

$$\frac{\mathrm{d}\boldsymbol{y}(x)}{\mathrm{d}x} = \boldsymbol{f}[x, \boldsymbol{y}(x)], \tag{12.12}$$

where x is the independent variable, $\boldsymbol{y} = (y_1(x), y_2(x), \ldots, y_n(x))$ is an array of dependent variables and $\boldsymbol{f}$ is an array of functions $f_j[x, \boldsymbol{y}(x)]$ with $j = 1, \ldots, n$.

We need to divide the interval into subintervals, and integrate the equation

$$\int_{x_n}^{x_{n+1}} \frac{\mathrm{d}\boldsymbol{y}}{\mathrm{d}x}\,\mathrm{d}x = \int_{x_n}^{x_{n+1}} \boldsymbol{f}[x, \boldsymbol{y}(x)]\,\mathrm{d}x \quad \Rightarrow \quad \boldsymbol{y}_{n+1} = \boldsymbol{y}_n + \int_{x_n}^{x_{n+1}} \boldsymbol{f}[x, \boldsymbol{y}(x)]\,\mathrm{d}x. \tag{12.13}$$

Depending on what method we choose to approximate the integral, we obtain the Runge-Kutta methods of different orders.

The workhorse of differential equation solvers is the *4th order Runge-Kutta* method

$$
\begin{aligned}
h &= x_{n+1} - x_n \\
\boldsymbol{k}_1 &= h\boldsymbol{f}\left(x_n, \boldsymbol{y}_n\right) \\
\boldsymbol{k}_2 &= h\boldsymbol{f}\left(x_n + h/2,\ \boldsymbol{y}_n + \boldsymbol{k}_1/2\right) \\
\boldsymbol{k}_3 &= h\boldsymbol{f}\left(x_n + h/2,\ \boldsymbol{y}_n + \boldsymbol{k}_2/2\right) \\
\boldsymbol{k}_4 &= h\boldsymbol{f}\left(x_n + h,\ \boldsymbol{y}_n + \boldsymbol{k}_3\right) \\
\boldsymbol{y}\left(x_{n+1}\right) &= \boldsymbol{y}\left(x_n\right) + \left(\boldsymbol{k}_1 + 2\boldsymbol{k}_2 + 2\boldsymbol{k}_3 + \boldsymbol{k}_4\right)/6 + O\left(h^5\right).
\end{aligned} \tag{12.14}
$$

It offers a nice balance between simplicity, speed, and accuracy. Still one question remains: how do we know if the 4-th order Runge-Kutta method provides the right solution? We generally don't know. We may repeat the calculation with different values of the step size h, and if the result does not depend on the step size within the required accuracy, we are on fairly safe ground.

12.2 Bound States in One-Dimensional Potential

Here we show the code that produced Fig. 3.1. We adopted units such that $\hbar = 1$ and $m = 1$. The one dimensional potential is given by

$$V(x) = 2[\tanh(2(x-2)) - \tanh(2(x+2))]. \tag{12.15}$$

The Schrödinger equation reads

$$-\frac{\hbar^2}{2m}\frac{\mathrm{d}^2\psi(x)}{\mathrm{d}x^2} + V(x)\psi(x) = E\psi(x), \tag{12.16}$$

with the normalization condition $\int_{-\infty}^{\infty} \mathrm{d}x|\psi(x)|^2 = 1$. The potential is symmetric and the Hamiltonian commutes with the parity operator. Consequently, the wave function should possess either even or odd parity. This means that $\psi(0) \neq 0$ and $\psi'(0) = 0$ for even parity solutions, while $\psi(0) = 0$ and $\psi'(0) \neq 0$ for odd parity solutions. At large x values the potential vanishes, and the wave function behaves like

$$\psi(x) \to \exp(-\kappa x) \quad \text{as} \quad x \to \infty, \tag{12.17}$$

where $\kappa = \sqrt{2m(-E)/\hbar^2}$.

By introducing $y_1(x) = \psi(x)$, $y_2(x) = \psi'(x)$ and $y_3(x) = \int_x^{\infty} \mathrm{d}x'|\psi(x')|^2$ we can write the Schrödinger equation as three coupled first order equations

$$\frac{\mathrm{d}}{\mathrm{d}x} y_1(x) = y_2(x),$$
$$\frac{\mathrm{d}}{\mathrm{d}x} y_2(x) = -\frac{2m}{\hbar^2}(E - V(x))y_1(x), \tag{12.18}$$
$$\frac{\mathrm{d}}{\mathrm{d}x} y_3(x) = -[y_1(x)]^2.$$

In solving these equations, we start at some large x_{max} value with boundary condition Eq. (12.17), integrate inwards until we reach $x = 0$, and obtain $\psi(0)\psi'(0)$ as a function of E. Then by using the Chebyshev method, we find energy values that make $\psi(0)\psi'(0)$ vanish. The bound states corresponding to the wave functions in Fig. 3.1 are −3.69387, −2.89273, −1.77154, and −0.61736.

We should note that this code can easily be modified for solving the Schrödinger equation for bound states in spherical potentials in three dimensions. Only the boundary condition at $r = 0$ has to be modified accordingly. Moreover, it is also relatively easy to modify the code to solve the Klein-Gordon equation or the Dirac equation.

Listing 12.3 Bound states in one-dimensional potential

```

module setup

    use numtype
    implicit none

    integer, parameter :: n_eq = 3
    real(dp), parameter :: hbar =1._dp, hbar2 = hbar**2, mass = 1._dp
    real(dp) :: energy, xmax, dstep
    real(dp), allocatable, dimension(:,:) :: wf
    integer :: imax

end module setup

program bs1d
!    1-d bound states

    use setup
    use chebyshev
    implicit none

    real(dp) :: emin, emax, e0, x
    real(dp), external :: psi0, potential
    integer :: nch, iz, i

    xmax = 5
    dstep = 0.001_dp       ! step size
    emin = -4              ! E_min
    emax = -0.0            ! E_max
    nch = 35               ! Chebyshev points

    imax = abs(nint(xmax/dstep))
    allocate( wf(-imax:imax, 2 ))
    do i = -imax, imax
        x = i*dstep
        write(unit= 1 , fmt='(2f15.5)') x, potential(x)
    end do

    call chebyex(psi0,nch,cheb,emin,emax)
    call chebyzero(nch,cheb,emin,emax,z0,iz0)

```

```
    do iz = 1, iz0
        e0 = z0(iz)
        call wavef(iz, e0)
    end do

end program bs1d

subroutine wavef (iw, e)
!   psi(x)

    use setup
    implicit none
    real(dp), intent(in) :: e

    real(dp) :: x, psi(n_eq), parity, kappa
    integer :: iw, i
    real(dp), external :: potential

    energy = e
    x = xmax
    kappa = sqrt( 2*mass/hbar2 * abs(e))
    psi(1) = exp(- kappa * x )
    psi(2) = - kappa* psi(1)
    psi(3) = exp(-2*kappa*x)/(2*kappa)

    do while ( x > 0 )
        call rk4step( x, -dstep, psi )
    end do

    if ( abs(psi(1)) > abs(psi(2)) ) then
        parity = 1
    else
        parity = -1
    end if

    x = xmax
    psi(1) = exp(- kappa *x ) / sqrt(2*psi(3))
    psi(2) = - kappa * psi(1)
    psi(3) = 1/(2*psi(3)) * exp(-2*kappa*x)/(2*kappa)
    i = imax+1
    do while ( x > 0 )
        i = i-1
        wf(i,1) = x;  wf(-i,1) = -x
        wf(i,2) = psi(1);  wf(-i,2) = parity * psi(1)
        call rk4step( x, -dstep, psi )
    end do

    print *,iw,' E = ', e, '  | psi | ', 2*psi(3)
    do i = -imax, imax
        write(unit= 20+iw, fmt='(2f15.5)') wf(i,1), energy
        write(unit= 120+iw, fmt='(2f15.5)') wf(i,1),  energy + wf(i,2)
    end do

end subroutine wavef

function psi0(e)
!   psi(0) * psi'(0)

    use setup
    implicit none
    real(dp), intent(in) :: e
    real(dp) :: psi0, kappa
    real(dp) :: x, psi(n_eq)

    energy = e
    x = xmax
    kappa = sqrt(2*mass/hbar2 * abs(e))
```

```
    psi(1) = exp(- kappa *x )
    psi(2) = - kappa * psi(1)
    psi(3) = exp(-2*kappa*x)/(2*kappa)
    do while ( x > 0 )
        call rk4step( x, -dstep, psi )
    end do
    psi0 = psi(1) * psi(2)

end function psi0

subroutine rk4step( x, h, y )

    use setup
    implicit none
    real(dp), intent(inout) :: x
    real(dp), intent(in) :: h
    real(dp), intent(inout), dimension(n_eq) :: y
    real(dp), dimension(n_eq) :: k1, k2, k3, k4, dy

    k1 = kv ( x,      h, y)
    k2 = kv ( x+h/2, h, y+k1/2)
    k3 = kv ( x+h/2, h, y+k2/2)
    k4 = kv ( x+h,    h, y+k3)
    dy = ( k1 + 2*k2 + 2*k3 + k4) / 6
    y = y + dy
    x = x + h

    contains

        function kv ( x, dx, y) result(k)

            use setup
            real(dp), intent(in) :: x, dx
            real(dp), intent(in), dimension(n_eq) :: y
            real(dp), dimension(n_eq) :: f, k
            real(dp), external :: potential

            f(1) = y(2)
            f(2) = - 2*mass/hbar2 * ( energy - potential(x) ) * y(1)
            f(3) = -y(1)**2

            k = dx * f

        end function kv

end subroutine rk4step

function potential (x)

    use setup
    real(dp) :: x, potential

    potential = 2 * (tanh(2*(x-2)) - tanh(2*(x+2)))

end function potential
```

12.3 Scattering in One Dimension

We consider a one dimensional scattering problem. We assume that the potential is such that $V(-\infty) = 0$ and $V(\infty) = V_0$. In these asymptotic regions, the unnormalized wave function is given by

$$\psi(x) = \begin{cases} \psi_-(x) = A\exp(ik_1x) + B\exp(-ik_1x), & \text{if } x < x_{min}, \\ \psi_+(x) = \exp(ik_2x), & \text{if } x_{max} < x, \end{cases} \tag{12.19}$$

where $k_1 = \sqrt{2m/\hbar^2\, E}$ and $k_2 = \sqrt{2m/\hbar^2\,(E - V_0)}$. Otherwise, $\psi(x)$ and $\psi'(x)$ are continuous everywhere. So, the numerical procedure is quite straightforward. We choose x_{min} and x_{max} such that the potential is already asymptotic in those regions. Then starting from x_{max}, with the boundary condition $\psi_+(x)$, we integrate the Schrödinger equation backwards to $x = x_{min}$. Then matching $\psi(x_{min})$ to $\psi_-(x_{min})$, we infer the amplitudes

$$A = \frac{\psi(x_{min}) + \psi'(x_{min})/(ik_1)}{2\exp(ik_1x_{min})} \quad \text{and} \quad B = \frac{\psi(x_{min}) - \psi'(x_{min})/(ik_1)}{2\exp(-ik_1x_{min})}, \tag{12.20}$$

and calculate the reflection and transmission coefficients

$$R = |\frac{B}{A}|^2 \quad \text{and} \quad T = \frac{k_2}{k_1}|\frac{1}{A}|^2. \tag{12.21}$$

The code in Listing 12.4 produced the wave function in Fig. 3.5. Similarly to the bound state code, this code can also easily be modified to calculate scattering in three dimensions and with Klein-Gordon or Dirac dynamics.

Listing 12.4 Fortran code for one dimensional scattering

```

module setup

    use numtype
    implicit none

    integer, parameter :: n_eq = 2
    real(dp), parameter :: hbar = 1._dp, hbar2 = hbar**2, mass = 1._dp
    real(dp), parameter :: xmax = 10._dp, dstep = 0.001_dp
    real(dp) :: energy

end module setup

program schroedinger_1d_sc
!    1-d scattering

    use setup
    implicit none
    real(dp) :: rr, tt, x
    complex(dp) :: psi(n_eq), en, k2, k1, aa, bb, cc
    real(dp), external :: potential
    integer :: n, i, imax

```

```
    imax = xmax/dstep
    do i = - imax, imax
        x = i * dstep
        write(10, * ) x, potential(x)
    end do

    do n = 0, 1

        energy = 1.75_dp + n * 0.5_dp
        en = energy

        x = xmax
        k2 = sqrt( 2*mass/hbar2 * (en - potential(x)) )
        psi(1) =   exp( iic * k2 * x )    ! psi
        psi(2) =    iic * k2 * psi(1)          ! psi'
        do while ( x > -xmax )
            call rk4step ( x, -dstep , psi )
        end do
        x = -xmax
        k1 = sqrt( 2*mass/hbar2 * (en - potential(x)) )
        aa = ( psi(1) + psi(2)/(iic*k1) )/( 2* exp(iic*k1*x) )
        bb = ( psi(1) - psi(2)/(iic*k1) )/( 2* exp(-iic*k1*x) )
        rr = (abs(bb/aa))**2;     tt = k2/k1 * (abs(1/aa))**2
        print *, energy, k1, k2, rr, tt, rr+tt

        cc = 1/aa
        x = xmax
        k2 = sqrt( 2*mass/hbar2 * (en - potential(x)) )
        psi(1) =   cc*exp( iic * k2 * x )    ! psi
        psi(2) =    iic * k2 * psi(1)          ! psi'
        do while ( x > -xmax )
            write(800+n, * ) x, energy + realpart(psi(1))
            write(70+n, * ) x, energy
            call rk4step ( x, -dstep , psi )
        end do
        x = -xmax
        k1 = sqrt( 2*mass/hbar2 * (en - potential(x)) )
        aa = ( psi(1) + psi(2)/(iic*k1) )/( 2* exp(iic*k1*x) )
        bb = ( psi(1) - psi(2)/(iic*k1) )/( 2* exp(-iic*k1*x) )
        rr = (abs(bb/aa))**2;     tt = k2/k1 * (abs(cc/aa))**2
        print *, energy, k1, k2, rr, tt, rr+tt
        print *, aa,bb

    end do

end program schroedinger_1d_sc

subroutine rk4step ( x, h, y )

    use setup
    implicit none
    real(dp), intent(inout) :: x
    real(dp), intent(in) :: h
    complex(dp), intent(inout), dimension(n_eq) :: y
    complex(dp), dimension(n_eq) :: k1, k2, k3, k4, dy

    k1 = kv( x, h, y )
    k2 = kv( x+h/2, h, y + k1/2 )
    k3 = kv( x+h/2, h, y + k2/2 )
    k4 = kv( x+h , h, y + k3 )
    dy = ( k1 + 2*k2 + 2*k3 + k4 ) / 6
    x = x + h
    y = y + dy

    contains

        function kv( x, dx, psi ) result(k)
```

```
            use setup
            implicit none
            real(dp), intent(in) :: x, dx
            complex(dp), intent(in), dimension(n_eq) :: psi
            complex(dp), dimension(n_eq) :: f, k
            real(dp), external :: potential

            f(1) = psi(2)
            f(2) = - 2*mass/hbar2 * ( energy - potential(x)) * psi(1)
            k = dx * f

        end function kv

end subroutine rk4step

function potential(x)

    use setup
    real(dp) :: x, potential
    real(dp), parameter :: v0 = 1._dp, x0 = 2._dp, a0 = 0.1_dp

    potential = v0/2 * (1 + tanh((x+x0)/a0) )*(1 - tanh((x-x0)/a0) )

end function potential
```

12.4 Solution of the Lippmann-Schwinger Equation

In this section we provide the Fortran code corresponding to the method presented in Sect. 10.3.3. We determine bound, resonant, and scattering states in a spherical Coulomb-like potential. As an example, we present a model for the ${}^{212}_{84}\text{Po} \to {}^{208}_{82}\text{Pb} + \alpha$ decay process. The α particle experiences a deep finite-range nuclear potential and a repulsive Coulomb potential. A typical parametrization is a Woods-Saxon plus a smeared Coulomb potential (Fig. 7.5)

$$V(r) = \frac{v_0}{1+\exp[(r-r_0)/a_0]} + \text{erf}(\gamma_0 r)\frac{2Ze^2}{r}, \tag{12.22}$$

where v_0 and r_0 are the depth and the range of the nuclear forces, respectively, a_0 is the "skin" of the nucleus, γ_0 is a smearing parameter for the Coulomb potential, Z is the charge of the "daughter" nucleus and the factor 2 stands for the charge of the α particle. The parameters $v_0 = -50$ MeV, $r_0 = 7$ fm, $a_0 = 0.4$ fm, $\gamma_0 = 0.1$ fm^{-1}, $e^2 = 1.44$ MeV fm and $Z = 82$ roughly correspond to the α decay of ${}^{212}_{84}\text{Po}$. The adopted units are such that $\hbar^2/m_n = 41.47$ MeV fm^2 where m_n is the mass of the nucleon. Consequently, $m_\alpha = 4$ for the mass of the α particle and $m_{Pb} = 208$ is the mass of the ${}^{208}_{82}\text{Pb}$ nucleus.

Listing 12.5 Fortran module for the inverse Coulomb-Green's matrix

```

module green

    use setup_cs_b_s
```

```
contains

    subroutine gri2b(e,bst,xm,z1,rl,nm,g)
    ! (< n l | g_l^C (e) | n' l>)^(-1) - inverse Coulomb-Green

        implicit none
        integer, intent(in) :: nm
        complex(dp), intent(in) :: e
        complex(dp), dimension(ndim,ndim), intent(out) :: g
        real(dp), intent(in) :: rl, bst, xm, z1
        integer :: n
        complex(dp) :: zk, zki, gami, w0, w1, znm, s0

        zk = sqrt(2*xm*e); if(dble(e) < 0._dp .AND. &
            dimag(zk) < 0._dp) zk = -zk
        zki = (0.0_dp,1.0_dp)*zk
        gami = (0.0_dp,1.0_dp)* z1*xm/zk
        w0 = (zk*zk-bst*bst)/(2*xm*bst)
        w1 = -(zk*zk+bst*bst)/(4*xm*bst)
        g(1:nm+1,1:nm+1) = (0._dp,0._dp)
        do n = 0, nm-1
            g(n+1,n+1) = w0*(n+rl+1) - z1
            s0 = w1*sqrt((n+1)*(n+2*rl+2))
            g(n+1,n+2) = s0
            g(n+2,n+1) = s0
        end do
        znm = nm+1
        g(nm+1,nm+1) = w0*(nm+rl+1) - z1 + &
            w1*w1*znm*(znm+2*rl+1) * &
            4*xm*bst/((znm+rl+1+gami)*(bst-zki)**2) * &
            f21u(-rl+gami,znm,znm+rl+1+gami,((bst+zki)/(bst-zki))**2)

    end subroutine gri2b

    function f21u(a,b,c,z) result(q1)
    !   2_F_1 ( a, b+1; c+1; z ) / 2_F_1 ( a, b; c; z )
    !   L. Lorentzen, H. Waadeland: Continued fractions
    !   with Applications, North-Holland, page 293
    !

        implicit none
        complex(dp), intent(in) :: a, b, c, z
        complex(dp) :: p0, q0, q1, r, p2, q2, qq
        real(dp), parameter :: eps = 1.e-14_dp
        integer, parameter :: nn = 800
        integer :: n

        if ( abs(a) == 0._dp .AND. abs(b) == 0._dp ) then
            q1 = (1._dp,0._dp)
        else
            p0=(1._dp,0._dp)
            q0=(0._dp,0._dp)
            q1=(1._dp,0._dp)
            qq=(0._dp,0._dp)
            do n = 0,nn
                r = -(a+n)*(c-b+n)/((c+2*n)*(c+2*n+1))*z
                p2 = 1+r*p0
                q2 = q1+r*q0
                p0 = 1/p2
                q0 = q1*p0
                qq = q2*p0
                r = -(b+n+1)*(c-a+n+1)/((c+2*n+1)*(c+2*n+2))*z
                p2 = 1+r*p0
                q2 = qq+r*q0
                p0 = 1/p2
                q0 = qq*p0
                q1 = q2*p0
                if ( abs(1-qq/q1) < eps ) exit
            end do
            if (n >= nn) &
                write(iout,'('' F_21 accuracy '',d16.9)') abs(1-qq/q1)
        end if

    end function f21u

    FUNCTION LnGamma(x) result(tmp)
```

```fortran
        !
        ! Based on W.H. Press, S.A. Teukolsky, W.T. Wetterling, B.P. Flannery:
        ! Numerical Recipes, The Art of Scientific Computing, Third Edition,
        ! Canbridge Univerity Press, code GammLn, page 257.
        !
            integer :: j
            real(dp), parameter :: coef(0:13)= &
                (/ 57.1562356658629235_dp,  -59.5979603554754912_dp, &
                    14.1360979747417471_dp, -0.491913816097620199_dp, &
                    0.339946499848118887e-4_dp, 0.465236289270485756e-4_dp, &
                    -0.983744753048795646e-4_dp, 0.158088703224912494e-3_dp, &
                    -0.210264441724104883e-3_dp, 0.217439618115212643e-3_dp, &
                    -0.164318106536763890e-3_dp, 0.844182239838527433e-4_dp, &
                    -0.261908384015814087e-4_dp, 0.368991826595316234e-5_dp /)
            complex(dp) :: x, tmp, y, ser

            IF ( real(x) <= 0 ) stop ' Re(x) < 0 in GammaLn '
            y = x
            tmp = x + 5.2421875_dp
            tmp = (x + 0.5_dp) * LOG(tmp) - tmp
            ser = 0.999999999999997092_dp
            do j = 0, 13
                y = y + 1._dp
                ser = ser + coef(j)/y
            end do
            tmp = tmp + LOG(2.5066282746310005_dp * ser / x)

        END FUNCTION

end module green
```

Listing 12.6 Fortran module for calculating the Fredholm determinant and the Coulomb-modified phase shift

```fortran

module fredholm

    use setup_cs_b_s
    use matrix
    use green
    contains

        function r2bfd(e) result(fdet)
        ! | D(E) |

            implicit none
            real(dp) :: e, fdet
            complex(dp) :: z

            z = e
            fdet = (abs(c2bfd(z)))**2

        end function r2bfd

        function c2bfd(ep) result(fdet)
        ! Fredholm determinant D(E)

            implicit none
            complex(dp), intent(in) :: ep
            complex(dp), dimension(ndim,ndim) :: g2
            complex(dp) :: det(2), fdet

            call gri2b(ep,bg,xm,z1,r1,nm,g2)
            g2(1:nm+1,1:nm+1) = g2(1:nm+1,1:nm+1) - vv2(1:nm+1,1:nm+1)
            call zsydet(nm+1,g2,ndim,det)

            if( det20 == 0 ) then
                fdet =  det(1); det20 = nint(realpart(det(2)))
            else
                fdet = det(1)*10._dp**int(det(2)-det20)
            endif
            write(iout,'(2x,2d15.8,5x,2d15.8)') ep,fdet
```

```

      end function c2bfd

      subroutine scphase(en,zk,eta,delta)
      ! eta :      Coulomb phase
      ! delta :    Coulomb-modified scattering phase

         implicit none
         real(dp) :: en, zk, gam, delta, eta
         integer :: i, info
         complex(dp) :: zz, t1
         real(dp), dimension(ndim) :: phi, vphi
         complex(dp), dimension(ndim,ndim) :: g2, gc
         complex(dp), dimension(ndim) :: psi

         zz = en;  zk = sqrt(2*xm*en); gam = z1*xm/zk
         call gri2b(zz,bg,xm,z1,r1,nm,g2)
         gc(1:nm+1,1:nm+1) = g2(1:nm+1,1:nm+1)
         call zsyi(nm+1,gc,ndim)
         gc(1:nm+1,1:nm+1) = zk/(-4*(0._dp,1._dp)*xm)* &
              (gc(1:nm+1,1:nm+1)-conjg(gc(1:nm+1,1:nm+1)))
         phi(1:nm+1) = gc(1,1:nm+1)/sqrt(gc(1,1))
         vphi(1:nm+1) = matmul(phi(1:nm+1),vv2(1:nm+1,1:nm+1))
         psi(1:nm+1) = matmul(g2(1:nm+1,1:nm+1),phi(1:nm+1))
         g2(1:nm+1,1:nm+1) = g2(1:nm+1,1:nm+1)-vv2(1:nm+1,1:nm+1)
         call zsys(nm+1,g2,ndim,psi)
         t1 = dot_product(vphi(1:nm+1),psi(1:nm+1))
         delta = log(1-2*(0.0_dp,1.0_dp)*zk*t1/en)/(2*(0.0_dp,1.0_dp))
         eta = dimag(lngamma(r1+gam*(0._dp,1._dp)+1))

      end subroutine scphase

end module fredholm
```

Listing 12.7 Coulomb-Sturmian matrix elements of the potential

```

module potential

     use numtype
     use setup_cs_b_s
     integer , parameter :: nlag = 50  ! Gauss-Laguerre poinst and abscissas
     real(dp), dimension(nlag) :: &
         xlag = (/ 2.8630518339377187E-002_dp, 0.15088293567693040_dp, &
         0.37094878153489474_dp, 0.68909069988104643_dp, &
         1.1056250235399128_dp, 1.6209617511025007_dp,   &
         2.2356103759151811_dp,  2.9501833666418351_dp,  &
         3.7653997744057808_dp,  4.6820893875592837_dp,  &
         5.7011975747848878_dp,  6.8237909097945497_dp,  &
         8.0510636693907873_dp,  9.3843453082583981_dp,  &
         10.825109031549140_dp,  12.374981608757464_dp,  &
         14.035754599829904_dp,  15.809397197844673_dp,  &
         17.698070933350262_dp,  19.704146535461568_dp,  &
         21.830223306578265_dp,  24.079151444411504_dp,  &
         26.454057841252997_dp,  28.958376011937386_dp,  &
         31.595880956622864_dp,  34.370729963090426_dp,  &
         37.287510610550456_dp,  40.351297573586045_dp,  &
         43.567720269995029_dp,  46.943043991603027_dp,  &
         50.484267963129874_dp,  54.199244880168571_dp,  &
         58.096828017248505_dp,  62.187054175688935_dp,  &
         66.481373878444842_dp,  70.992944826619464_dp,  &
         75.737011547727292_dp,  80.731404802477627_dp,  &
         85.997211136463221_dp,  91.559690412533840_dp,  &
         97.449565614850570_dp,  103.70489123669230_dp,  &
         110.37385880764033_dp,  117.51919820311112_dp,  &
         125.22547013347359_dp,  133.61202792272863_dp,  &
         142.85832548925399_dp,  153.26037197260368_dp,  &
         165.38564331668255_dp,  180.69834370921444_dp  /)
     real(dp), dimension(nlag) :: &
          wlag = (/ 7.1404726135184851E-002_dp, 0.14714860696458673_dp, &
         0.18567162757483005_dp, 0.18438538252735392_dp,  &
         0.15420116860635602_dp, 0.11168536990226884_dp,  &
         7.1052885490195783E-002_dp, 4.0020276911508341E-002_dp, &
```

```
        2.0050623080071741E-002_dp, 8.9608512036462896E-003_dp, &
        3.5781124153156691E-003_dp, 1.2776171567890505E-003_dp, &
        4.0803024498372074E-004_dp, 1.1652883223097317E-004_dp, &
        2.9741704936941894E-005_dp, 6.7778425265420190E-006_dp, &
        1.3774795031713664E-006_dp, 2.4928861817200803E-007_dp, &
        4.0103543504277839E-008_dp, 5.7233317481414088E-009_dp, &
        7.2294342491825992E-010_dp, 8.0617101422817482E-011_dp, &
        7.9133930999436231E-012_dp, 6.8157366176767403E-013_dp, &
        5.1324267165894947E-014_dp, 3.3656247624378764E-015_dp, &
        1.9134763269650927E-016_dp, 9.3855897818274410E-018_dp, &
        3.9500699645033818E-019_dp, 1.4177495178275269E-020_dp, &
        4.3099702762923835E-022_dp, 1.1012575198456037E-023_dp, &
        2.3446177556090356E-025_dp, 4.1185441546381148E-027_dp, &
        5.9022467635963069E-029_dp, 6.8120089165532410E-031_dp, &
        6.2374494988122300E-033_dp, 4.4524405796836503E-035_dp, &
        2.4268623522505048E-037_dp, 9.8529714810500406E-040_dp, &
        2.8910788723183865E-042_dp, 5.9061627081121127E-045_dp, &
        8.0128745975035743E-048_dp, 6.7895754243968155E-051_dp, &
        3.3081730108487511E-054_dp, 8.2509648764409800E-058_dp, &
        8.8487281282991855E-062_dp, 3.0648948898441319E-066_dp, &
        1.9887082293307360E-071_dp, 6.0495671522392335E-078_dp /)

contains

    function pot1(r) result(v)
    !   v^(s) - short-range part of the potential

        implicit none
        real(dp), intent(in) :: r
        real(dp) :: v

        v =  v0* 1/(1+exp((r-r0)/a0)) +  erf(g0*r)*z1/r - z1/r

    end function pot1

    subroutine vpot(rl,nm,b,vp)
    !     < n l ; b | v^(s) | n' l ; b >

        implicit none
        integer, intent(in) :: nm
        real(dp), intent(in) :: b, rl
        real(dp), dimension(ndim,ndim) :: vp
        real(dp), allocatable, dimension(:,:) :: st
        real(dp), dimension(nlag) :: pf
        real(dp) :: x, r, potf
        integer :: n, i, j
        allocate (st(ndim,nlag))

        do n=1,nlag
            x = xlag(n)
            r = x/(2*b)
            call csturm(x,rl,nm,st(1,n))
            potf = pot1(r)
            pf(n)  = potf * wlag(n)/(2*b)
        end do
        forall (i=1:nm+1, j=1:nm+1 )
            vp(i,j)  = sum(st(i,1:nlag)*st(j,1:nlag)*pf(1:nlag))
        end forall
        deallocate (st)

    end subroutine vpot

    subroutine csturm(x,rl,nm,st) ! Coulomb-Sturmian basis
    !   st(n+1) = exp(x)*S_n (x) = [n!/(n+2l+1)!]^(1/2) x^(l+1) L_n^(2l+1!(x)

        implicit none
        integer, intent(in) :: nm
        real(dp), intent(in) :: x, rl
        real(dp), dimension(ndim) :: st
        integer :: i
        real(dp) :: x10, df
        st(1) = x**(rl+1)/sqrt(gamma(2*rl+2))
        x10 = 0._dp
        do i = 1, nm
            df = sqrt(i*(i+2*rl+1))
            st(i+1) = ((2*i+2*rl-x)*st(i)-x10)/df
```

```
                    x10 = df*st(i)
                end do

            end subroutine csturm

end module potential
```

Listing 12.8 The main program for bound and resonant state energies and phase shifts for the $\alpha - {}^{208}_{82}$PB system

```

module setup_cs_b_s

    use NumType
    implicit none
    save
    integer , parameter :: inp=5, iout=6, ndim=41
    real(dp), parameter :: coul2 = 1.44_dp, hb2 = 41.47_dp
    real(dp), parameter :: m1 = 4, m2 = 208, rm = m1*m2/(m1+m2)

    real(dp), parameter :: z1 = 2*82*coul2, xm = rm/hb2
    real(dp), parameter :: v0 = -50._dp, g0 = 0.1_dp, r0 = 7._dp, a0 = 0.4_dp

    integer :: nm, l1
    real(dp) :: bg, r1
    integer :: det20
    real(dp), dimension(ndim,ndim) :: vv2

end module setup_cs_b_s


program Lippmann_Schwinger_CS_basis
    ! Solution of the Lippmann-Schwinger equation on CS basis
    ! for bound and resonant state energies and for scattering phase shift

    use setup_cs_b_s
    use matrix
    use potential
    use fredholm

    implicit none
    real(dp) :: emin, emax, en, en0, de, zk, eta, delta
    integer :: ifail, nch, iz, maxf, nm7, nm0, i

    l1     = 0        ! angular momentum
    bg   =   4        ! basis parameter
    nm0 =   30        ! basis size

    write(iout,'(/a)')  '          '
    write(iout,'((a,i4,a,f12.3),a,i8)') ' l =',l1,' bg =',bg,' nm   =',nm0
    nm = nm0;     nm7 = nm + 7
    call vpot(r1,nm7,bg,vv2)
    call dsyi(nm7+1,vv2,ndim)
    call dsyi(nm +1,vv2,ndim)

    emin    = -25  ! energy range
    emax    = -18
    if ( emin > emax ) stop ' emin > emax '
    call pole_search(emin, emax)

    emin    = -18  ! energy range
    emax    = -10
    if ( emin > emax ) stop ' emin > emax '
    call pole_search(emin, emax)

    emin    = -10  ! energy range
    emax    = -2
    if ( emin > emax ) stop ' emin > emax '
    call pole_search(emin, emax)

```

```
    emin    = 0   ! energy range
    emax    = 6
    if ( emin > emax ) stop ' emin > emax '
    call pole_search(emin, emax)

    emin    = 6   ! energy range
    emax    = 12
    if ( emin > emax ) stop ' emin > emax '
    call pole_search(emin, emax)

    emin    = 12  ! energy range
    emax    = 20
    if ( emin > emax ) stop ' emin > emax '
    call pole_search(emin, emax)

    print *,'-------   Scattering   ------------'
    en0 = 5._dp
    de = 1._dp
    do i = 0, 20
        en = en0 + i*de   ! Energy in MeV
        call scphase(en,zk,eta,delta)
        write(iout,'(4(a,f12.8))')' energy =',en,'   k =', &
                zk,'   eta =',eta,'   delta =',delta
    end do

end program Lippmann_Schwinger_CS_basis


subroutine pole_search(emin, emax)

    use setup_cs_b_s
    use chebyshev
    use fredholm

    implicit none
    real(dp) :: emin, emax, eps, e0
    complex(dp) :: e2, de2
    integer :: nch, iz, maxf, minf, it

    nch = 13      ! Chebyshev points
    eps = 1.e-13_dp
    minf = 3
    maxf = 10

    call chebyex(r2bfd, nch, cheb, emin, emax)
    call chebyderiv(nch,cheb,chder,emin,emax)
    call chebyderiv(nch-1,chder,chder2,emin,emax)
    call chebyzero(nch-1, chder, emin, emax, z0, iz0)
    print *,'---------'

    de2 = abs(emax-emin)/50
    do iz = 1, iz0     ! root polish
        e0 = z0(iz)
        if( emin < e0 .and. e0 < emax .and. &
                cheby(e0,nch-2,chder2, emin, emax) > 0 ) then
            det20 = 0
            e2 = e0
            call root_polish(c2bfd,e2,de2,eps,maxf,minf,it)
            if ( it < maxf ) then
                write(iout,'(/a)')  '        '
                write(iout,'(a,i5,a,2f20.15,a)') ' nm =',nm, &
                        ' energy =',e2,'  MeV'
                write(iout,'(/a)')  '        '
            end if
        end if
    end do

end subroutine pole_search
```

The $^{208}_{82}\mathrm{Pb} - \alpha$ model system has three $l = 0$ bound states at $E = -19.75120$ MeV, $E = -14.34965$ MeV, $E = -7.25114$ MeV and three resonant states at $E = 1.210795 - 0.62 \cdot 10^{-20}i$ MeV, $E = 10.55372 - 0.463 \cdot 10^{-7}i$ MeV and $E = 19.40508 - 0.098336i$ MeV. At $E = 20$ MeV the Coulomb phase shift is $\eta_0 = 17.333$ rad, while the Coulomb modified nuclear phase shift is $\delta_0 = 1.005$ rad.

Further Reading

1. J.P. Boyd, Computing the zeros, maxima and inflection points of Chebyshev, Legendre and Fourier series: solving transcendental equations by spectral interpolation and polynomial rootfinding. J. Eng. Math. **56**(3), 203–219 (2006)
2. J. Zinn-Justin, Strong interactions dynamics with Padé approximants. Phys. Rep. **1**(3), 55–102 (1971)

Index

Z. Papp, *Mastering Quantum Mechanics*,
https://doi.org/10.1007/978-3-032-09011-9

www.ingramcontent.com/pod-product-compliance
Lightning Source LLC
Chambersburg PA
CBHW070936180226
39774CB00016B/12
9783032090102